ELECTRONIC DRAFTING AND PRINTED CIRCUIT BOARD DESIGN

Second Edition

ELECTRONIC DRAFTING AND PRINTED CIRCUIT BOARD DESIGN

James M. Kirkpatrick

Drafting Instructor and Program Coordinator
Dallas (Texas) Community College District

Second Edition

Australia • Brazil • Japan • Korea • Mexico • Singapore • Spain • United Kingdom • United States

Electronic Drafting and
Printed Circuit Board Design
Second Edition
James M. Kirkpatrick

Associate Editor: **Joan Gill**
Editing Manager: **Barbara A. Christie**
Project Editor: **Lawrence T. Main**

Library of Congress Control Number: 88016145

ISBN-13: 978-0-8273-3285-0

ISBN-10: 0-8273-3285-8

Delmar Cengage Learning
5 Maxwell Drive
Clifton Park, NY 12065-2919
USA

Cengage Learning products are represented in Canada by Nelson Education, Ltd.

For your lifelong learning solutions, visit **delmar.cengage.com**

Visit our corporate website at **www.cengage.com**

Printed in the United States of America
18 19 20 21 22 12 11 10 09 08

Contents

Preface

Many manufactured products are developed from drawings. Drawings precisely describe each part of an assembly, the relationships of the parts to one another, and the relationships of subassemblies to whole systems. Electronic assemblies are described by a variety of drawing types. Many electronic assemblies are built around printed circuit boards. Printed circuit boards are manufactured by reproducing special printed circuit board drawings on the actual board. ELECTRONIC DRAFTING AND PRINTED CIRCUIT BOARD DESIGN, second edition, explains each of the kinds of drawings used to manufacture electronic assemblies.

Many students of printed circuit board design have limited knowledge of drafting techniques. Chapter 1 of the text is included as an aid to those students by briefly reviewing the essential drafting tools and drafting techniques. This chapter is not a course in drafting, but it does serve as a refresher for those who do not have recent drafting experience.

Chapter 4 is an overview of common electronic symbols and the devices they represent. This chapter will be especially useful to drafters who have not had previous courses or experience in electronics. The electronic devices are described and their functions in circuits are covered very briefly. Electronic drafters are not usually required to select appropriate electronic devices. Neither are they expected to design the operating parameters of the circuit. So, the explanations of devices in this chapter are brief. The focus of the chapter is on producing final drawings from engineering sketches, not circuit design. Readers who have a strong background in electronics may not need to dwell on this chapter.

Chapters 8, 9, and 10 are of particular importance to the drafter in electronics. These are the chapters that cover the actual layout and drafting for printed circuit boards. Although the other areas of electronic drafting are important, it is in printed circuit board design and drafting that the greatest growth is taking place. An understanding of the content in these chapters will prepare a competent drafter to become a valuable printed circuit board designer.

References to Computer-Aided Drafting and Design are included throughout the text. Chapter 12 presents an introduction to the use of computers in electronic drafting. CAD has replaced manual drafting in many areas of drafting. Printed circuit design is one of them. All of the drawings in this book may be done either manually or on a personal computer using AutoCAD or a PCB software package. The information on manual drafting techniques is presented to allow students who do not have access to a personal computer equipped with appropriate software to complete this study. The information in this book is basic to manual or computer drafting.

The final chapter, Chapter 13, describes how the reader may make his own printed circuit boards. Taking the printed circuit board from a rough schematic diagram to a finished printed circuit board which performs a function in an electronic circuit is a valuable experience. Many students will find this to be an extremely useful chapter.

Like drafters in other fields, electronic drafters are required to use many references. ELECTRONIC DRAFTING AND PRINTED CIRCUIT BOARD DESIGN, second edition, includes an extensive appendix. This appendix includes references of the type the drafter will use frequently.

Each chapter consists of three major parts: objectives, development of content, and exercises. The objectives at the beginning of each chapter tell the reader what to expect from its content. The objectives are written in measurable terms, so they can be used without revision in competency-based programs. The body of the chapter develops the content necessary to achieve marketable drafting skills. The emphasis is on following industry standards and using the best techniques available. An important feature of the book is its use of quality illustrations. Some of the illustrations are reproduced with permission from industrial sources. These illustrations serve as excellent examples of work done by experienced drafters. Other illustrations were drawn especially for this book in order to illustrate a point most clearly. Where appropriate to chapter content, a number of drafting exercises is included.

Some exercises require the student to put into practice what has been learned from the content. Other exercises are designed to test for achievement of the chapter objectives.

James M. Kirkpatrick is an experienced drafter and drafting instructor. He has worked for 22 years in industry as a drafter, technical illustrator, and technical writer in both permanent and contract capacities. He has been a drafting instructor and program coordinator in the Dallas (Texas) Community College District for 21 years. He holds a Bachelor of Arts degree, Master of Education degree, and Doctor of Education degree. He is the author of six other textbooks.

The instructional content and exercises have been tested in the classroom by the electronic drafting students at six of the campuses of the Dallas County Community College District. The text was reviewed by:

Noel M. Smith, Sr.
Central Texas College

John Brady
Orange Coast College

William G. Clapp
Utah Technical College

Frank Ling
Santa Fe Community College

Howard Whited
San Jacinto College

David L. Goetsch
Okaloosa-Walton Junior College

Paul Keicher
Asheville-Buncombe Technical College

Chapter 1

Basic Drawing Tools, Materials, and Skills for Electronic Drafting

OBJECTIVES

After completing this chapter, you will be able to

- ✓ list the materials used for electronic drafting and describe how each is used.
- ✓ name the tools commonly used for electronic drafting, and describe the function of each.
- ✓ draw pencil lines of acceptable density and width or plot them on a computer driven plotter.
- ✓ draw pencil lettering of acceptable density, height, and slant or plot them on a computer driven plotter.

INTRODUCTION

In general, the drawing skills needed to make most types of electronic drawings are less complex than many types of drafting. In addition to electronic layout principles, the four main characteristics of most electronic drawings are:

1. proper line weight
2. uniform and legible lettering
3. pleasing page arrangement
4. accuracy

The increasing use of computer graphics and the explosion of low priced personal computers and useable software has greatly lessened the need for manual drawings. Many preliminary drawings, however, are still drawn manually. Therefore, to become most marketable, the drafter must develop skills in both manual and computer graphics.

Some electronic drawings are done in pencil on drafting paper (vellum) or drafting film. Most electronic drawings are very high-quality photographic or ink artwork having accurately drawn symbols, circuits, and computer plotted lettering. This chapter describes the materials and tools required to produce electronic drawings. It also includes exercises to develop the skills needed to draw good lines and lettering.

MATERIALS

The basic materials used for electronic drawings are:

- Plain and gridded paper
- Plain and gridded Mylar®*
- Preprinted patterns (transfer letters and patterns, and adhesive letters, symbols, and tape) or symbol libraries stored on a computer.

Plain and Gridded Paper

Several varieties of vellum make good drawing papers. The paper selected should be durable, and translucent enough to trace through. It should have a surface which will accept either pencil or ink. Vellum is manufactured with and without an imprinted but nonreproducible grid. The nonreproducible grid size most adaptable for electronic drafting work is $\frac{1}{10} \times \frac{1}{10} \times 1''$. This means that a major 1″ grid contains 100 subdivisions of one-tenth inch square grids within each 1″ square.

Plain and Gridded Mylar®

Drafting film, commonly called Mylar®, is also used for electronic drawings, particularly for printed circuit artwork. Mylar® is preferred over paper because it resists stretching and shrinking when the temperature or humidity changes. This film has a frostlike coating on one side, with the other side clear. The frosted side, called the *matte* side, is the drawing surface. Some films are frosted on two sides so that either side or both may be used for drawing. Ink or plastic pencil (a crayonlike lead) is usually used on drafting film. Graphite pencil tends to smear easily on film and should not be used for a permanent drawing unless the drawing is sprayed with a fixative upon completion.

Paper and Sheet Sizes • Papers and films come in standard sizes, identified by letters. The drafter must memorize these sizes because they are commonly used and referred to by their letters. Table 1–1 lists the standard sizes and gives an easy way to remember them.

TABLE 1–1 Standard Sizes of Drawing Papers

Size Designation	Dimensions	How to Remember
A	$8\frac{1}{2}'' \times 11''$	Memorize the A size. Then, use the long dimension of the A size for the short dimension of the B size and double the short dimension of the A size for the long dimension of the B size. The long dimension of the B size is the short dimension of the C size; twice the B size short dimension is the C size long dimension; and so on.
B	11″ × 17″	
C	17″ × 22″	
D	22″ × 34″	
E	34″ × 44″	
R[a]	34″ wide × any length[b]	

[a] This size designation is often shown with other letters or words.
[b] Any length can be taken from a roll of paper.

®* duPont registered trademark

Preprinted Patterns

Preprinted patterns and tape are available for many often repeated symbols, letters, numbers, and parts. Figure 1–1 shows several of those most commonly used. Tape is also used for connectors between pads on artwork.

The patterns are available either on transfer sheets (rub-on type) or as adhesives (stick-ons). Rub-ons are applied by rubbing on the top of the sheet with a pencil lead or a blunt object, Figure 1–2A. These same patterns and connections are available in many software packages for personal computers. The symbol or letter is transferred from the back of the sheet onto the drawing, Figure 1–2B. Stick-ons are cut out and placed on the drawing using nonreproducible alignment marks, Figure 1–3. It is usually best to cut stick-ons fairly close to the image, using a square cut. This allows other lines to be added easily, and edges to be removed easily if they show when the drawing is reproduced.

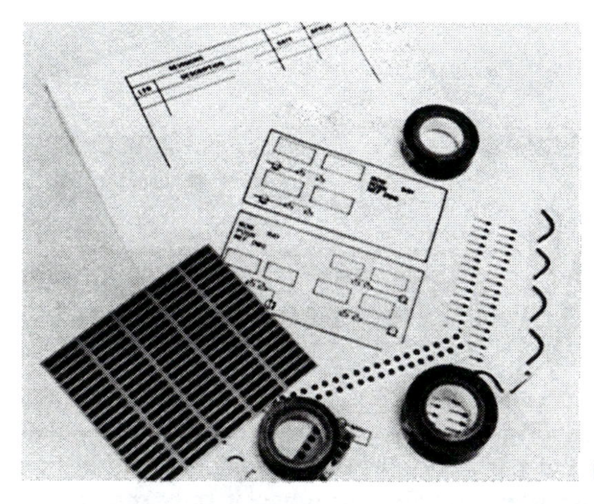

Figure 1–1 Typical preprinted patterns and tape

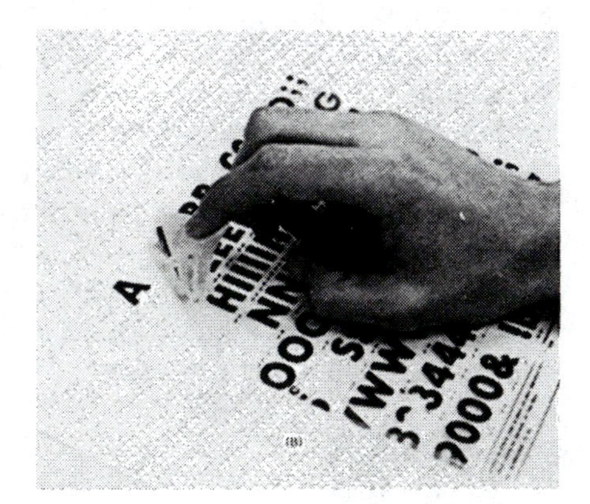

Figure 1–2 Using a transfer sheet

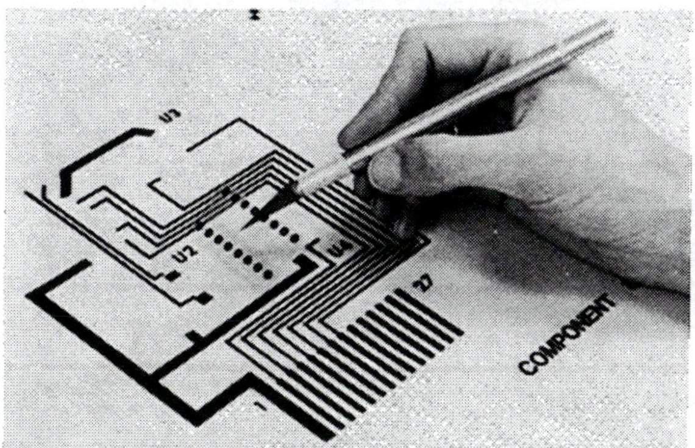

Figure 1–3 Adhesive (stick-on) patterns

TOOLS

The tools used for electronic drawings include:

Manual

- Drawing tables or boards, and light tables
- Horizontal straight edges (T-squares, drafting machines)
- Tape
- Mechanical pencils and lead
- Nonreproducible lead
- Pencil pointers
- Drafting powder
- Dusting brush
- Triangles
- Templates
- Erasers and erasing devices
- X-acto© knife
- Reservoir inking pens
- Pen cleaners
- Measuring scales and other devices

CAD

- A computer (PC or mainframe) of sufficient capacity, appropriate software, and a plotter of sufficient size and accuracy.

Drawing Tables or Boards

Although some drafting drawings are very large and require special-size boards, an 18″ × 24″ board is big enough for the assignments in this book. You need a surface you can wash or change when it becomes dirty. Some people use poster board to cover the table. The poster board can then be discarded when it is dirty or cut. The pale green vinyl covers sold by drafting or art supply stores are good surfaces. They are easily cleaned and last a long time provided they are not used as a cutting surface. Figure 1–4 shows two types of boards that can be used for drafting.

A light table, Figure 1–5, is used extensively in printed circuit design. Some light tables have a gridded image composed of small dots to allow highly accurate placement of lines and patterns. Horizontal straight edges, such as T-squares and drafting machines, are often not necessary when this type of light table is used.

(A)

(B) Courtesy of J. S. Staedtler, Inc.

Figure 1–4 Two types of drawing boards

Figure 1-5 One type of light table

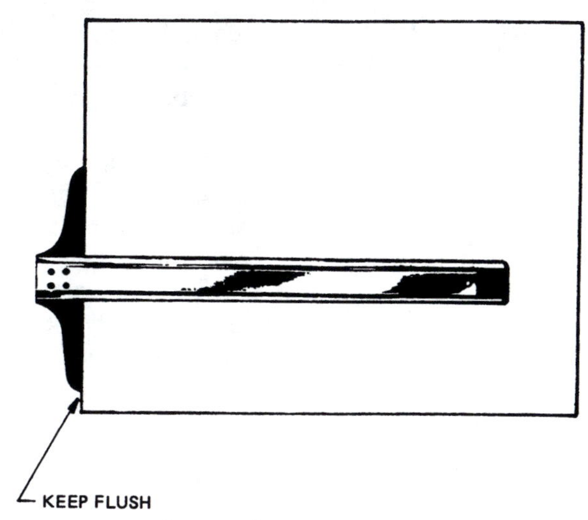

Figure 1-6 Using a T-square

Horizontal Straight Edges

In electronic drafting, a T-square or drafting machine is used to make horizontal lines and to keep these lines straight and square with the paper. The T-square is the original straight edge tool, and is the least expensive.

One of these two types of tools is used for all drawings.

T-Square • A T-square is shown in Figure 1–6. The most important thing to remember about using a T-square to draw horizontal lines is to keep the head of the T-square flush with the edge of the drawing board. Use the left hand to move the T-square, and always check to be sure that the head is snug against the board edge before drawing. (Left-handed people should reverse these directions.) Occasionally, check the head to be sure it is securely fastened and is square. Clean the blade often to keep drawings clean.

Drafting Machines • The two types of drafting machines are the arm type and the track type, Figure 1–7. Both types are used to draw horizontal lines or lines at any angle. All drafting machines have locking devices and releases to set angles. There are different types of locks and releases. The operating manual should be read carefully in order to learn how to operate a drafting

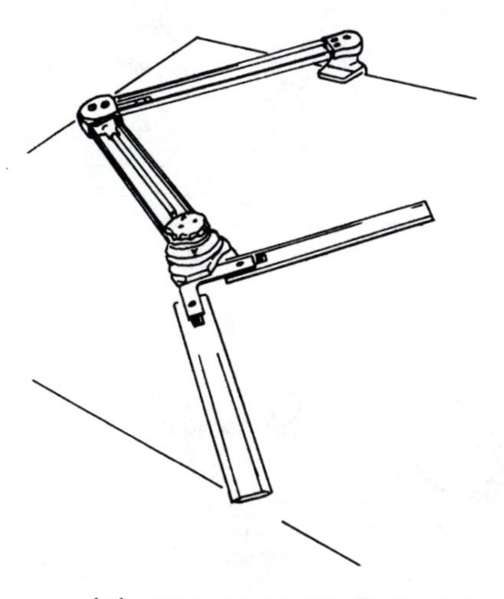

(A) ARM DRAFTING MACHINE

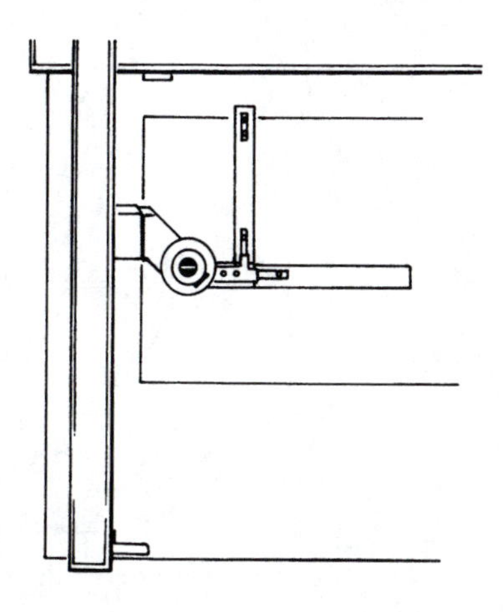

(B) TRACK DRAFTING MACHINE

Figure 1-7 Two types of drafting machines

machine before using it. If the machine does not move freely, it should not be forced—a brake or locking device is probably in place. These devices should be handled with care as they are easily broken or damaged beyond repair.

The blade of the machine must be kept clean, and operation of the machine should be checked occasionally. This is done by drawing a line near the bottom of the board, then moving the machine to its extreme upper position and then back to the line that was drawn. If the blade does not align with the original line, a connection is loose or a band is broken. Loose connections should be tightened first before assuming that the machine must be sent out for repair.

Tape

Drafting tape or masking tape is used to fix paper in position for drawing. Drafting tape is preferred, however, because it does not stick as firmly to paper as does masking tape and, thus, is not as likely to tear the drawing when it is removed. Drafting dots (small, circular pieces of drafting tape) are used by many drafters in addition to the more common roll of tape. These dots are available at most drafting supply stores.

Mechanical Pencils and Lead

Several types of pencils work well for pencil drawings. The thin-lead mechanical pencil, Figure 1–8A, is the drafting pencil used most by drafters. Its advantage is that it does not have to be sharpened. Mechanical pencils should be held straight up and down to get an even line and to avoid breaking the lead, Figure 1–8B. Lead for these pencils is made in several degrees of hardness and in several widths. Leads with .5 mm and .7 mm diameters are used for most types of electronic drawings. A .3 mm lead is often used for drawing centerlines, dimension lines, and other thin lines.

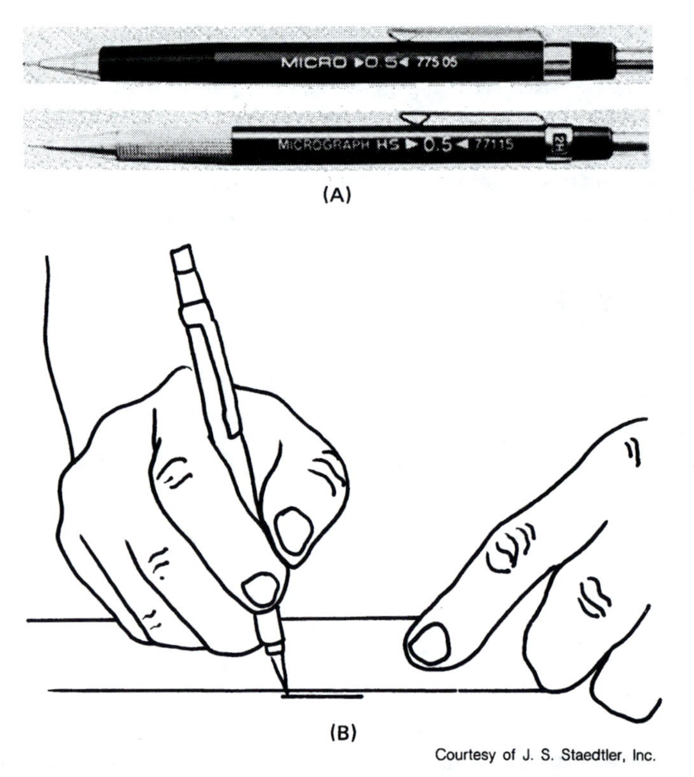

Courtesy of J. S. Staedtler, Inc.

Figure 1–8 Drawing with a mechanical pencil

The drafting lead holder, Figure 1–9, uses thicker lead than the mechanical pencil, and will take more pressure. The lead in this type of holder can be pointed to make thin, dark lines, as well as thicker ones. Its disadvantage is that it must be pointed often. The holder will hold either graphite or plastic leads.

Graphite lead hardness categories are shown in Figure 1–10. The hard leads are used for construction lines and for very precise work. The medium leads are used for drawing and lettering. The soft leads are not used in drafting.

Plastic lead comes in grades similar to those of graphite lead. It is more like a crayon, however, and makes a blacker line than does graphite lead. It works well on drafting film, and its reproduction qualities are almost as good as ink. Working with plastic as compared to graphite lead, requires a slightly different skill. To acquire this skill, the drafter should practice drawing with plastic lead.

Wooden pencils may also be used, but sharpening is more of a problem with them than with any of the other pencils. Whichever type of pencil is chosen, it is necessary to be able to make dark lines of even width. Wooden pencils and the drafting lead holder should be held at about a 60° angle to the paper, Figure 1–11. They should be rolled slightly between the fingers while drawing in order to keep a uniform line width. A rag or tissue should be kept on hand to wipe off the points of wooden pencils and lead holders after they have been sharpened.

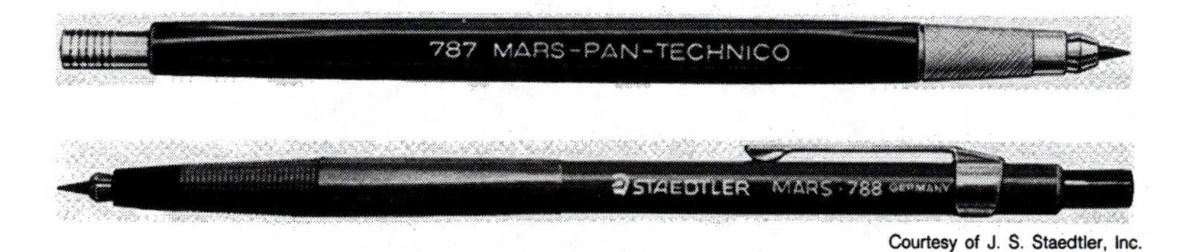

Courtesy of J. S. Staedtler, Inc.

Figure 1–9 Lead holders

9H —— 4H | 3H 2H H F HB | B —— 7B
HARD | MEDIUM | SOFT

Courtesy of *Basic Industrial Drafting Skills*, Kirkpatrick. Breton Publishers

Figure 1–10 Graphite lead hardness categories

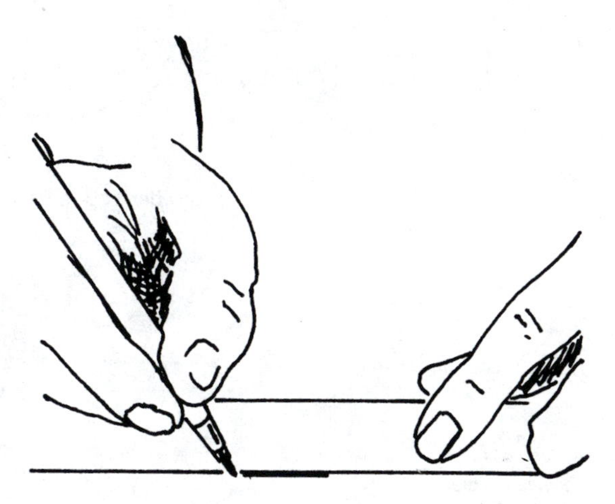

Figure 1–11 Drawing with a wooden pencil or a drafting lead holder

Many professional drafters prefer to use nonreproducible lead for construction lines and layout. This lead will not reproduce by most reproduction methods (unless the lines are drawn too dark). Construction and layout lines drawn with this lead do not have to be erased and, as a result, drawing time is saved.

Courtesy of J. S. Staedtler, Inc.

Figure 1–12 Rotary-type pencil pointer

Pencil Pointers

Many good pencil pointers are on the market. Figure 1–12 shows the common rotary-type pencil pointer preferred by many drafters. It is fairly fast, is not messy, and is very durable. A sandpaper pad sharpens well but it is slow. Figure 1–13 shows the points needed for good line work.

Drafting Powder

The drafting powder bag or can is an excellent aid for keeping drawings clean. The bag is made of a coarsely woven material, Figure 1–14A. The powdered particles fall out of it when the bag is kneaded, Figure 1–14B. The can-type container resembles a saltshaker. Both types are used in the same way. The powder is sprinkled over the paper before a drawing is started. Most of the excess lead is then picked up by the powder as the drawing is made. After the drawing is completed, the powder is rubbed over the drawing *very lightly* to pick up the excess graphite, Figure 1–15. The powder is then brushed off the drawing.

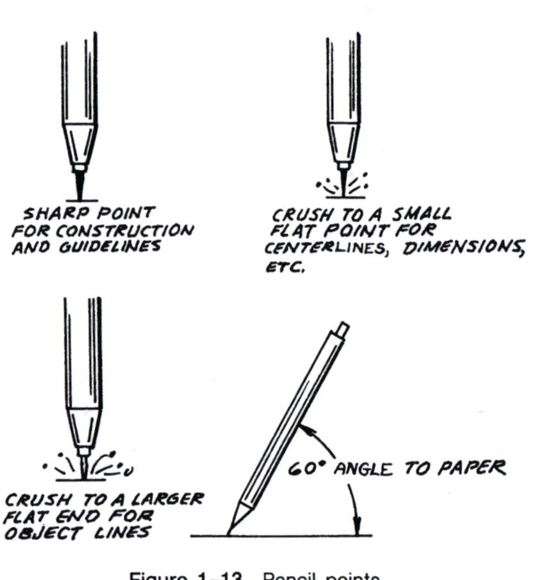

Figure 1–13 Pencil points

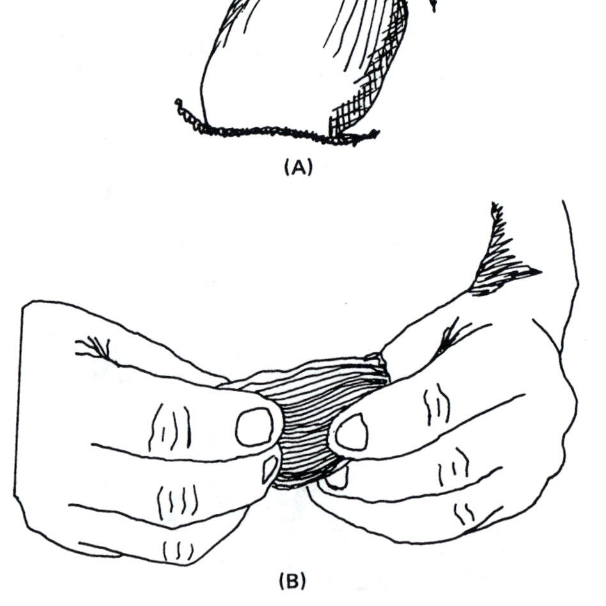

Figure 1–14 A drafting powder bag is used to keep drawing clean

Figure 1–15 Drafting powder picks up excess graphite

Dusting Brush

A dusting brush is often needed at the drafting table to brush away eraser crumbs and drafting powder. A brush commonly used is shown in Figure 1–16.

Triangles

The 30° - 60° and 45° triangles are both needed for electronic drafting, Figure 1–17. Either 8″ or 10″ triangles are excellent sizes, but triangles larger than that are awkward to use for some types of drafting.

Some inexpensive triangles are quite acceptable for drafting, whereas the thick, expensive ones are often hard to use when inking. The best triangles are about $\frac{1}{16}$″ thick, having some means of keeping the drawing edge off the paper, thus preventing the ink from running underneath the triangle. If this kind is not available, the edges of an ordinary triangle can be beveled with an X-acto© knife, Figure 1–18. Some drafters put tape on the edge of their triangle to raise it off the paper, but the tape gets dirty easily and is difficult to remove. Small plastic circles with adhesive, known as *risers,* are available to accomplish the same purpose. They can be cleaned when they become dirty.

To draw vertical lines, place the triangle on the straight edge. Hold the triangle firmly with one hand as you draw upward with the other, Figure 1–19. You will soon learn to return downward over a line to improve its density. Notice that the pencil is slanted in the direction of the line, but is not tilted in relation to the edge of the triangle. As mentioned previously, a mechanical pencil should be held perpendicular to the paper.

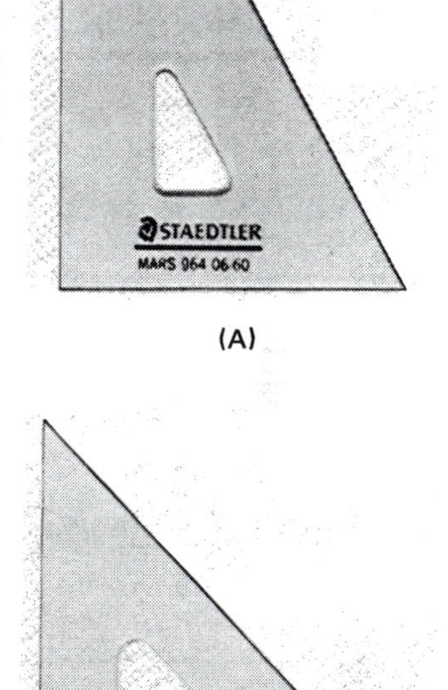

(A)

(B)

Courtesy of J. S. Staedtler, Inc.

Figure 1–17 (A) 30°–60° and (B) 45° triangles

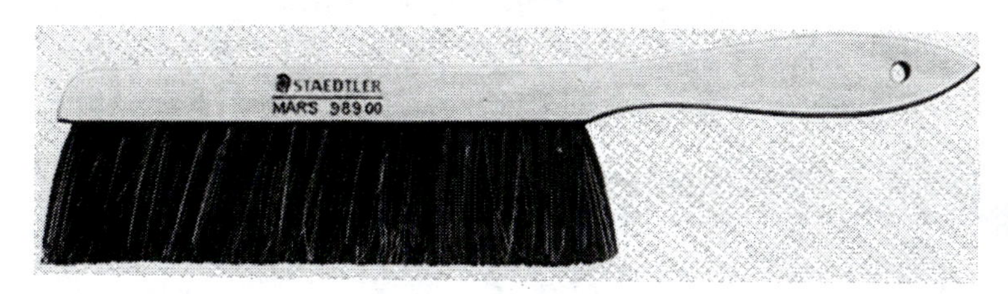

Courtesy of J. S. Staedtler, Inc.

Figure 1–16 Drafting table dusting brush

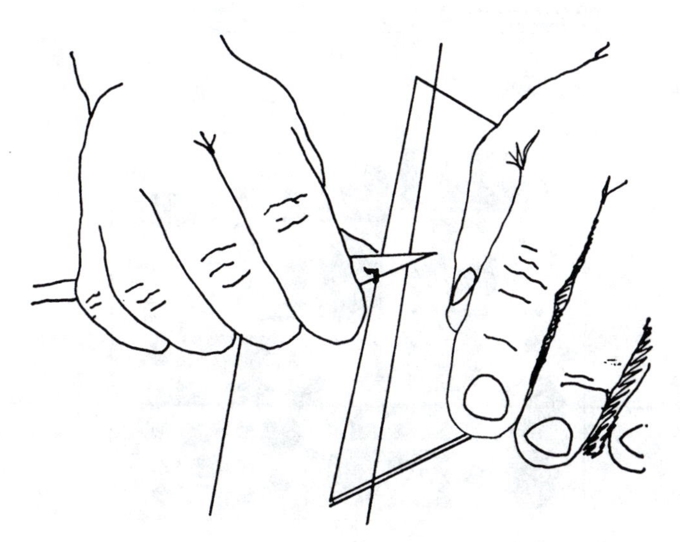

Figure 1–18 Beveling the edges of a triangle

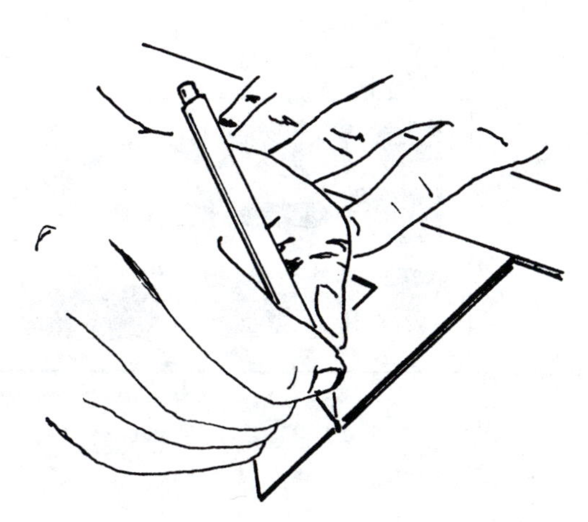

Figure 1–19 Drawing vertical lines with a triangle

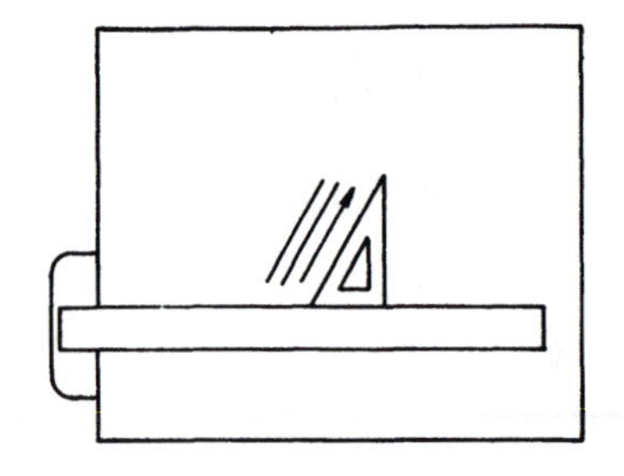

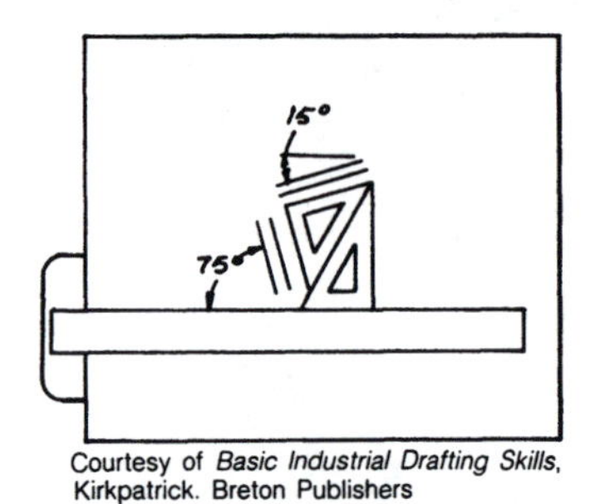

Courtesy of *Basic Industrial Drafting Skills,* Kirkpatrick. Breton Publishers

Figure 1–20 Drawing inclined lines with a triangle

Figure 1–20 shows how triangles are used to draw inclined lines. Using a drafting machine eliminates the need for triangles to draw unusual angles.

Templates

Templates are available for drawing just about anything that needs to be repeated often. A few of the most commonly used templates are those for hexagonal nuts and bolts, springs, circles, ellipses, and many kinds of symbols. The ones needed for electronic drafting are circle templates, $\frac{1}{8}$'' to $1\frac{3}{8}$'' diameter, Figure 1–21; electronic symbol templates, Figure 1–22; logic symbol templates, Figure 1–23; and printed circuit board component symbols, Figure 1–24.

To use the circle template or any other similar template, align the cross hairs of the template on the center of where the circle is to be drawn. Draw the circle with the pencil held perpendicular to the paper, Figure 1–25. Bear down firmly to produce a dense line.

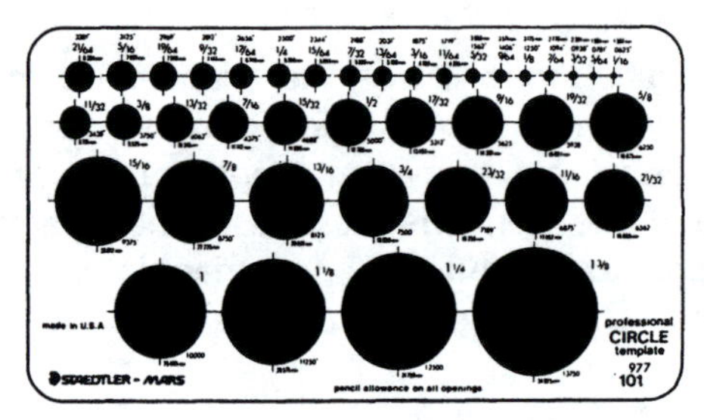

Courtesy of J. S. Staedtler, Inc.

Figure 1–21 Circle template

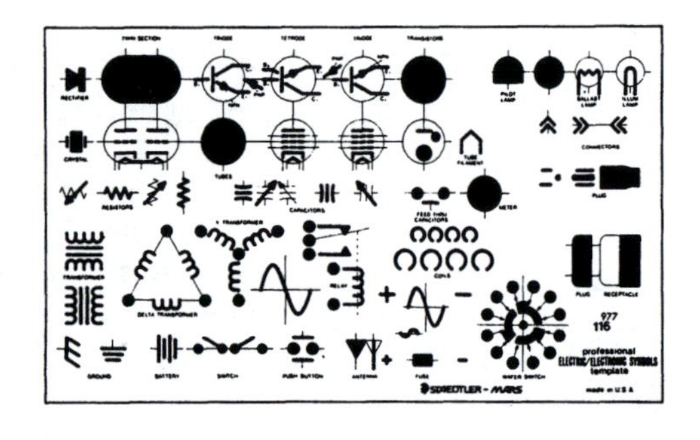

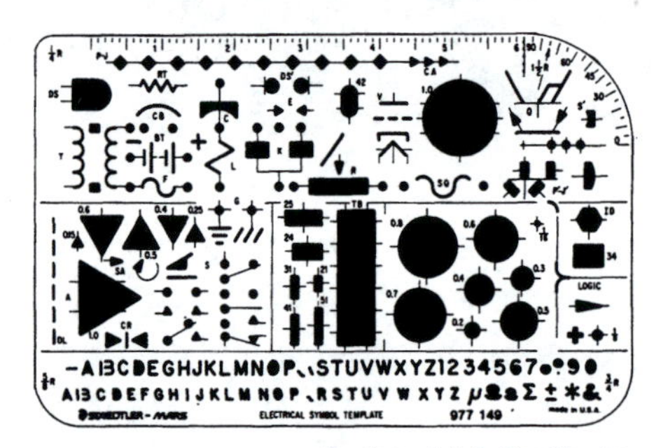

Courtesy of J. S. Staedtler, Inc.

Figure 1–22 Electronic symbol templates

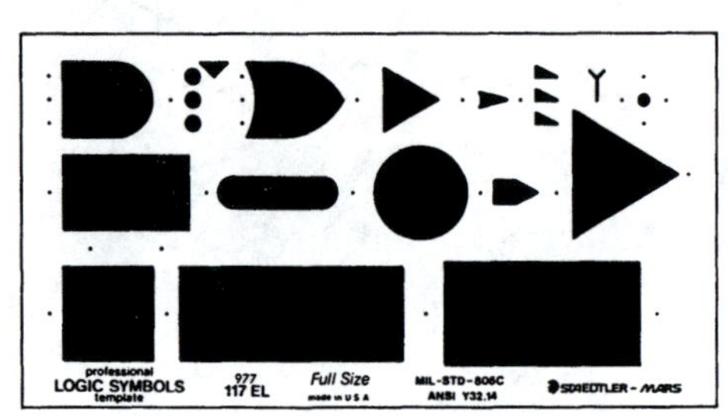

Courtesy of J. S. Staedtler, Inc.

Figure 1–23 Logic symbol template

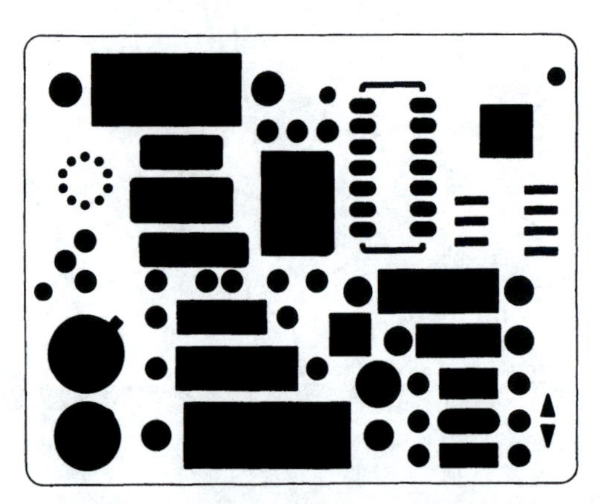

Figure 1–24 Printed circuit board component template

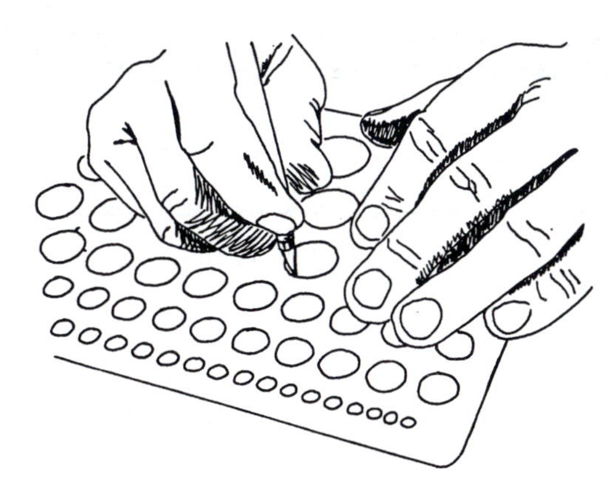

Figure 1–25 Using a circle template

Erasers and Erasing Devices

A Pink Pearl* eraser is best for most types of drafting vellum. Drafting film requires a soft, white eraser which will remove lines without damaging the drawing surface. An art gum eraser works best for removing smudges. The Pink Pearl eraser and the white eraser are available in block form, Figure 1–26, or in thin, pencil-shaped form. Plug or strip forms are used in electric erasers.

Courtesy of J. S. Staedtler, Inc.

Figure 1–26 Block eraser

An electric eraser is almost a necessity when there is a great deal of drafting to be done. Ink and pencil both erase easily with this tool. The eraser motor must be heavy duty to take the constant use and abuse it often gets; lightweight machines simply do not hold up for long. Figure 1–27 shows the heavy-duty eraser. Cord-type erasers last longer, but cordless ones are sometimes handier. Cordless types will hold up for at least two or three years before they must be replaced. When using an electric eraser, it must be kept moving so that it does not rub a hole in the paper.

An erasing shield, as shown in Figure 1–28, is often helpful to avoid erasing correct lines or letters. Many drafters use a slightly moistened round, white eraser in a plastic holder to remove small ink and pencil mistakes from drafting film, Figure 1–29.

Courtesy of J. S. Staedtler, Inc.

Figure 1–27 An electric eraser

Courtesy of J. S. Staedtler, Inc.

Figure 1–28 An erasing shield

Courtesy of J. S. Staedtler, Inc.

Figure 1–29 Erasing tool used with drafting film

* Trade name of Eberhard Faber

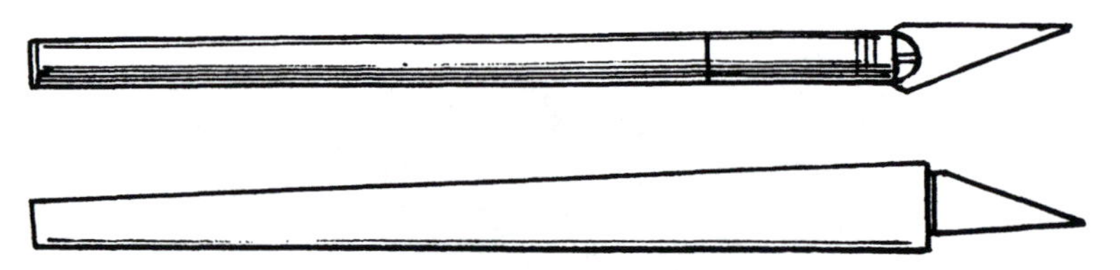

Figure 1–30 X-acto© knives

X-acto© Knife

The X-acto© knife is used a great deal in electronic drafting. *Care must be used with an X-acto© knife because the blade is surgically sharp.* The handle of the knife can be either cylindrical—about the same diameter as a pencil—or flat. The blades are shaped as shown in Figure 1–30. For scratching off ink, the blade works best when held at a low angle to the paper.

Reservoir Pens

Reservoir inking pens, Figure 1–31, have become a necessity for many drafters. Many are on the market now, some excellent and some poor. In general, the more expensive pens are the best. They flow easily and deliver a clean, dense line when they are used correctly. Some of the less expensive pens work well, but must be purchased with care. Working with a pen that clogs can be very frustrating.

Reservoir inking pens range in size from very thin (5 × 0 or .13 mm) to very thick (6 or 2.0 mm) and larger. A size 0 or .35 mm pen can be used to do most of the work done by beginners. Other pens can be added later as needed. Three or four sizes are all that most drafters use.

Pen points, Figure 1–32, are replaceable when they wear out or are damaged. They should not be removed for cleaning more often than necessary because the points are easily damaged. Figure 1–33 shows two styles of pens. The pen point with the cylindrical shoulder can be used on both types of pens, and is, by far, the most commonly used.

Courtesy of J. S. Staedtler, Inc.

Figure 1–31 Reservoir inking pen

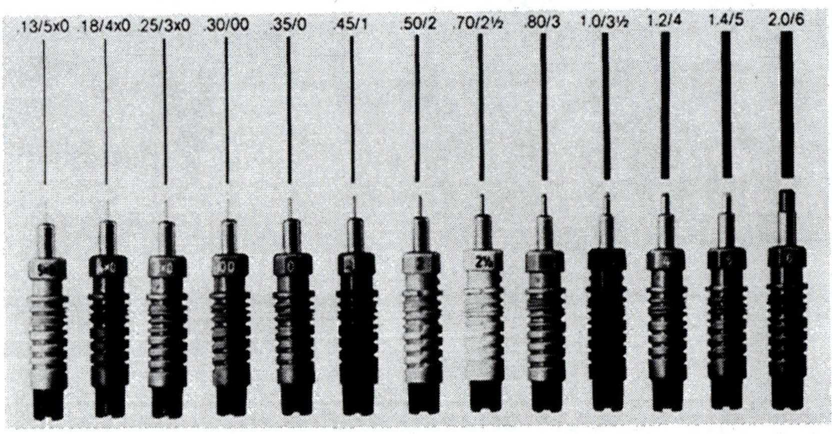

Courtesy of J. S. Staedtler, Inc.

Figure 1–32 Pen point sizes for reservoir inking pens

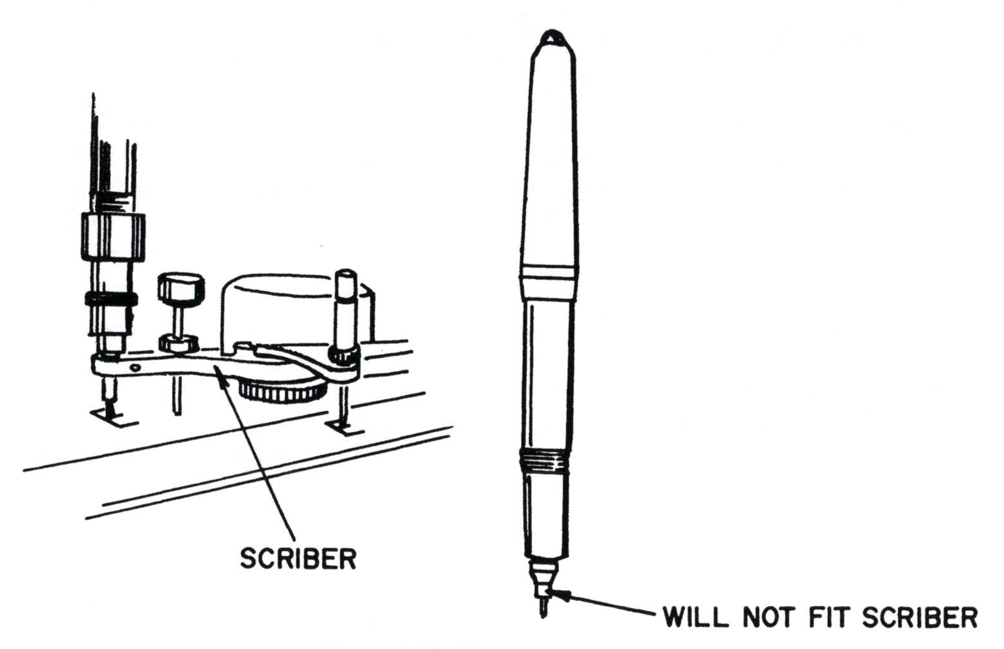

Figure 1–33 Two styles of pens

Pen Cleaners

Pen cleaning is an important part of working with ink. Pens should be tightly capped and stored upright so that the ink empties out of the point. When this is done, the pens need not be cleaned as often as those that are not given good care. Pens that are not severely clogged can be placed, completely assembled, in an ultrasonic pen cleaner where they are cleaned in just a few minutes, Figure 1–34.

Courtesy of Koh-i-noor Rapidograph, Inc.

Figure 1–34 Ultrasonic pen cleaner

If a pen is severely clogged, it should be completely disassembled, as shown in Figure 1–35, and soaked in the ultrasonic cleaner for several minutes. Use great care when unscrewing the point from the barrel because the plastic barrel is easily broken. Use care also when handling the flow regulating device (plunger). If the wire part of the plunger becomes bent, it is likely that the whole point will have to be replaced.

If an ultrasonic pen cleaner is not available, pens can be disassembled and soaked in detergent overnight for deep cleaning or washed in water for light cleaning. Washing by hand requires that the pens be thoroughly rinsed and carefully dried with a paper towel. It is also helpful to blow the water out of the barrel so that the ink does not become diluted when the reservoir is refilled.

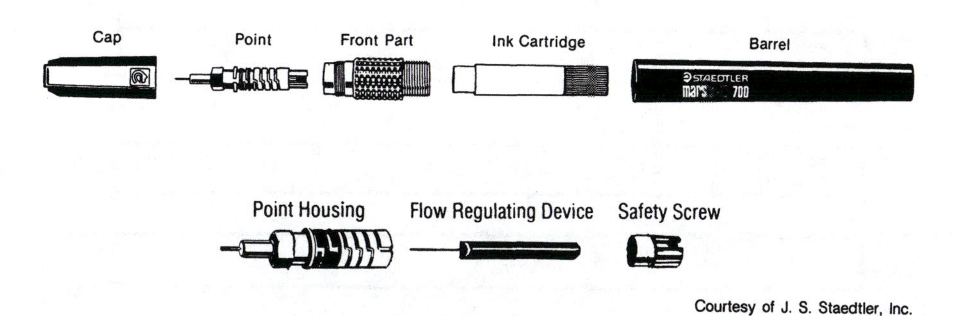

Courtesy of J. S. Staedtler, Inc.

Figure 1–35 Reservoir pen disassembly

Measuring Scales and Other Devices

The measuring scales used in electronic drafting differ considerably from those used in most other types of drafting. Drawing to scale is not used for schematics, block diagrams, logic diagrams, cabling diagrams, and similar drawings. Such drawings must fit a particular page or paper size. The ability to read a standard English ruler or metric ruler is all that is necessary, Figure 1–36.

Half-size Scales • Occasionally, an architect's or civil engineer's half-size scale is helpful. When a reduced scale is used, it is necessary only to select the correct scale and measure with it. No arithmetic is necessary. For example: the one-half scale on the architect's scale and the 20-parts-to-the-inch scale on the civil engineer's scale produce the same size drawing. The drawing produced when measured with either of these scales will be half size. Figure 1–37 shows these two scales.

Twice Full-size Scales and Greater • Scales greater than full size are often used on printed circuit board and integrated circuit drawings. Often these drawings are done on an oversized grid which automatically produces a drawing to the desired scale. If grids such as these are not available, all full-size dimensions must be doubled to produce a drawing that is twice full size. The scale for a twice full-size drawing is shown as 2×. Integrated circuit drawings are often drawn to a scale of 100 to 1 or greater. The following are some common scale references:

2× (twice full size)
100× or 100:l (One hundred times full size)
2500× or 2500:1 (Two thousand five hundred times full size)

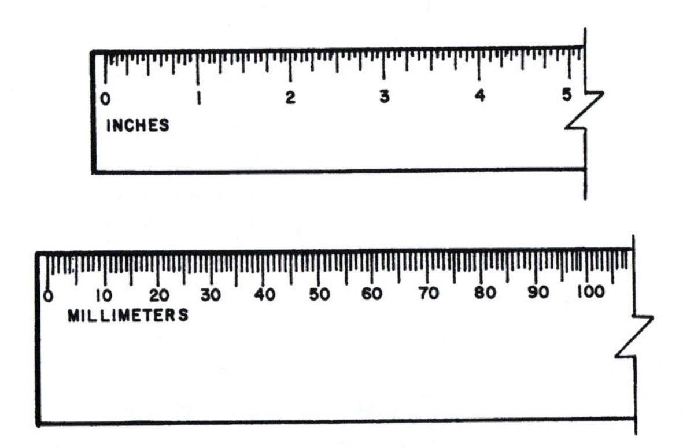

Figure 1–36 Standard English and metric rulers

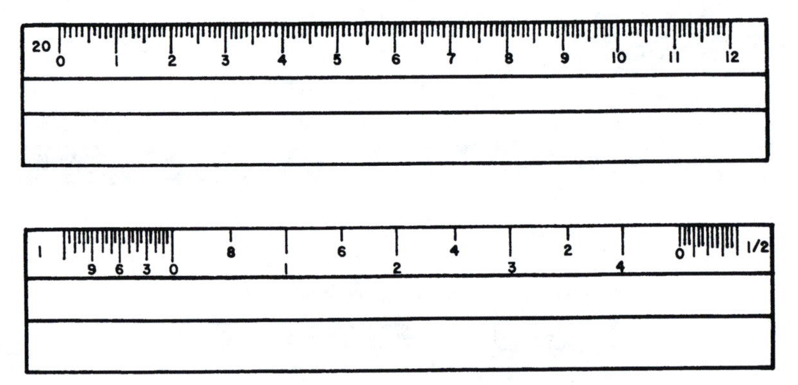

Figure 1–37 Half-size scales

The Steel Rule (Hundredths of an Inch)

The steel rule is also helpful in many instances. The most common form of the steel rule is six inches in length and is a very accurate measuring device in many cases. Figure 1–38 shows an enlarged picture of the left end of a steel rule. Notice that the top edge of this rule is the decimal rule and is marked off in hundredths of an inch (.01). The bottom edge is the fractional rule and is marked off in 64ths of an inch.

To read the decimal edge, Figure 1–39,

- Study the largest divisions (those marked with the numbers 1 and 2. These are inches and would be written as 1, 1.0, or 1.00.
- Study the divisions marked with the whole numbers 1 through 9. These are tenths of an inch ($\frac{1}{10}$ or .1).
- Study the smallest divisions. These are hundredths of an inch ($\frac{1}{100}$ or .01).
- Study the longer marks which make it easier to locate and count the smaller divisions. For example, the longest mark between the divisions marked 1 and 2 is halfway between the two divisions and is .05 or $\frac{5}{100}$ of the distance between .1 and .2.
- To locate a dimension of .67, find the center point between 6 and 7 on the scale and count over 2 more small marks.

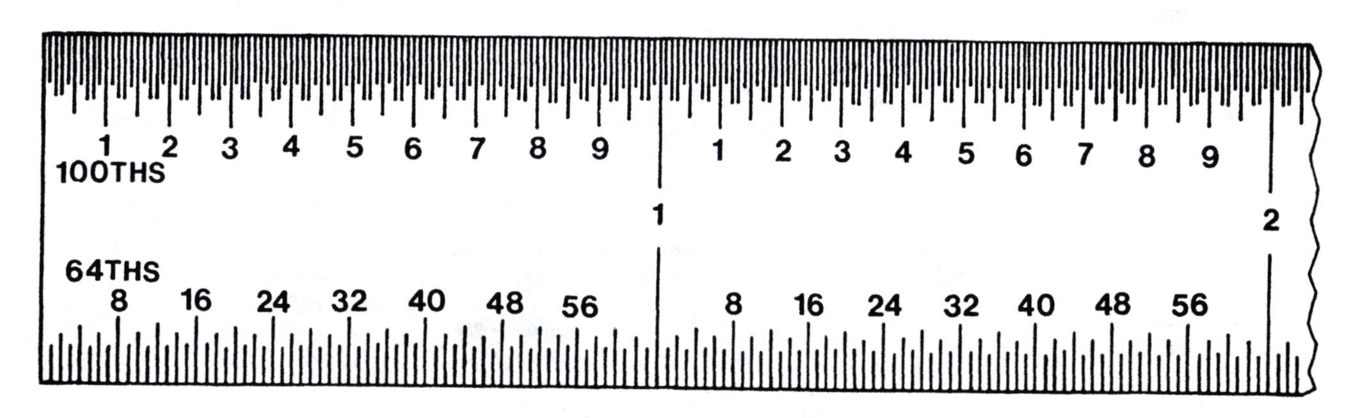

Figure 1–38 Enlarged picture of the left end of a steel rule

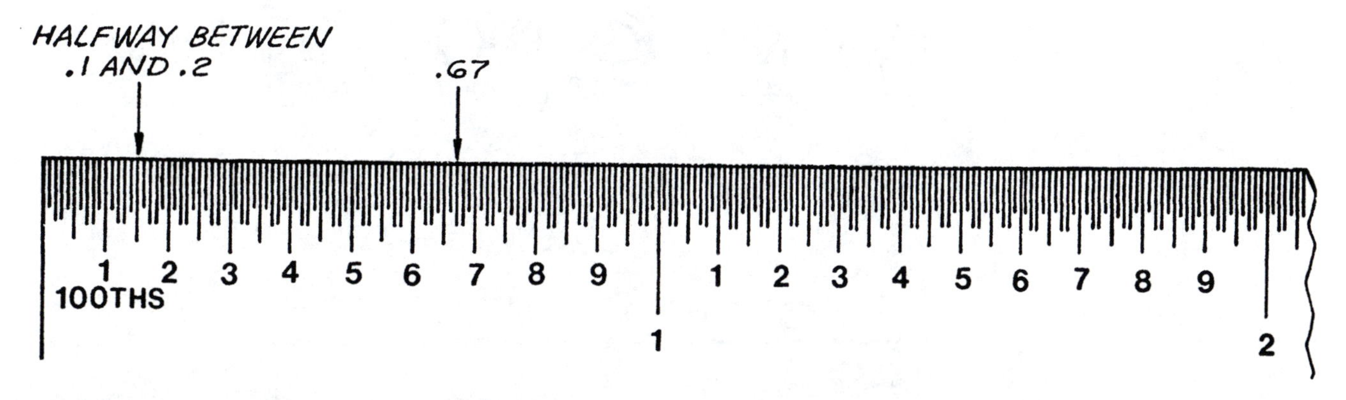

Figure 1–39 The decimal edge of a steel rule

LETTERING

Many methods are used to put written descriptions on a drawing. Lettering may be typewritten or produced by using mechanical lettering devices, stick-on or rub-on letters, or lettering templates. However, freehand lettering is still the most popular method. Not only is good freehand lettering most used on the job, but it makes a valuable first impression when it is used on a job application. Many inexperienced applicants have been hired for good drafting jobs because they had developed a neat lettering style and used it on an employment application.

Some companies use a particular style of lettering, which is learned on the job. Most drafting room standards, however, require lettering that is legible, uniform in height and slant, and carefully formed so that letters or numbers are not mistaken for others. After meeting these requirements, slight variations in style are allowed.

The Drafting Alphabet

The alphabet used in drafting is a single-stroke, sans-serif, Gothic style. This means that each part of a letter is made with a single stroke, there are no short, cross-lines (serifs) on the ends of the letters (*sans* meaning without), and all strokes are the same width. The letter R is given as an example in Figure 1–40. Most companies use only uppercase or capital letters. Very few use lowercase letters, because it is difficult to get good reproductions of reduced lowercase letters.

The sequence of strokes for each letter and number in the vertical drafting alphabet is shown in Figure 1–41. The forms of the letters are very important to a professional style. Study each character so that you know

Courtesy of *Basic Industrial Drafting Skills*, Kirkpatrick. Breton Publishers

Figure 1–40 An example of the drafting alphabet

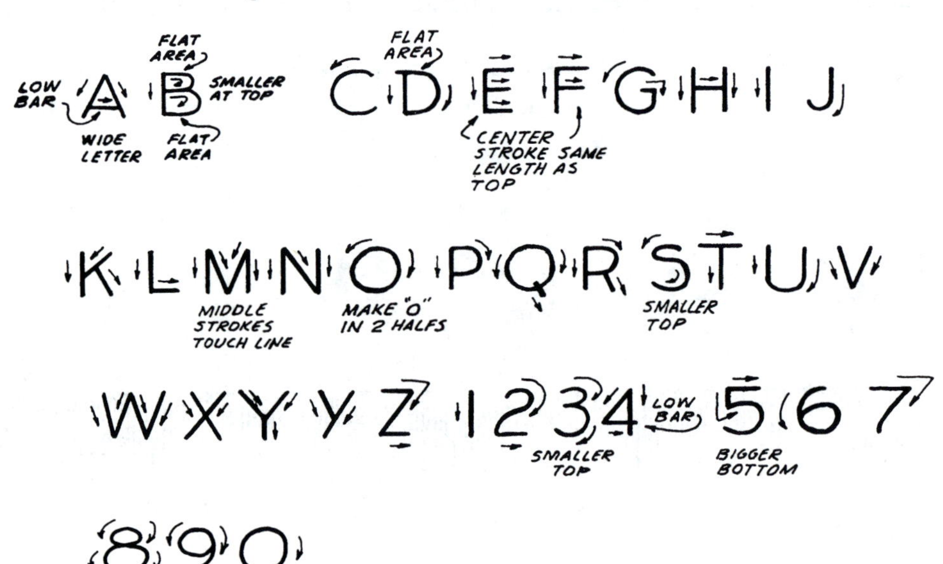

Courtesy of *Basic Industrial Drafting Skills*, Kirkpatrick. Breton Publishers

Figure 1–41 Forming letters and numbers in the vertical drafting alphabet

how each letter should look. It is important that you practice only well-formed letters. After you have drawn a letter, compare it to its form in Figure 1–41. If your letter does not look as good as the one in the figure, or better, analyze what is wrong and correct it. Many students waste valuable time practicing poor lettering. Go slowly at first and do it correctly. You can pick up speed later.

It is very important that fractions be large enough to be microfilmed or reproduced at a reduced size and still be legible. Fractions should be twice the size of whole numbers when they are vertical, Figure 1–42A. Do not allow fraction numbers to touch the bar because that can make the fraction illegible, Figure 1–42B. Vertical fractions are preferred in electronic drafting. However, if the fraction must be horizontal to fit a given space, each number of the fraction should be nearly the same size as whole numbers, Figure 1–43.

Letters can be vertical or inclined; either style is acceptable. The style used usually depends on the drafter's natural ability, but some companies do require one style or the other. The most important characteristics of a good lettering style are that all letters are uniform in height and slant, and that they are bold enough to reproduce well. Figure 1–44 shows the inclined alphabet.

Notice that the curved stroke of the letter D in Figures 1–41 and 1–44 extends a bit beyond the vertical stroke. This is done so that it is not mistaken for a capital letter O or a zero.

Guidelines should be used for all letters and numbers. These lines can be drawn with a 2H or 4H lead, .5 mm or .3 mm thick. Guidelines must be very thin and light—so light that they are barely visible. H or HB lead is used for lettering.

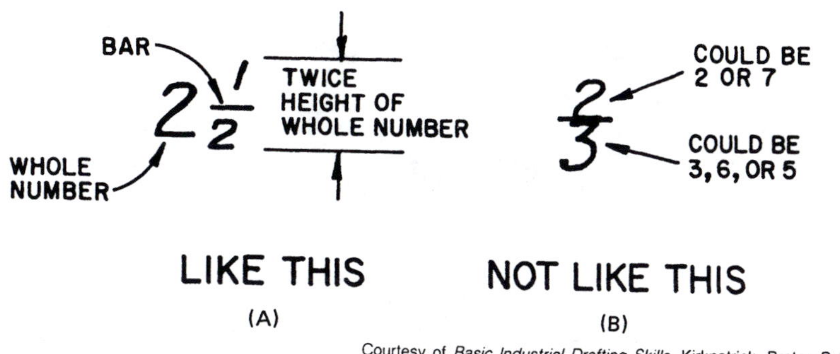

Courtesy of *Basic Industrial Drafting Skills*, Kirkpatrick. Breton Publishers

Figure 1–42 Lettering vertical fractions

2 1/2

Courtesy of *Basic Industrial Drafting Skills*, Kirkpatrick. Breton Publishers

Figure 1–43 Lettering a horizontal fraction

ABCDEFGHIJKLMN
OPQRSTUVWXYZ
1234567890

70°
APPROXIMATE
SLANT

Courtesy of *Basic Industrial Drafting Skills*, Kirkpatrick. Breton Publishers

Figure 1–44 The inclined drafting alphabet

Once you have developed a good lettering style, you can begin to use some of the aids that professional drafters use. But first you must have a good style before you begin using the aids so that your curved lettering form does not become flattened at the top or bottom of a stroke. The lettering aid shown in Figure 1–45 can be used to draw guidelines when practicing at first. Later it can be used, as most professional drafters use it, for lettering within the slots, as shown in Figure 1–46.

When letters are put together to form a word, they should be placed as close together as possible. Because letter widths vary, equal areas of space can result in an unbalanced appearance. The spacing between letters should be about equivalent to the space taken by the capital letter I. The space between words should be equivalent to the area taken up by the capital letter O. The space between lines of lettering should be about two-thirds the height of the letters. Examples of letter, word, and line spacing are shown in Figure 1–47.

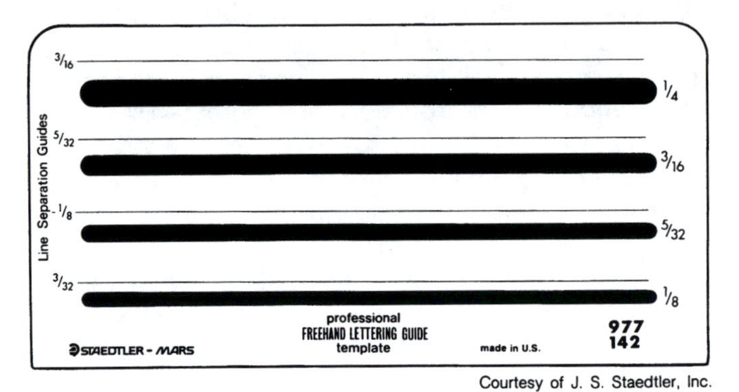

Courtesy of J. S. Staedtler, Inc.

Figure 1–45 Lettering guide

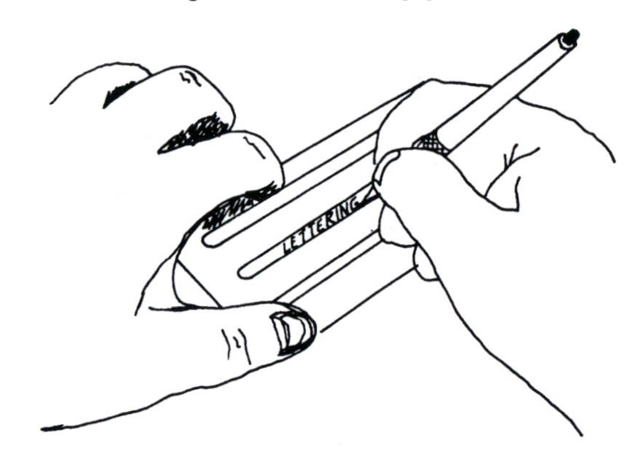

Figure 1–46 Using a lettering guide

Courtesy of *Basic Industrial Drafting Skills*, Kirkpatrick. Breton Publishers

Figure 1–47 Letter, word, and line spacing

Developing a Good Lettering Style

Legibility is the main characteristic of a drafter's lettering. Lettering must also be consistent and neat, but it must be legible so that numbers and values cannot be misread. The alphabet styles shown in Figures 1–41 and 1–44 can be lettered legibly with a little practice. With even more practice, the letters can be developed into an attractive, personalized style. The following are the rules for developing a good lettering style.

- Make sure the form of the letter is correct. Do not mix uppercase and lowercase letters. Most drafting companies use all uppercase (capital) letters.
- Use guidelines either by drawing them or using ruled paper. When drawing guidelines, make sure they are very light and thin.
- Keep the slant of the letters the same. Use either a vertical or a slanted stroke, but keep the stroke consistent.
- Keep the areas between letters the same. Some letters must be spaced farther apart or closer together to balance the areas in between, as is shown in Figure 1–47. An easy way to accomplish this is to draw all the letters as close as possible to one another while maintaining a pleasing appearance.
- Concentrate on keeping the characters open and easily read.
- Do not make letters too tall for the thickness of the stroke. Using a slightly blunted pencil point, letters $\frac{1}{8}''$ to $\frac{3}{16}''$ high are best.
- Work to increase your speed as soon as you have the correct form, density, and slant.
- Make the space between words approximately the width of the capital letter O (see Figure 1–47).
- Do not allow letters or numbers to touch any object line, border, or fraction bar. Letters and numbers should have clear space all around them—sides, top, and bottom.
- Begin by drawing letters and numbers. Students who are lettering for the first time tend to hurry. They finish a drawing on which they have spent a great deal of time, only to ruin it by rushing through the lettering. If you do not have time to complete the drawing using a good lettering technique, set it aside and go back to it later.
- Make all letters very dark. If you must repeat a stroke to improve its density, do so. Many professional drafters use a technique that looks almost as if they make a quick double stroke first and then move on through the letter. You will find a technique that suits you if you work on good lettering form, consistent slant, and very dense letters.

A Device for Inked Letters

The instrument most commonly used to make letters in ink is the device shown in Figure 1–48. This device, manufactured by several com- companies, is commonly called Leroy equipment, after its original manufacturer. It is available at any good drafting supply store.

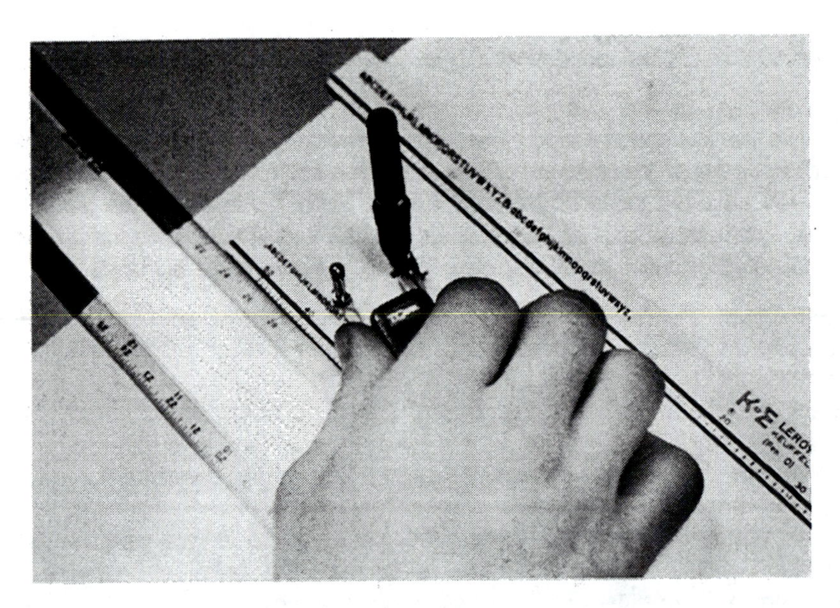

Courtesy of Keuffel and Esser

Figure 1–48 Leroy lettering device

The device has two major components: a scriber and a template. The scriber has several parts, all of which must be kept tight when in use. One of these parts, called the *tracer point,* has a fine point on one end and a blunt point on the other. The blunt point is used for all templates larger than an 80 guide. The fine point is used for the 80 guide and all lettering smaller than .080. (All template sizes are labeled in thousandths of an inch; therefore, an 80 guide is .080, or 80 thousandths of an inch.)

An adjustable arm is used for making slanted letters. Not all scribers have this feature, but, when one does, it must be tight. The penholder for the technical pen must also be tight.

Every template has an identifying number, such as 120CL, for example, which means that the numbers and letters on the template are .120″ high, and that it contains both capital and lowercase letters. The template also recommends a pen size to use.

Using the scriber and the template is very simple, but some practice is necessary to become skilled at it. Start by putting the pen in the hole and tightening all parts of the scriber. Then place the template on your drafting machine or T-square so that you can keep all the letters in a straight line. Put the back pin of the scriber in the long slot of the template. Put the scriber in the letter you select and, using no downward pressure whatsoever, guide the scriber through the letter. The pen should trace a perfect letter.

If the pen does not flow properly, jiggle it until it does and try again. Using force on the equipment is the worst mistake you can make. Use a light touch for the best results. Space the letters in a word evenly. Leave a space about two-thirds of the letter height between lines of lettering. As in freehand lettering, the correct space between words is equivalent to the space taken by the letter O.

COMPUTER AIDED TOOLS

Drawings for electronics may be done on a personal computer (Figure 1–49) or a mainframe computer. Plotters which produce photographic or ink drawings of excellent quality are readily available. Pens for these plotters are similar to the technical drawing pens used for manual drafting and require similar care. A relatively new product is a disposable pen which rarely clogs and is discarded when it runs out of ink. Consult any good CAD textbook for further information on how to use the computer to make drawings.

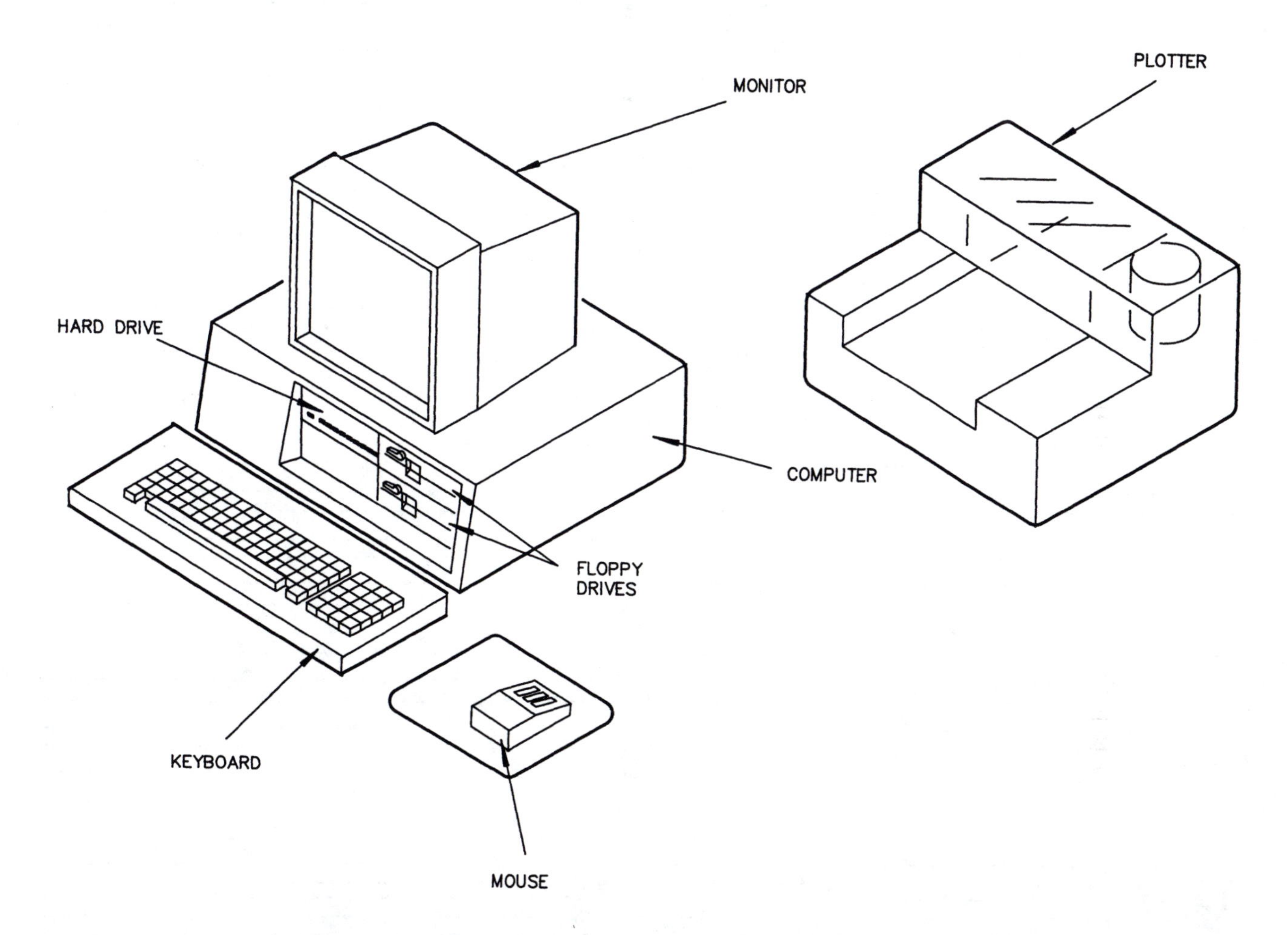

Figure 1–49 Drawing for electronics may be done on a personal computer or a mainframe computer

EXERCISE 1-1

Using the following guide as a sample, draw the vertical style of lettering the stated number of times for each of the letters, numbers, fractions, words, and sentences.

DO LETTERS AND NUMBERS 10 TIMES

A B
C D
E F
G H
I J
K L
M N
O P
Q R
S T
U V
W X
Y Z
1 2
3 4
5 6
7 8
9 0

$1\frac{1}{2}$ 5 TIMES

$2\frac{3}{4}$ 5 TIMES

GOOD LETTERING MAKES A GOOD DRAWING BETTER

G
G
G
G
G

5 TIMES

EXERCISE 1–2

Identify drawing tools labeled (A) through (G).

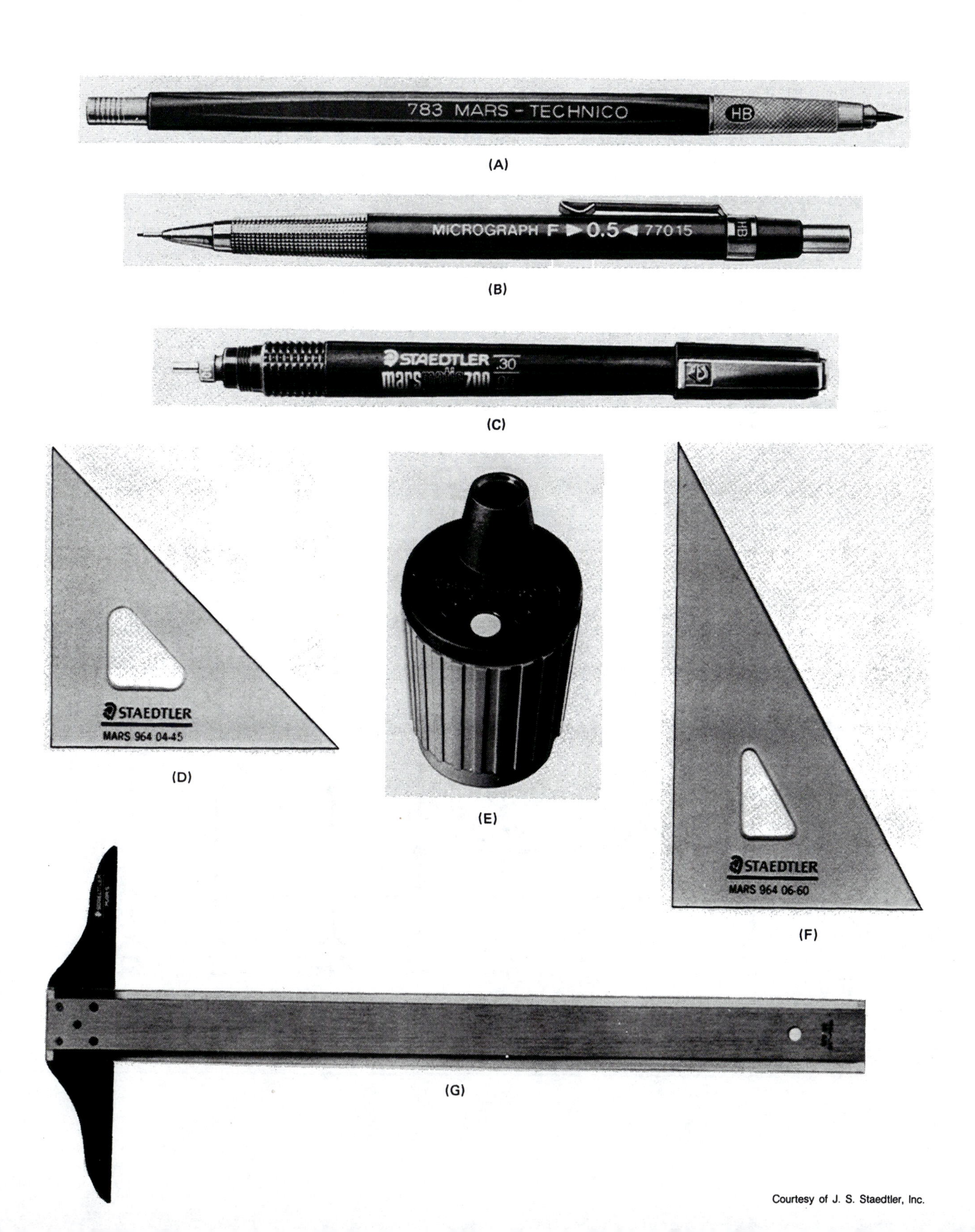

(A)

(B)

(C)

(D)

(E)

(F)

(G)

Courtesy of J. S. Staedtler, Inc.

EXERCISE 1-3

Trace or copy the following drawing on vellum using an electronic symbol template. Make sure all of your lines are very dark and the full width of a .5 mm HB lead.

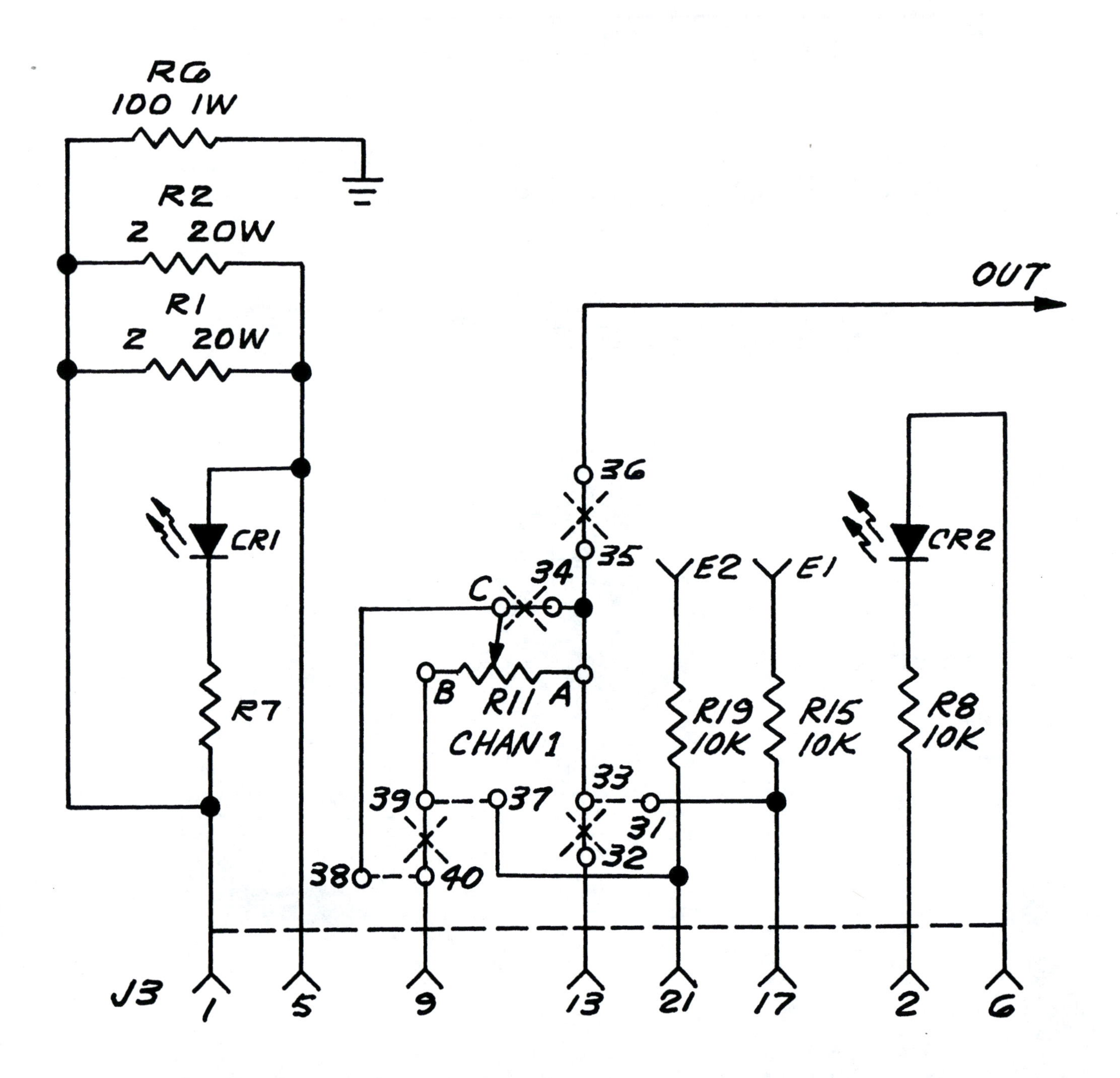

EXERCISE 1-4

Trace or copy the following drawing on vellum using a logic symbol template.

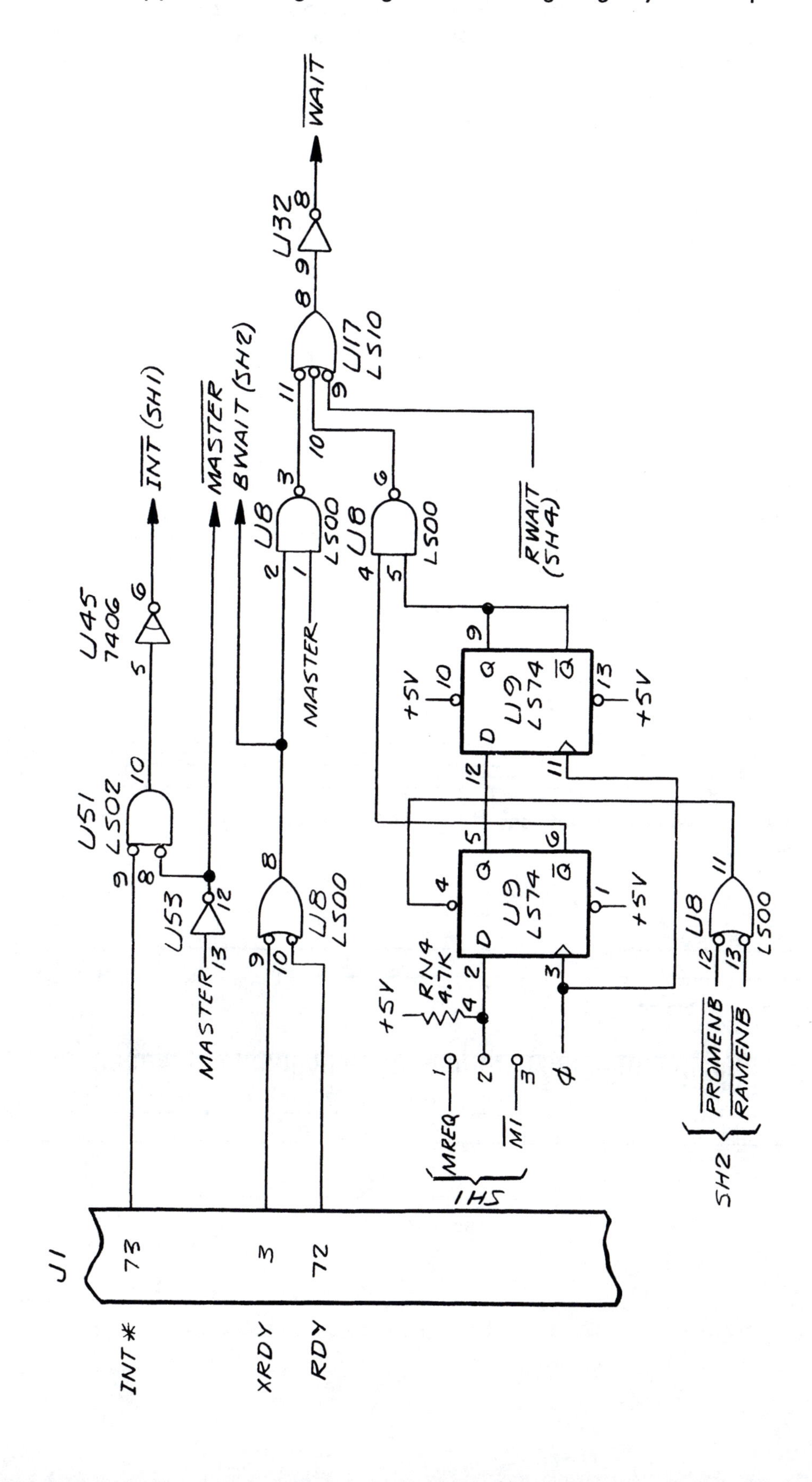

EXERCISE 1-5

Trace the following drawing on vellum using a PC component template. Use a triangle, straight edge, and circle template if a component template is not available. If you plan to use a personal computer for the drawings in this book, construct these parts to the sizes shown and place them in drawing files on your hard disk. They can be used for later drawings.

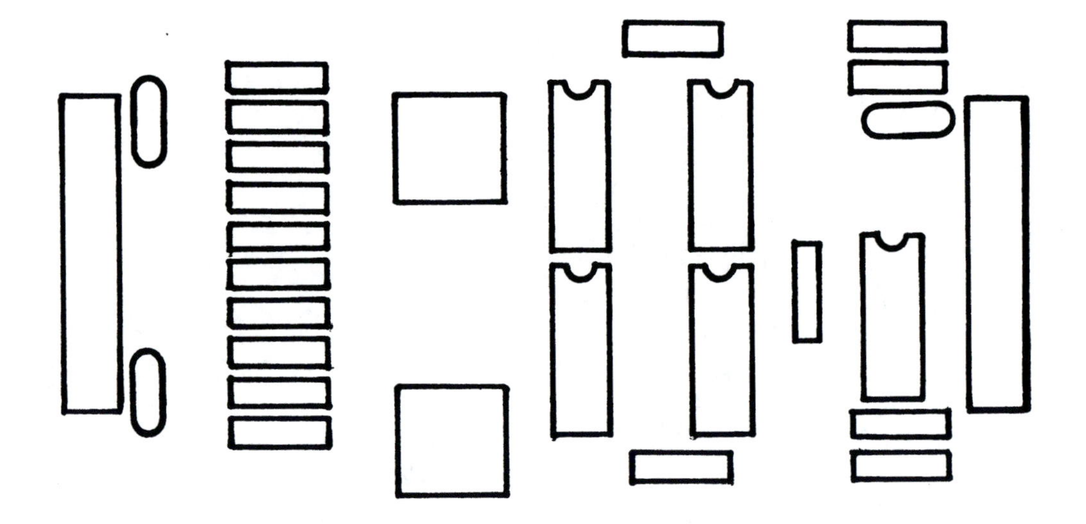

EXERCISE 1-6

Write the correct measurement in the blank on each dimension line for the scale shown.

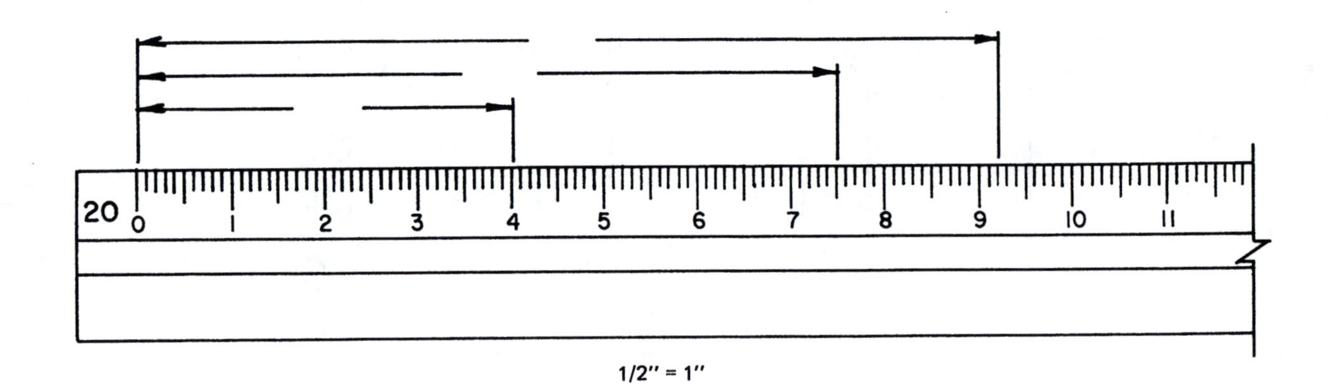

1/2" = 1"

EXERCISE 1-7

Write in the correct answer for each of the measurements shown on the ruler that follows.

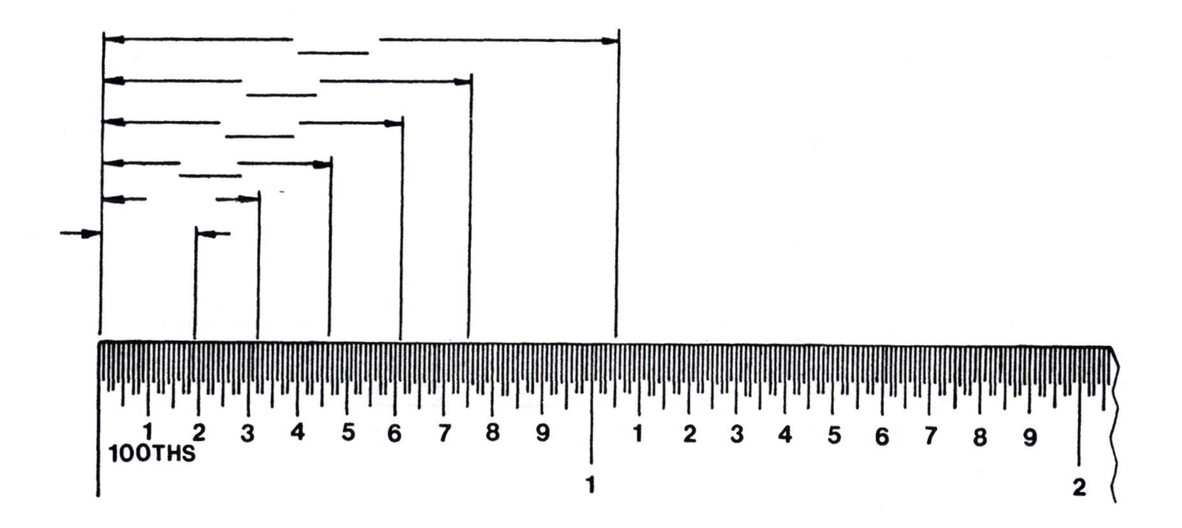

EXERCISE 1-8

Write the correct measurement in the blank on each dimension line. Write the dimension as a two-place decimal, ex: 5.25.

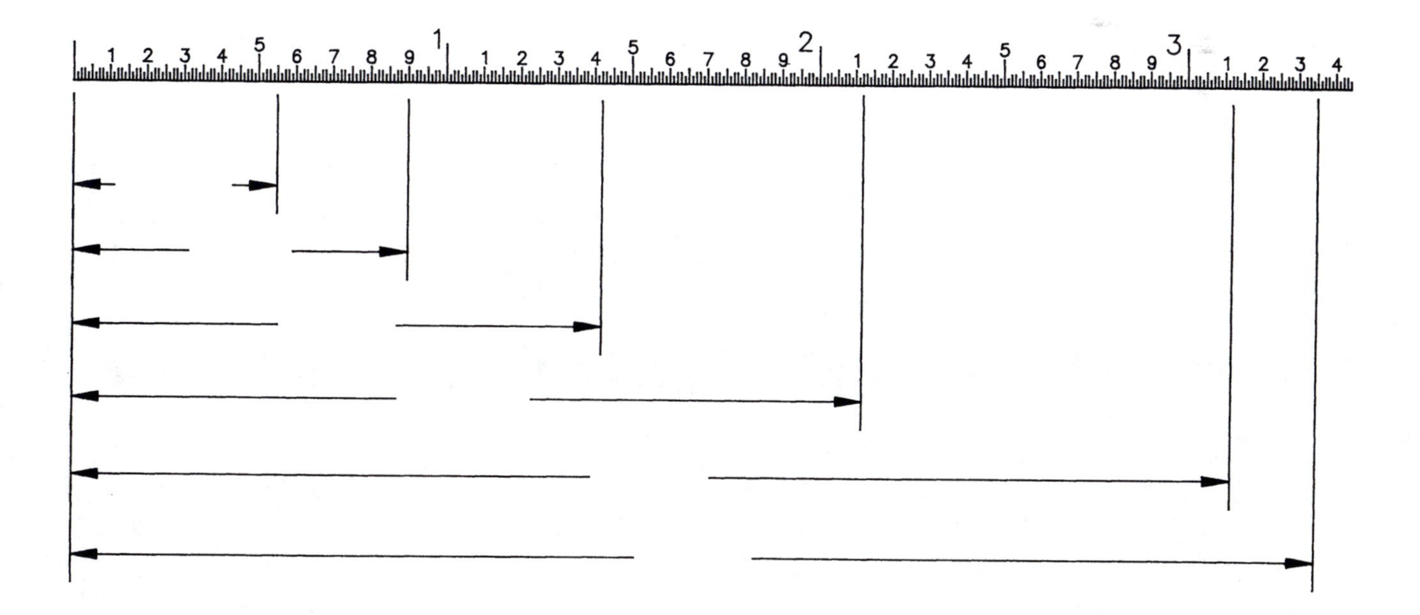

EXERCISE 1-9

Trace the following drawing including the lettering. Be careful to make your lettering exactly the same as shown on the drawing. Draw the box outlines with a .7 mm lead and everything else with a .5 mm lead.

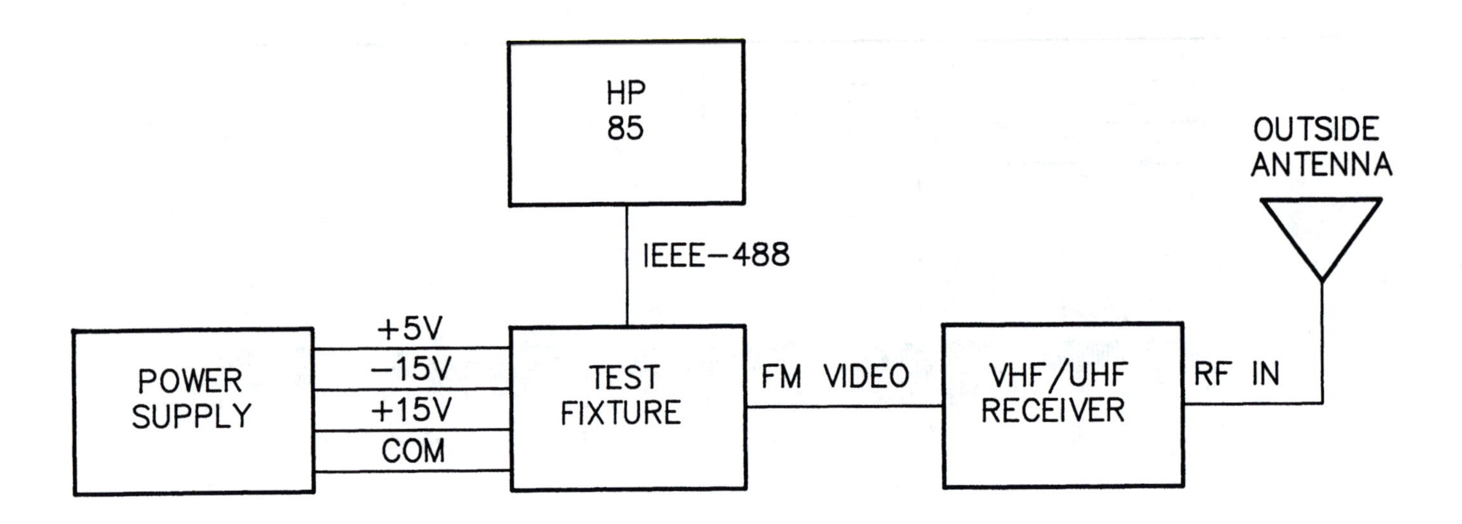

Chapter 2

Types of Electronic and Mechanical Drawings

OBJECTIVES

After completing this chapter, you will be able to

✓ list eight major types of electronic drawings, and explain how each is used.

MAJOR TYPES OF DRAWINGS

Eight major types of drawings are commonly produced in the electronics industry. They are:

- Block diagrams
- Logic diagrams
- Schematic diagrams
- Cabling diagrams
- Component assembly drawings
- Printed circuit artwork
- Drill plans or tapes
- Mechanical drawings

A type of mechanical drawing called a *specification control drawing* is used often to ensure that purchased parts meet the required standards. All of these types of drawings are covered in detail in this text, except for mechanical drawings which are better discussed in a basic drafting text.

Block Diagrams

All of the drawings shown in Figure 2–1 are block diagrams. Figure 2–1A is a block diagram showing how an engineering group is organized. Figure 2–1B is a functional block diagram showing how and when work is performed throughout an engineering project. Figure 2–1C is a functional block diagram which shows how functional units are related in a piece of equipment. The functional block diagram is by far the most commonly used type of block diagram in the electronics industry.

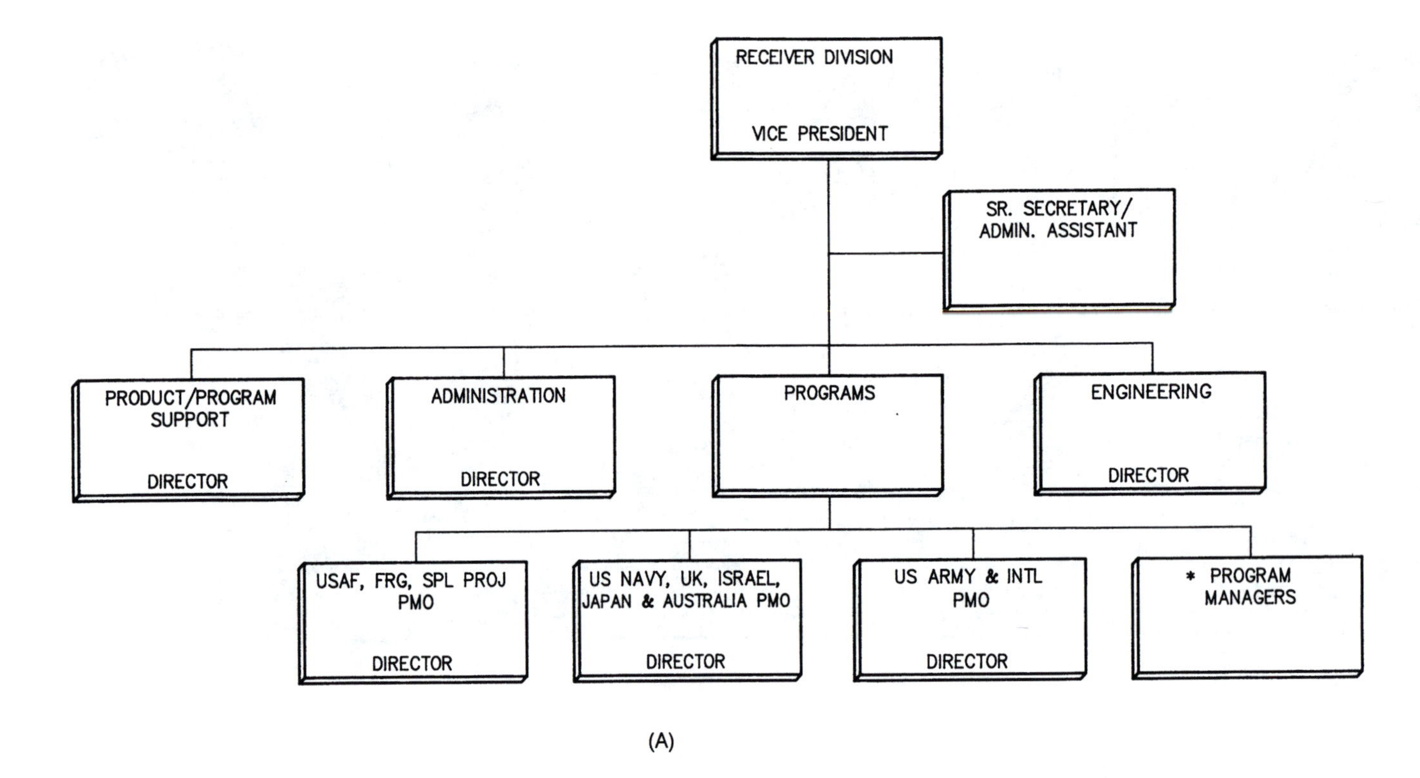

(A)

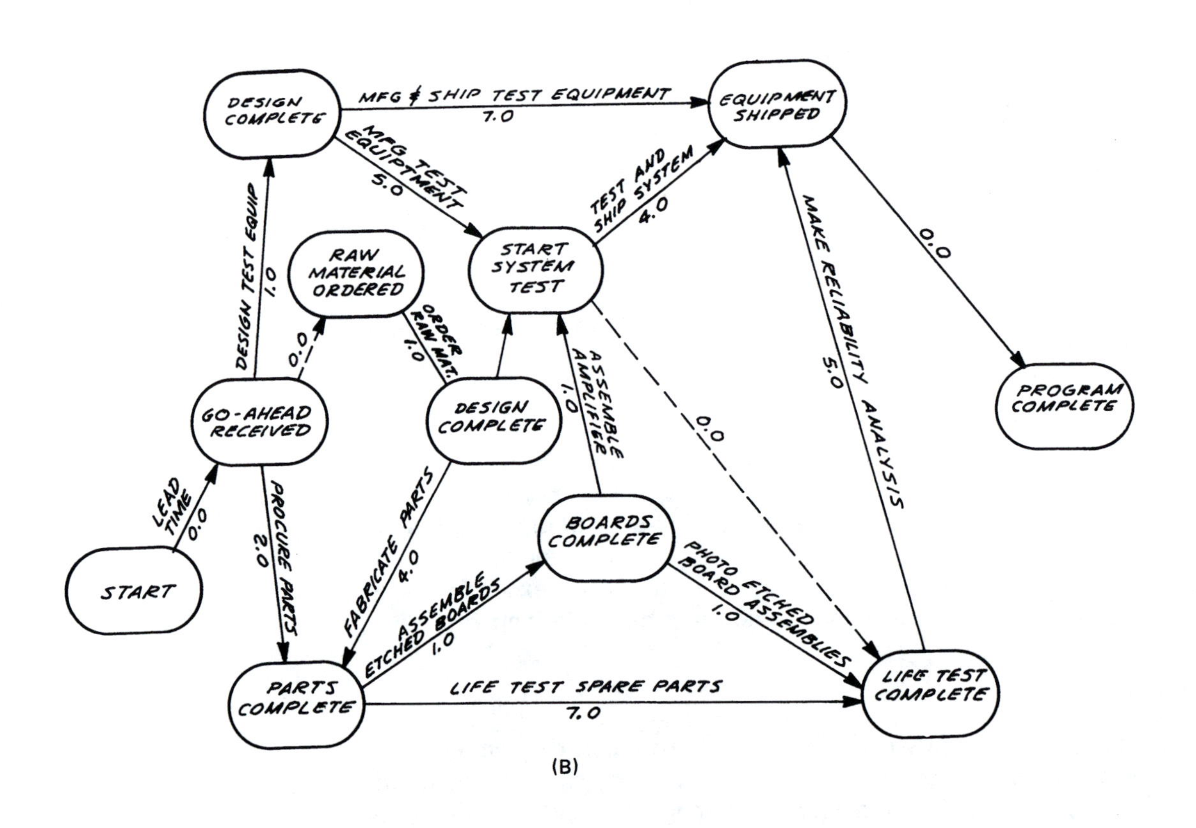

(B)

Figure 2-1 All of these drawings are block diagrams

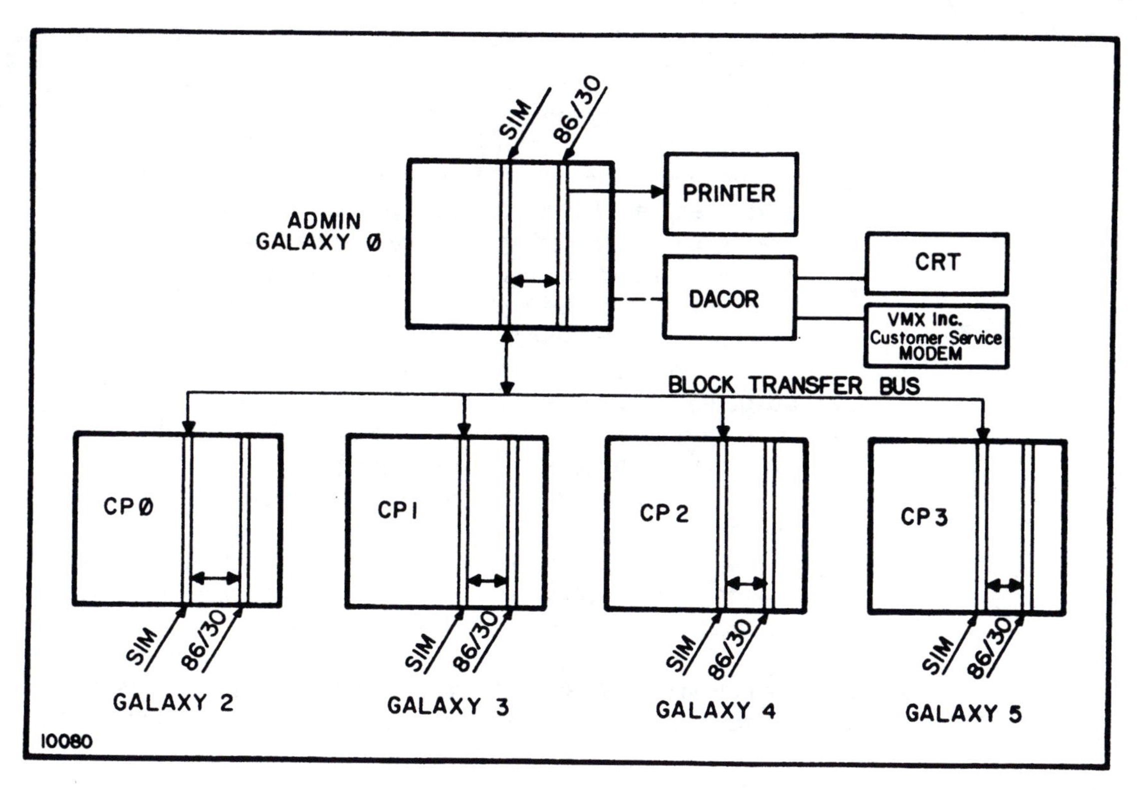

Figure 2–1 (continued)

Block diagrams can be identified by the blocks which are used to illustrate units, people, functions or various other factors. For whatever the purpose, these diagrams always use blocks and usually use arrows to present the desired concepts. Electronic symbols are often intermixed with the blocks on functional block diagrams, shown in Figure 2–2. Block diagrams can be very simple or very complex depending on what ideas are being illustrated. Block diagrams are closely related to logic diagrams.

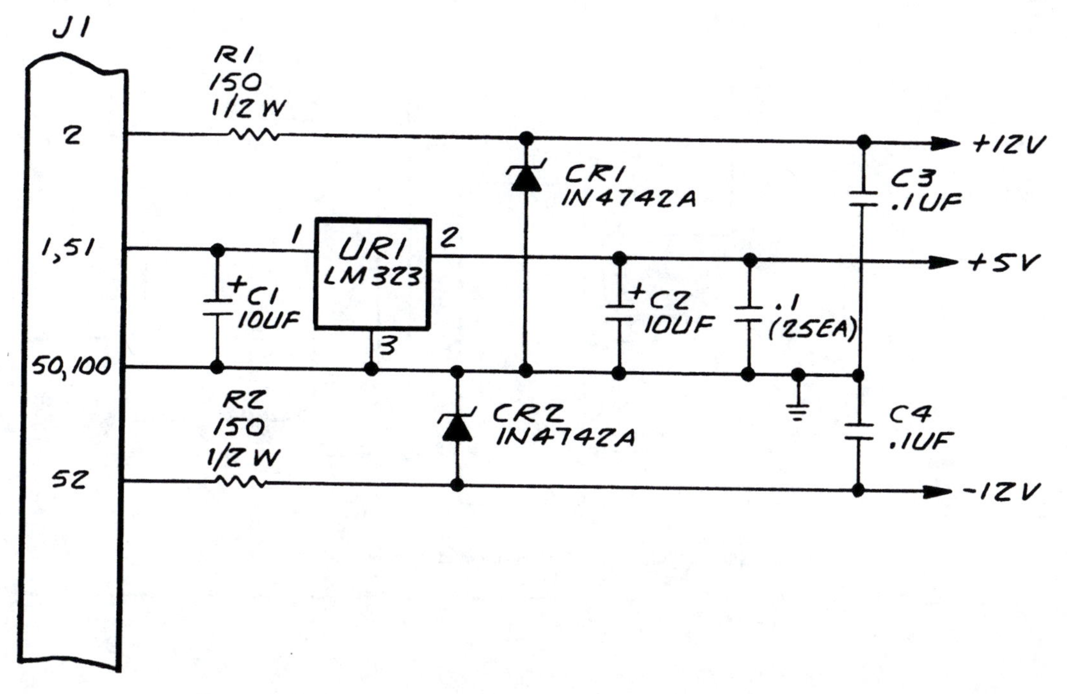

Figure 2-2 Electronic symbols are often intermixed with the blocks

Logic Diagrams

Three logic diagrams are shown in Figure 2–3. These are diagrams of switching circuits such as those commonly used in computers. In Figure 2–3A, each one of the rounded box-type symbols, called *gates,* represents several different electronic components; other boxes represent complete integrated circuits. Logic diagrams are readily identified by the logic gates such as those shown in this figure, and they often have gates or blocks that represent a complete integrated circuit package.

The wedge-type symbols in Figure 2–3B represent connecting metal strips on the edge of a printed circuit board. Notice in the figure that there are two gates identified as I5. This labeling indicates that both gates are physically contained within the same package. Symbols I2 and I3 in Figure 2–3A represent complete integrated circuits.

Figure 2–3C is another example of a logic diagram.

Schematic diagrams are closely related to block diagrams and logic diagrams. Schematic diagrams use symbols to identify electronic or electrical components. A symbol is used instead of a picture of the part. Schematic symbols are often mixed with logic symbols and blocks to produce a complete schematic diagram for a piece of equipment which involves both simple electronic parts and integrated circuits. In many types of circuits the terms *logic diagram* and *schematic diagram* are used interchangeably.

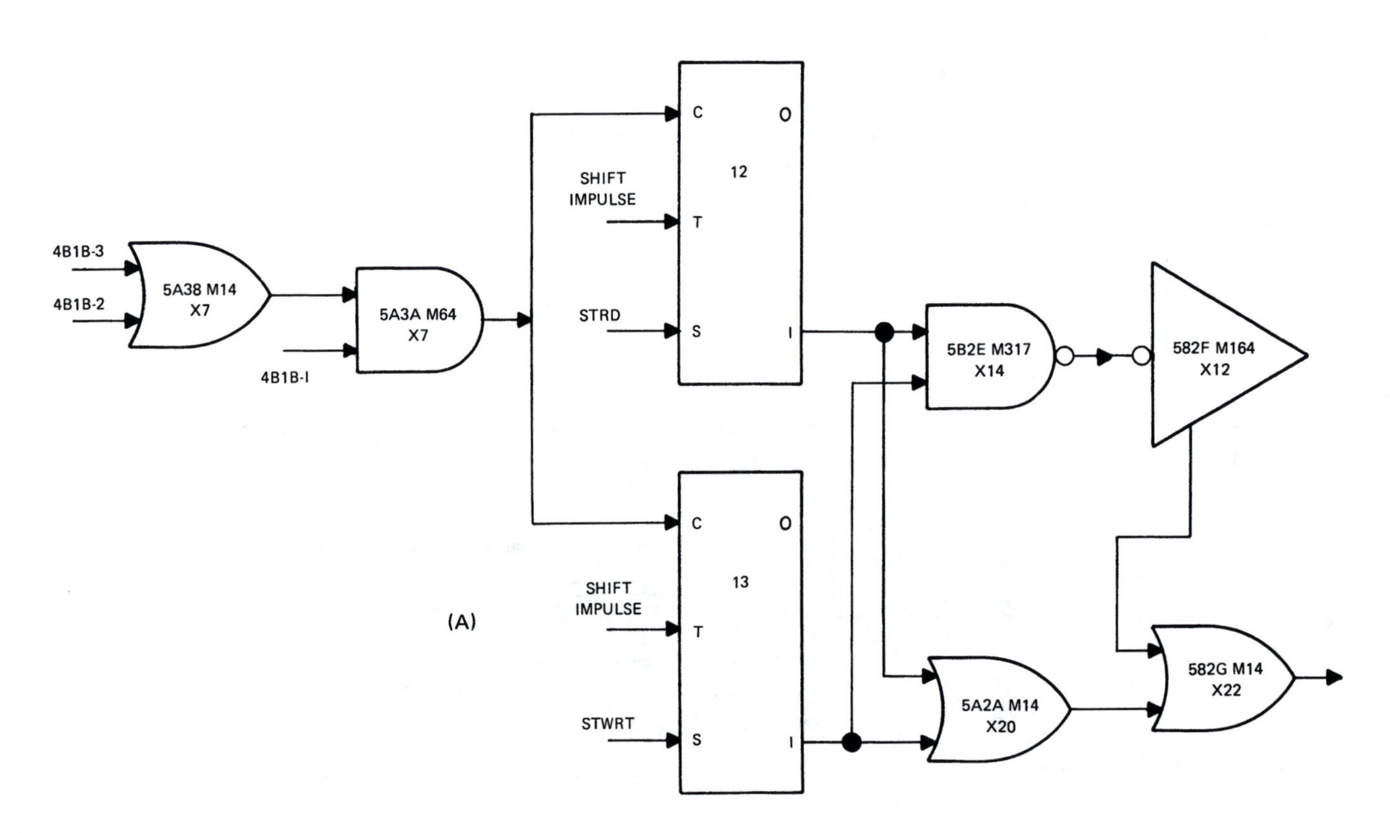

Figure 2–3 Examples of logic diagrams

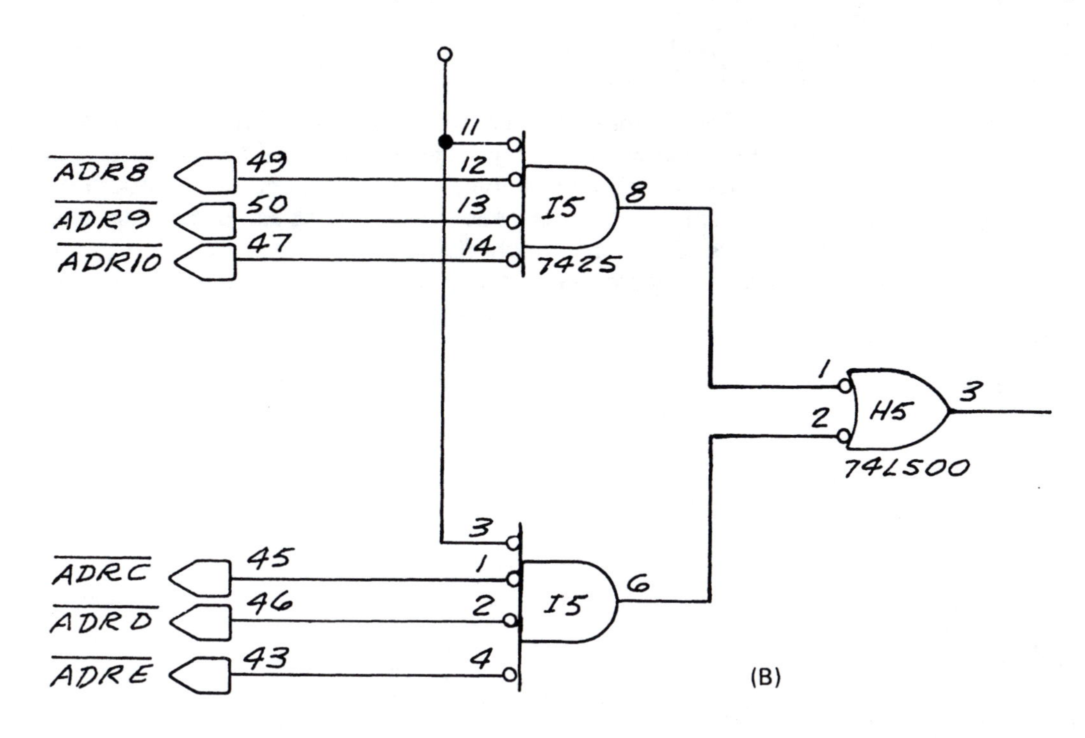

(B)

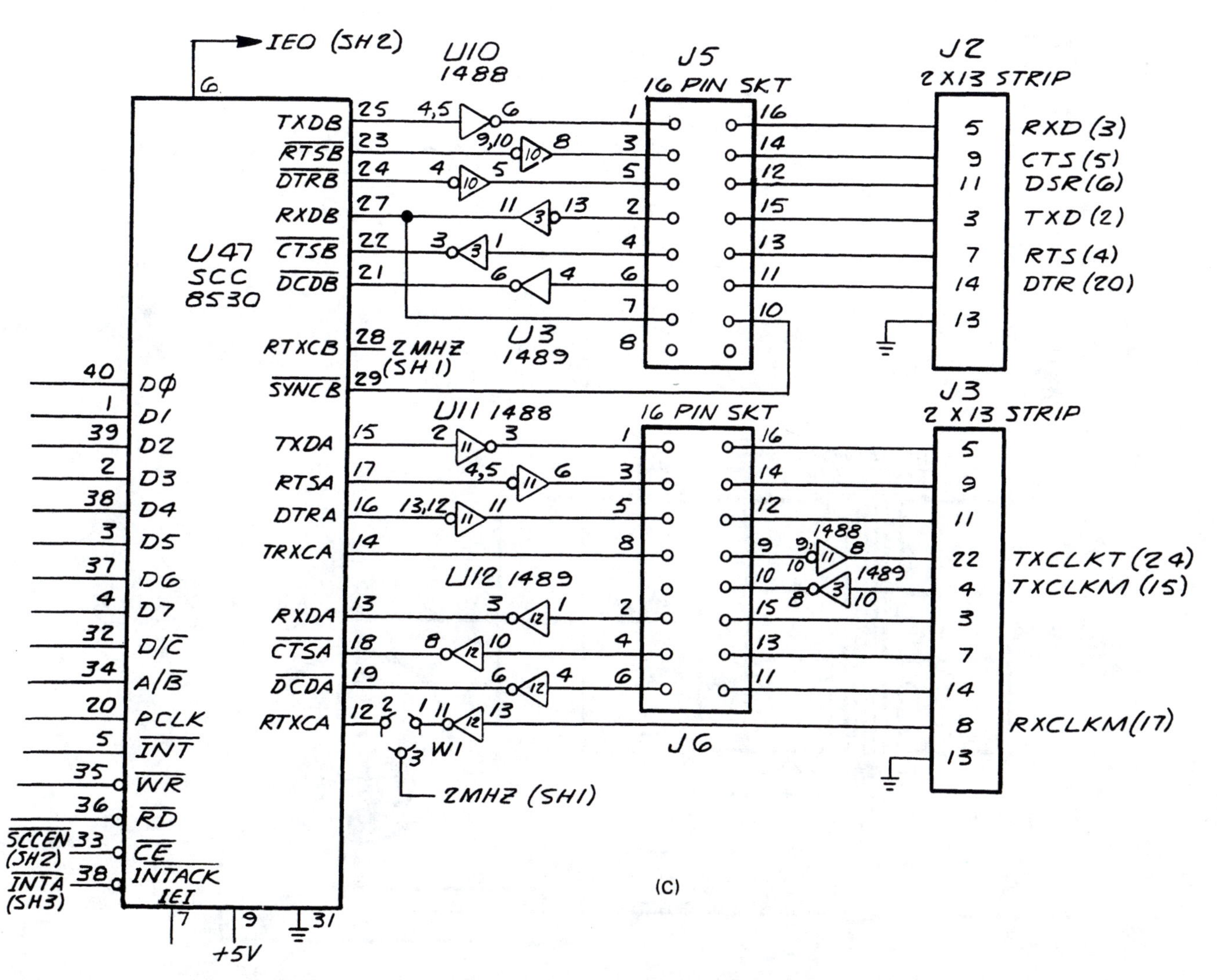

(C)

Figure 2-3 (continued)

Schematic Diagrams

Three types of schematic diagrams are shown in Figure 2–4. Schematic diagrams show which components are connected together, but do not show the physical position of the part on a printed circuit board or in a piece of equipment. More information is given about schematic diagrams in later chapters. Schematic diagrams usually precede cabling diagrams in the order in which they are drawn (the individual electronic units are usually completed before they are cabled together).

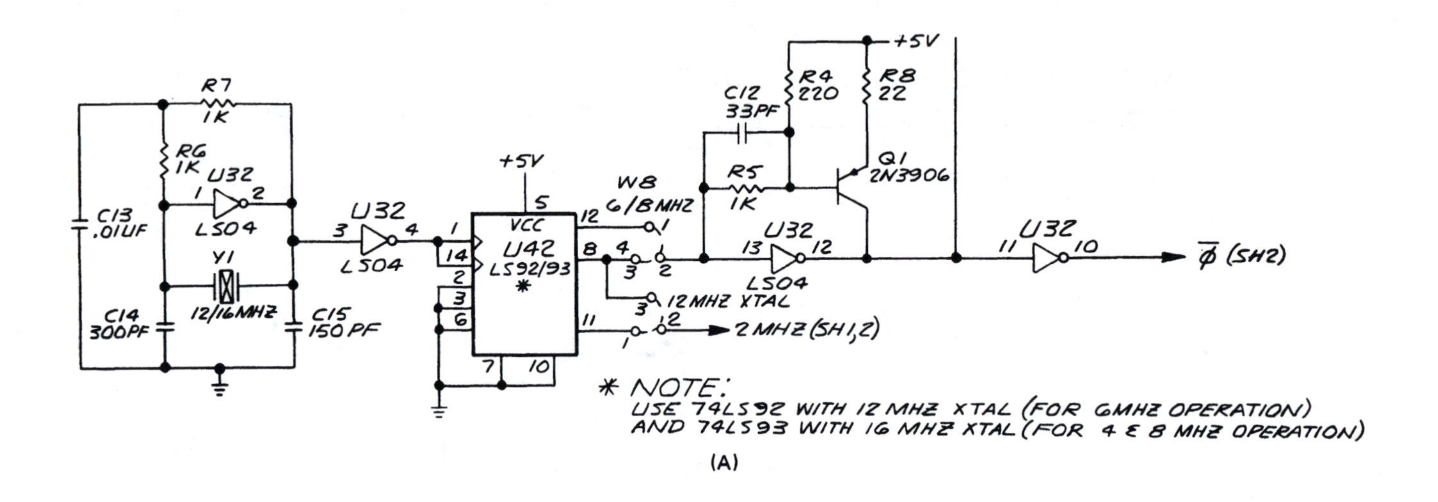

(A)

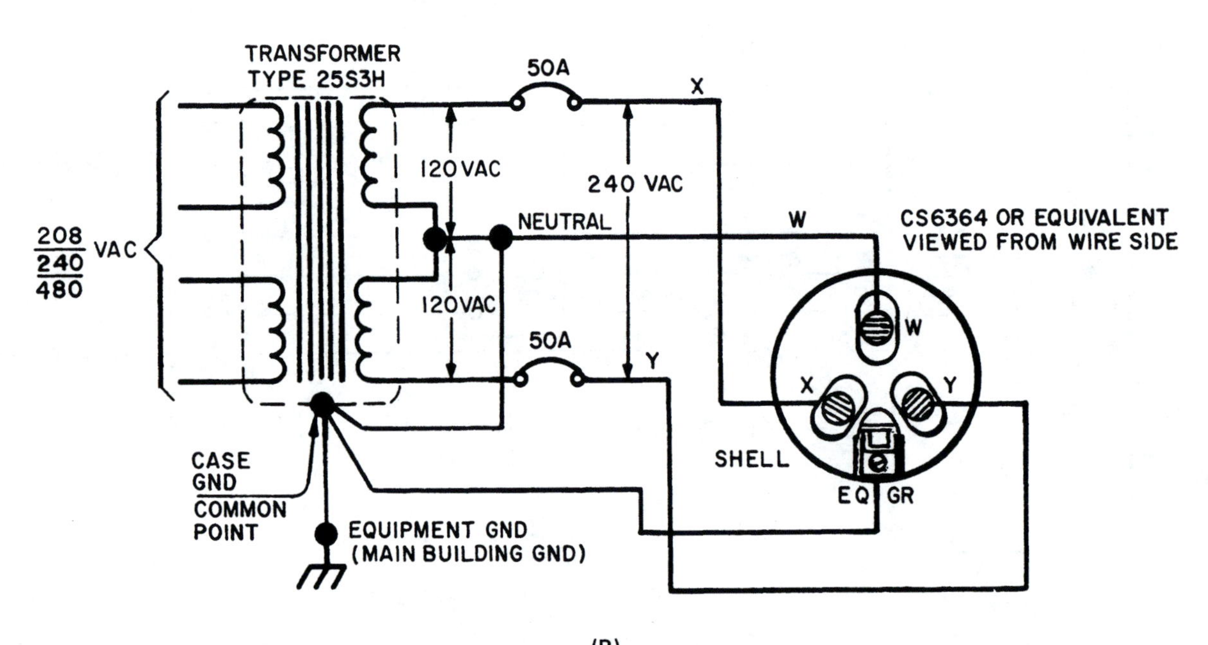

(B)

Figure 2–4 Three types of schematic diagrams

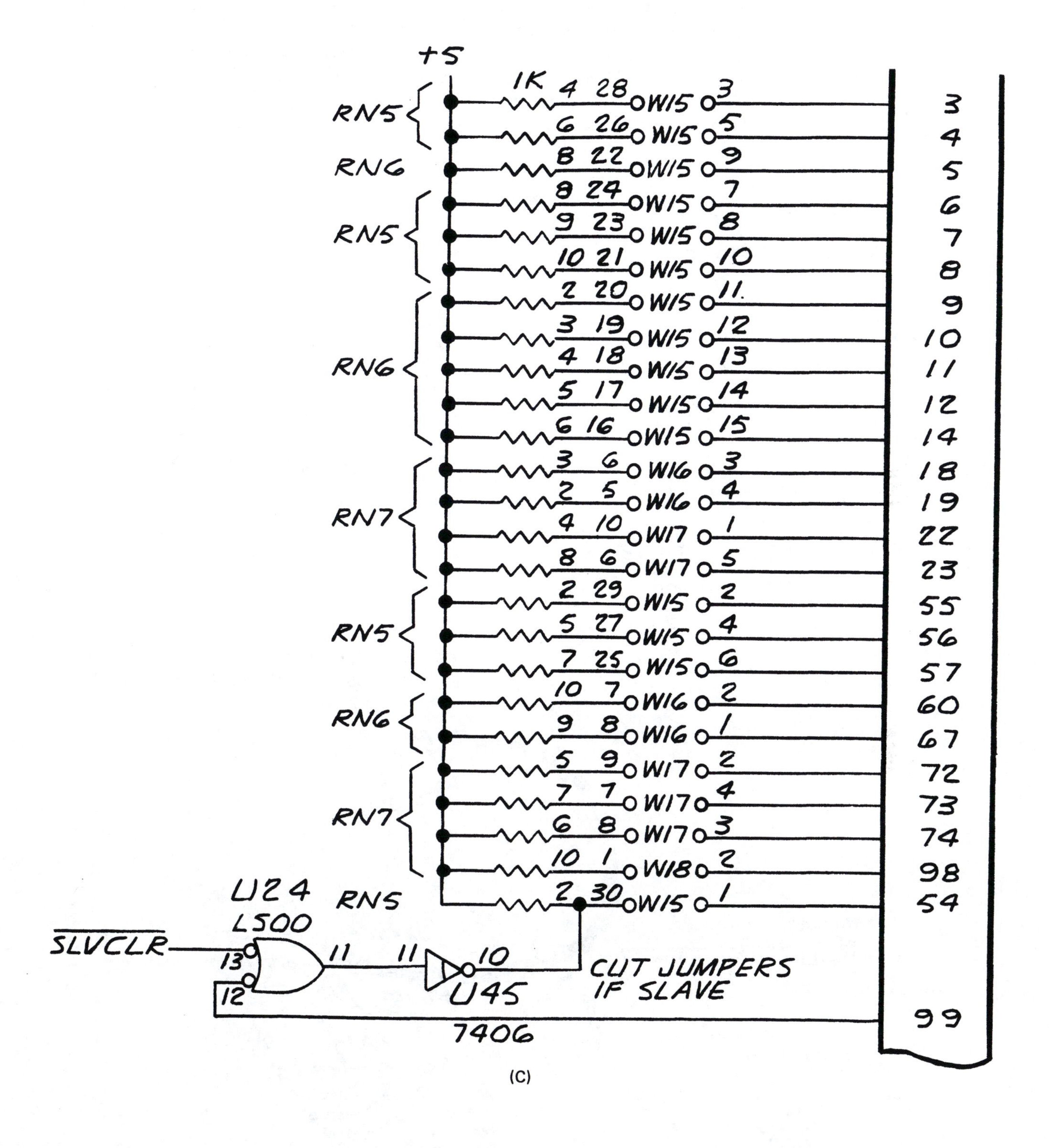

(C)

Figure 2–4 (continued)

Cabling Diagrams

Three common types of cabling diagrams are shown in Figure 2–5. Figure 2–5A is a *highway diagram* which shows how cables are routed among units in a system, the destinations of the wires (where they go), and the wire colors or functions. The highway diagram is used to reduce the number of lines drawn on a cabling diagram. Figure 2–5B shows a *point-to-point wiring diagram.* Figure 2–5C is a drawing of a wiring harness which shows how to build a wiring harness that connects parts of a system.

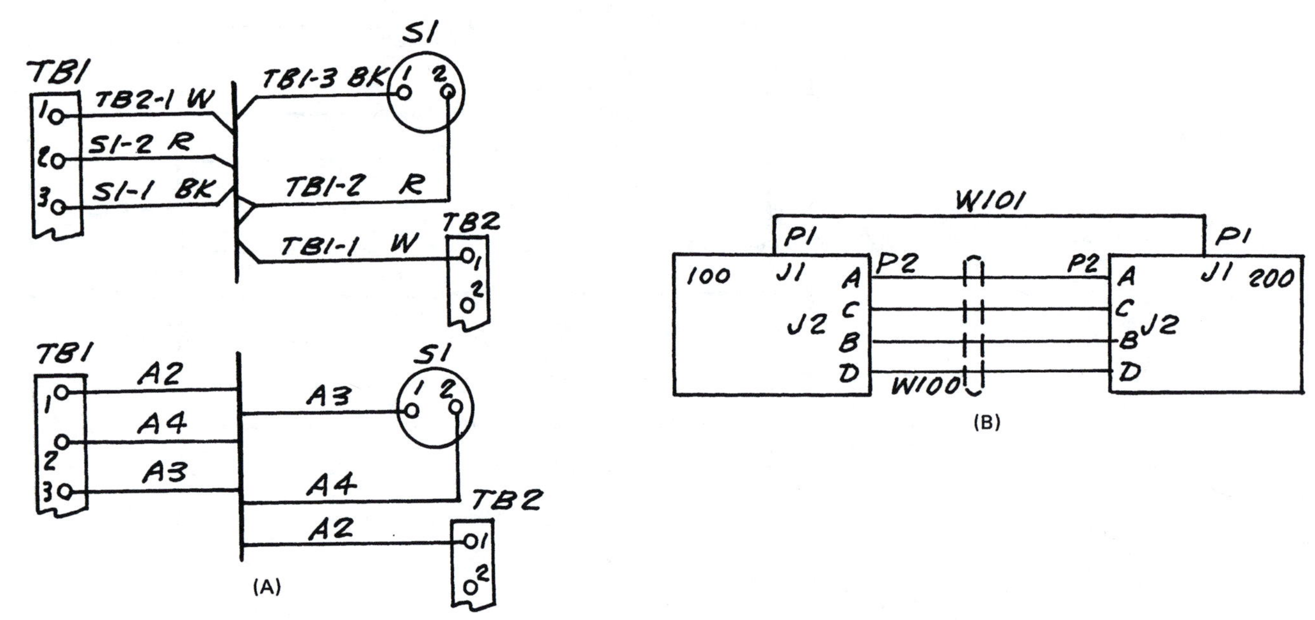

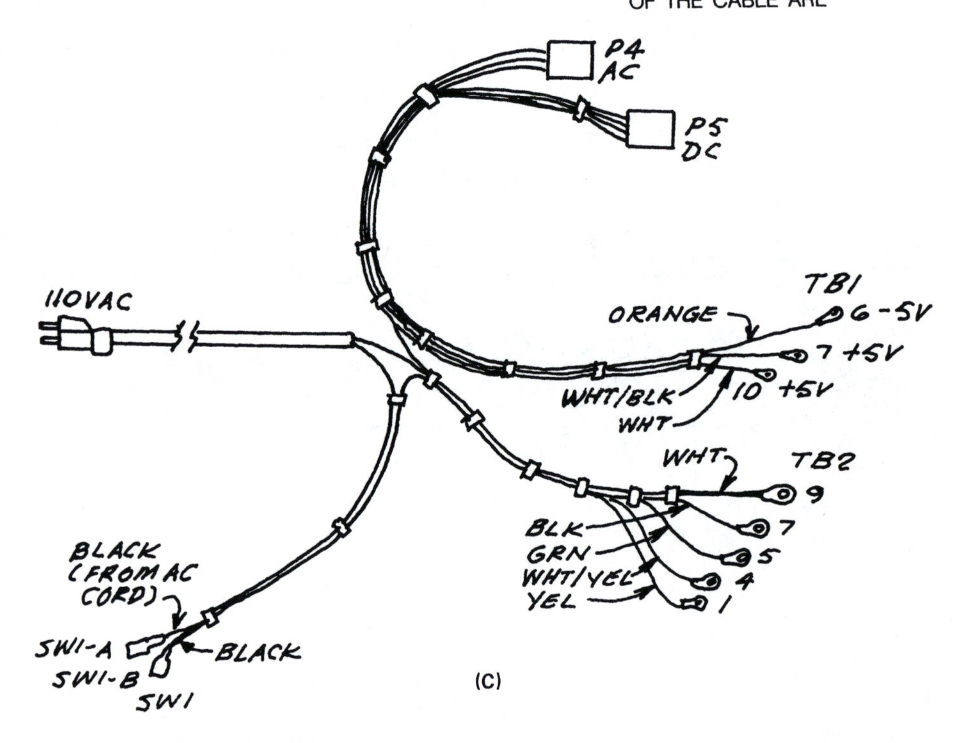

Figure 2–5 Common types of cabling diagrams

Component Assembly Drawings

Component assembly drawings (or component layout drawings) are used to assemble electronic parts. They show how to assemble components on a printed circuit board, and often show how the components are connected. Examples of component assembly drawings are shown in Figure 2–6.

Component assembly drawings are easily recognized because they show the actual size of the components and their physical appearances. A label, called a *reference designator,* is also shown for each component. In assembly #1 in Figure 2–6A, R1 and R2 are resistors, C2 is a capacitor, and Q1 is a transistor. These are known as discrete components. Methods for identifying these components are given in a later chapter. Integrated circuit packages are shown in assembly #2, Figure 2–6B, and assembly #3, Figure 2–6C.

The artwork used to etch the circuit on the printed circuit board is often used on the assembly drawing to show the hole patterns for components and how the components are connected. This artwork is often used for checking.

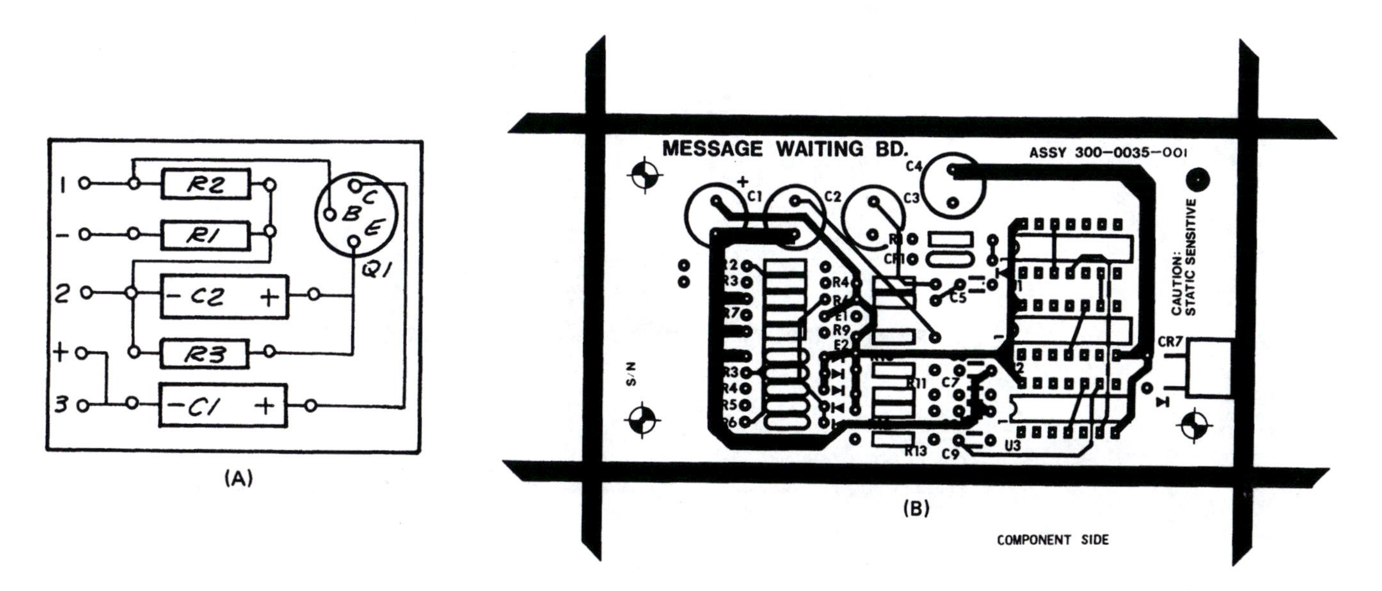

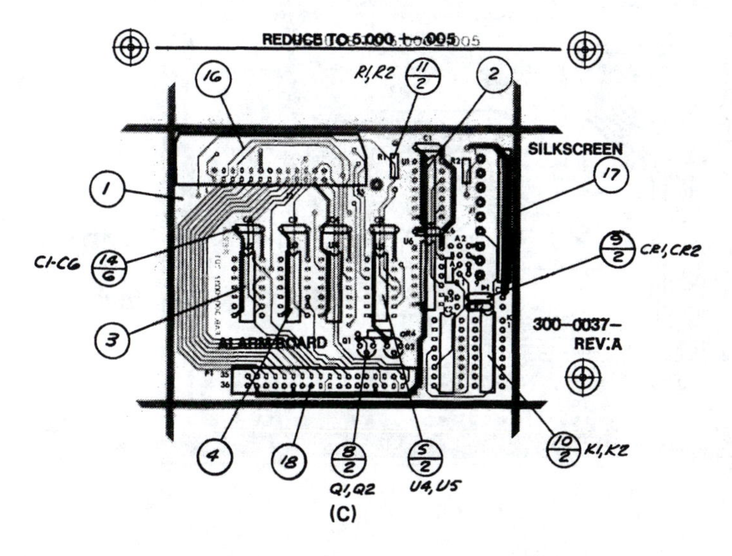

Figure 2–6 Examples of component assembly drawings

Artwork

Figure 2–7 shows two different examples of artwork used for printed circuit boards or printed wiring boards (they are identified by both names). Artwork is used to etch the circuit on the board. This type of drawing shows no components. Artwork is easily identified by its long lines of etched circuits.

More information about how artwork is used in manufacturing is given in a later chapter.

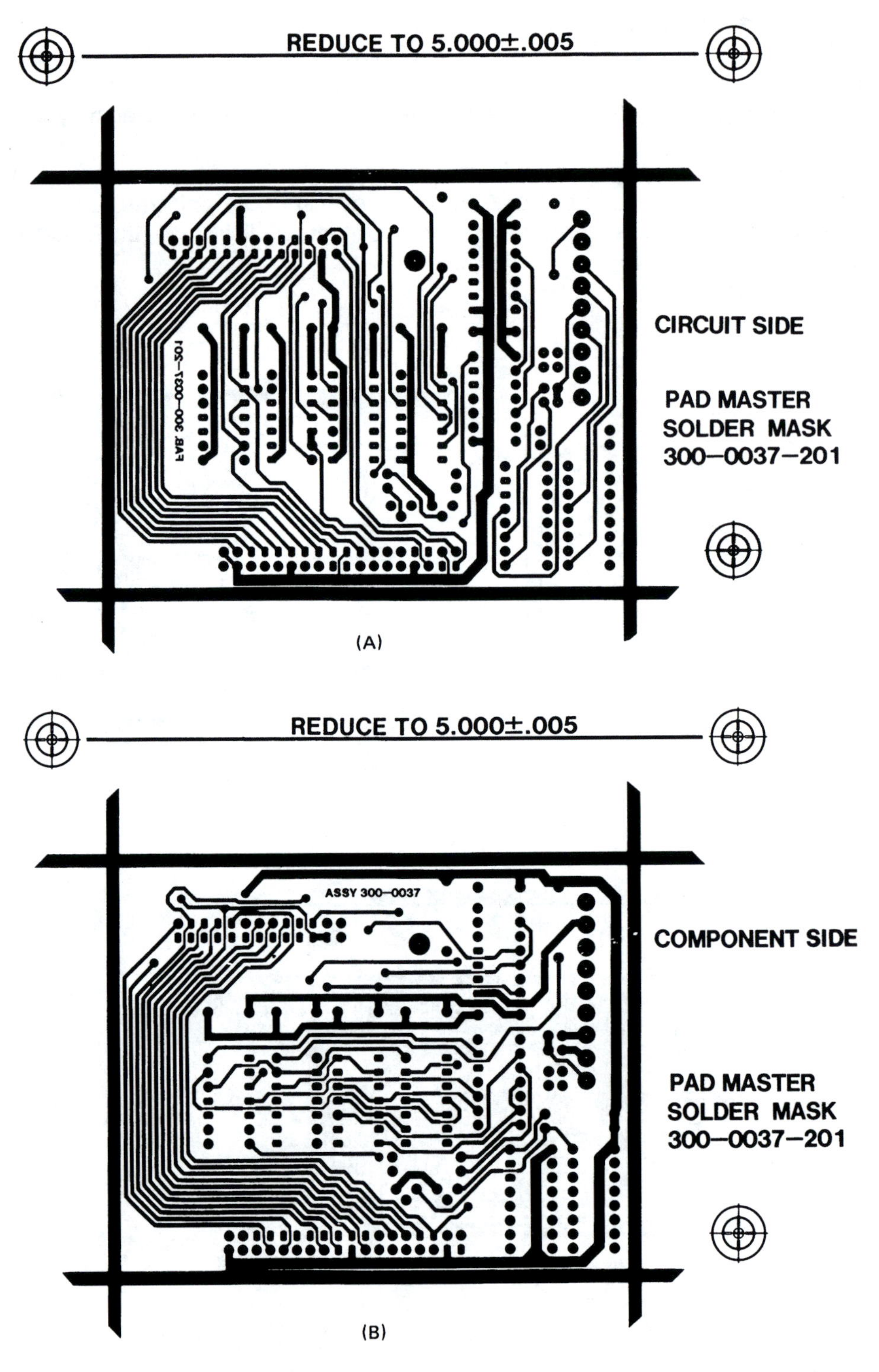

Figure 2–7 Artwork is used to etch the circuit on the board

Drill Plans or Tapes

A drill plan is a drawing used to manufacture the bare printed circuit board (without components mounted on it). These drawings show the size and shape of the board, hole sizes and where the holes are to be drilled on the printed circuit board. Components are not shown on a drill plan. Figure 2–8 shows two different drill plans. These drawings show hole sizes and where the holes are to be drilled on the printed circuit board. Holes allow components to be mounted on the printed circuit board. The table shown on the drill plan, called the *hole schedule,* shows how the hole is identified (usually by a letter), the hole size, and how many of each hole is required. The quantity of holes, at a certain cost per hole, is often used in determining the cost of the board.

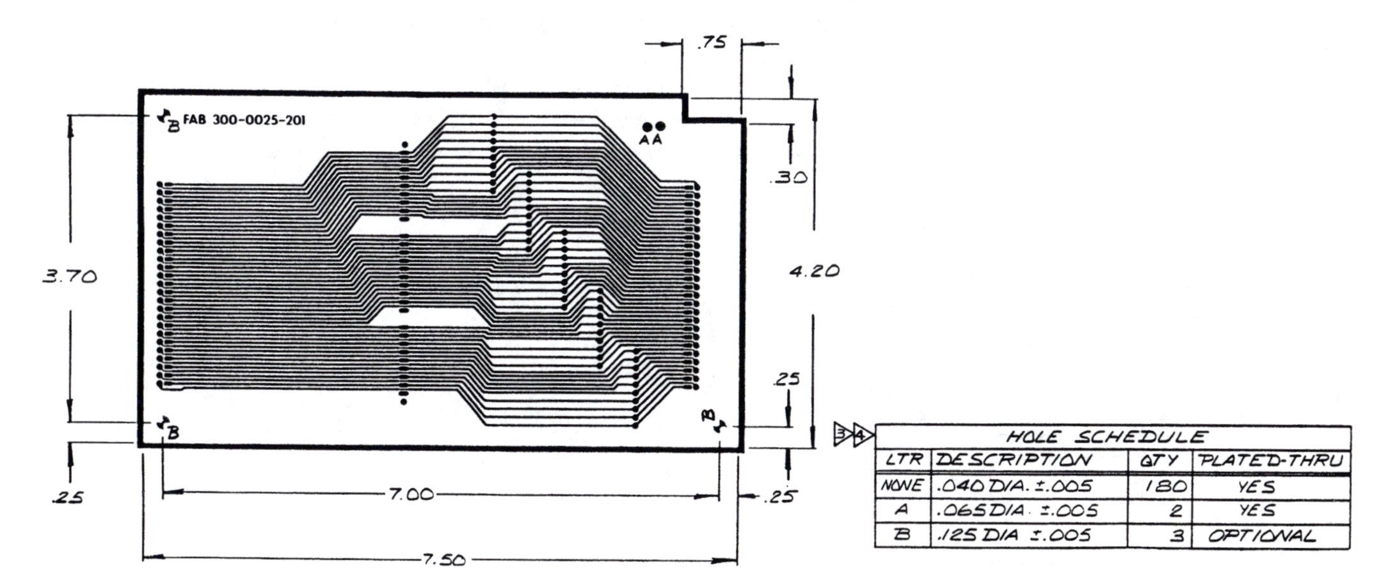

HOLE SCHEDULE			
LTR	DESCRIPTION	QTY	PLATED-THRU
NONE	.040 DIA. ±.005	180	YES
A	.065 DIA ±.005	2	YES
B	.125 DIA ±.005	3	OPTIONAL

DRILL PLAN A

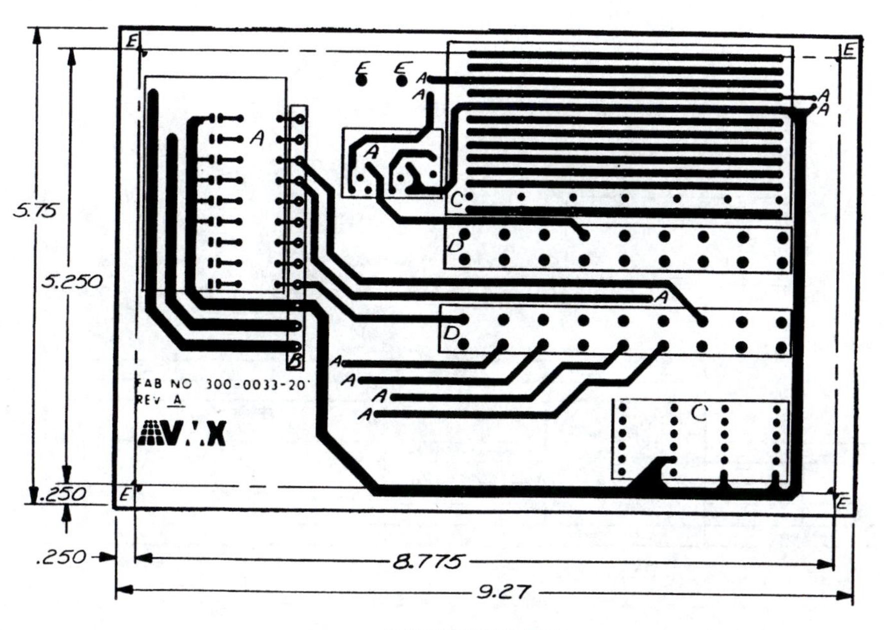

NOTES:

1. MATERIAL-EPOXY GLASS LAMINATE, FOIL CLAD NEMA GRADE FR4, .062 THK, TYPE FLGEN 1 OZ CU BOTH SIDES.
2. FINISH-ELECTROPLATED SOLDER, 60/40 TIN LEAD PER QQ-S-291, .0005 IN. THK MIN.
3. PLATE-THRU HOLES MUST HAVE MINIMUM TOTAL WALL THICKNESS OF .001 INCHES.
4. TOLERANCES APPLY AFTER PLATING.
5. SILKSCREEN USING YELLOW EPOXY INK. APPLY SIDE 1 TO COMPONENT SIDE.
6. SOLDER MASK SHALL BE APPLIED TO BOTH SIDES IN ACCORDANCE WITH STANDARD IPC-SM-840 COLOR: GREEN

HOLE SCHEDULE			
LTR	DESCRIPTION	QTY	REMARKS
A	.043 DIA ±.005	62	
B	.060 DIA ±.005	12	
C	.065 DIA ±.005	115	
D	.070 DIA ±.005	36	
E	.166 DIA ±.005	6	

CIRCUIT SIDE

DRILL PLAN B

Courtesy of VMX, Inc.

Figure 2–8 Drill plans

Drill tapes, rather than drill plans, can be used with numerically controlled manufacturing equipment to make holes in boards. Drill tapes perform the same function as drill plans. Drill plans can be identified by the letters placed next to the holes, and the hole schedule on the drawing. Components are not usually shown on a drill plan.

Mechanical Drawings

Once electronic equipment and printed circuit boards have been designed, they are usually held together by parts made of sheet metal, plastic or wood. The drawing of such parts comes under the general category of mechanical drawing. Figure 2–9 shows three examples of mechanical drawings which are used to build mechanical (metal, plastic, or wooden) parts, and assemble mechanical and electronic parts into a usable unit. Figure 2–9A is a detail drawing showing a single part and calling out all of its dimensions so that it can be built. Figure 2–9B is an assembly drawing which shows how parts are to be fitted together, and identifies all parts with a number in a circle. Figure 2–9C is a parts list which gives an accurate description of each part, and tells how many of each part is needed to build the equipment.

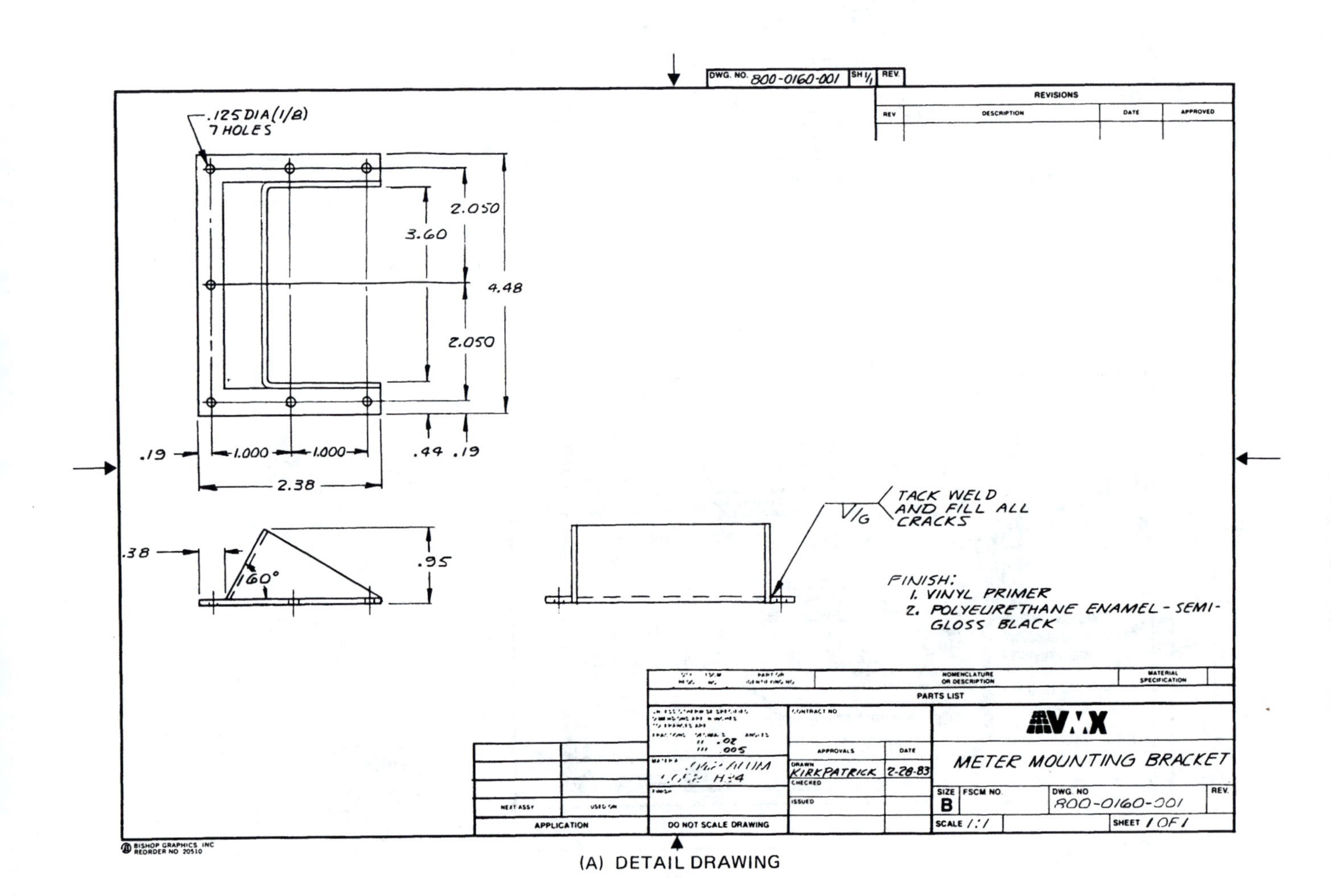

(A) DETAIL DRAWING

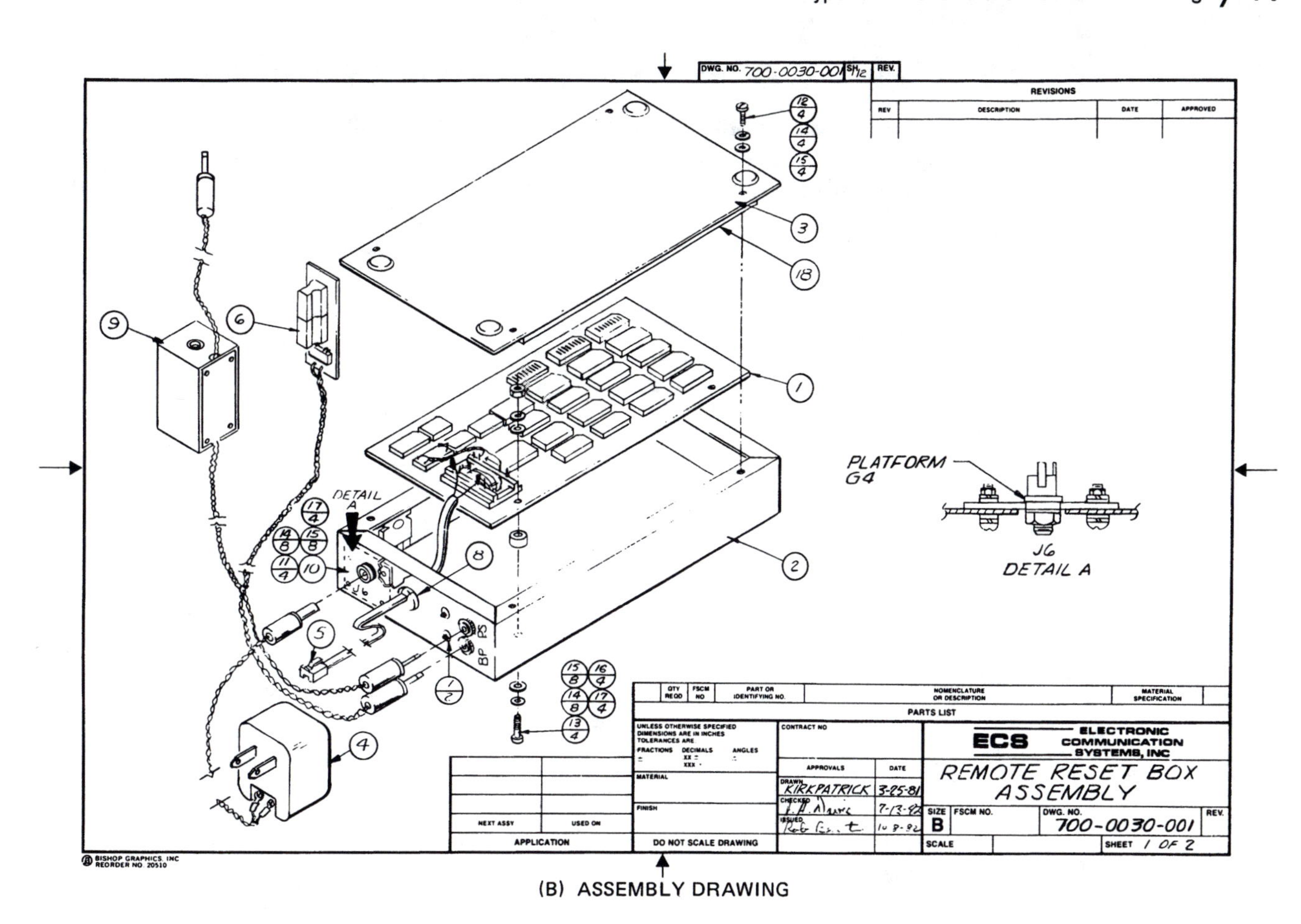

(B) ASSEMBLY DRAWING

LM 700-0030 6-30-82

ITEM	QTY REQD -001	QTY REQD -002	DESCRIPTION	REF DESIG	PART NUMBER	MANUFACTURER PART NUMBER	MANUFACTURER	CODE IDENT
1	1		RESET PCB ASSY		300-0027-001			
2	1		CHASSIS (ALUMINUM)		800-0137-001	AC1408	BUD	
3	1		CHASSIS BOTTOM PLATE (ALUMINUM)		407-0001-001	BPA1508	BUD	
4	1		TRANSFORMER ASSEMBLY		700-0031-001			
5	25'		PHONE CABLE ASSEMBLY		500-0020-001			
6	1		BACK PLANE CABLE ASSEMBLY		530-0004-001			
7	2		MOUNT, LED		477-0001-002	515-0004	DIALIGHT	
8	1		STRAIN RELIEF		230-0006-002	1077	HEYCO	
9	1		POWER SIFTER RELAY BOX		700-0105-001			
10	1		PLATE, INSULATING		800-0142-001			
11	4		SCREW, PH, PH, SS, #4-40 X 9/16 LG		432-0030-80			
12	4		SCREW, PAN HEAD, PH, SS #4-40 X 5/8 LG		432-0030-050			
13	4		" " " " " #4-40 X 5/16"		432-0030-002			
14	20		WASHER, SPLIT LOCK NO. 4 SS		432-0040-020			
15	20		WASHER, FLAT, NO. 4 SS		432-0040-001			
16	4		SPACER, 1/4 OD NYLON, (#6) X 1/4" LG		476-0002-012	314-1431-008	EF JOHNSON	
17	8		NUT, HEX, BRASS, CAD PL, 4-40 STD		432-0041-040			
18	1		FOAM RUBBER, 7" X 3-3/4" X 1/2"		465-0002-001			

ECS ELECTRONIC COMMUNICATION SYSTEMS, INC	REVISION NUMBER	REV. APPROVAL	DATE	DRAWN BY	NEXT ASSY	SHEET
APPROVED BY: / DATE	TITLE REMOTE RESET BOX ASSEMBLY				USED ON / ASSY SIZE LM 700-0030	REV A

(C) PARTS LIST

Courtesy of VMX, Inc.

Figure 2–9 Mechanical drawings

Specification Control Drawings and Altered Drawings

A specification control drawing, used to ensure that purchased parts meet the company's requirements for that part, is shown in Figure 2–10. The drawing also shows an approved source (or sources) from which the part may be purchased. Figure 2–11 shows an altered item drawing; in this case a purchased connector was modified for a particular application.

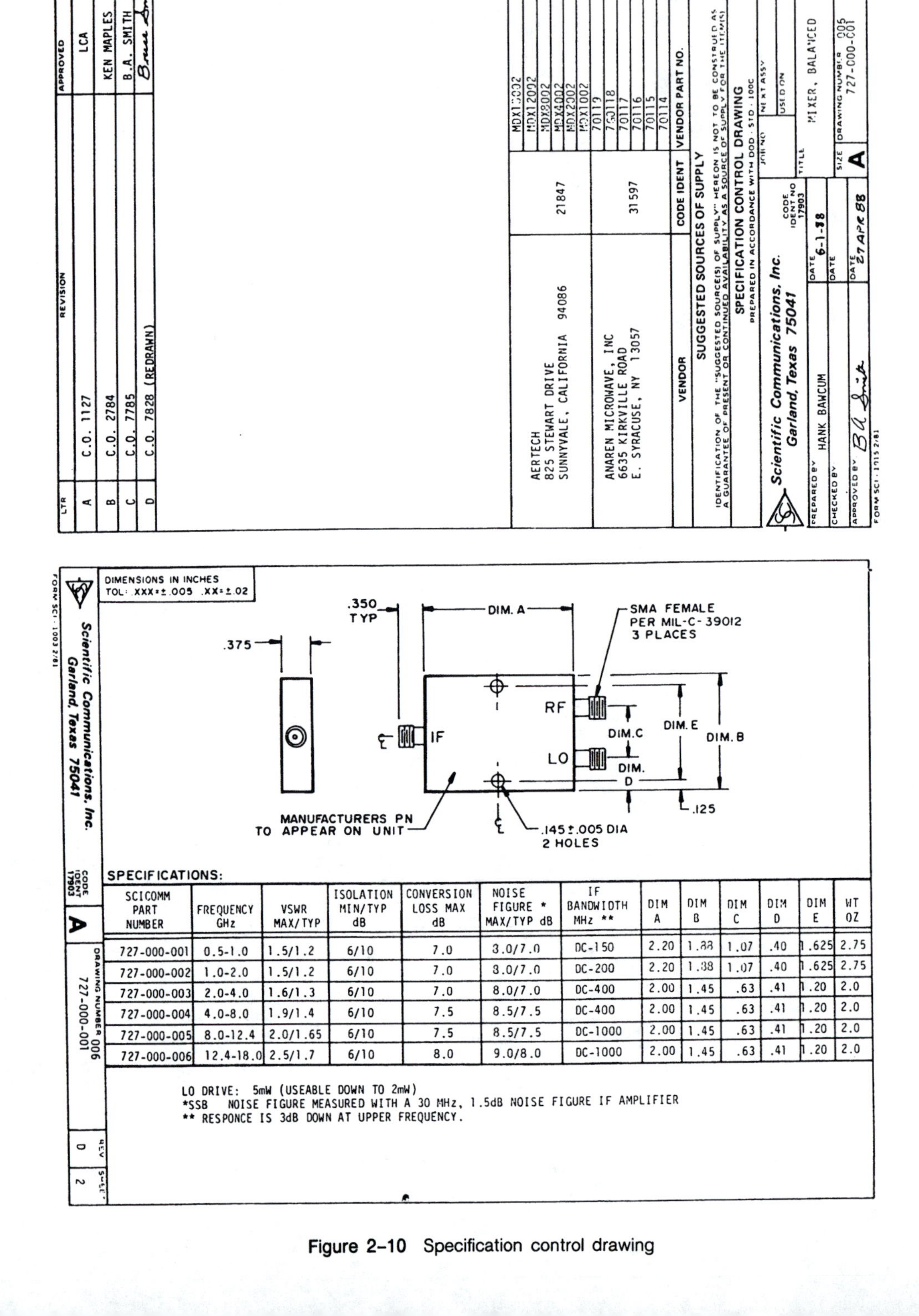

SPECIFICATIONS:

SCICOMM PART NUMBER	FREQUENCY GHz	VSWR MAX/TYP	ISOLATION MIN/TYP dB	CONVERSION LOSS MAX dB	NOISE FIGURE * MAX/TYP dB	IF BANDWIDTH MHz **	DIM A	DIM B	DIM C	DIM D	DIM E	WT OZ
727-000-001	0.5-1.0	1.5/1.2	6/10	7.0	3.0/7.0	DC-150	2.20	1.38	1.07	.40	1.625	2.75
727-000-002	1.0-2.0	1.5/1.2	6/10	7.0	3.0/7.0	DC-200	2.20	1.38	1.07	.40	1.625	2.75
727-000-003	2.0-4.0	1.6/1.3	6/10	7.0	8.0/7.0	DC-400	2.00	1.45	.63	.41	1.20	2.0
727-000-004	4.0-8.0	1.9/1.4	6/10	7.5	8.5/7.5	DC-400	2.00	1.45	.63	.41	1.20	2.0
727-000-005	8.0-12.4	2.0/1.65	6/10	7.5	8.5/7.5	DC-1000	2.00	1.45	.63	.41	1.20	2.0
727-000-006	12.4-18.0	2.5/1.7	6/10	8.0	9.0/8.0	DC-1000	2.00	1.45	.63	.41	1.20	2.0

LO DRIVE: 5mW (USEABLE DOWN TO 2mW)
*SSB NOISE FIGURE MEASURED WITH A 30 MHz, 1.5dB NOISE FIGURE IF AMPLIFIER
** RESPONCE IS 3dB DOWN AT UPPER FREQUENCY.

Figure 2–10 Specification control drawing

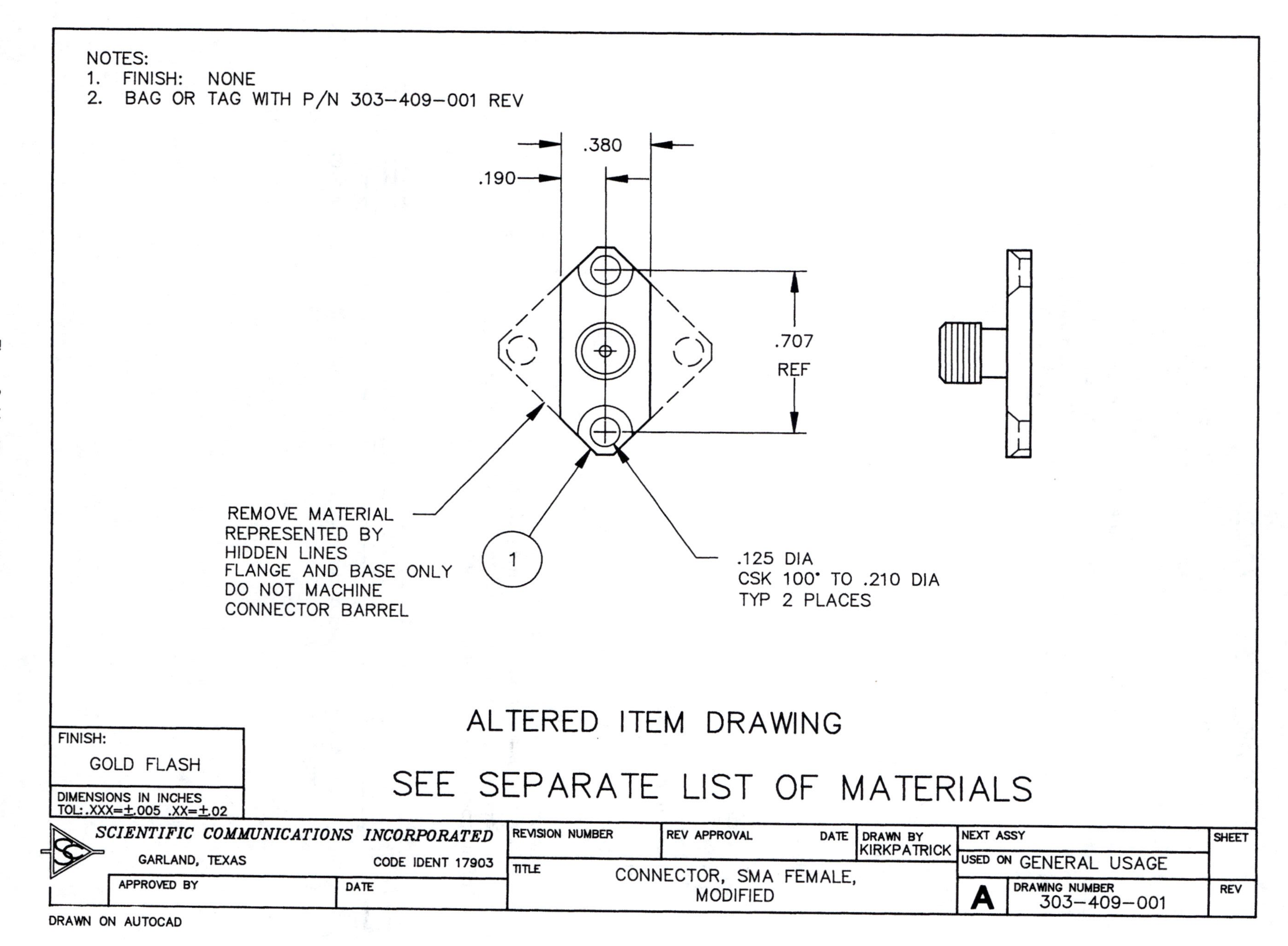

Figure 2–11 Altered item drawing

COMPUTER-AIDED DESIGN (CAD)

Some of the major computer-aided design (CAD) systems in use today are described in Chapter 12. These systems can produce all of the drawings described in this text. However, an understanding of what is being drawn is necessary in order to operate this computer graphics equipment. Electronic drafters must have the basic knowledge of the drawings described in this text before beginning to use a computer to design printed circuit boards and other parts. Chapter 12 describes the components of the computerized system, how the equipment is operated, many of the commands used, and the drawings produced.

SUMMARY

Eight major types of drawings are commonly used in the electronics industry. They are:

- Block diagrams
- Logic diagrams
- Schematic diagrams
- Cabling diagrams
- Component assembly drawings
- Artwork
- Drill plans
- Mechanical drawings

Block diagrams, logic diagrams, schematic diagrams, and most cabling diagrams provide the information needed to troubleshoot, service, design, and repair electronic equipment. Component assembly drawings, artwork, drill plans, mechanical drawings, and some types of cabling drawings are used in manufacturing the equipment. Each drawing type has a specific function and serves a purpose that no other drawing serves.

An electronic drafter is expected to be proficient at drawing each type of diagram discussed in this chapter.

EXERCISE 2–1

List eight major types of drawings used in the electronics industry and the function of each type.

EXERCISE 2-2

Identify each of the drawing types shown.

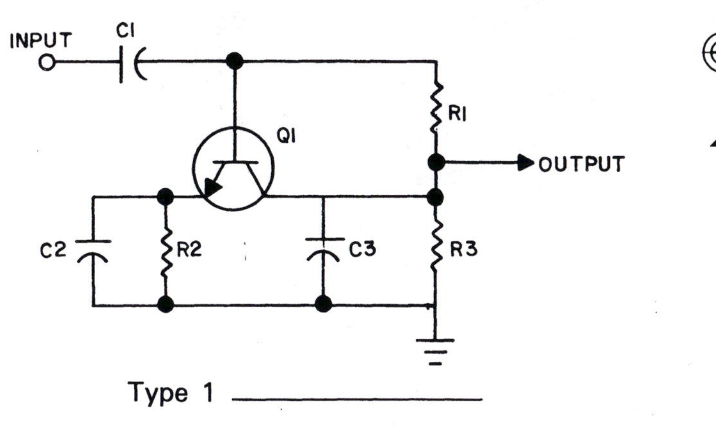

Type 1 ______________

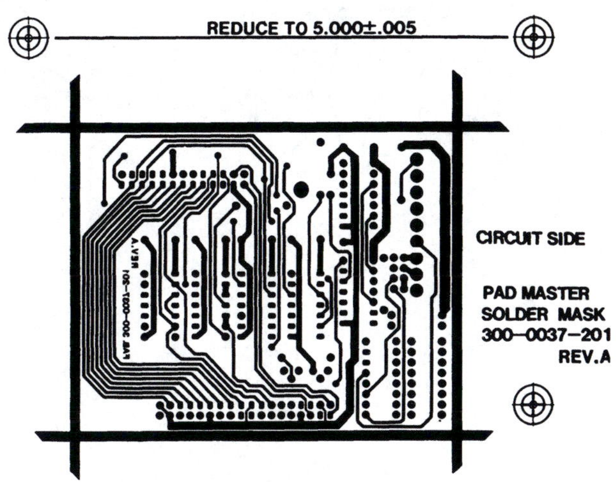

Type 2 ______________

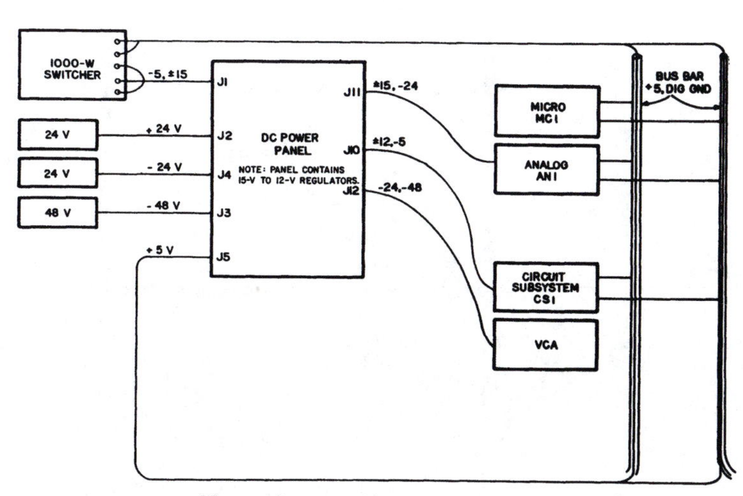

Type 3 ______________

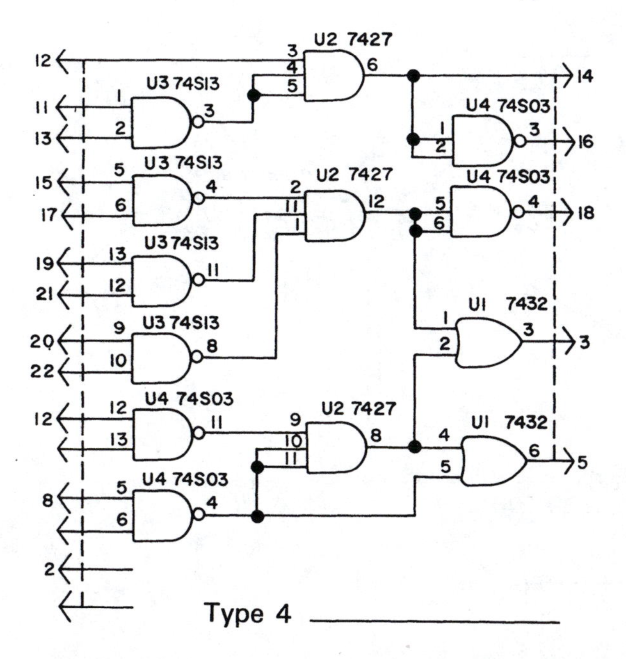

Type 4 ______________

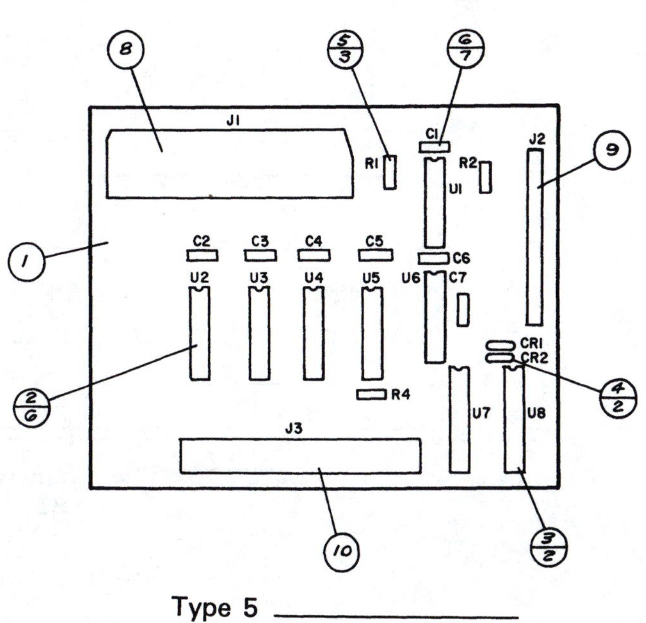

Type 5 ______________

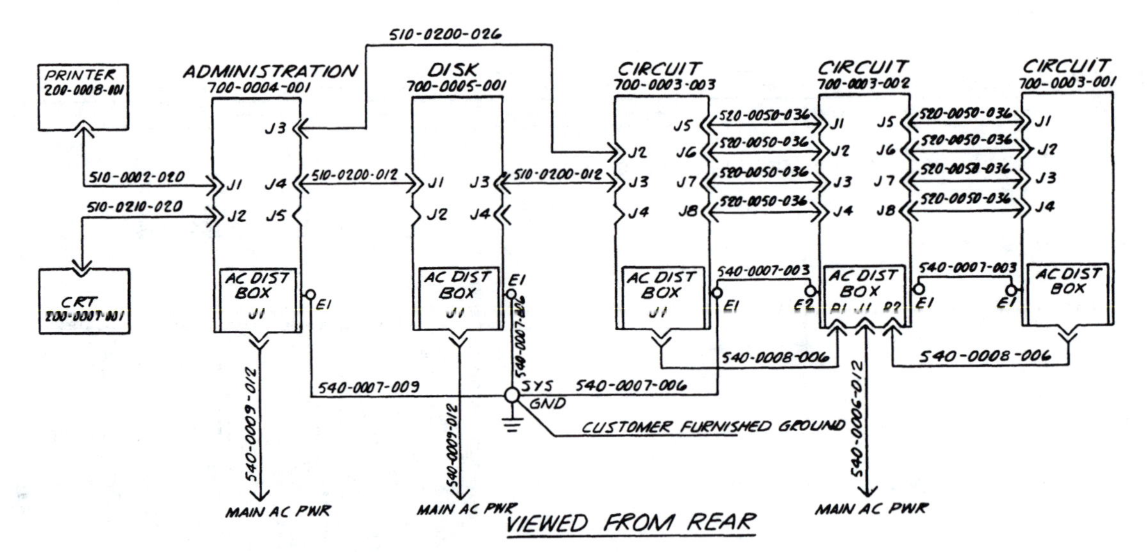

Type 6 ________________

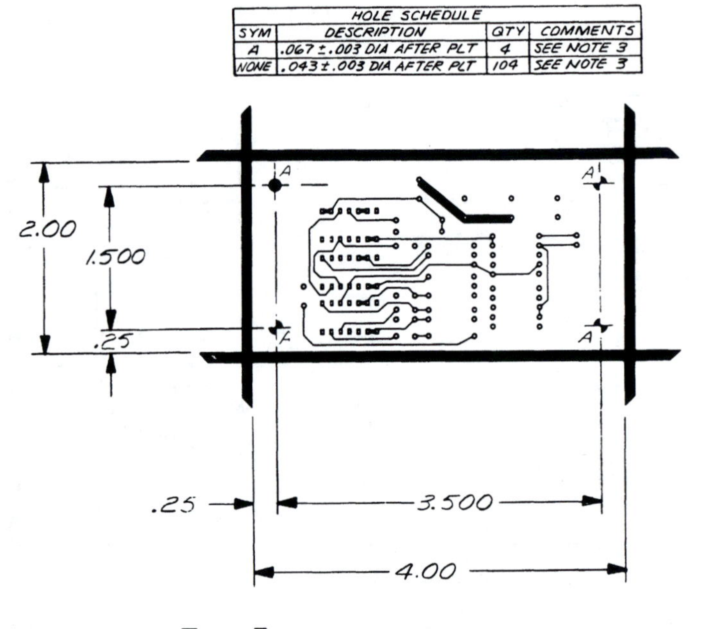

Type 7 ________________

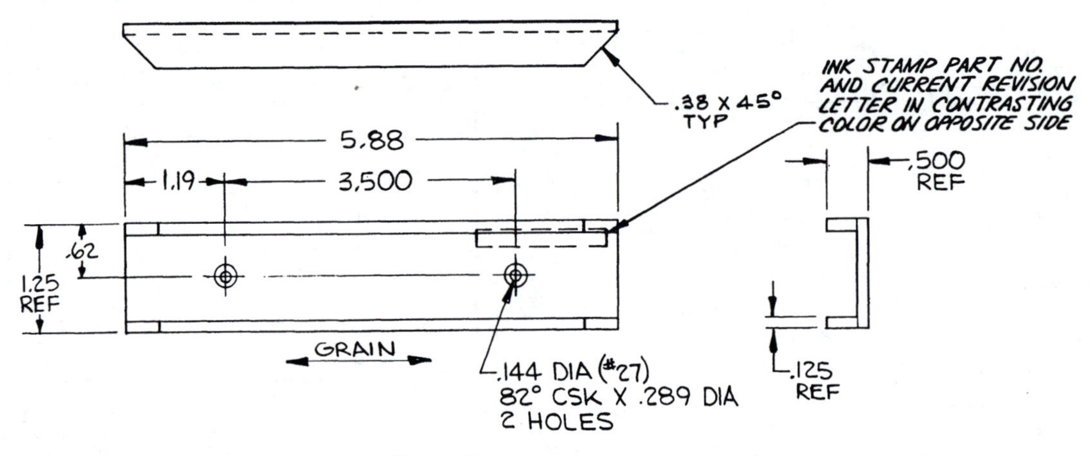

Type 8 ________________

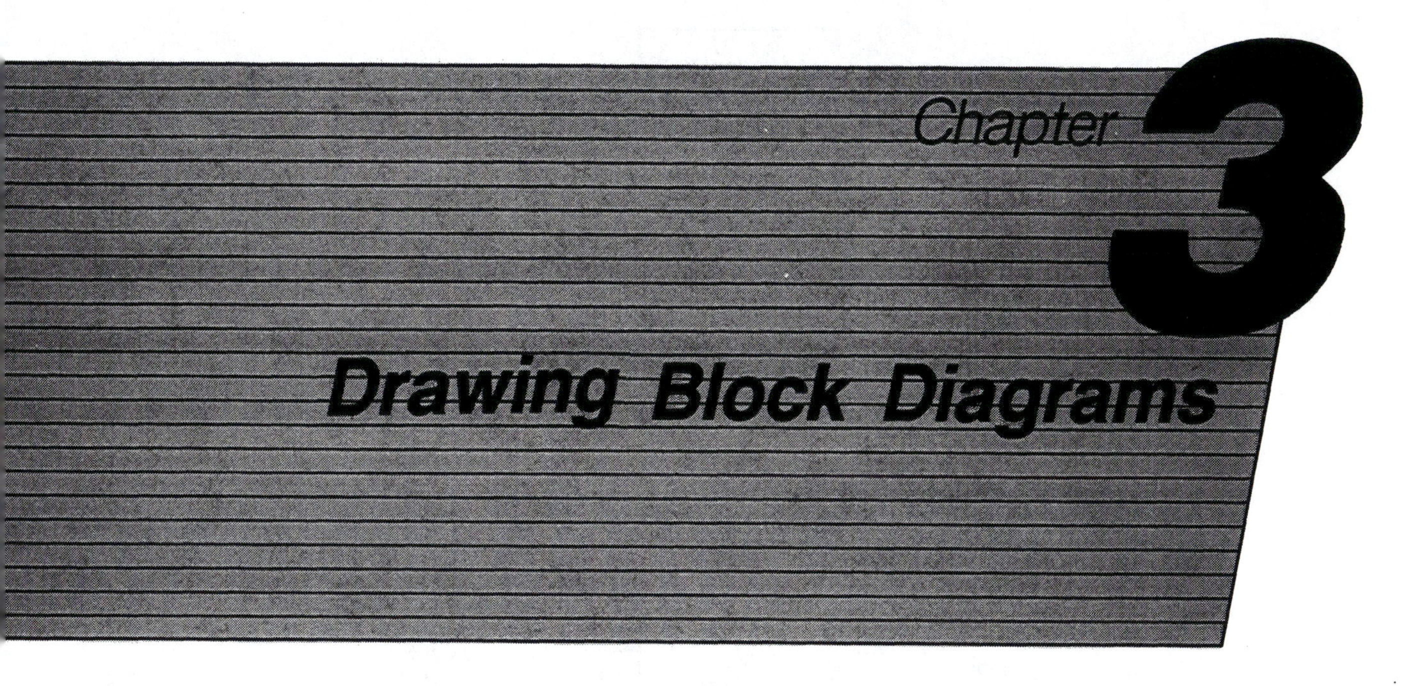

Chapter 3 Drawing Block Diagrams

OBJECTIVES

After completing this chapter, you will be able to
- ✓ list three functions of block diagrams.
- ✓ draw in pencil two block diagrams using dense, uniform, unbroken lines following the rules for good layout.
- ✓ complete the block diagrams, using sans serif, Gothic lettering of uniform slant and height.

TYPES OF BLOCK DIAGRAMS

Block diagrams are relatively simple drawings that are used to show many different concepts or ideas. Block diagrams serve three major functions, as shown in Figure 3–1. Figure 3–1A is an *organizational chart* showing how an organization is structured. Figure 3–1B shows a manufacturing sequence or a *project flow chart.* The drawing in Figure 3–1C is called a *functional block diagram.* This type of diagram shows how units within a system function.

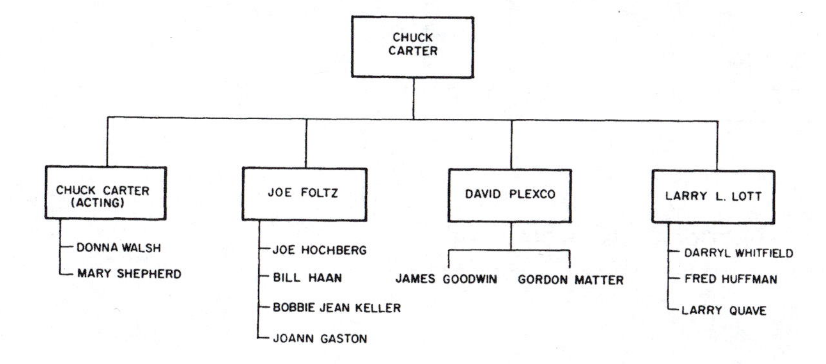

Figure 3–1 Common types of block diagrams

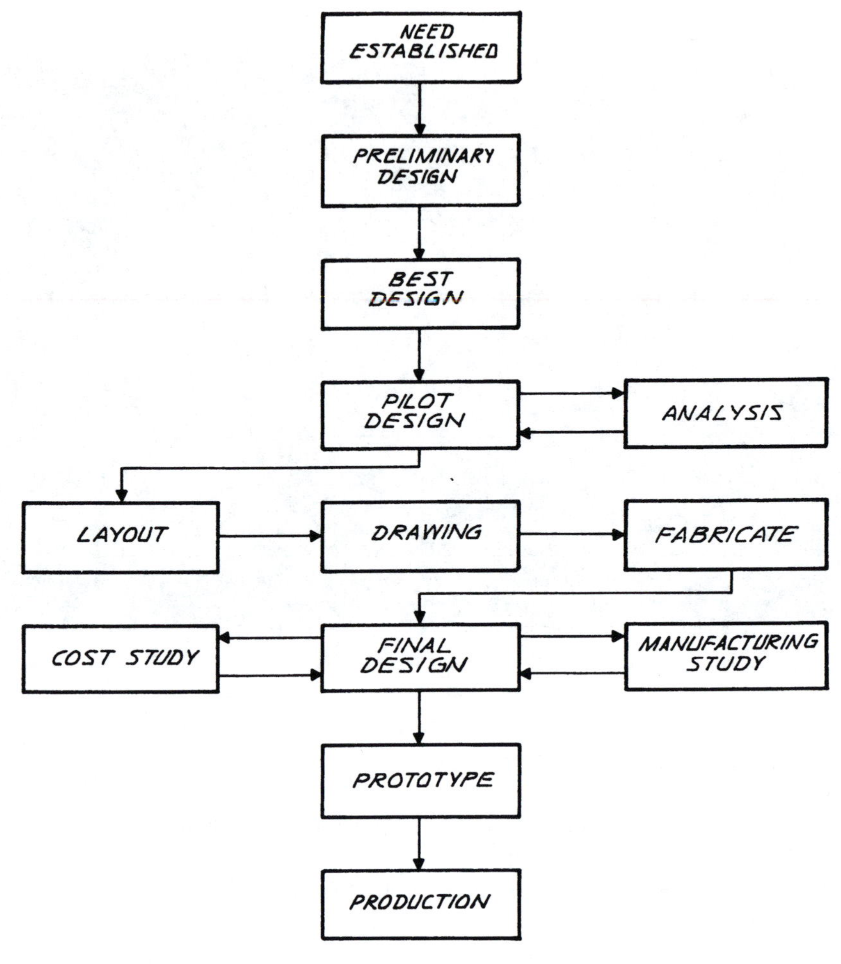

(B) PROJECT FLOW CHART

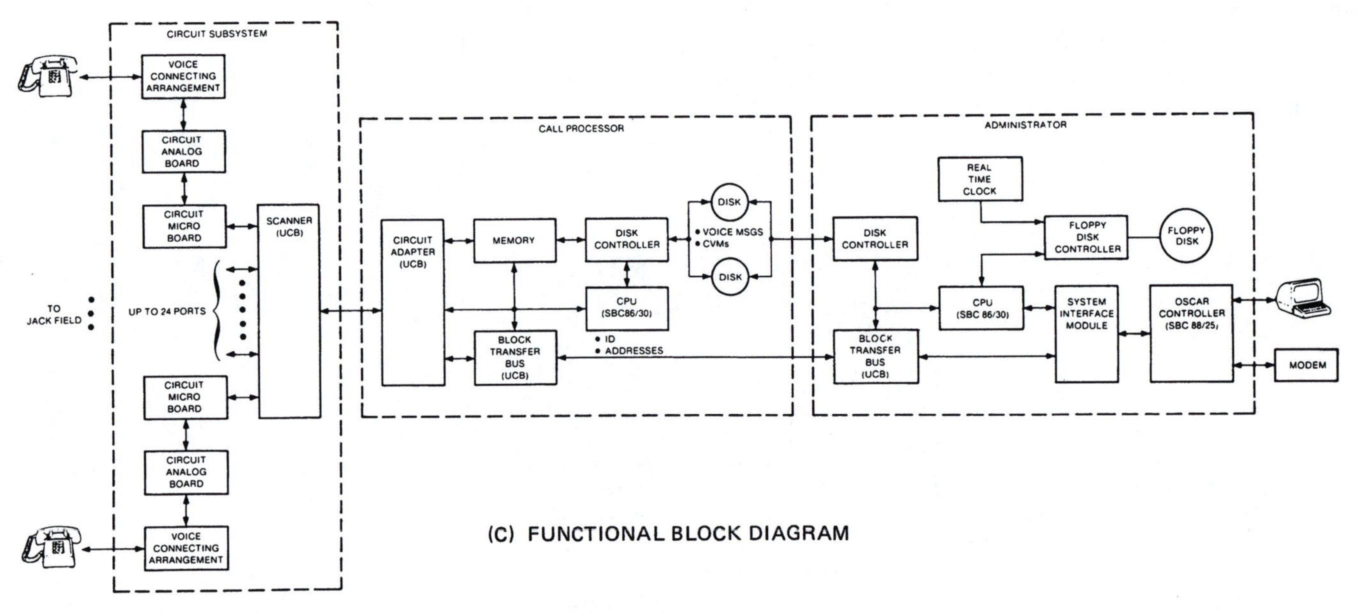

(C) FUNCTIONAL BLOCK DIAGRAM

Courtesy of VMX, Inc.

Figure 3–1 (continued)

DRAWING BLOCK DIAGRAMS

Layout Steps

A logical, organized sequence of steps should be followed when drawing block diagrams. These steps save time overall, and result in better layouts and more consistent appearances. These steps are:

Manual	*CAD*
1. Determine box size.	1. Approximate box size as necessary.
2. Lightly draw a layout.	2. Enlarge or shrink boxes using a command such as the AutoCAD stretch command.
3. Draw all horizontal lines.	3. Copy boxes or other repetitious sections such as lines and arrowheads throughout the drawing to complete the layout.
4. Draw all vertical lines.	4. Add connecting lines and move boxes and lines when necessary.
5. Draw arrowheads and circles.	5. Add arrowheads, circles, lettering, and other symbols.
6. Complete lettering.	6. Plot the drawing.

Figure 3–2 shows the first step in making a formal block diagram: determine box size. The engineering rough sketch from which the layout is to be made is shown at the top left of the figure. If practical, the boxes should all be the same size, although many diagrams require boxes of more than one size. Look carefully at the sketch to determine which box contains the most lettering. Size the box large enough to accommodate the lettering. Letter height and space between lines of letters is used to determine the vertical height of the box. The longest line of wording is used to determine the horizontal length of the box. All the boxes should then be drawn to the same size whenever practical.

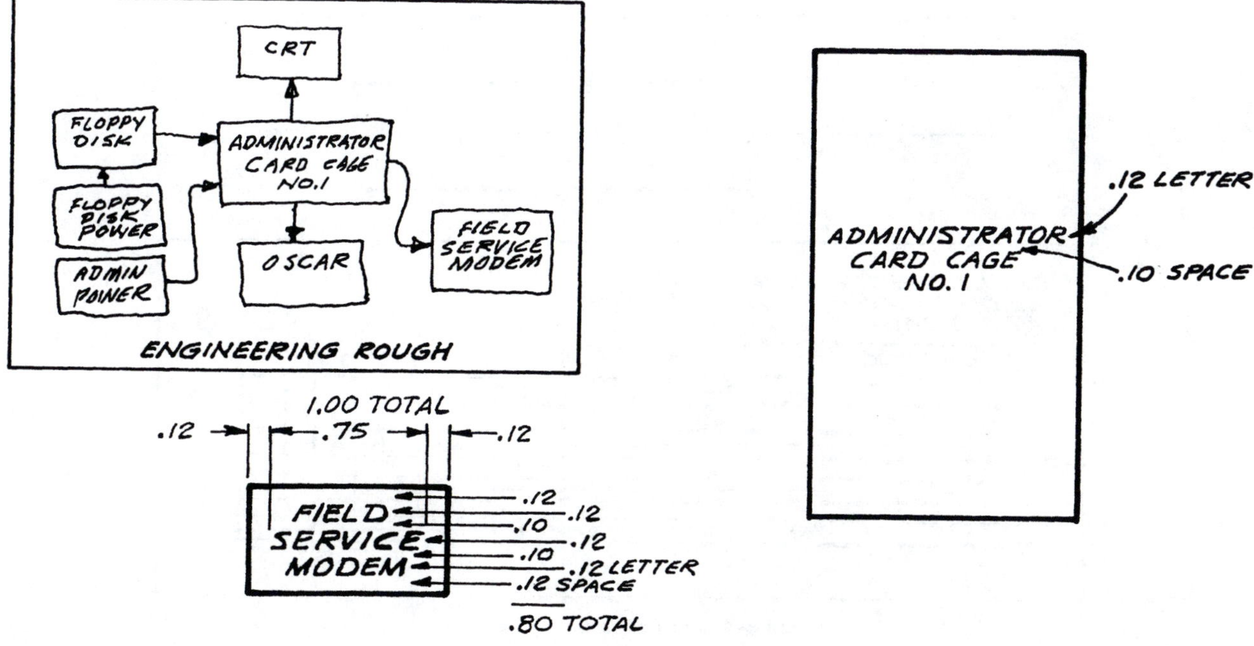

Figure 3–2 Determine box size

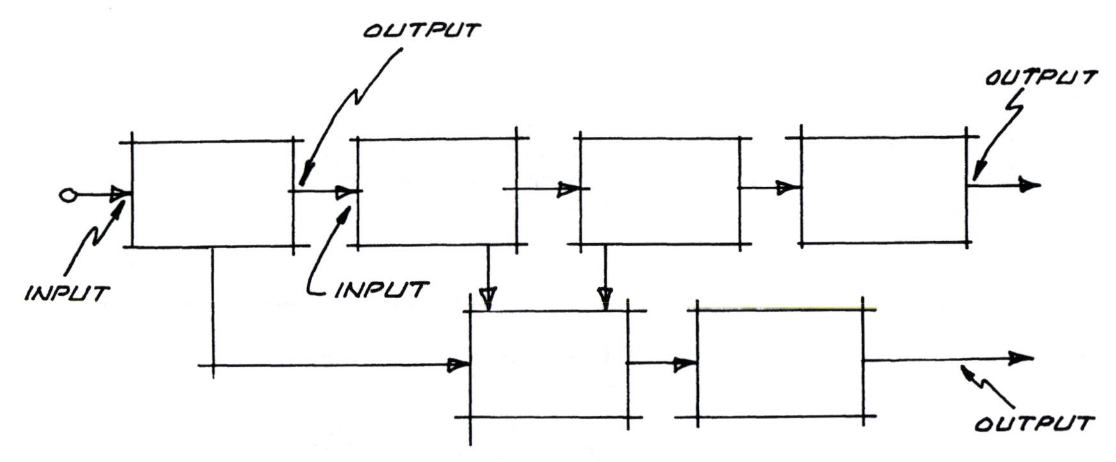

Figure 3–3 Sketch a layout

In Figure 3–3, a sketch has been lightly drawn with nonreproducing pencil or light pencil lines. Another way to accomplish this second step is to draw the sketch using dark lines, then making a copy of the sketch and tracing the copy. Notice in the figure that all of the boxes are the same size and the same distance apart. However, this is not always possible. Sometimes the drawing will require boxes of different sizes and unequal spacing. But, in general, all boxes should be drawn to the same size, if possible, and spaced equally apart.

The arrows shown in Figure 3–3 are called *input* and *output* arrows. The first input into this diagram is indicated by the arrow with a circle on the end at the far left. This arrow meets the first box. On the other side of the same box, the arrow leading out of it is an output, which then becomes an input into another box. In this way, the sequence and flow of information is established. The flow is generally from left to right. The output on the far right is terminated with an arrow.

The next step following the layout is to draw all horizontal lines, as shown in Figure 3–4. Start at the top of the page and work downward,

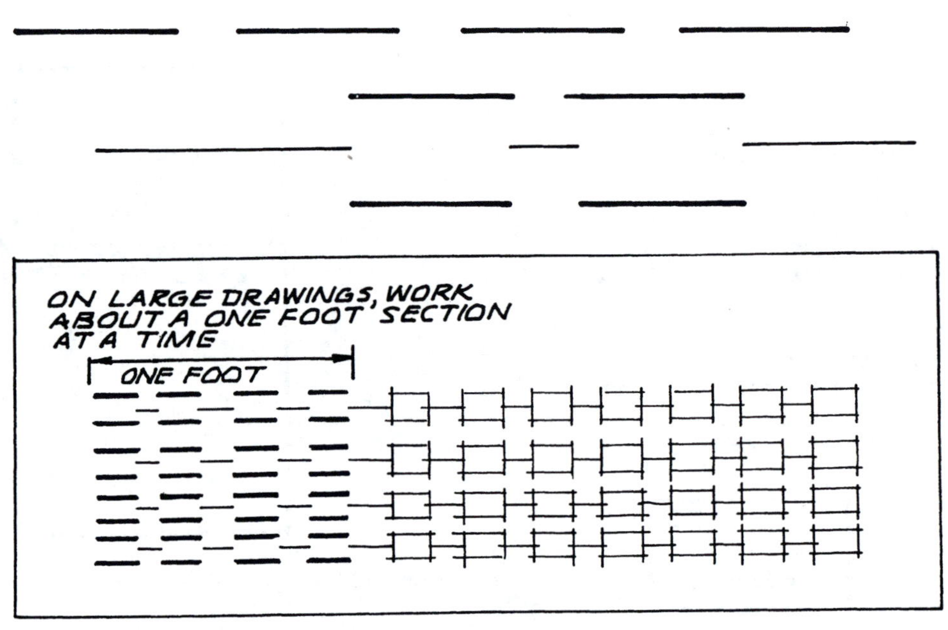

Figure 3–4 Draw all horizontal lines

drawing all horizontal lines as they occur. If the drawing is very large, it is best to draw one section at a time, so that hand movements can be better controlled. All the lines should be dark. The box outlines are usually darker than the connecting lines. This is discussed further later in the chapter.

Step 4 is to draw all vertical lines, as shown in Figure 3–5. This is done from left to right (or right to left for some left-handed people). When drawing a small diagram, all vertical lines can usually be drawn in one pass. Large diagrams require that all vertical lines be drawn within the height of the triangle all the way across the page. Then, the drafter returns to the far left and draws the next set of vertical lines until all vertical lines have been drawn.

The next step, step 5, is to draw arrowheads and circles, and other symbols if any, Figure 3–6. Templates should be used to keep all arrowheads and circles uniform in size.

The final step is to add lettering, keeping it consistently spaced within the boxes, Figure 3–7. Do not touch any lines with lettering and maintain spacing of approximately two-thirds the letter height between lines of lettering. This results in a neat, consistent appearance, and makes the drawing easier to read and reproduce.

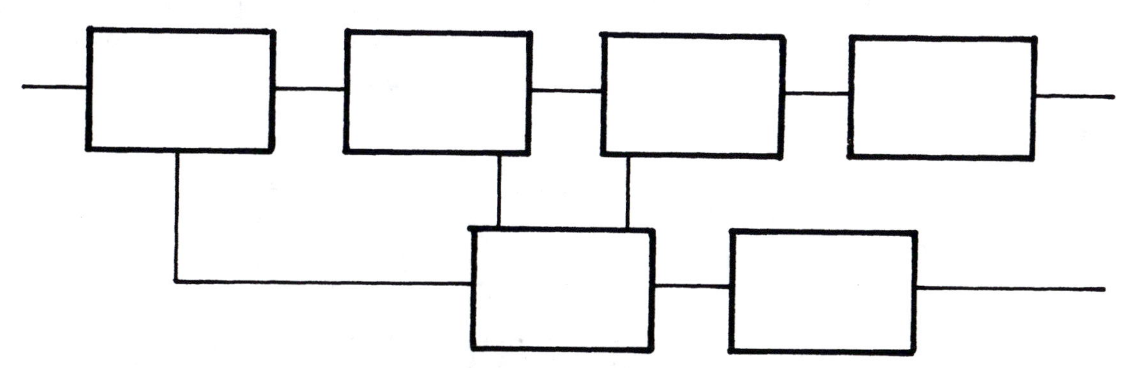

Figure 3–5 Draw all vertical lines

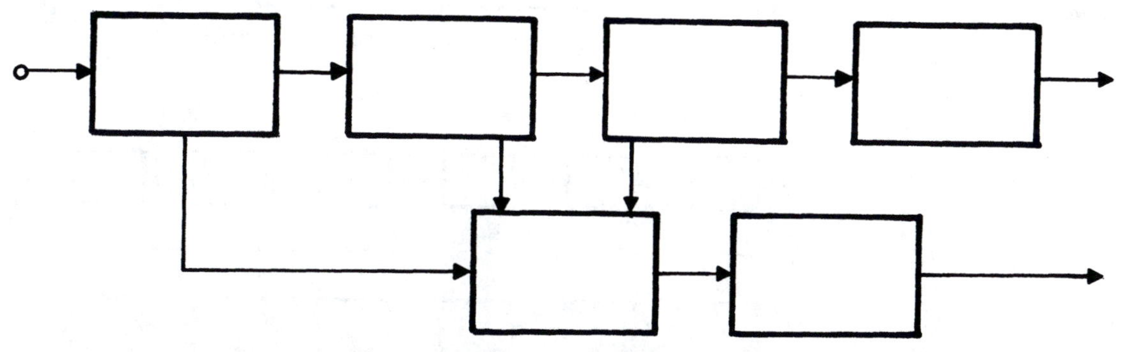

Figure 3–6 Add arrowheads, circles, and other symbols

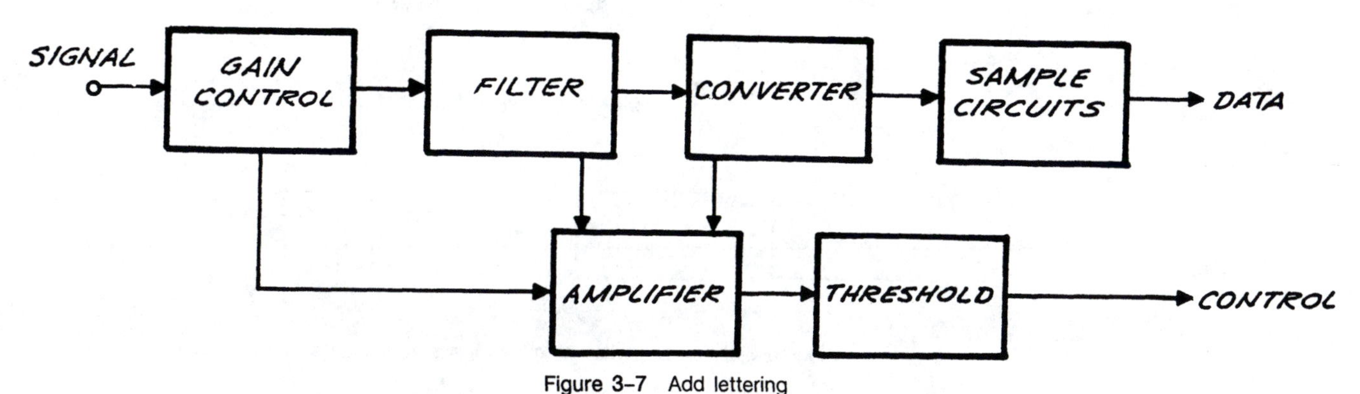

Figure 3–7 Add lettering

Inputs and Outputs

Figure 3–8 shows several fundamental layouts. The best arrangement is to lay out the boxes so that the flow is from left (input) to right (output), Figure 3–8A. Often, however, that cannot be done. So, the next best arrangement is the flow from top to bottom, Figure 3–8B. The layout of a double row of boxes, shown in Figure 3–8C, is used occasionally.

If the drawing is highly complex, inputs may have to come into the right-hand side of the box and outputs from the left, unless there is a great deal of time to solve all of the problems involved. Whether or not it is worth the time required depends on the customer.

Flow arrows are shown in Figure 3–9. The flow arrow is different from a dimensioning arrow or other arrows that you may have encountered in other types of drafting. The flow arrow has an arrowhead with a relationship of one to one; that is, its width is the same as its height. Arrowheads that are about $\frac{1}{8}''$ are adequate for most drawings. Many templates contain arrowheads, some of which may vary slightly from the $\frac{1}{8}''$ size. This is acceptable.

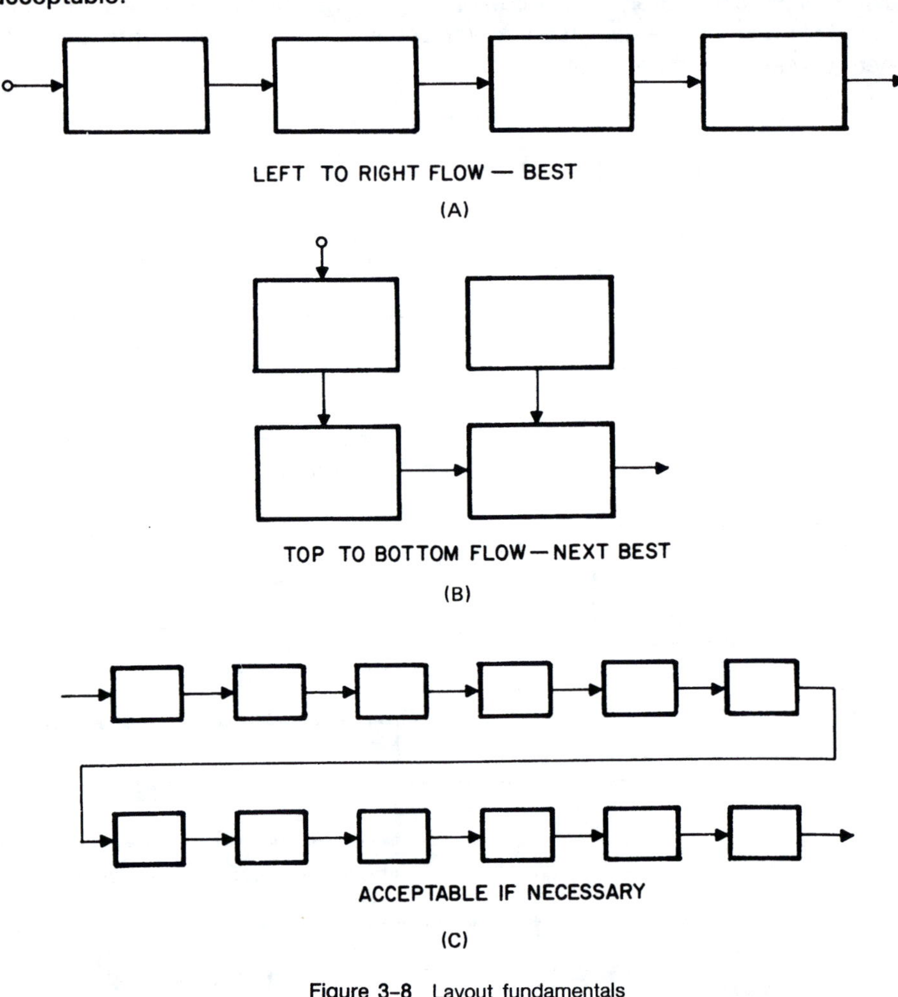

Figure 3–8 Layout fundamentals

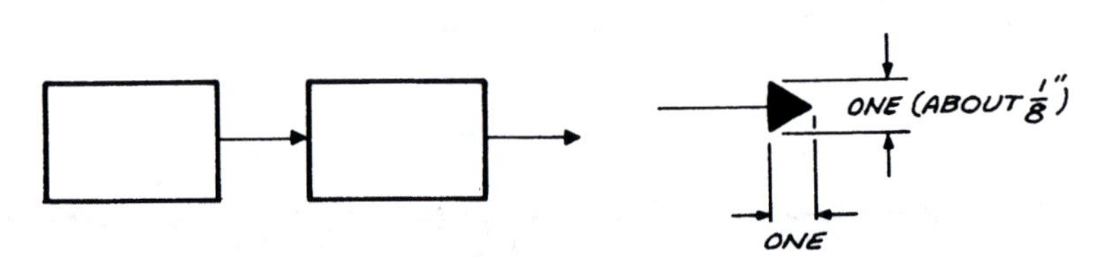

Figure 3–9 Arrowheads on flow arrows are as wide as they are high

Figure 3–10 Identifying inputs and outputs

Figure 3–10 shows how the inputs and outputs into a box or block diagram are often identified. An engineer who submits a sketch that does not identify the inputs and outputs may have left them off deliberately. But, then the question should be raised as to whether or not they should be included.

Lettering Arrangement

Figure 3–11 shows how lettering and line spacing should be arranged in the boxes. In the box at (A), the lettering is centered both vertically and horizontally. Each word is centered around the centerline of the box, and then all lines are centered within the box horizontally. The box at (B) shows that all the lettering is centered vertically, and that all the letters start the same distance from the left side of the box. This arrangement is called *flush left.* The instructions at (C) for drawing the block diagram specify the arrangement of lettering. Notice that the large block still keeps lines of lettering the same distance apart. The illustration at (D) shows that lettering should never touch a line. Notice that the spacing between lines of lettering is approximately ⅔ the letter height, (full letter height spacing is okay).

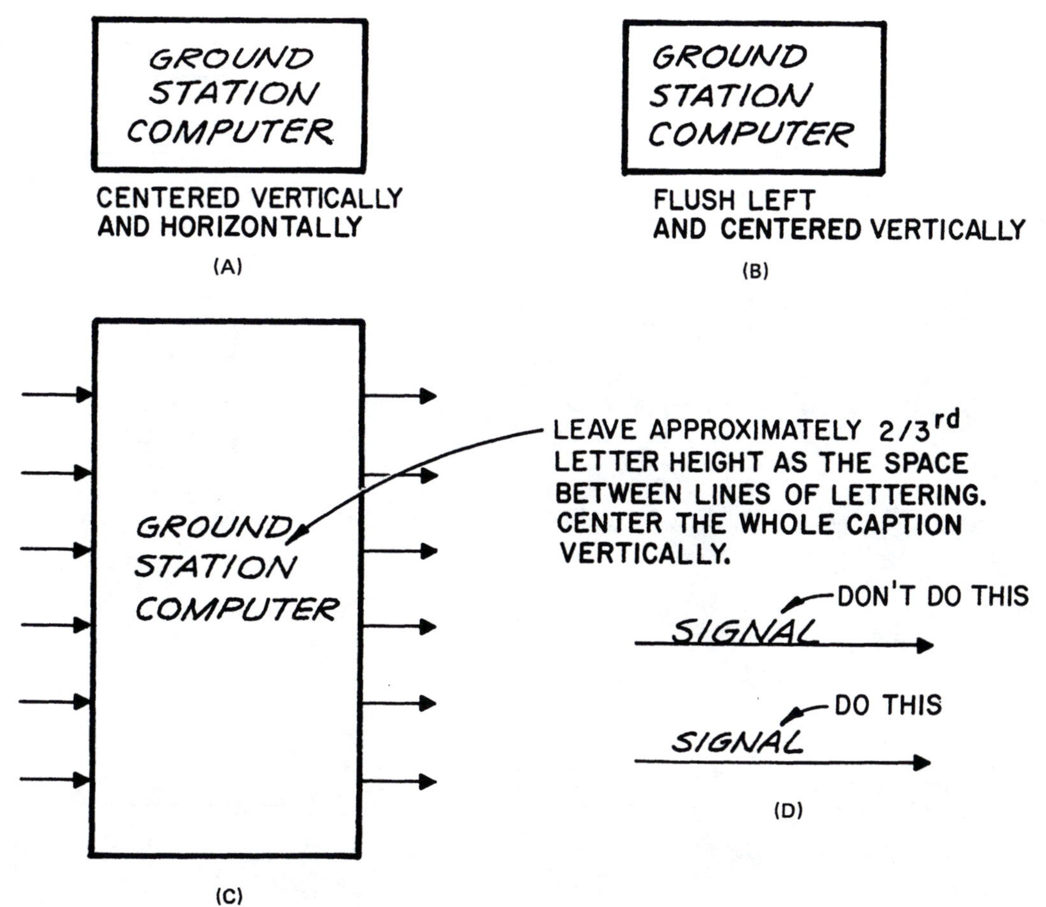

Figure 3–11 Spacing of lettering

Crossovers and Doglegs

Crossovers and doglegs are two pitfalls to be avoided in laying out block diagrams. These are described in the two before and after illustrations of Figure 3–12. The diagram shows that a *Crossover* results when a line crosses another line leading out of box A. This crossover is eliminated by bringing one of the outputs out of the bottom of box A. Crossovers can be a real problem in complex diagrams, however, since not all of them can be avoided. The student should keep in mind that a much cleaner layout will be made and the drawing will be much more legible than it would be with many lines crossed unnecessarily.

As shown in Figure 3–12, a *dogleg* is a line that has at least two bends in it, and sometimes more. In this example, the doglegs were eliminated by moving box C down, and bringing the output from box A in to the top of box C instead of to the left side. The output from box B is then drawn straight across to box C. A much cleaner layout results by eliminating the doglegs. Sometimes doglegs cannot be avoided. Complicated diagrams may contain lines with three or four bends. But, if the drafter avoids doglegs whenever possible, a much better layout is produced.

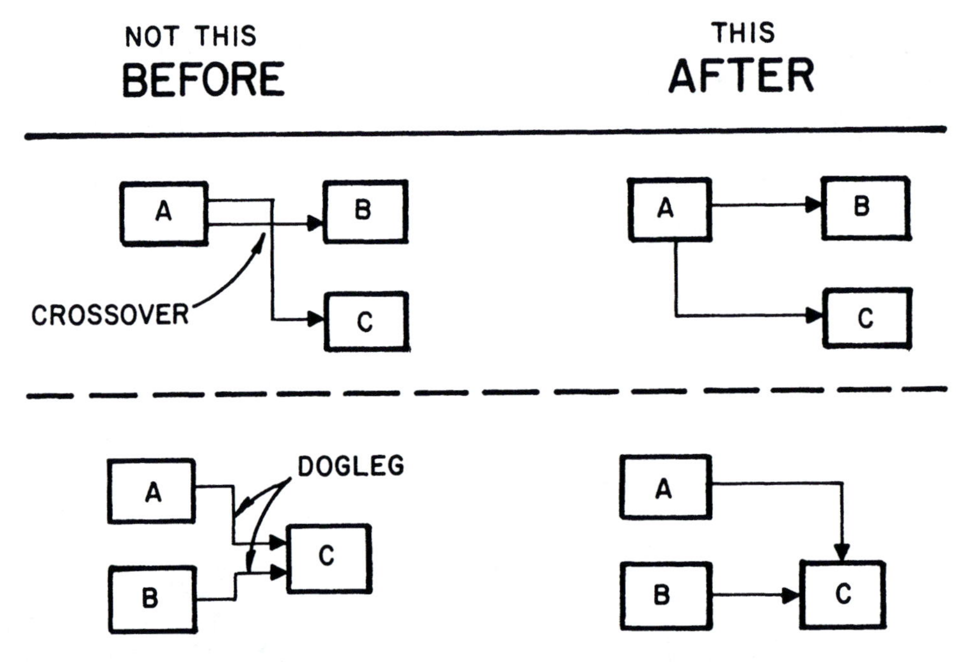

Figure 3–12 Avoid crossovers and doglegs

Line Thickness

The approximate relationship of the thickness of the box outline to the thickness of the interconnecting lines is shown in Figure 3–13. Notice that the box outline is about three times as thick as the interconnecting lines. These are the line weights to use on the diagrams assigned in the exercises at the end of this chapter.

Completed Block Diagram

A completed block diagram is shown in Figure 3–14.

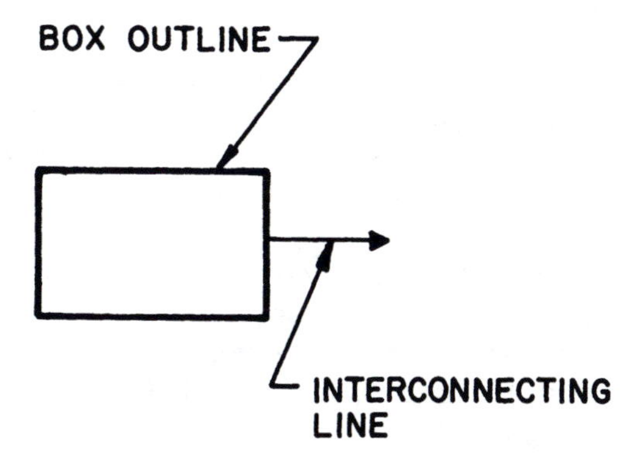

Figure 3-13 Line thickness

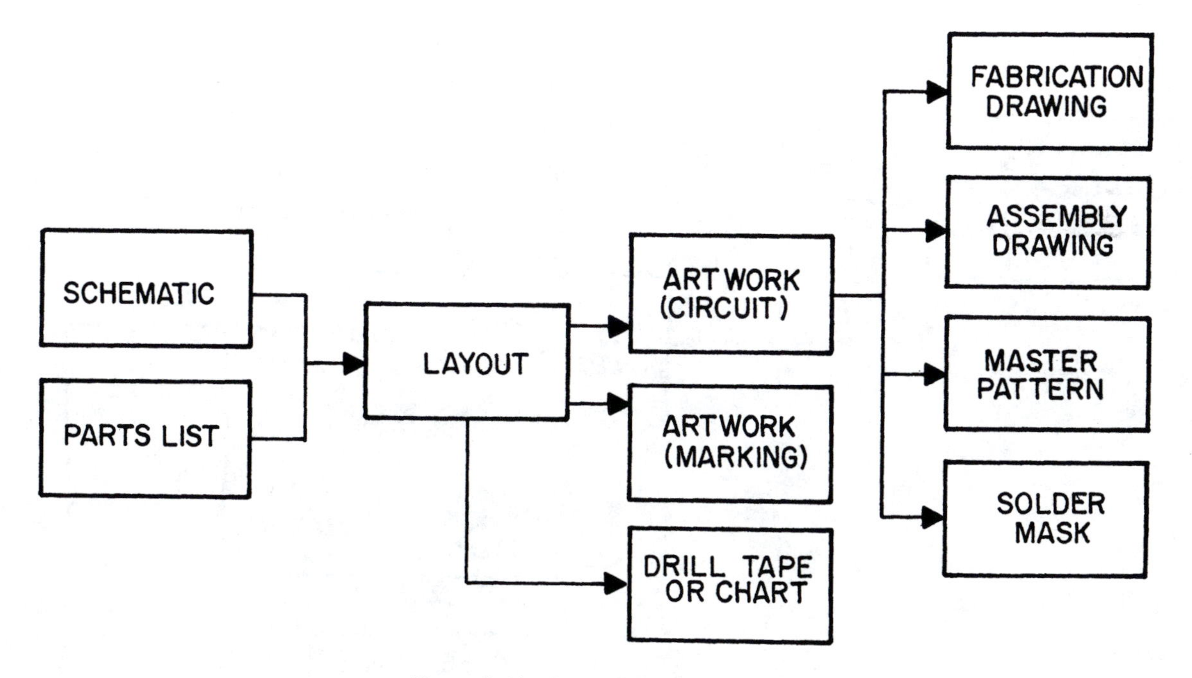

Figure 3-14 Completed block diagram

EXERCISE 3-1

Redraw the following sketch in pencil using the steps for *correct* layout and drawing of block diagrams.

Center the drawing lengthwise on an $8\frac{1}{8}'' \times 11''$ sheet of grid paper. Leave at least 1″ of white space all around the drawing. Make all lettering $\frac{1}{8}''$ high, flush left, $\frac{1}{10}''$ from the left edge of the box, and center it vertically. Use a flow arrow with a $\frac{1}{8}'' \times \frac{1}{8}''$ arrowhead.

Remove all doglegs and as many crossovers as possible. All inputs should go into the left side or the top of the boxes. All outputs should leave from the right side or the bottom of the boxes.

Begin your diagram by deciding the size boxes to use. Notice that the box for the analyzer is larger than the other boxes. Draw this box larger than the other boxes. Draw all the smaller boxes the same size. The tie dots where lines connect to one another.

After you have your drawing pretty well laid out, you can place another piece of grid paper over it to center it on the sheet. Trace your layout using clear, sharp, dark lines with box outlines about three times the thickness of the interconnecting lines. Remember to make all your lettering very dark, and keep it at the same slant and height.

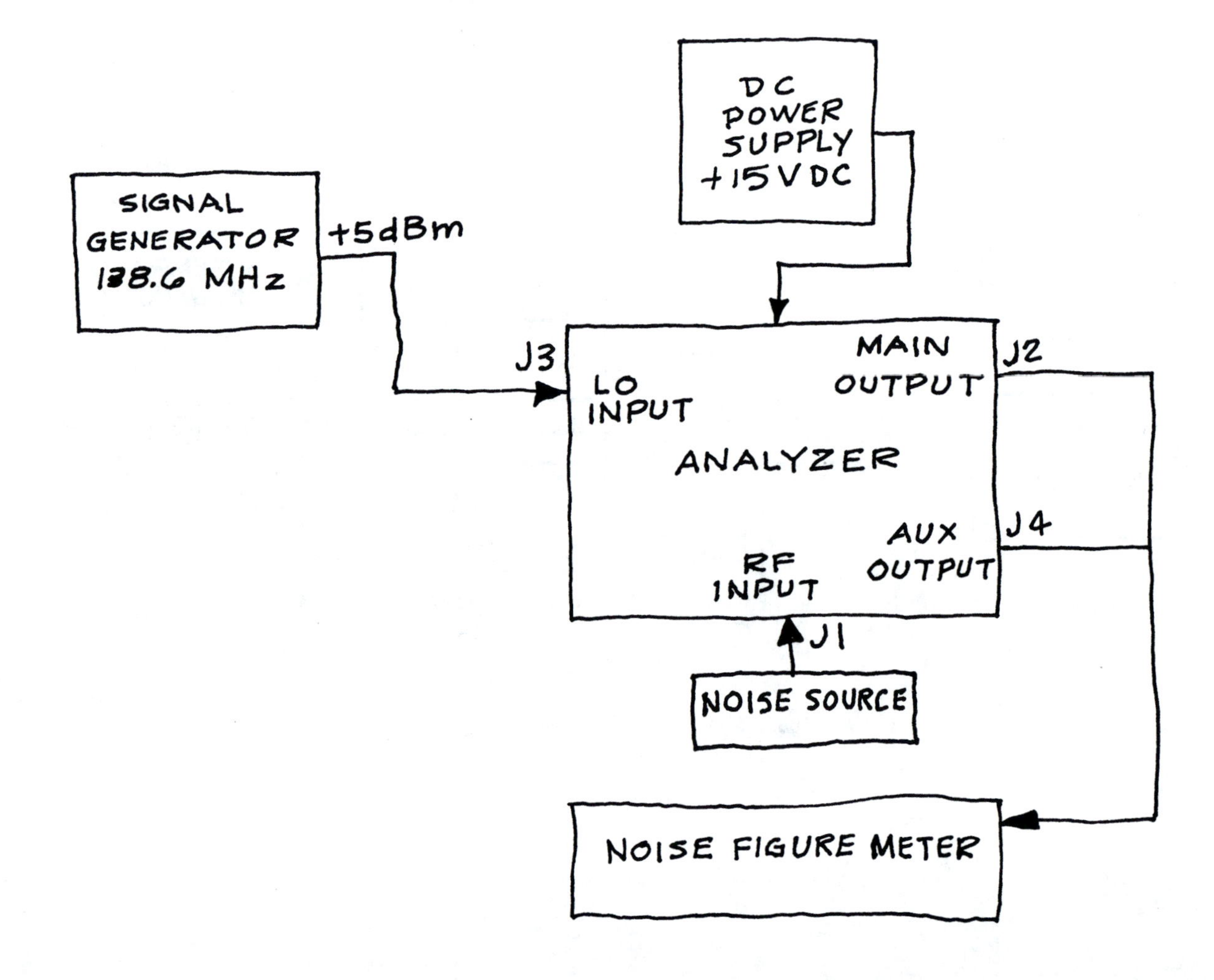

EXERCISE 3-2

Redraw the following sketch in pencil using the steps for *correct* layout and drawing of block diagrams. You may elect to draw it on the computer if you like.

Center the drawing lengthwise on an $8\frac{1}{2}''$ × 11″ sheet of grid paper. Leave a 1″ margin all around the drawing. Do not draw a border. Make all lettering $\frac{1}{10}''$ high, flush left, $\frac{1}{10}''$ from the left edge of the box, and center it vertically. Leave $\frac{1}{8}''$ between lines of lettering. Use a flow arow with a $\frac{1}{8}''$ × $\frac{1}{8}''$ arrowhead.

Remove all doglegs and as many crossovers as possible. All inputs should go into the left side or the top of the boxes. All outputs should leave from the right side or the bottom of the boxes.

Begin your diagram by deciding the size boxes to use. Notice that the boxes for the regulator control unit and the radar control panel are larger than the other boxes. Draw these two boxes the same width (the length may vary) and draw all the smaller boxes the same size. The tie dots where lines connect to one another should be made by using a $\frac{1}{8}''$ diameter closed circle, blackened in.

After you have your drawing pretty well laid out, you can place another piece of grid paper over it to center it on the sheet. Trace your layout using clear, sharp, dark lines with box outlines about three times the thickness of the interconnecting lines. Remember to make all your lettering very dark, and keep it at the same slant and height. In this drawing you will be able to eliminate all doglegs and most crossovers. You may have two crossovers, but not more than that.

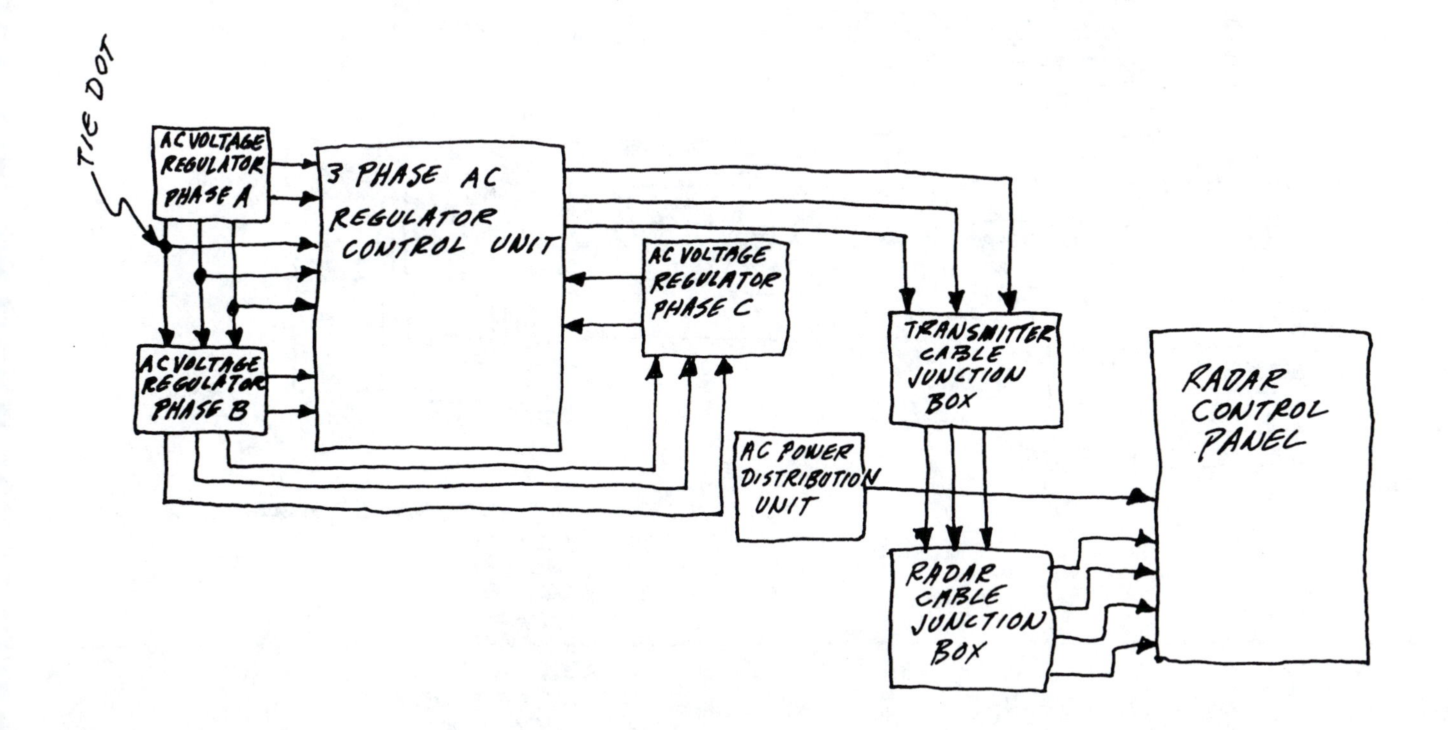

EXERCISE 3-3

Redraw the following sketch using the guidelines for *correct* layout and drawing of block diagrams. You may elect to draw it on the computer if you like.

Center the drawing lengthwise on a 17″ × 22″ sheet of grid paper. Leave at least a 1″ margin all around the drawing. Do not draw a border. Make all lettering flush left, $\frac{1}{10}$″ from the left edge of the box, and centered vertically. Use a flow arrow with a $\frac{1}{8}$″ × $\frac{1}{8}$″ arrowhead. All input circles should be a $\frac{1}{8}$″ diameter open circle. Make all lettering $\frac{1}{8}$″ high. *Do not allow lettering to touch any line or box.* Notice that you are asked to spell out all abbreviated words, and that the complete spellings of those abbreviations are given at the bottom of the sketch.

Remove as many doglegs and crossovers as possible. If possible, inputs should go into the left side or the top of the boxes, and outputs should leave from the right side or the bottom of the boxes.

This drawing has several major components. These components are the signal level detector, the signal conditioner, and the signal modifier. Show these major components by using dashed lines that are approximately $\frac{1}{4}$″ long with $\frac{1}{16}$″ between dashes. The dashed lines should go around these boxes as shown on the sketch. For example, in the box for the signal level detector, you will include the calibrate signal generator, the detector, the speaker, the A to D converter, the signal data regulator, and the buffer circuits. All of these items should be within the same dashed box. All the other boxes will be solid. Make all lettering $\frac{1}{8}$″ high. Lettering should be sharp and dark, and drawn to the same slant. Do not allow lettering to touch any line or box.

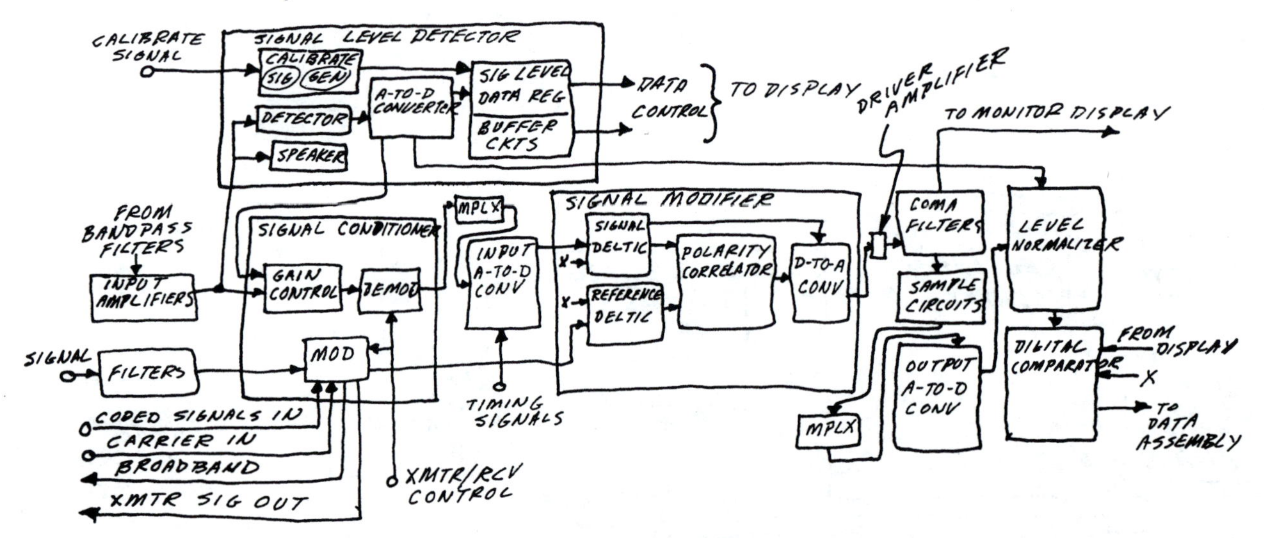

NOTE: SPELL OUT ALL ABBREVIATED WORDS:

RCV- RECEIVER
CONV- CONVERTER
MPLX- MULTIPLEX
SIG - SIGNAL
GEN- GENERATOR
REG- REGULATOR
DEMOD- DEMODULATOR
MOD- MODULATOR
CKTS- CIRCUITS
XMTR- TRANSMITTER

EXERCISE 3–4

Lay out and draw a formal block diagram from the following sketch. Use rules for good layout: inputs left or top, outputs right or bottom. On this diagram, remove all doglegs. Make all of the boxes the same size. Arrange the drawing so that it fits into a 6″ × 9″ area centered across the 11″ side of an $8\frac{1}{2}$″ × 11″ sheet. Make all lettering approximately $\frac{1}{8}$″ high, flush left, centered vertically in the boxes with approximately $\frac{1}{10}$″ between lines of lettering.

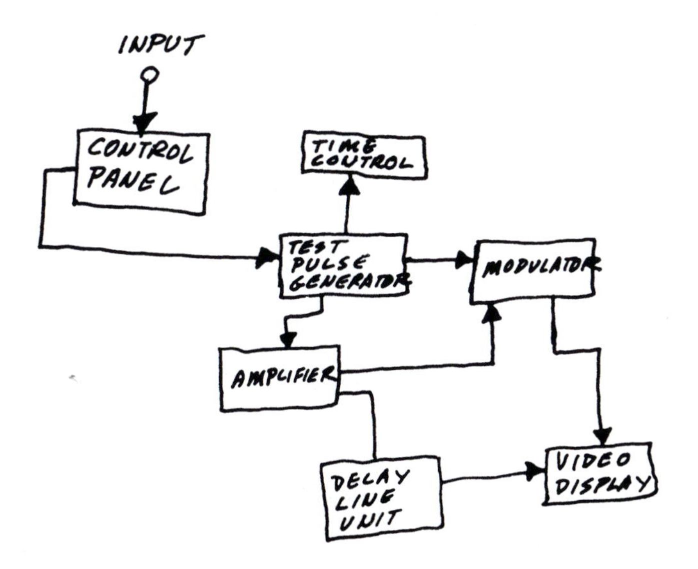

EXERCISE 3–5

Follow the same instructions as for Exercise 3–4, using the following sketch.

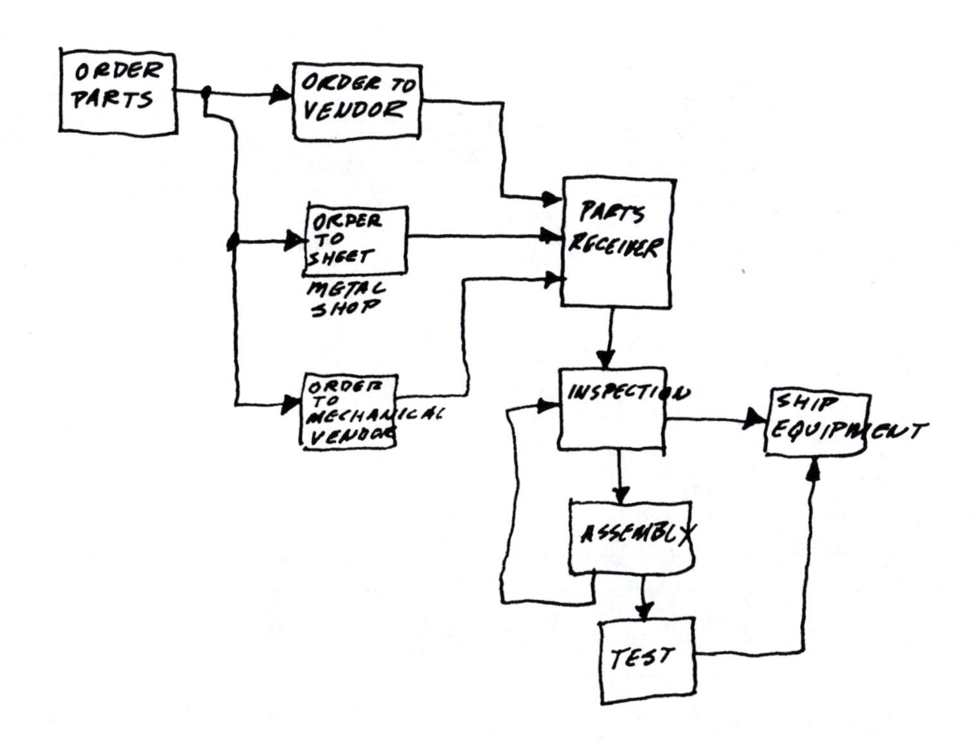

EXERCISE 3–6

Layout and draw a formal block diagram from the following sketch. Use rules for good layout described in this chapter. Remove doglegs wherever possible. Make all of the boxes approximately the same size except the 3-PHASE AC REGULATOR CONTROL UNIT. Draw this box as tall as necessary to allow all inputs to come into it from the left. Notice that some boxes have lettering overflowing the box. This lettering should be inside the box. Center the drawing lengthwise on an 11" × 17" sheet. Make all lettering approximately $\frac{1}{8}$" high with $\frac{1}{16}$" between lines of lettering. Lettering should be flush left, centered vertically in the boxes.

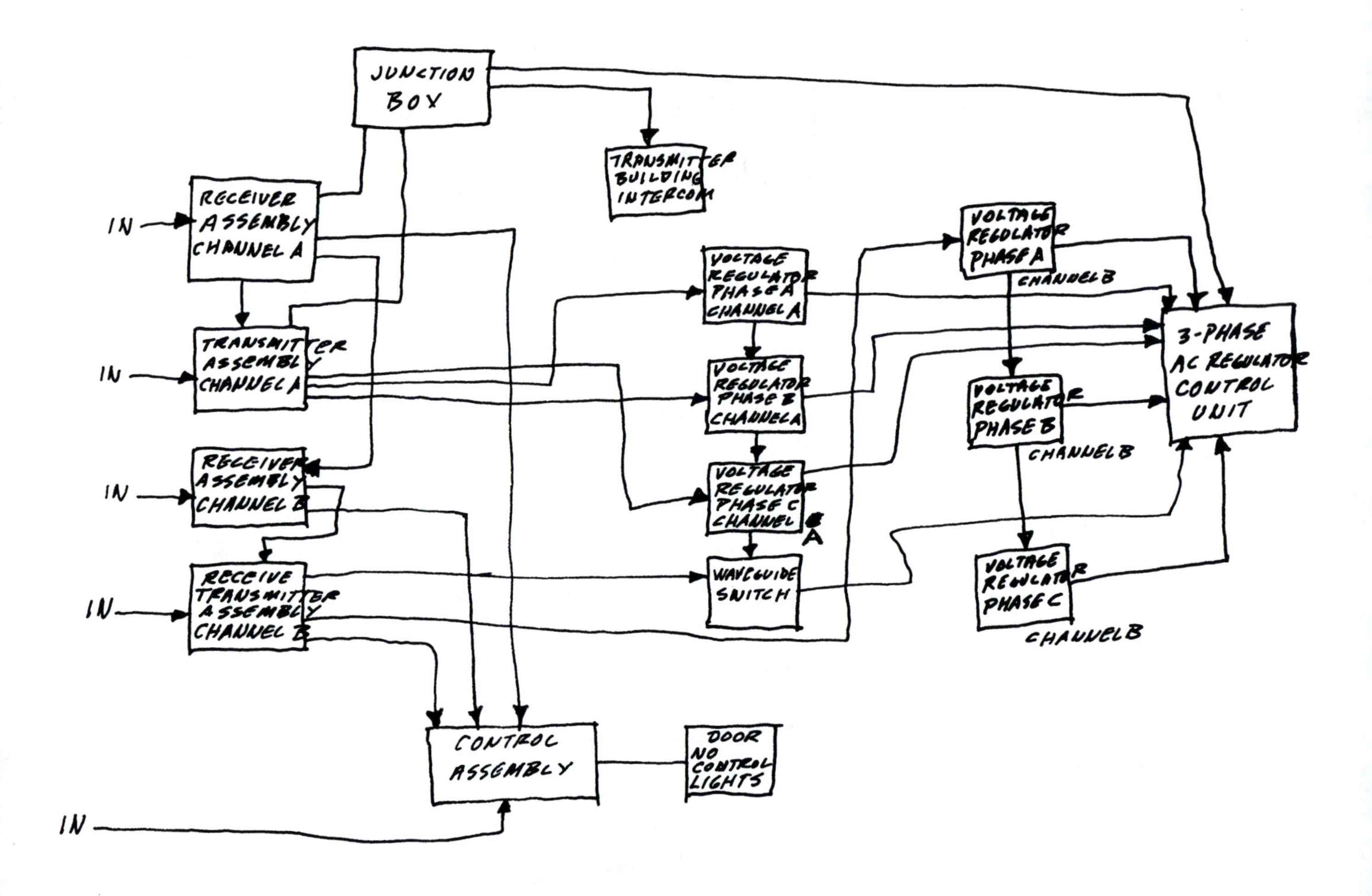

EXERCISE 3-7

Redraw the following sketch using the guidelines for *correct* layout and drawing of block diagrams.

Center the drawing vertically on a 17″ × 22″ sheet of grid paper. Leave at least a 1″ margin all around the drawing. Make all lettering flush left, $\frac{1}{10}$″ from the left edge of the box, and centered vertically. Use a flow arrow with a $\frac{1}{8}$″ × $\frac{1}{8}$″ arrowhead. All input circles should be a $\frac{1}{8}$″ diameter open circle.

Remove as many doglegs and crossovers as possible. If possible, inputs should go into the left side or the top of the boxes, and outputs should leave from the right side or the bottom of the boxes.

This drawing has three major components. These components are the RF distribution unit, antenna and the extra wide band tuner. Show these major components by using dashed lines that are approximately $\frac{1}{4}$″ long with $\frac{1}{16}$″ between dashes. The dashed lines should go around these boxes as shown on the sketch. For example, in the box for the extra tuner you will include the amplifier, the power divider and the remote tuner. All of these items should be within the same dashed box. All the other boxes will be solid. Make all lettering $\frac{1}{8}$″ high. Lettering should be sharp and dark, and drawn to the same slant. Do not allow lettering to touch any line or box.

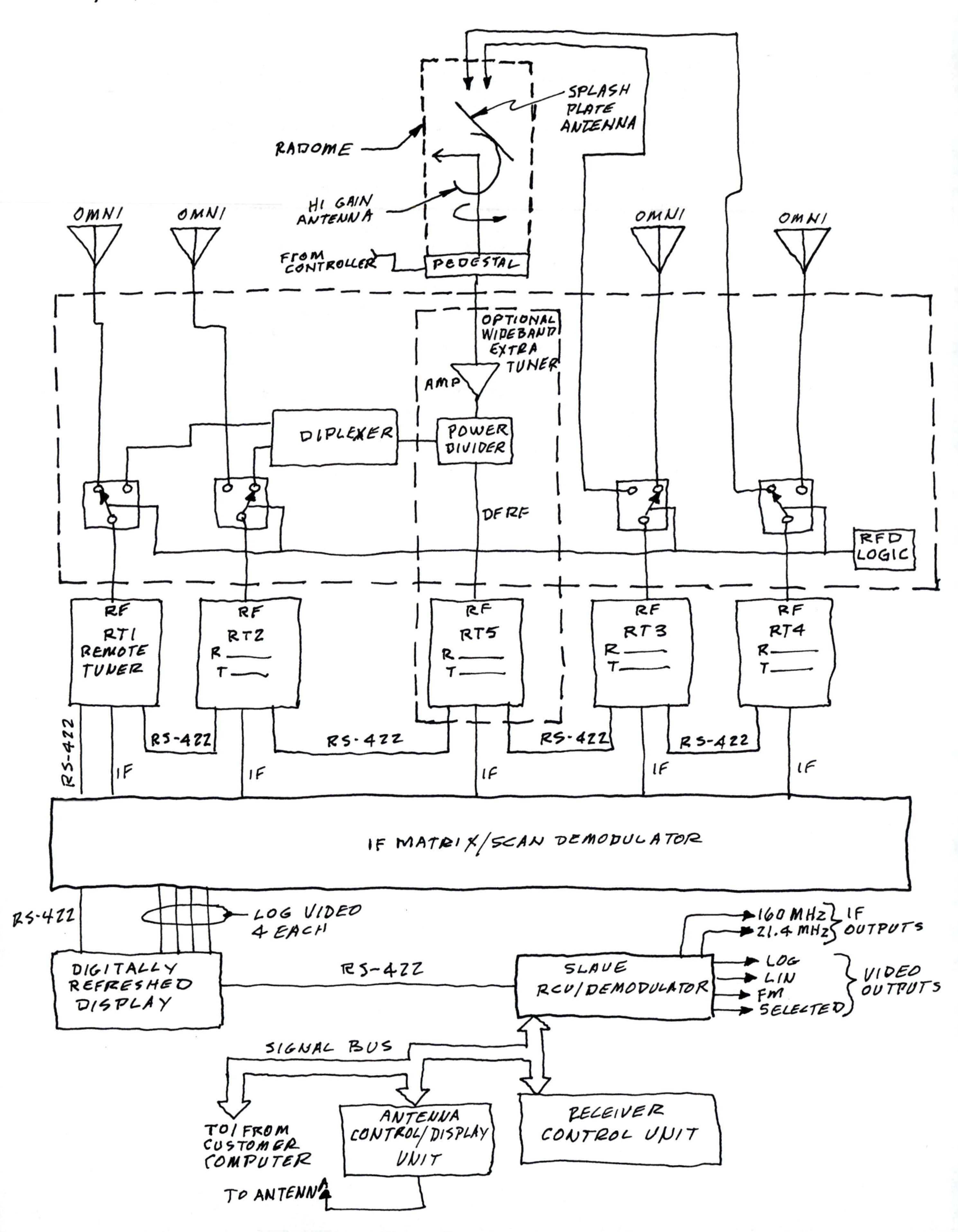
SPLASH PLATE ANTENNA
RADOME
HI GAIN ANTENNA
FROM CONTROLLER
PEDESTAL
OMNI
OMNI
OMNI
OMNI
OPTIONAL WIDEBAND EXTRA TUNER
AMP
DIPLEXER
POWER DIVIDER
DF RF
RFD LOGIC
RF
RT1 REMOTE TUNER
RF
RT2
R
T
RF
RT5
R
T
RF
RT3
R
T
RF
RT4
R
T
RS-422
RS-422
RS-422
RS-422
RS-422
IF
IF
IF
IF
IF
IF MATRIX/SCAN DEMODULATOR
RS-422
LOG VIDEO 4 EACH
160 MHz
21.4 MHz
IF OUTPUTS
DIGITALLY REFRESHED DISPLAY
RS-422
SLAVE RCU/DEMODULATOR
LOG
LIN
FM
SELECTED
VIDEO OUTPUTS
SIGNAL BUS
TO/FROM CUSTOMER COMPUTER
ANTENNA CONTROL/DISPLAY UNIT
RECEIVER CONTROL UNIT
TO ANTENNA

Chapter 4

Symbols for Schematic Diagrams

OBJECTIVES

After completing this chapter, you will be able to

- ✓ identify and draw the most commonly used electronic parts and symbols.
- ✓ identify the correct reference designator for each electronic symbol.

IDENTIFYING ELECTRONIC PARTS

Being able to identify correctly electronic parts, their symbols, values, names, and appearances is an important part of electronic drafting. The most common electronic parts are illustrated and discussed in this chapter. The actual shape of each part is shown, along with its symbol. The reference designator or the letter(s) by which the part is known is given after the symbol. The arrangement of the symbol on a page, with its value, is then shown. The student should become familiar with how these parts are labeled, what values are given to them, and how these items are arranged on an electronic schematic.

Resistors

Various types of resistors are shown in Figure 4–1. Figure 4–2 shows the resistor color code charts. Color bands on many resistors indicate the value of the resistor, and the percent of accuracy to which it is manufactured. Two types of variable resistors are shown in Figure 4–3. Variable resistors are manufactured in many different sizes and shapes. Some variable resistors can be mounted on printed circuit boards. They have small tabs that fit through the board and are soldered on the back.

Resistor–A part which resists the flow of electric current in a circuit.

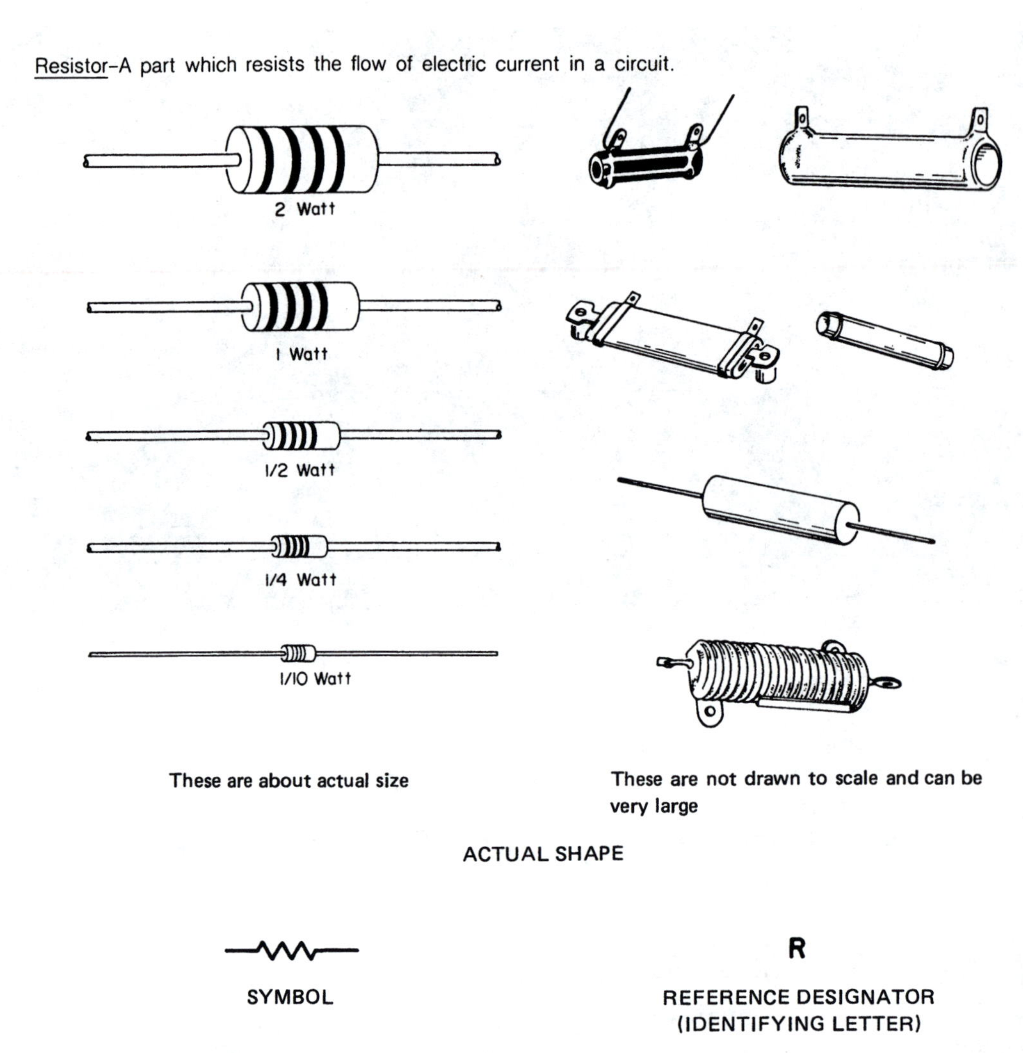

Resistance values are given in ohms, kilohms (1,000 ohms), or megohms (1,000,000 ohms). Power ratings are given in watts. A resistor symbol with its description looks like this:

R1 150   R2 5K   R3 1 MEG

Courtesy of Kelvin Electronics, Inc.

Figure 4–1 Resistors

1st digit — RED 2
2nd digit — YELLOW 4
3rd digit (multiplier) — ORANGE 3
4th digit (tolerance) — GOLD 5%
failure rate level (established relativity types only) — BROWN M FAILURE RATE LEVEL

ANSWER: 24000 OR 24K, ±5%

(K=1000, M=10000000)

Color	ABV	First Figure	Second Figure	Multiplier	Tolerance		Failure Rate Level
					Percent	LTR	
Black	BLK	0	0	1	±20	M	L
Brown	BR	1	1	10	± 1	F	M
Red	R	2	2	100	± 2	G	P
Orange	O	3	3	1,000			R
Yellow	Y	4	4	10,000			S
Green	G	5	5	100,000			T
Blue	BL	6	6	1,000,000			
Purple (Violet)	V	7	7				
Gray	GY	8	8				
White	W	9	9				
Silver	SIL			0.01	±10	K	
Gold	GLD			0.1	± 5	J	

Courtesy of Texas Instruments, Inc.

Figure 4–2 Resistor color code chart

Variable Resistor - A wirewound or composition resistor used to vary the resistance in a circuit. Rheostats and potentiometers are variable resistors.

ACTUAL SHAPE

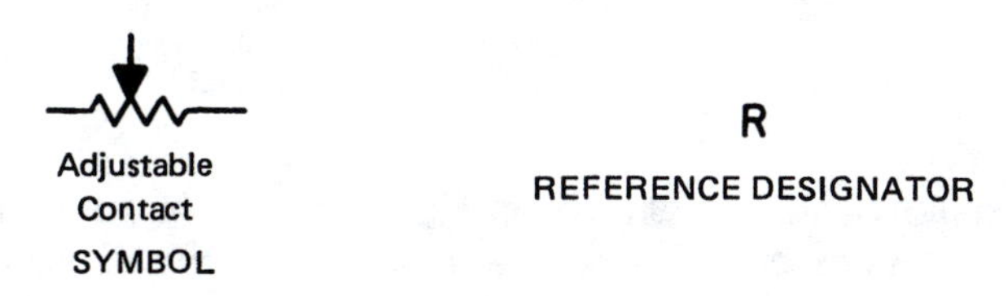

A variable resistor symbol with its description looks like this:

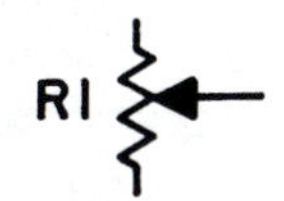

Courtesy of Kelvin Electronics, Inc.

Figure 4–3 Variable resistors

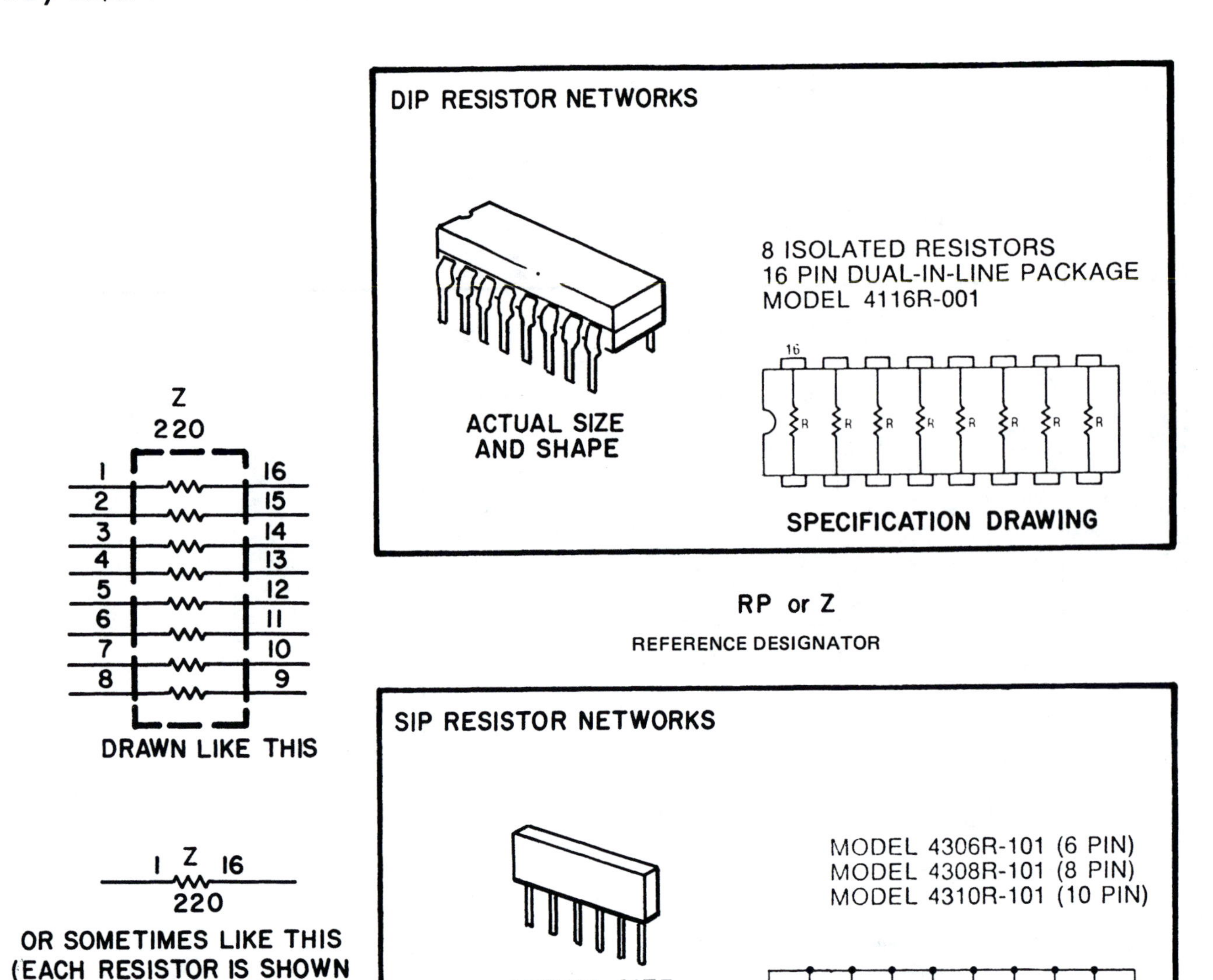

Figure 4–4 Resistor packs

Resistor packs are shown in Figure 4–4. They are shown actual size and shape. A pack contains several resistors. If the pack were pulled apart, one would not actually see a number of small resistors. Rather, the resistances of the value indicated are built into the pack by depositing a resistive material on a thin layer and connecting the pins as shown in the diagram. For example, pin 1 is on one end of the resistor and pin 16 is on the other end. The diagram at the lower left shows how these resistances are connected.

The figure shows only two of the many types of resistor packs. Resistor packs are available in several different shapes and arrangements.

Capacitor - A part which stores electrical energy, blocks the flow of direct current, and permits the flow of alternating current. Capacitors may be either variable or fixed.

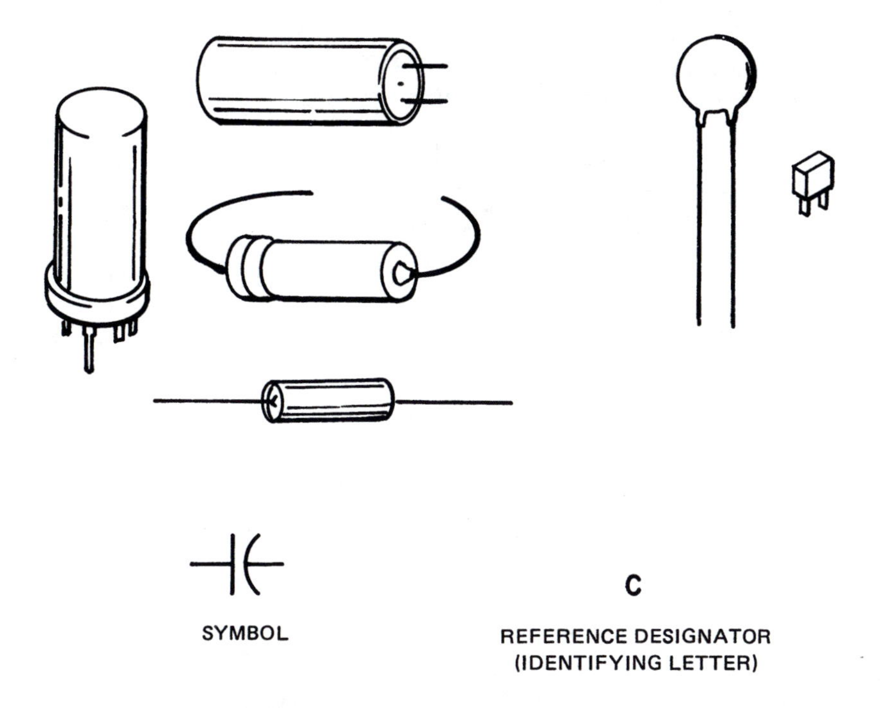

Capacitance values are given in farads. Microfarads (1/1,000) or picofarads (1/1,000,000) are most commonly used. Capacitor symbols with their descriptions appear below:

C1
10PF
+

Fixed Capacitor

C2
10 PF

Fixed Capacitor

C3

Variable Capacitor

Figure 4–5 Capacitors

Capacitors

Several forms of capacitors are shown in Figure 4–5. These are very common parts. Often, the positive side of the capacitor must be shown on a drawing. When a sketch shows a plus sign on one side of the capacitor, the drafter must be certain to include that sign since it is an important part of the symbol. Furthermore, the curved portion of the symbol typically faces circuit ground, which is discussed later.

Diode - A part which conducts electricity more easily in one direction than in the other.
Zener diode (breakdown diode) - This part has a small reverse current, as in a standard diode, until the voltage reaches a certain level at which point it suddenly increases rapidly.

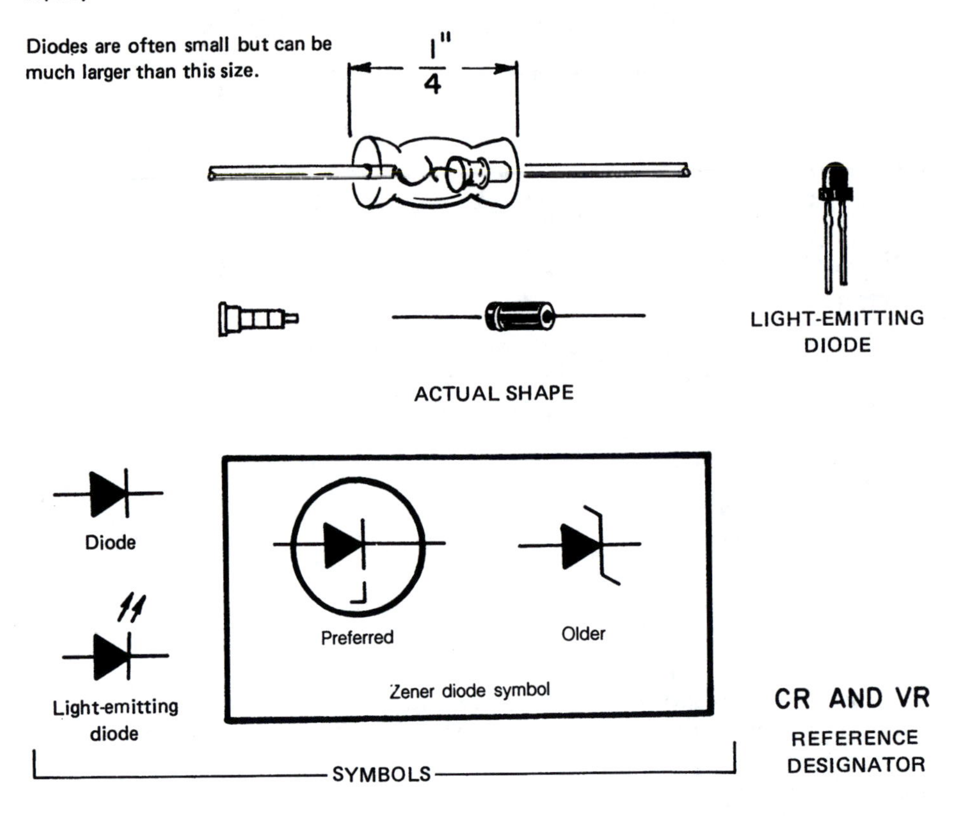

Diodes are identified on the drawing by type. No values are given for diodes on a schematic drawing. Diode symbols with their desciptions appear below:

Courtesy of Kelvin Electronics, Inc.

Figure 4–6 Diodes

Diodes

Several types of diodes are available, the standard type, the light-emitting diode, and the zener diode are most commonly encountered, Figure 4–6. Each has a different symbol.

Transistors

Figure 4–7 shows transistors. A drafter must be concerned about the tab location of the transistor, particularly when doing a printed circuit

<u>Transistor</u> - A part that has three or more electrodes. Transistors are used for many different purposes depending on how they are built.

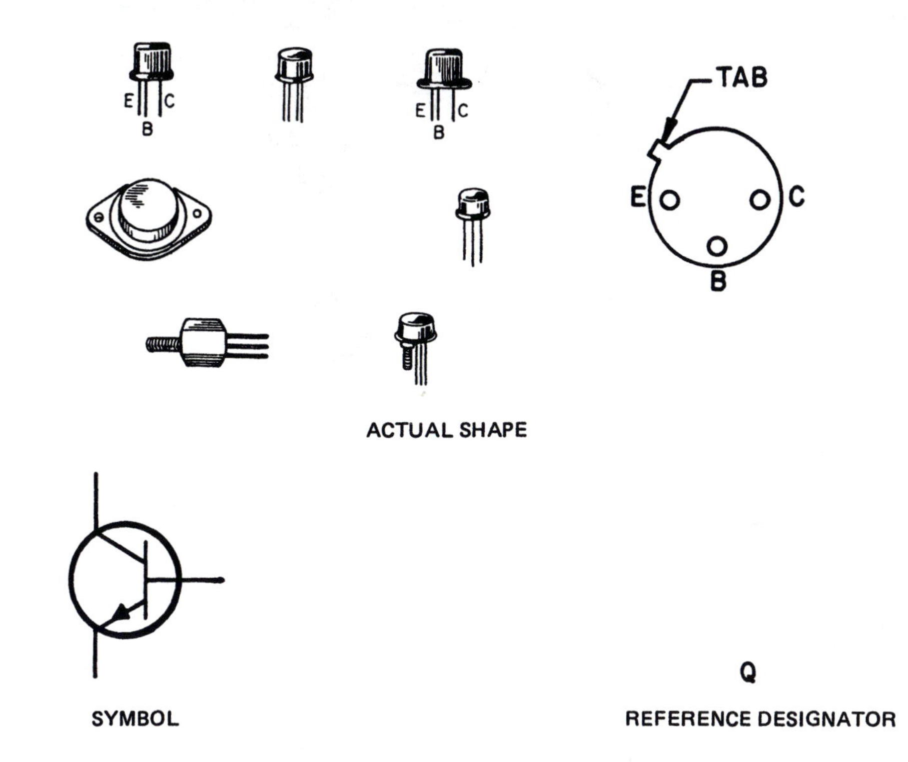

Transistors are identified on the drawing by type. No values are given for transistors on the schematic diagram. Transistor symbols with their descriptions appear below:

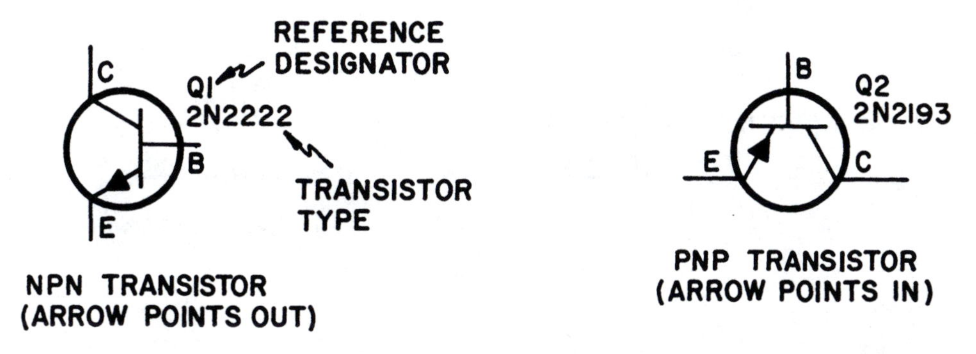

NOTE: THE LETTERS B,E,C IDENTIFYING THE BASE, EMITTER, AND COLLECTOR ARE NOT SHOWN ON THE SCHEMATIC

Courtesy of Kelvin Electronics, Inc.

Figure 4–7 Transistors

board. It is necessary to be sure the emitter, base, and collector (E, B, and C) terminals are identified correctly. These terminals on the transistor are arranged just as they are shown in Figure 4–7. The tab is located between the emitter and the collector, next to the emitter. The emitter is the leg with the arrow, the collector is the leg without the arrow, and the base is the line to which all the others attach.

Transformer - A part that changes electrical voltage from one or more circuits to one or more other circuits at the same frequency, but usually at a different voltage and current value.

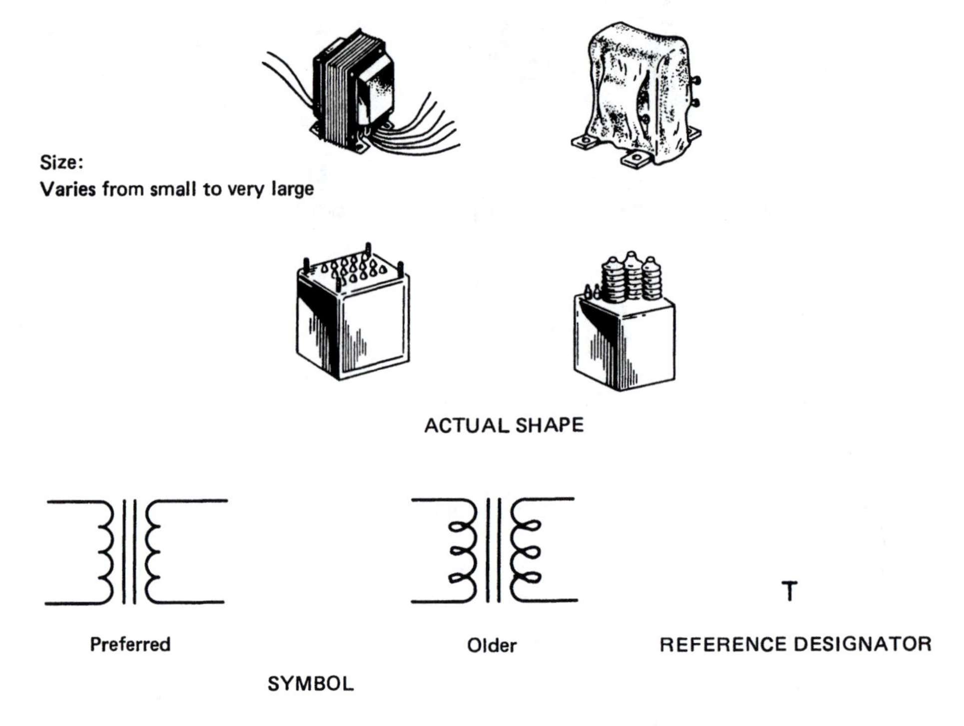

Values for transformers are often shown on the schematic, and the leads are often identified by color or number. Transformer symbols with their descriptions appear below:

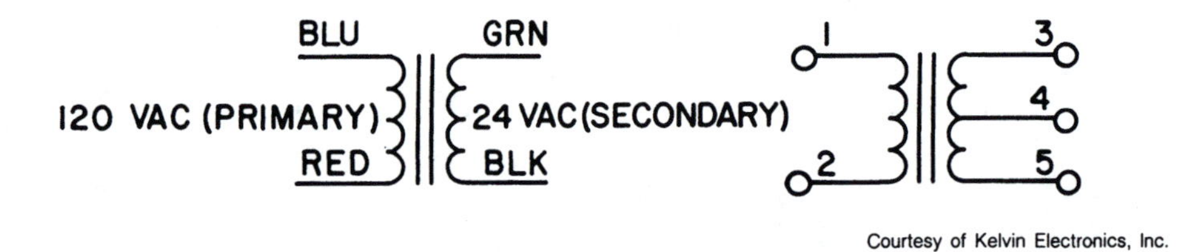

Courtesy of Kelvin Electronics, Inc.

Figure 4–8 Transformers

Transformers

Figure 4–8 shows several types of transformers and the transformer symbol. Often the secondary and the primary of the transformer must be shown, depending on the sketched instructions from the engineer or technician.

Inductors

Figure 4–9 describes an inductor or coil.

Fuses and Circuit Breakers

Figure 4–10 shows two different types of fuses and fuseholders. One type of circuit breaker is also shown. The fuseholders on the left accommodate the fuse inside by unscrewing the cap in the front. The fuseholder on the right is one that is commonly found in older automobiles. The circuit breaker shown is one that is found on some types of electronic equipment. It is equipped with a reset button. Notice that fuses are rated in amperes.

<u>Inductor or Coil</u> - An inductor (sometimes called a choke) is used for introducing inductance into an electrical circuit. Inductance is the property which opposes any change in the existing circuit.

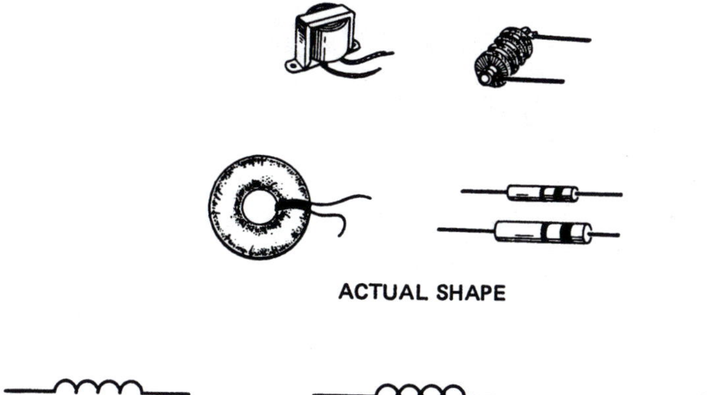

ACTUAL SHAPE

Preferred — Older — SYMBOL

L
REFERENCE DESIGNATOR

Inductance values are given in henrys. Millihenrys or microhenrys are often used for the inductor value. Inductor symbols with their descriptions appear below.

Courtesy of Kelvin Electronics, Inc.

Figure 4–9 Inductors

<u>Fuse</u> - A protective device, usually a short piece of wire (although sometimes a chemical compound) which melts, interrupting the current when it exceeds the rated value.

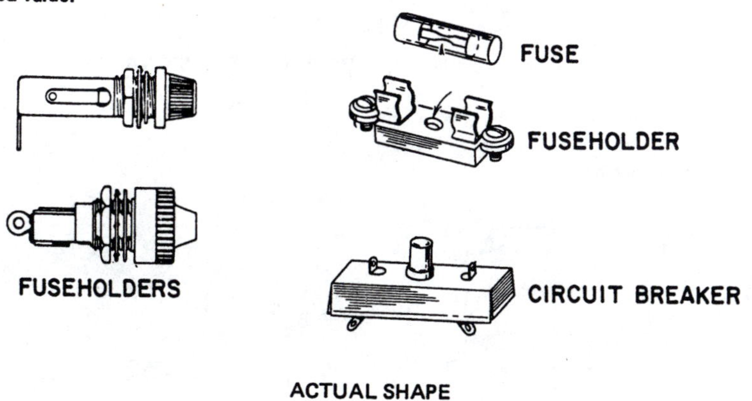

ACTUAL SHAPE

Preferred — Older — SYMBOL

F
REFERENCE DESIGNATOR

Values for fuses are given in amperes. A fuse symbol with its description appears below:

FI
I AMP

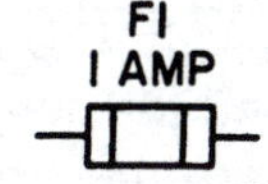

Figure 4–10 Fuses

Relay - A part which opens or closes contacts in a circuit in response to variations in that circuit or another circuit. It has a coil or winding which, when activated, opens or closes the contacts of a switch.

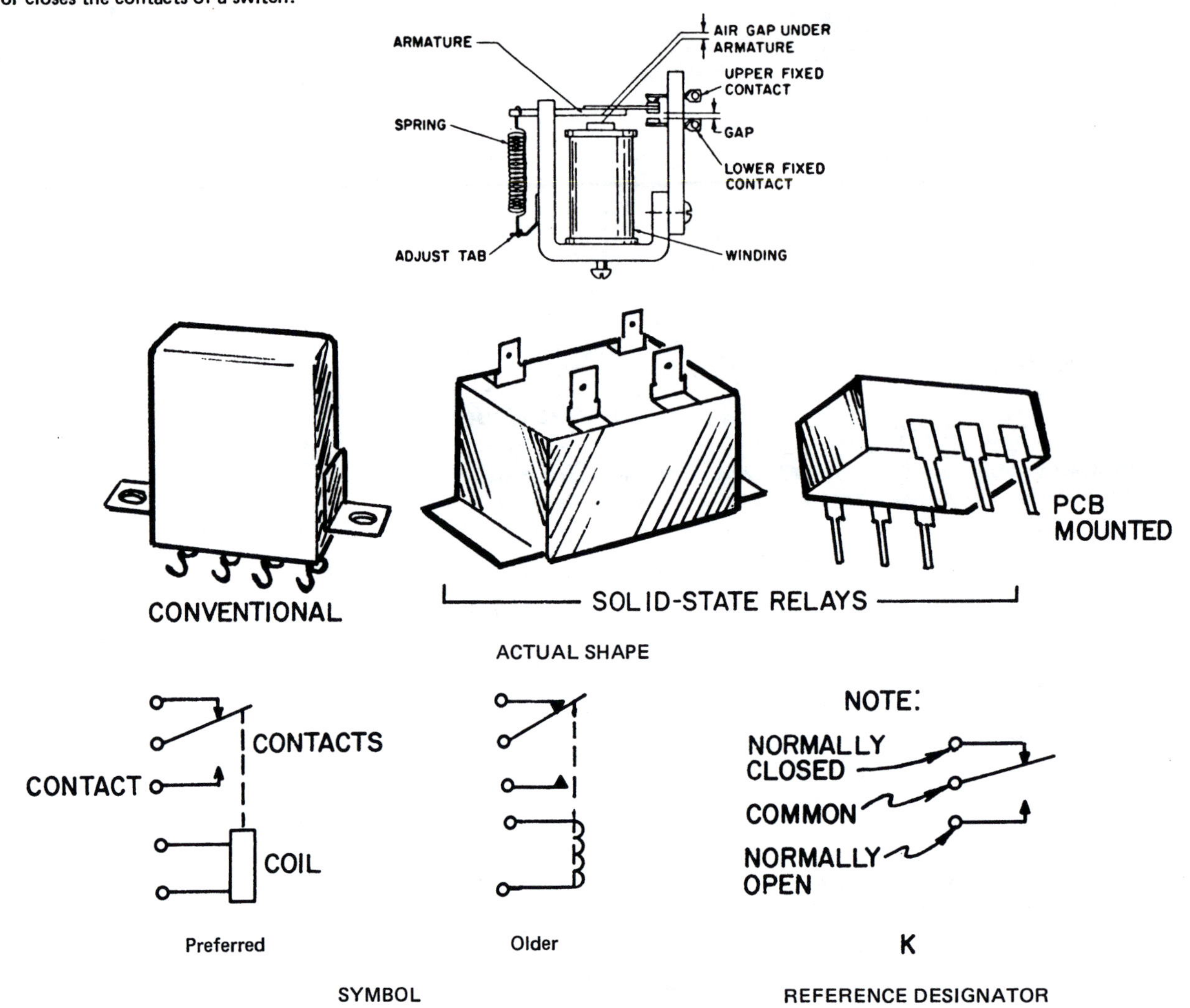

Values for a relay are not given on a schematic. The connections are numbered or labeled as they appear on the relay. The symbol may be rearranged to make a clearer layout. Relay symbols with their descriptions are shown below:

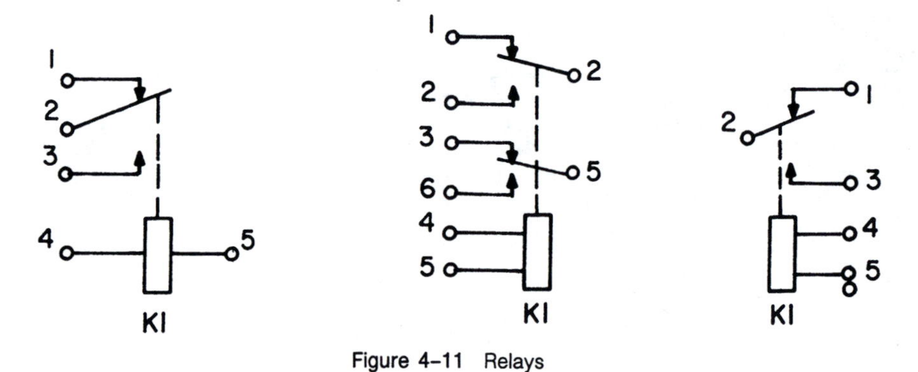

Figure 4–11 Relays

Relays

Relays are shown in Figure 4–11. The note shows a normally closed contact, a normally open contact, and the common contact where the wiper is located. A drafter should be aware that the wiper touches the normally closed contact, and does not touch the normally open contact.

Connector - A part which makes an electromechanical connection between two cables, between a cable and an electrical part or between two parts.

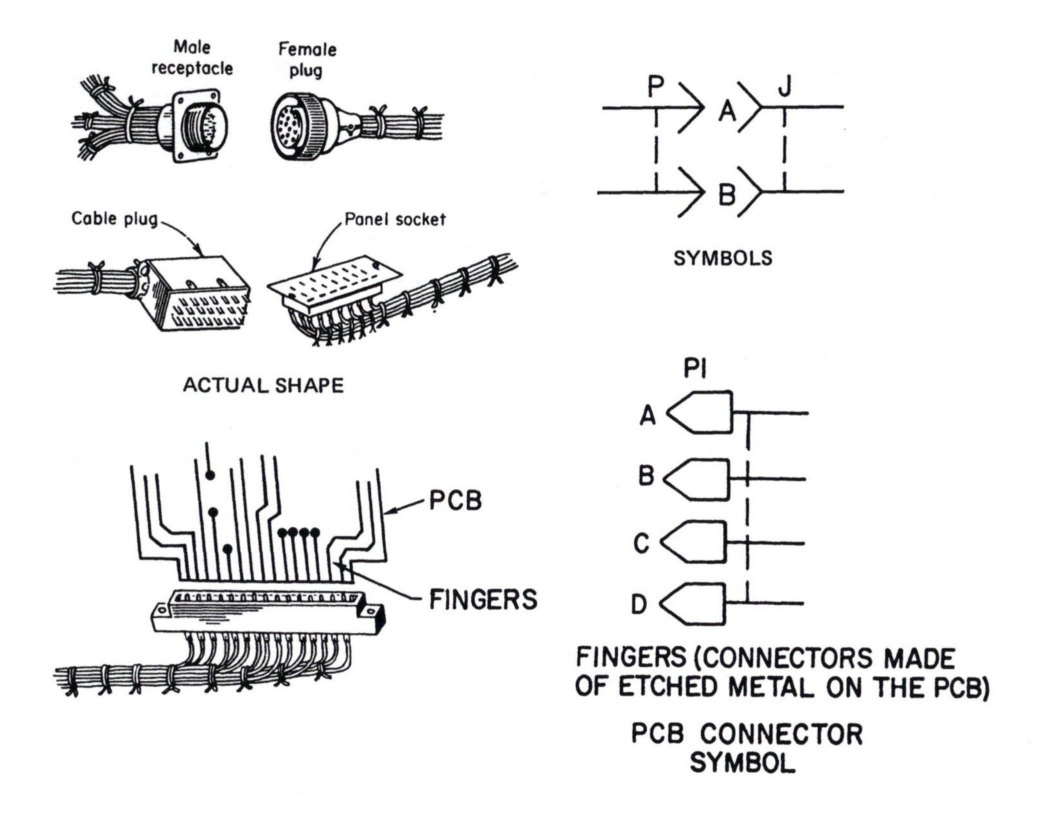

P IS FOR PLUG
J IS FOR RECEPTACLE

REFERENCE DESIGNATOR

The reference designator of the connector and the number of each connection (pin number) is shown on a schematic diagram. Pins may be labeled with either a number or a letter, depending on the manufacturer. A connector may have from one to sixty or more pins. Connector symbols with their descriptions are shown below:

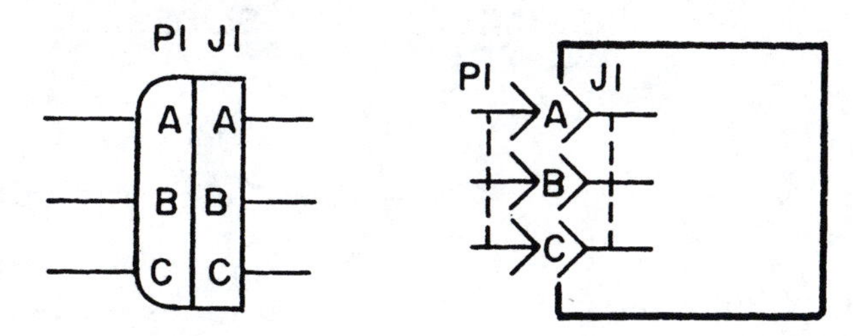

Figure 4–12 Connectors

Connectors

Figure 4–12 shows several different types of connectors and their symbols.

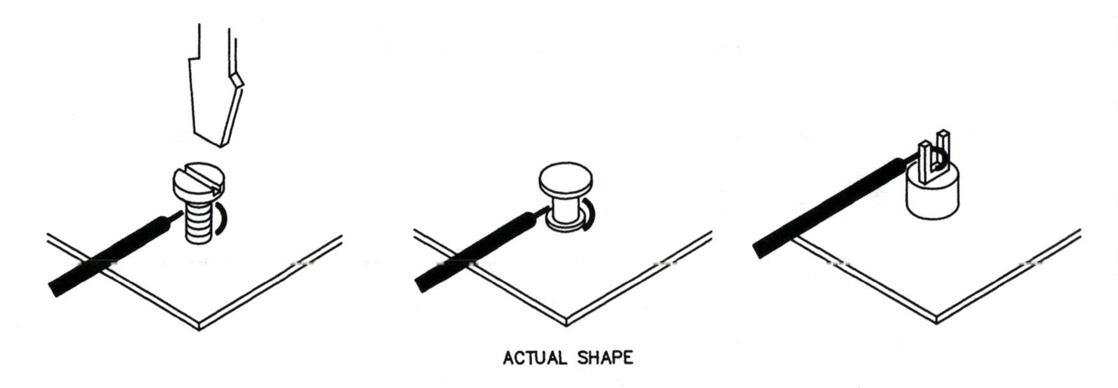

Figure 4–13 Terminals

Terminals

Figure 4–13 shows terminals or mechanical connections which are labeled. Terminals are given a reference designator but are not given a value.

Switches

Several different switches and their reference designators are shown in Figure 4–14.

Switch - A device used to open or close one or more electrical circuits.

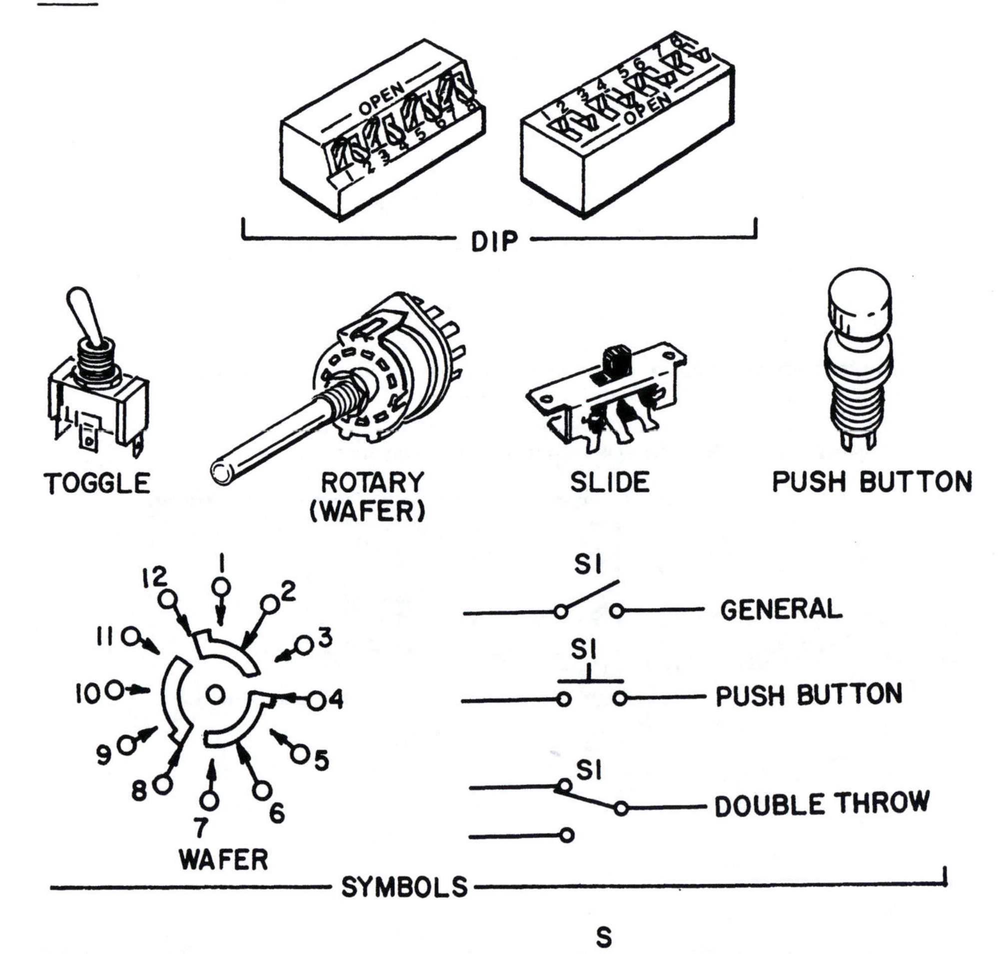

Figure 4–14 Switches

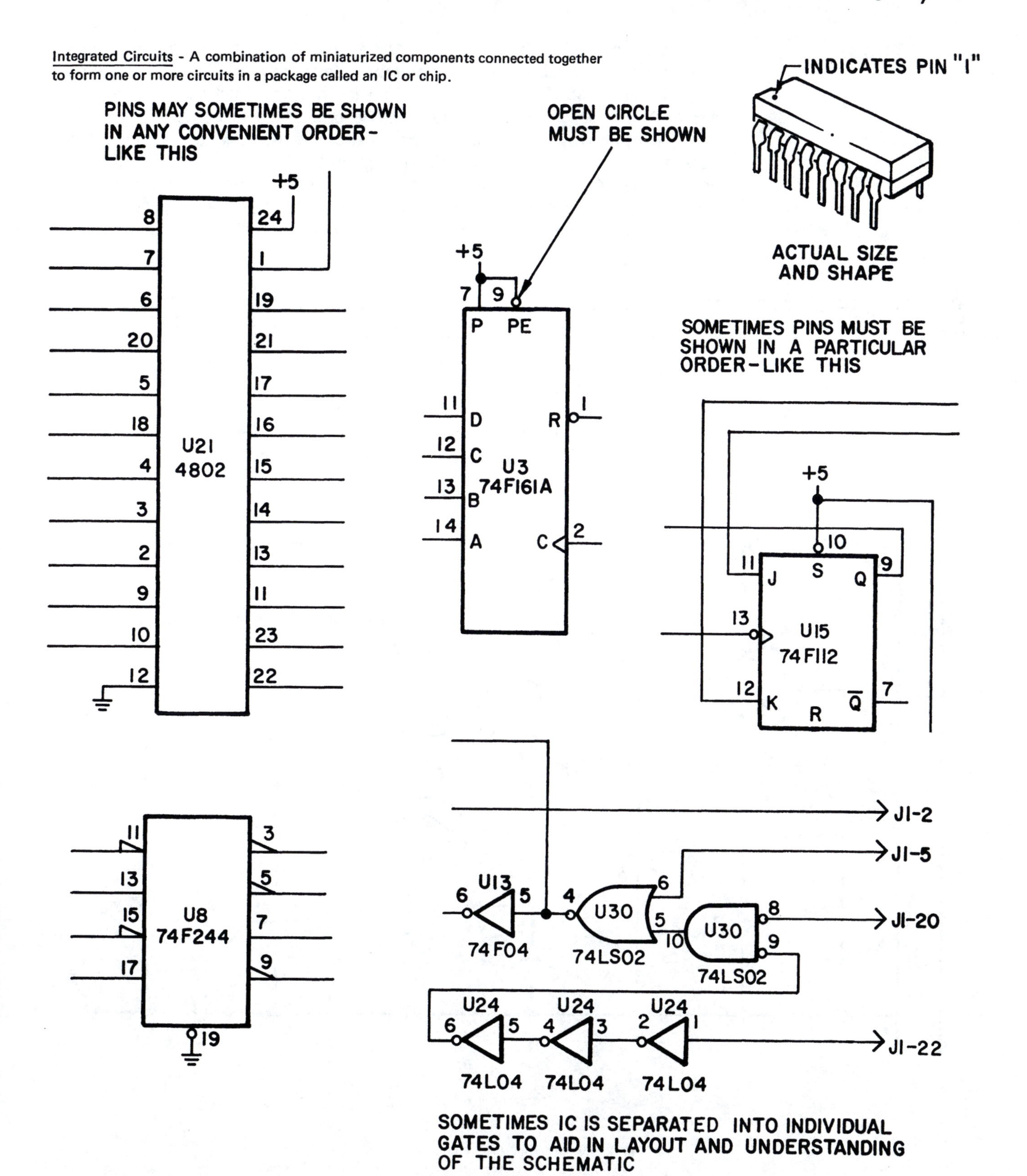

Figure 4–15 Integrated circuits (IC)

Integrated Circuits

Figure 4–15 shows several integrated circuits (IC). They are generally contained in dual-in-line (DIP) packages. They can also be of different shapes and sizes other than the ones shown.

Looking at Figure 4–15, note for example that the integrated circuit can be drawn as a box with its reference designator (U–21) and the type of device it is (4802) underneath it. The integrated circuit contains several *gates*. All of the gates are included inside one of these boxes.

Another way to draw an integrated circuit is by drawing each individual gate, as shown in the lower right of Figure 4–15. For example, U30 is a device type 74LS02. Each one of the gates is shown with its pin numbers. The device type and reference designator are shown on each gate.

Integrated circuits are described in greater detail in Chapter 6.

EXERCISE 4–1

Trace the schematic shown. Label the components correctly. Label the connectors and open circles, and determine what each of these symbols represents. Put the correct reference designator on each symbol, and number them left to right and top to bottom. For example, C1, C2, C3, C4, R1, R2, Q1, Q2, and so on. Be sure to put pin numbers on the connector.

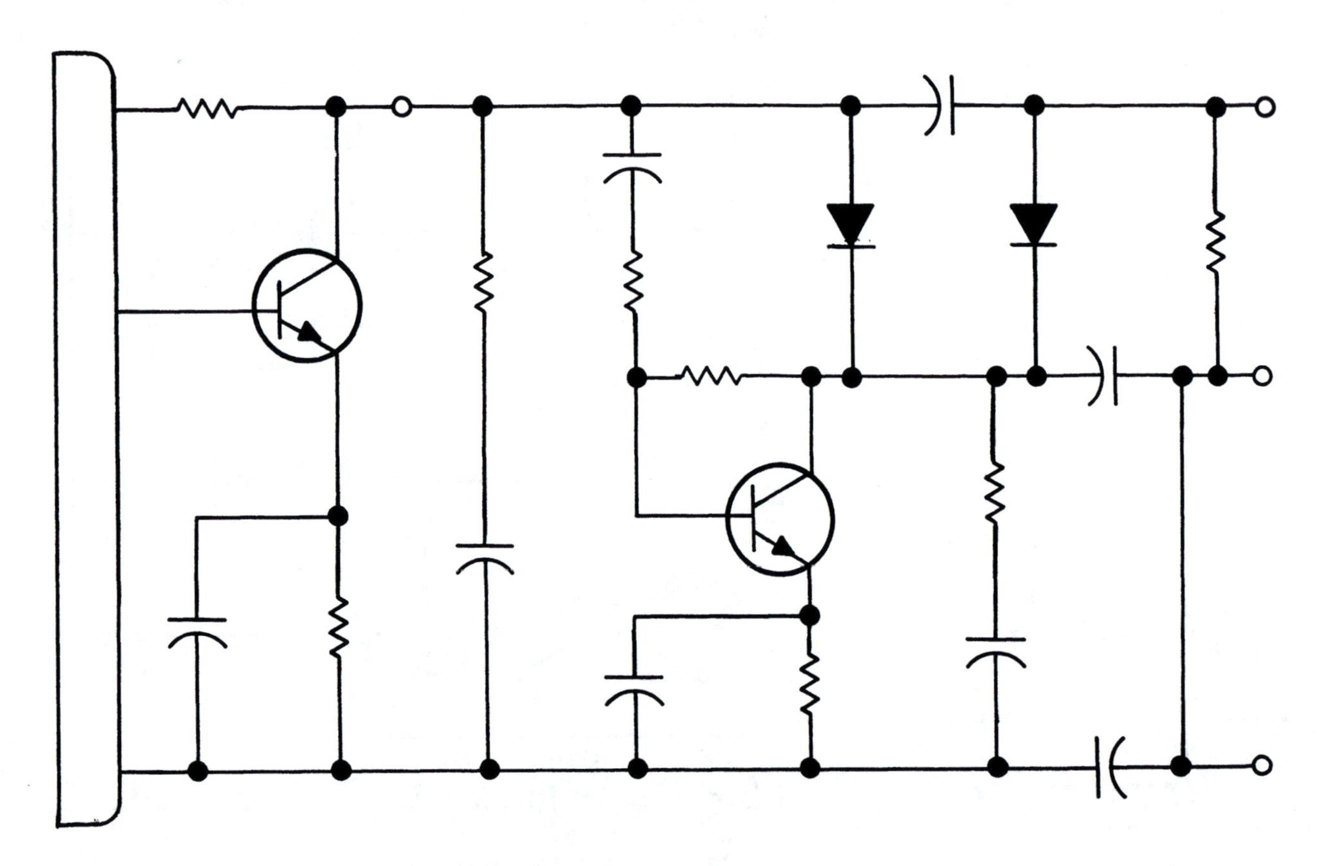

EXERCISE 4-2

On a separate sheet of paper, draw the correct symbol and write the correct reference designator for the following parts:

- Variable resistor
- Fuse
- Switch
- Inductor
- Relay
- Capacitor
- Integrated circuit
- Transistor
- Light emitting diode (LED)

EXERCISE 4-3

Redraw the following sketch as a formal schematic diagram. Use a schematic template for all of the symbols. Renumber all components using left to right, top to bottom numbering scheme. Use the correct symbol for each component. Make all lettering approximately $\frac{1}{8}''$ high with $\frac{1}{16}''$ between lines of lettering. Center the drawing lengthwise on an $8\frac{1}{2}'' \times 11''$ sheet.

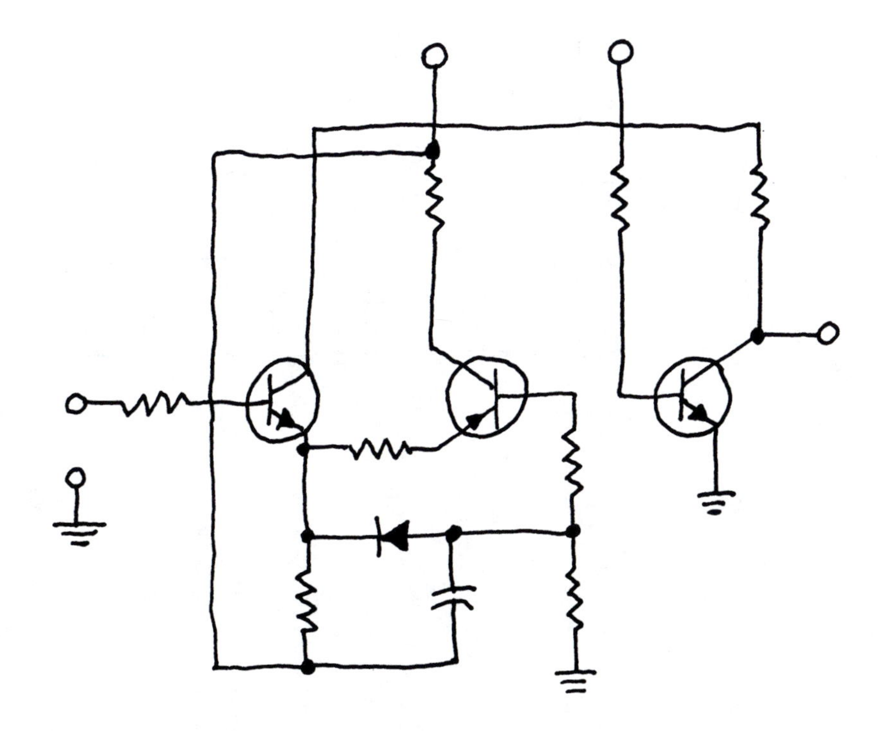

EXERCISE 4–4

Redraw the following sketch as a formal schematic diagram. Use a schematic template for all of the symbols. Renumber all components using left to right, top to bottom numbering scheme. Use the correct symbol for each component. Make all lettering approximately $\frac{1}{8}''$ high with $\frac{1}{16}''$ between lines of lettering. Center the drawing lengthwise on an $8\frac{1}{8}'' \times 11''$ sheet.

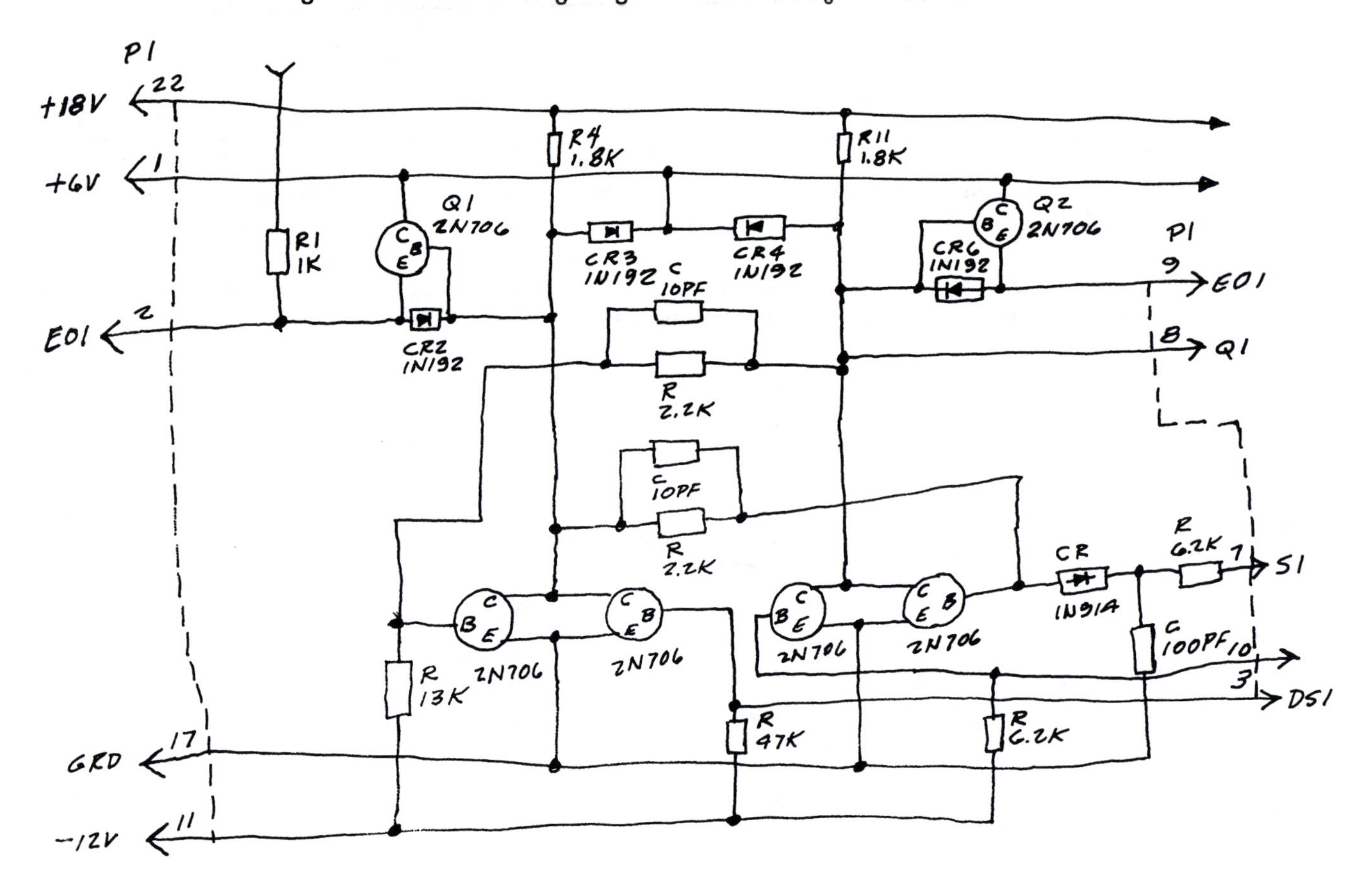

EXERCISE 4-5

Redraw the following sketch as a formal schematic diagram. Use a schematic template for all symbols. Assign reference designators to all components using left to right, top to bottom lettering scheme. Use the correct symbol for each component. Make all lettering approximately $\frac{1}{8}''$ high with $\frac{1}{16}''$ between lines of lettering. Center the drawing lengthwise on an $8\frac{1}{2}'' \times 11''$ sheet.

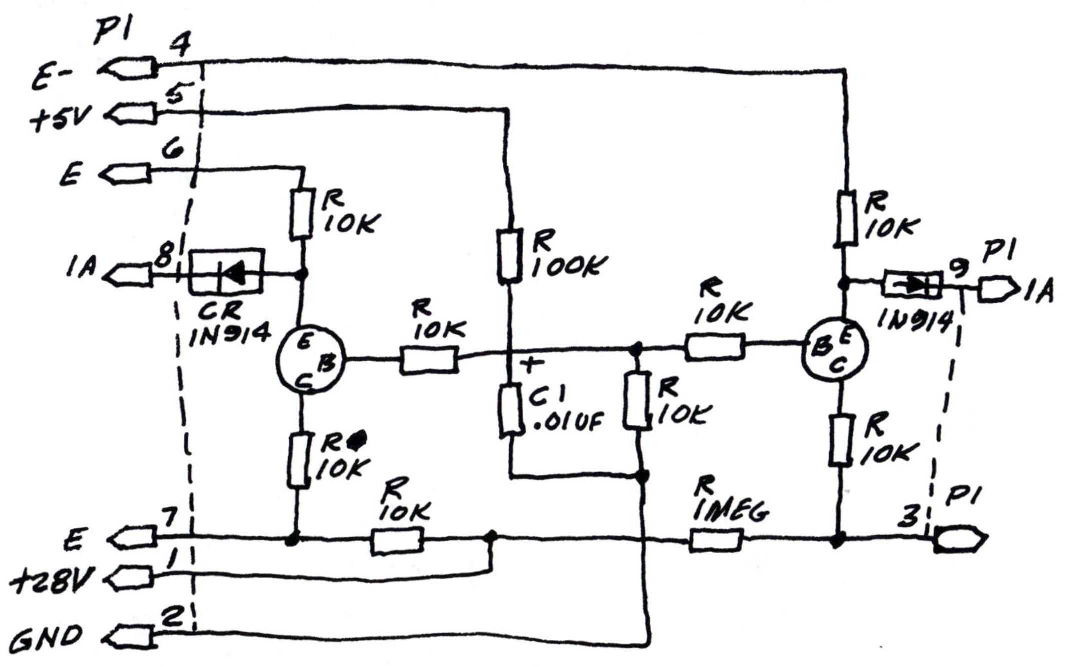

EXERCISE 4-6

Follow the same instructions as for Exercise 4–5, using the following sketch.

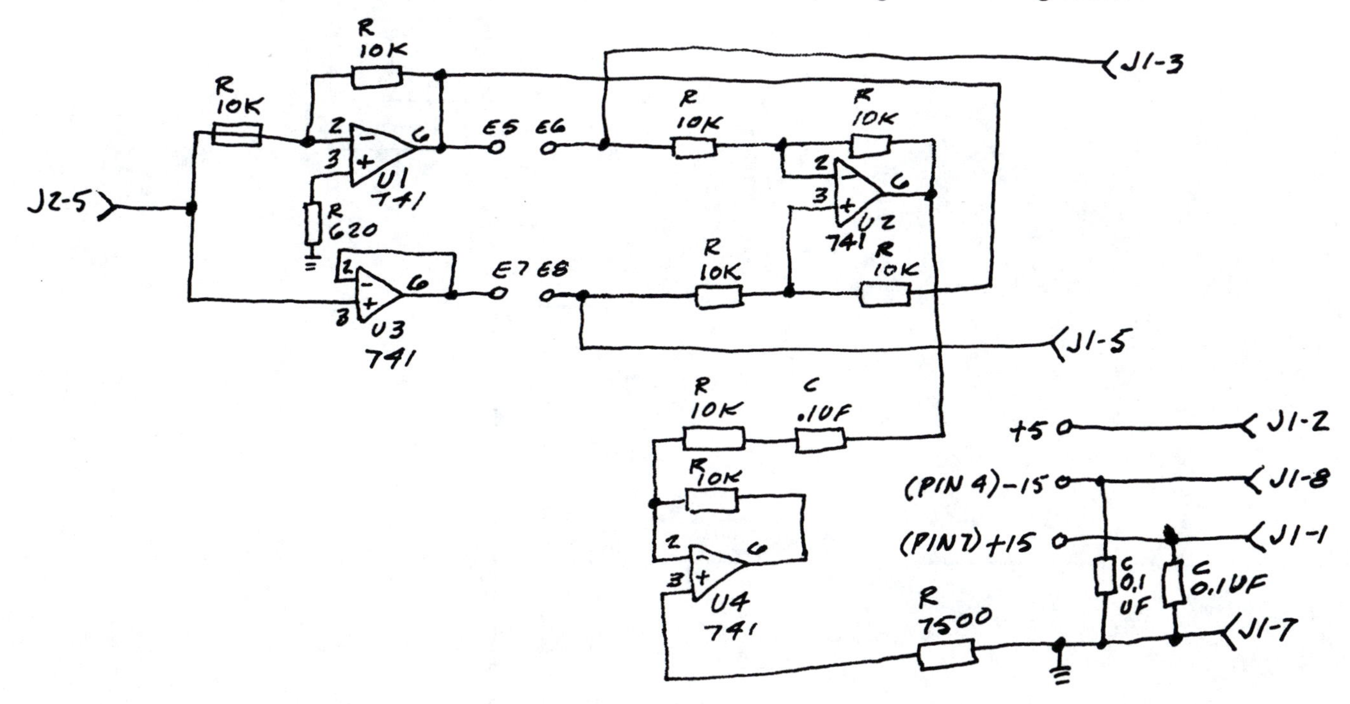

EXERCISE 4–7

Redraw the following wiring diagram as a schematic diagram. Use the correct symbols as described in the chapter. Do not show the wire colors. Use the reference designators as shown on the wiring diagram. An example of the correct symbols for switches S1 and S2 is shown in the upper right of the illustration to help you in drawing the schematic. DS1 shown on the sketch is an internal part of the switch and must be included on the schematic even though it does not appear on the wiring diagram. E1, E2, and E4 will be terminals and should be shown as open circles. Make all lettering approximately $\frac{1}{8}''$ high with $\frac{1}{16}''$ between lines of lettering. Center the drawing lengthwise on an $8\frac{1}{2}'' \times 11''$ sheet.

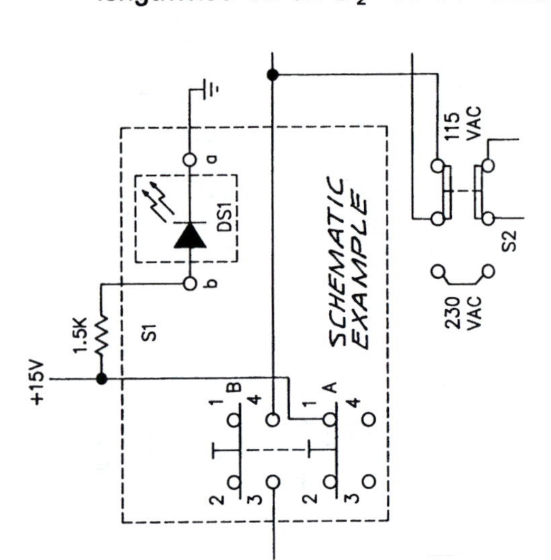

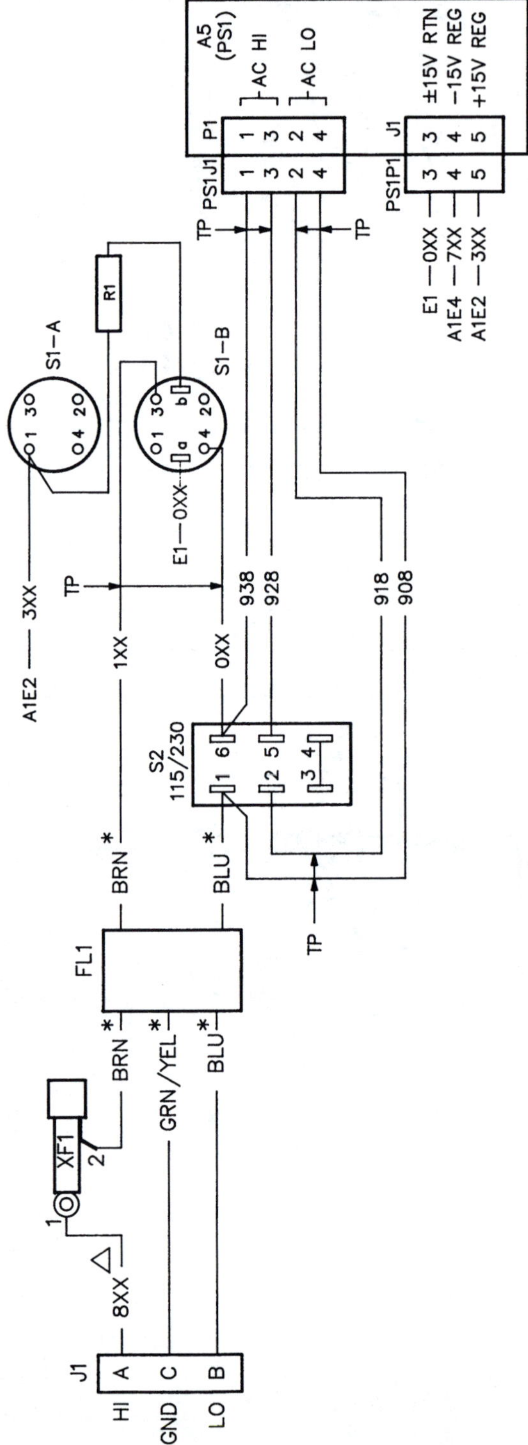

Chapter 5

Drawing Schematic Diagrams from Rough Sketches

OBJECTIVES

After completing this chapter, you will be able to

- ✓ list ten basic rules for the correct layout of component symbols and conducting lines.
- ✓ layout correctly, draw and letter in pencil, four schematic diagrams from rough sketches, and assign reference designators to one of them.

LAYING OUT SCHEMATIC DIAGRAMS

The basic rules for laying out schematic diagrams are described in the first ten figures of this chapter. Later figures show examples of good layouts. Then, typical engineering sketches are shown, along with examples of good schematic diagrams that were made from them.

The ten basic rules described are:

1. Inputs left, outputs right
2. Break connectors and jumble pins
3. Point grounds downward
4. Point curved side of capacitor symbol to ground
5. Arrange relay symbols
6. Remove doglegs
7. Avoid crossovers
8. Observe minimum spacing
9. Group leads
10. Eliminate 4-way ties

Inputs Left, Outputs Right

Figure 5–1 describes Rule 1: inputs on the left, and outputs on the right. It is not always possible to identify inputs and outputs from a rough sketch. Usually, then, the engineer must provide that information or the drafter has to make a judgment. However, the inputs and outputs are identified for most of the sketches given in this chapter.

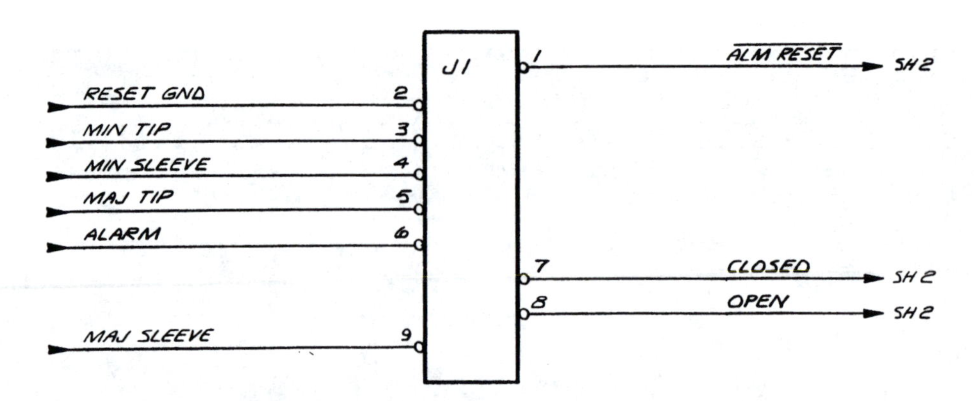

Figure 5–1 Rule 1—Inputs left, outputs right

Break Connectors and Jumble Pins

Figure 5–2 describes Rule 2: break connectors and jumble pins. *Pins* are similar to the connections of the common electrical wall plug. The plug usually has three prongs. Each one of the prongs is a pin. The connectors in this figure have ten pins. In this diagram, the pins are identified by a number. Often they are identified by a letter.

BEFORE and AFTER parts are shown. In the BEFORE part of the figure, connectors J1 and P1 are whole and all of their ten pins are shown on those connectors. Notice that the lines are jumbled considerably. The AFTER part shows the connector split. The abbreviation P/O has been placed above the reference designators J1 and P1 to show that this is just part of the connector. Notice also that the pins are no longer in order as they were before. This can often be done with an engineering sketch, unless the customer will not allow it for some reason. The drafter must consult the customer to determine if the connectors may be split and its pins jumbled.

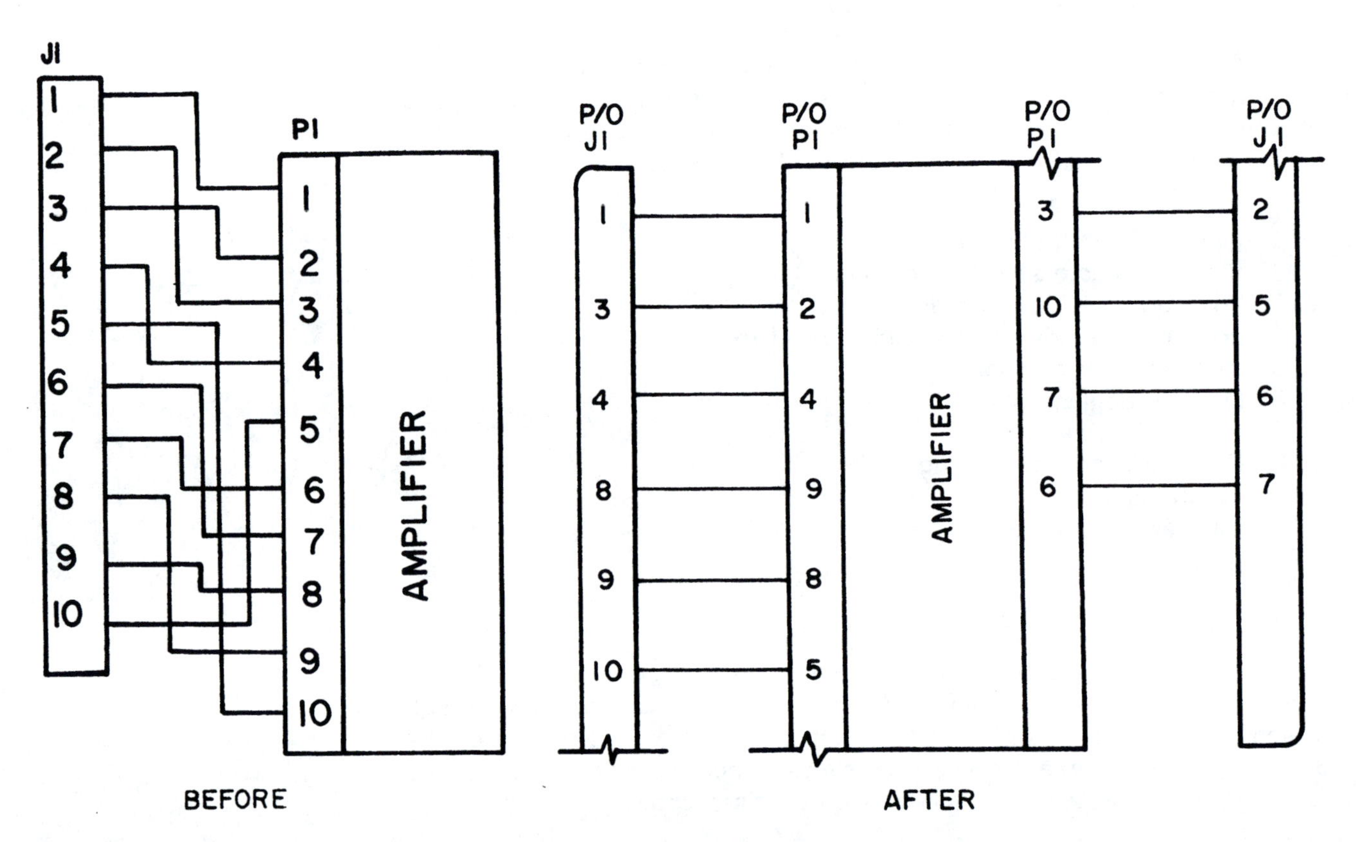

Figure 5–2 Rule 2—Break connectors and jumble pins

Point Grounds Downward

Figure 5–3 illustrates Rule 3: point grounds downward. Occasionally, grounds are pointed to the side on a schematic, but most specifications are very clear about pointing grounds downward.

Point Curved Side of Capacitor Symbol to Ground

Figure 5–4 illustrates Rule 4: point the curved side of the capacitor symbol to ground. The ground side should be clearly identified so that the drafter will know in what direction to point the curve. The straight side of the capacitor is the positive side and often (but not always) is identified with a plus (+). If the sketch shows a plus, it must be placed on the schematic.

Arrange Relay Symbols

Figure 5–5 illustrates Rule 5: arrange relay symbols. Notice that the mechanical linkage never touches the contacts. The drafter must be sure

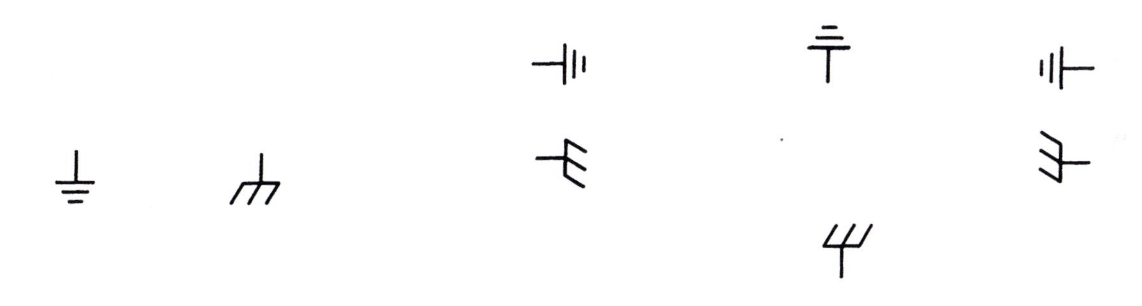

ALWAYS THIS

NEVER THIS

Figure 5–3 Rule 3—Point grounds downward

THIS

NOT THIS

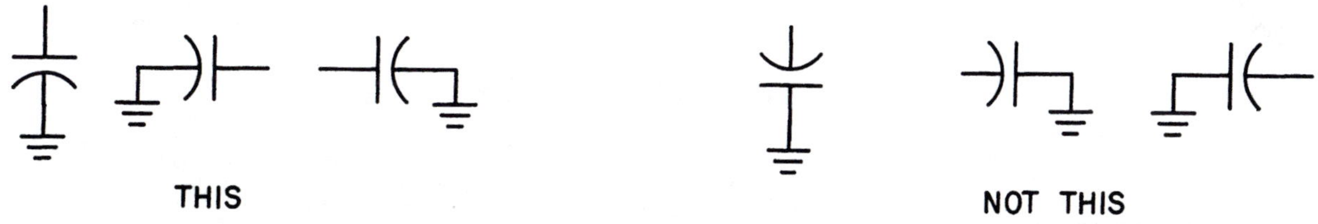

Figure 5–4 Rule 4—Point curved side of capacitor symbol to ground

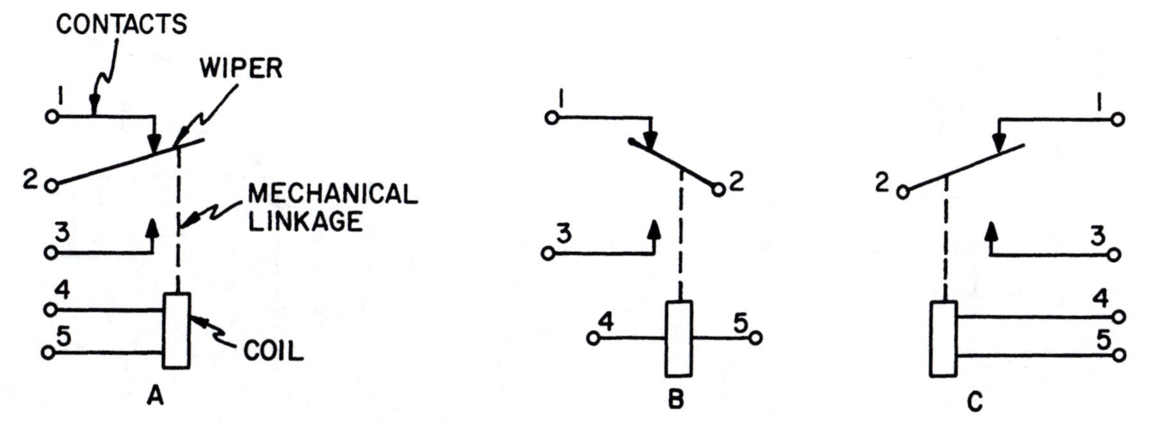

Figure 5–5 Rule 5—Arrange relay symbols

to draw it that way. The mechanical linkage touches only the wiper. The wiper is the movable contact. Occasionally, a symbol will have two sets of fixed contacts and two sets of wipers which are stacked on top of each other.

In Figure 5–5A the pins on the movable and fixed contacts have been identified with a number. The wiper can be switched over to the other side, as shown in Figure 5–5B. This figure also shows that the contacts of the coil may be placed with one on one side and one on the other. The other arrangements are to put all of the contacts on the right as shown in Figure 5–5C.

Remove Doglegs

Figure 5–6 shows Rule 6: remove doglegs. Occasionally, a situation arises where a dogleg must be used, but doglegs should be eliminated whenever possible. By moving pin 1 from its position in Figure 5–6A to where it is shown in Figure 5–6B, the dogleg has been removed. Although a dogleg is an unnecessary bend in a line, one 90° bend is not a dogleg. Two 90° bends produce a dogleg.

Avoid Crossovers

Figure 5–7 illustrates Rule 7: avoid crossovers. Recall that a crossover is a line crossing one or more other lines. Not all crossovers can be avoided,

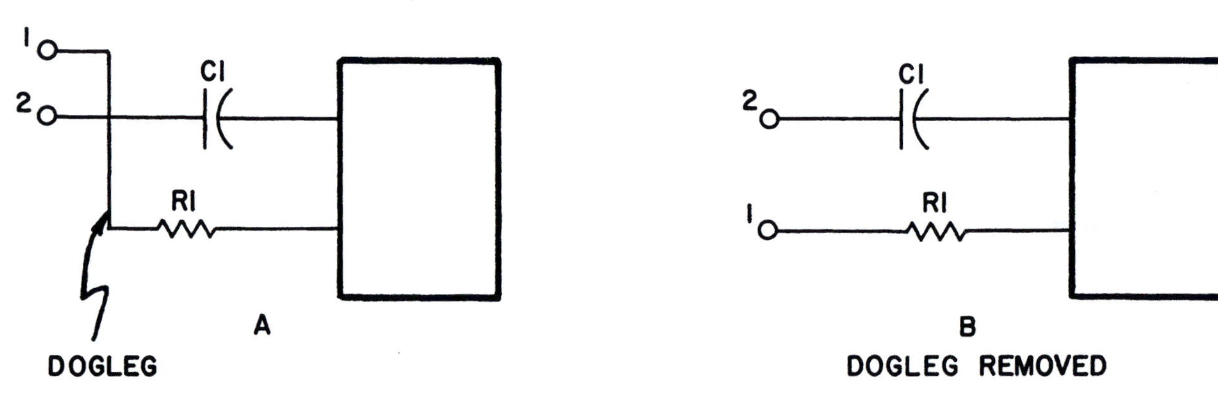

Figure 5–6 Rule 6—Remove doglegs

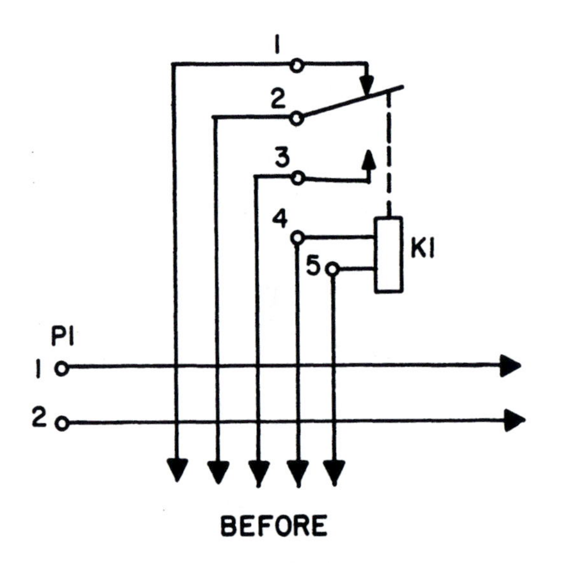

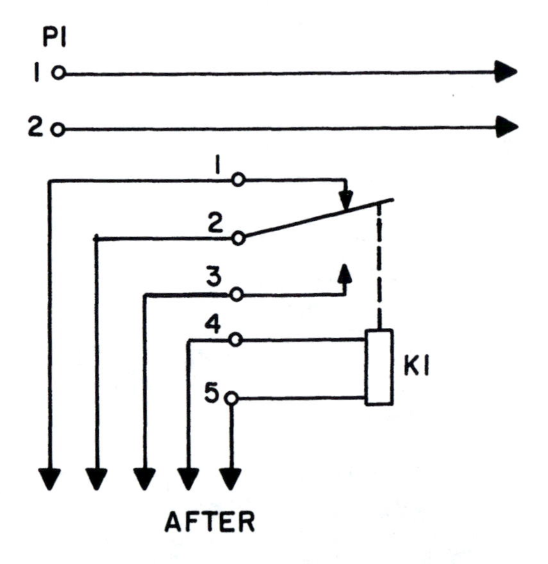

Figure 5–7 Rule 7—Avoid crossovers

but the drawing is clearer when they are eliminated. Some of the exercises at the end of this chapter require that all crossovers be eliminated. When the instructions state that all crossovers should be eliminated, it is possible to do so.

Observe Minimum Spacing

Figure 5–8 shows Rule 8: observe minimum spacing. In this schematic the minimum spacing is shown to be $\frac{1}{8}''$. This is fairly typical minimum spacing. When $\frac{1}{10}''$ grid paper is being used, the minimum spacing would necessarily be $\frac{1}{10}''$.

Group Leads

Figure 5–9 shows Rule 9: group leads. The leads in this figure would probably not be grouped because it is a simple diagram. In a diagram this small, there is no point in grouping leads. The example is given only to simplify the illustration of Rule 9.

Grouping of leads normally occurs only on large diagrams where leads are drawn two or three feet across a page. Then leads are grouped so it will be easier to follow them. Group leads in threes if more than six leads travel together for a distance of 10″ or greater.

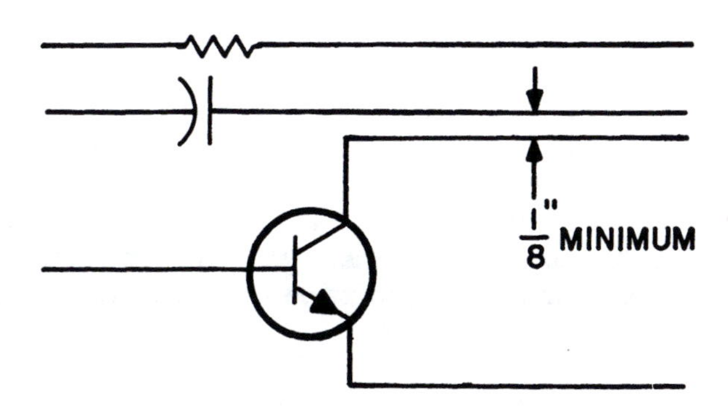

Figure 5–8 Observe minimum spacing

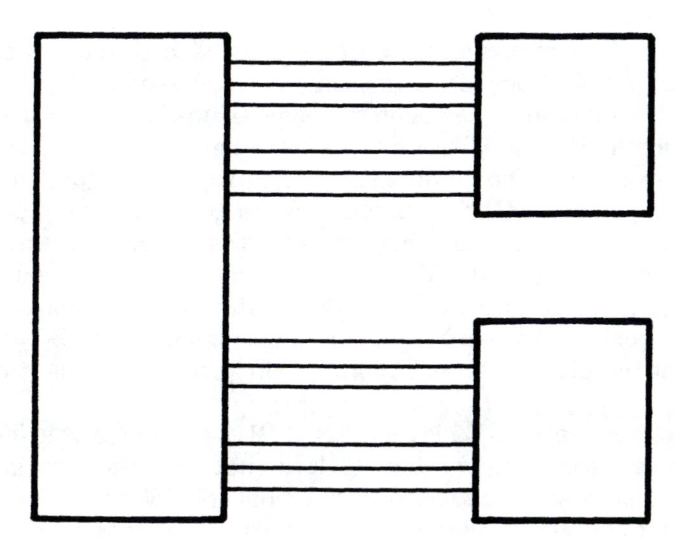

Figure 5–9 Group leads

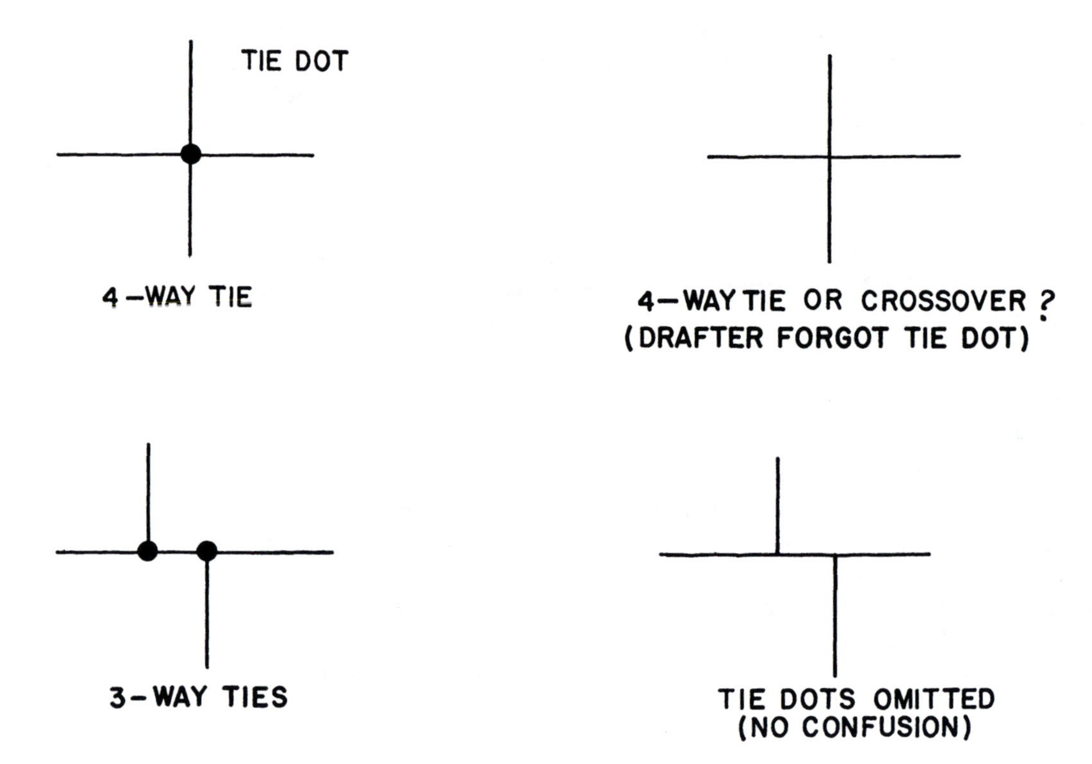

Figure 5–10 Eliminate 4-way ties

Eliminate 4-way Ties

Figure 5–10 shows Rule 10: eliminate 4-way ties. Many companies still allow the use of 4-way ties, but more and more companies are doing away with them. A 4-way tie can be confusing if the drafter has left off the tie dot and the lines cross each other; it is unclear if there should be a 4-way tie or if it is a crossover. If there were no such thing as a 4-way tie and the lines cross, there is no confusion as to whether or not the lines connect.

SCHEMATIC EXAMPLES

Tie Dots

Figure 5–11A shows a sketch of a schematic diagram. Usually, the sketches are in much rougher form than this. The sketch shown is poorly arranged. This arrangement has been improved by the layout in Figure 5–11B. Notice that the tie dots in Figure 5–11A have been arranged along the 12-V line in Figure 5–11B. The tie dots in this drawing are arranged in numerical order, but they need not be in order. The tie dots may be moved along that line anywhere as long as they do not cross a component. Tie dots may cross other tie dots, but they must not cross a component, such as a resistor or capacitor. In Figure 5–11B, tie dot number 2 could be drawn in the same position as number 6, and vice versa. Figure 5–11B merely indicates that the plus 12 volts occurs all along that line until it crosses a component.

For example, refer to R1 in Figure 5–11B. As the +12-volt line crosses R1, it becomes some other voltage. It is not important to understand electronics to be able to understand this concept. What the drafter must know is that the voltage value on one side of the component becomes a

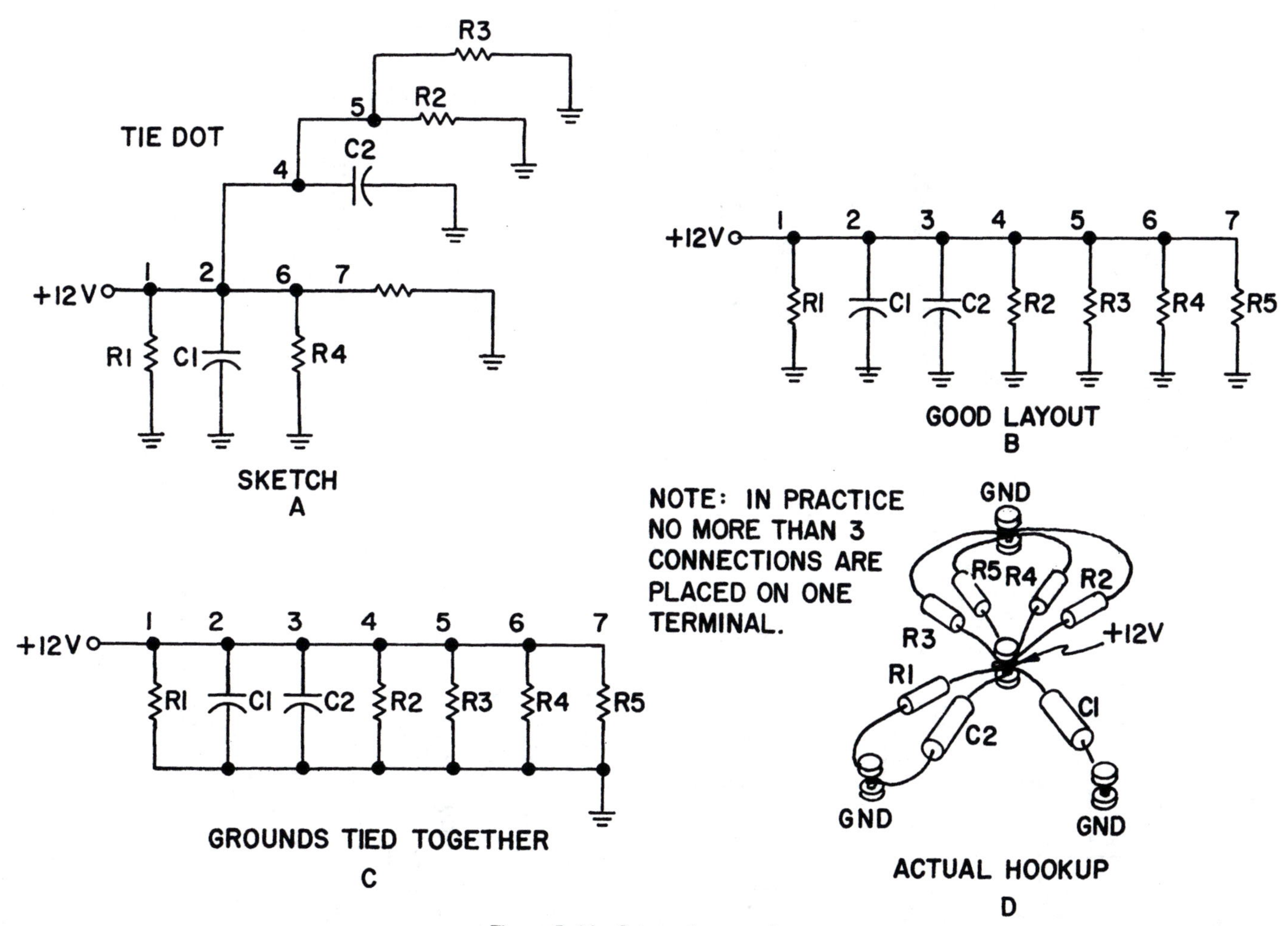

Figure 5–11 Schematic example

different value on the other side of the component. Tie dots may be moved and components rearranged in various ways, as long as the function of the circuit or the circuit itself is not changed in any way. Every component that must receive +12 volts on the sketch must receive +12 volts in the formal schematic. As long as a component is not placed in the +12-volt line, the line remains +12 volts.

Figure 5–11C shows all the grounds tied together. The diagram shows that R1 goes to ground, C1 goes to ground, C2 goes to ground, and R2 goes to ground. Individual grounds may be shown or the grounds may be tied together showing one ground. Some drawings show both methods. Individual grounds may be shown in one area of the schematic, whereas all grounds tied together may be shown in another area in order to improve the layout.

Space for Lettering

Space for symbols and lettering must be well planned. Consult the sketch to identify the lettering that must be placed by each symbol and allow adequate space for it. Recall that lettering should be all the same height and slant, and that the space between lines of lettering should be approximately two-thirds of the letter height. Figure 5–12 is a good example of well-spaced lettering on a relatively crowded schematic.

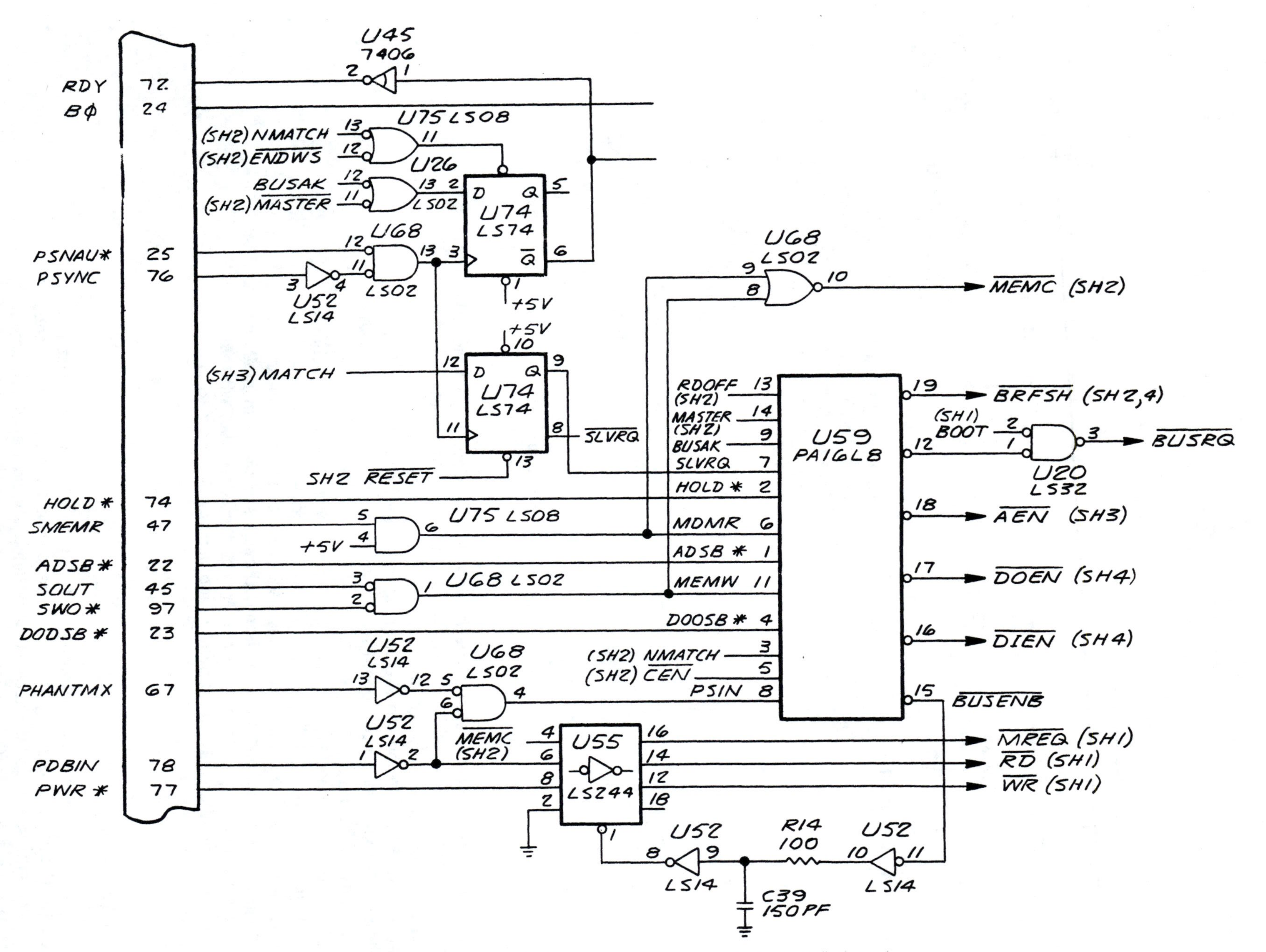

Figure 5-12 Adequate space for symbols and lettering should be well planned

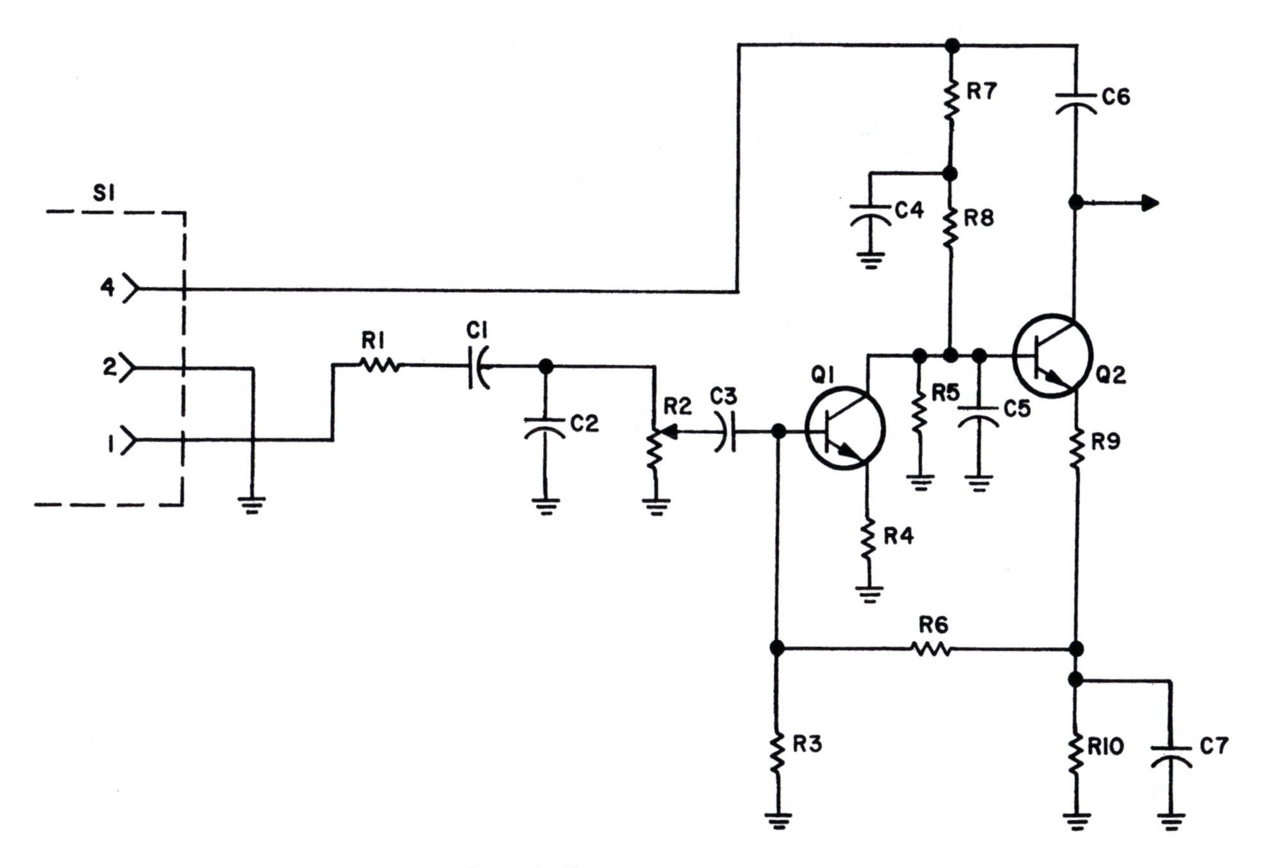

Figure 5–13 Numbering of symbols

Numbering of Symbols

Figure 5–13 shows how symbols are numbered: from left to right, and from top to bottom. Often, the symbols are not numbered on the engineering sketch of the schematic diagram, so it is up to the drafter to assign reference designators to those symbols. Starting at the top of the drawing on the left-hand side, the symbols are numbered from left to right. Start with R1 on the left-hand side. R2, R3, R4 will be labeled as they are encountered from left to right or top to bottom. If a new component is placed in the circuit sometime after the reference designators are assigned, it is preferable not to renumber the schematic sequentially. The drafter should just assign the next sequential number to that component.

Integrated Circuit Symbol

Figure 5–14 is an example of a diagram containing a symbol that the student possibly may not have seen before. The four triangles shown in the schematic are part of a device called an integrated circuit (IC). Integrated circuits are described in more detail later, but, for now, it is necessary only to know that these triangles represent a type of integrated circuit, and that the Us are the reference designators. Near the U1, U2, U3 and U4, note the designation 741. This designation identifies the type of part. Notice that the lines leading out of the triangles are numbered. These numbers are the pin numbers on that part.

A later chapter describes integrated circuits in detail, and how they are numbered.

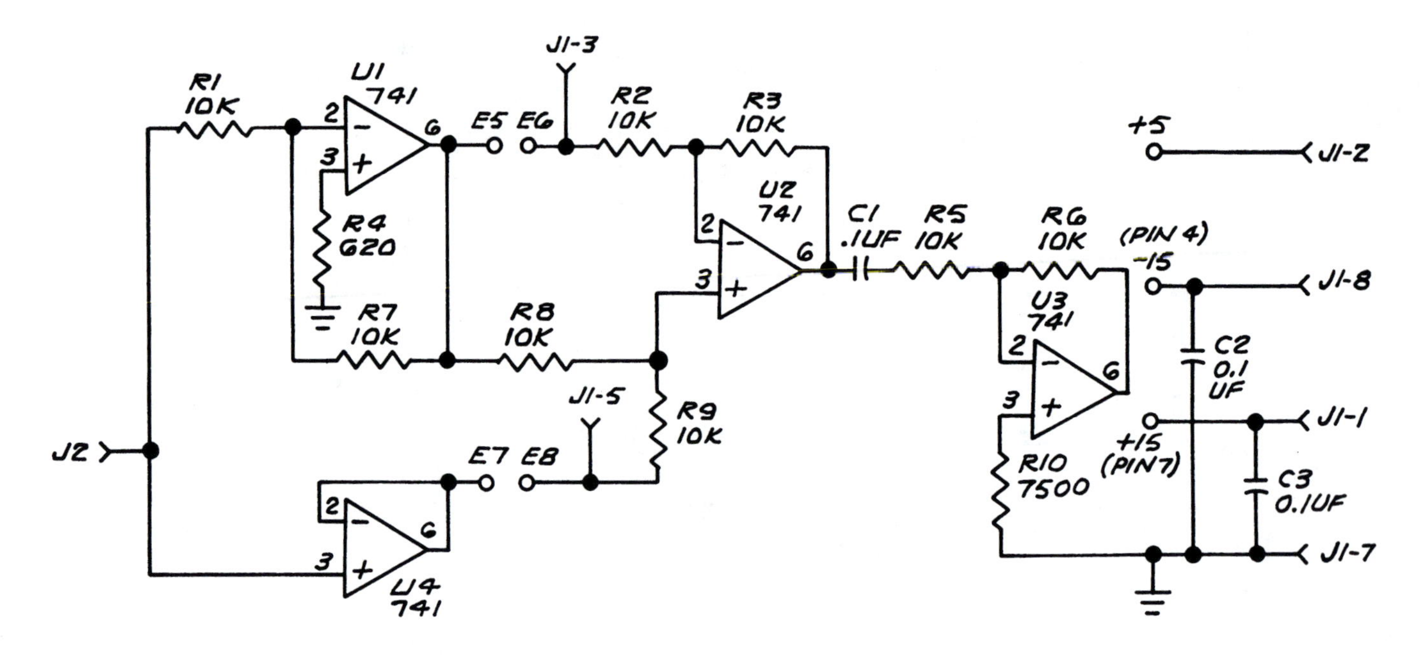

Figure 5–14 Example of symbol for an analog integrated circuit

EXAMPLES OF SKETCHES AND FINISHED SCHEMATICS

Example 1

Figure 5–15A is a sketch. Figure 5–15B is the finished schematic. Recall that inputs are on the left. The inputs in the sketch have been identified. The schematic shows the inputs on the left and the outputs on the right. Some unnecessary bends in the lines have been eliminated, and all the grounds have been tied together. Compare the two figures carefully. Notice that the ground with the minus shown at the top of the sketch has been tied to the ground in the finished schematic. This layout is acceptable in most cases. After studying the two diagrams, the student will note that the schematic, the way the diagram is hooked up, has not changed at all. The lines have just been turned around and the schematic has been simplified.

Example 2

Now compare Figures 5–16A and 5–16B. In 5–16B some doglegs shown in the sketch have been eliminated. The inputs have been placed on the left, and the outputs on the right. The symbols are still hooked up exactly as they are in the sketch. The circuit has just been turned over end to end.

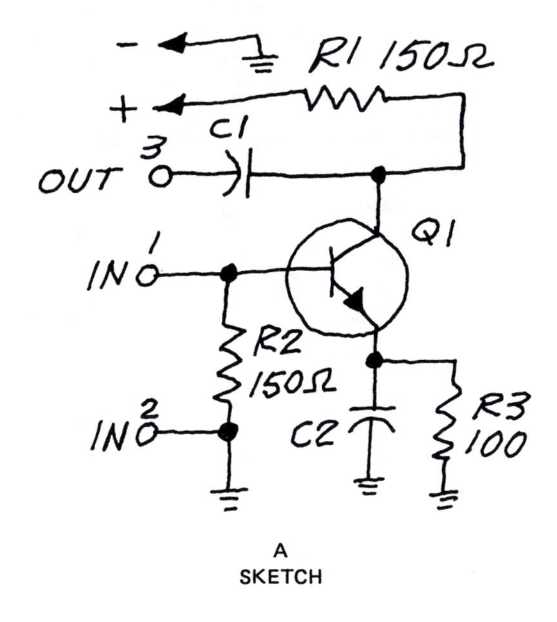

A
SKETCH

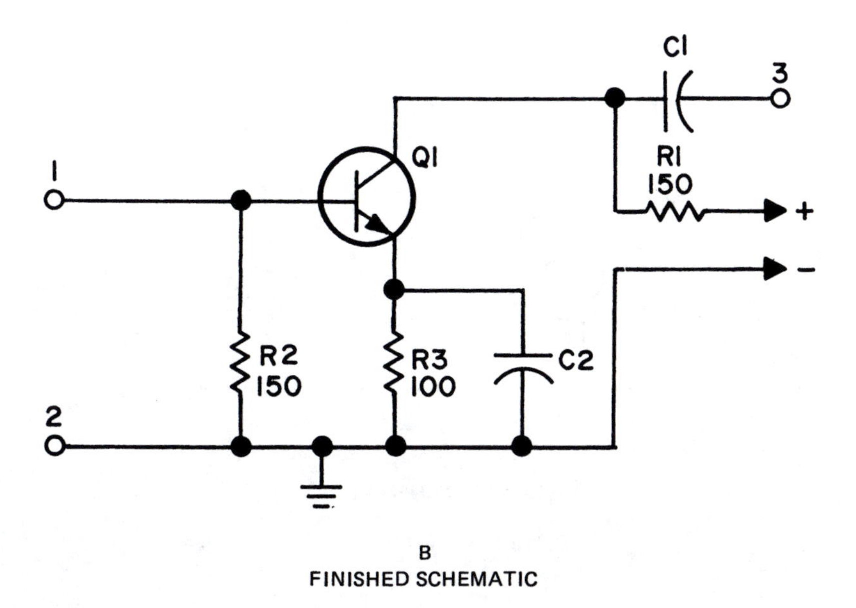

B
FINISHED SCHEMATIC

Figure 5–15 Example 1

Example 3

The sketch in Figure 5–17A is the rough drawing from which the finished schematic of 5–17B was drawn. The finished schematic in Figure 5–17B shows that several doglegs and crossovers have been eliminated, and that inputs have been placed on the left and outputs on the right. Study the sketch carefully to identify what has been done and what has been changed.

The triangle at the top of Figure 5–17B shows an arrow with the number 14 alongside labeled $+V_{CC}$. This is similar to splitting up a common

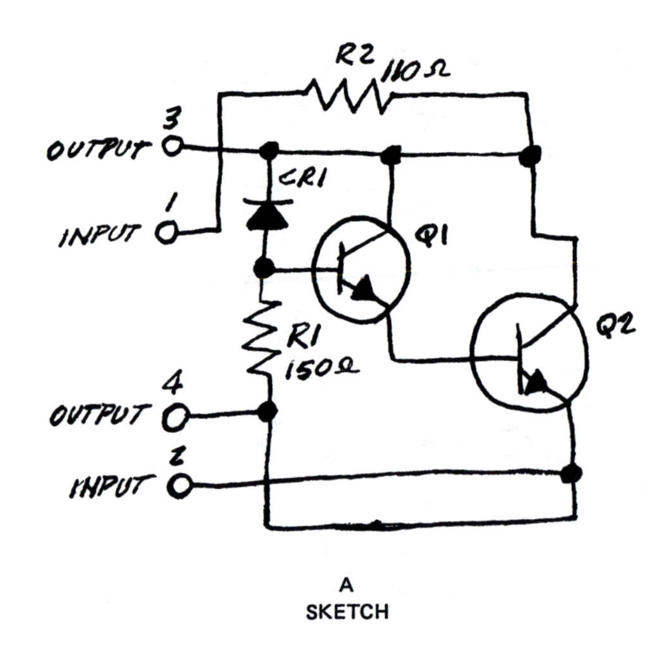

A
SKETCH

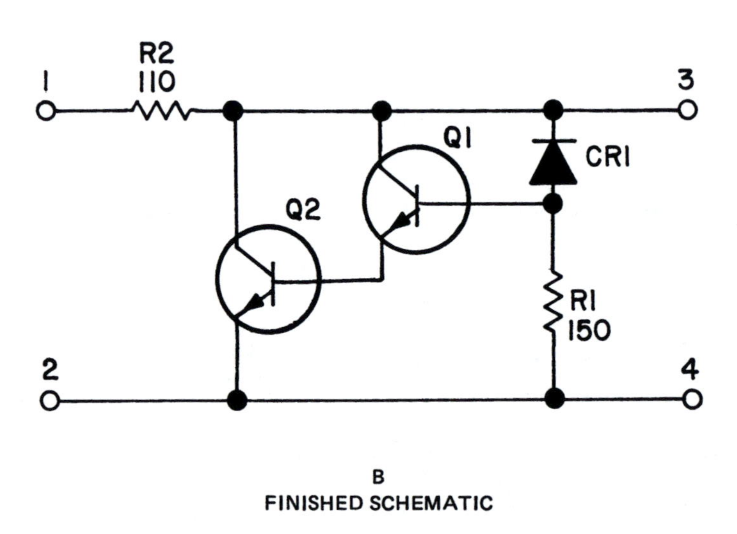

B
FINISHED SCHEMATIC

Figure 5–16 Example 2

ground and making individual grounds. In this case, the V_{CC} has been split and identified individually, instead of tying them all together as was done in the sketch. The same thing has been done on the $-V_{CC}$. Instead of tying all of these together, individual $-V_{CC}$ arrows are drawn coming out of the triangles. Then, the $+V_{CC}$ and the $-V_{CC}$ connections to L3 and L4 are shown at the lower portion of the diagram.

Example 4

Figure 5–18B illustrates a good example of splitting the connectors and jumbling the pins in order to simplify the layout given in the sketch at

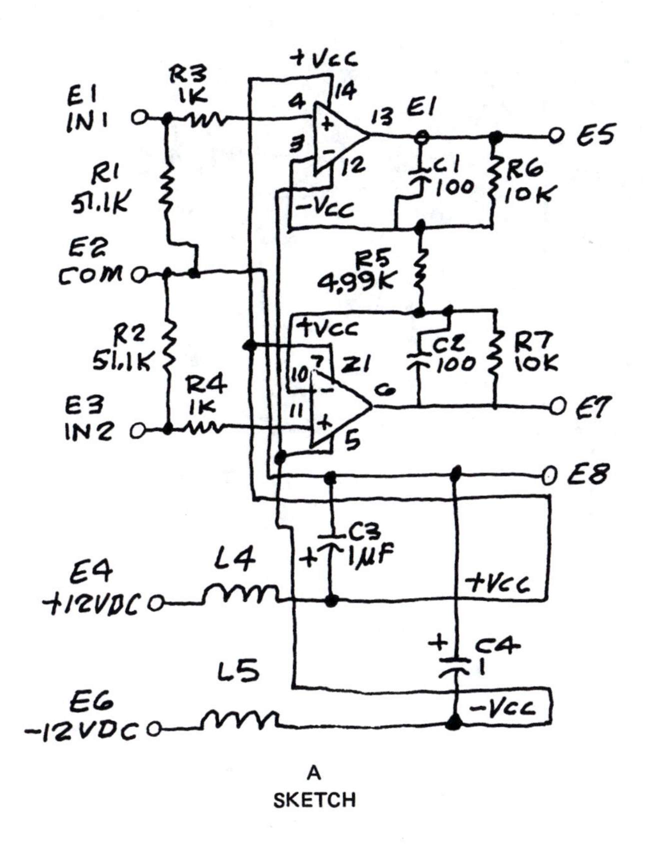

A
SKETCH

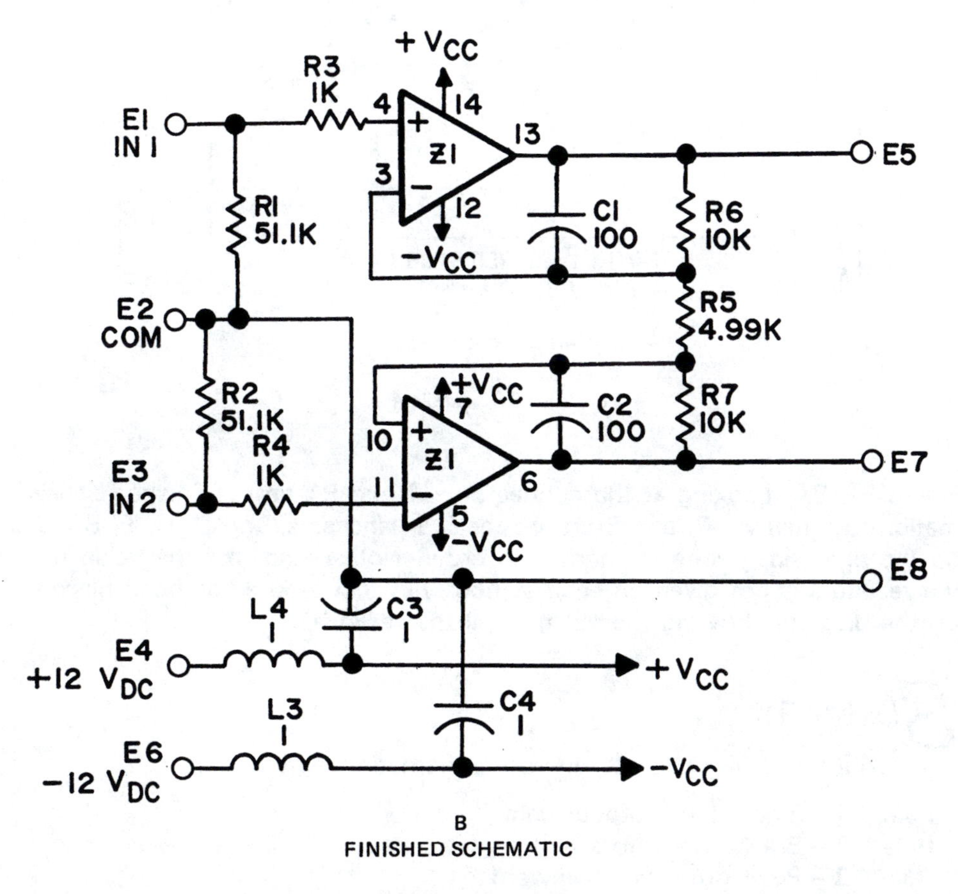

B
FINISHED SCHEMATIC

Figure 5–17 Example 3

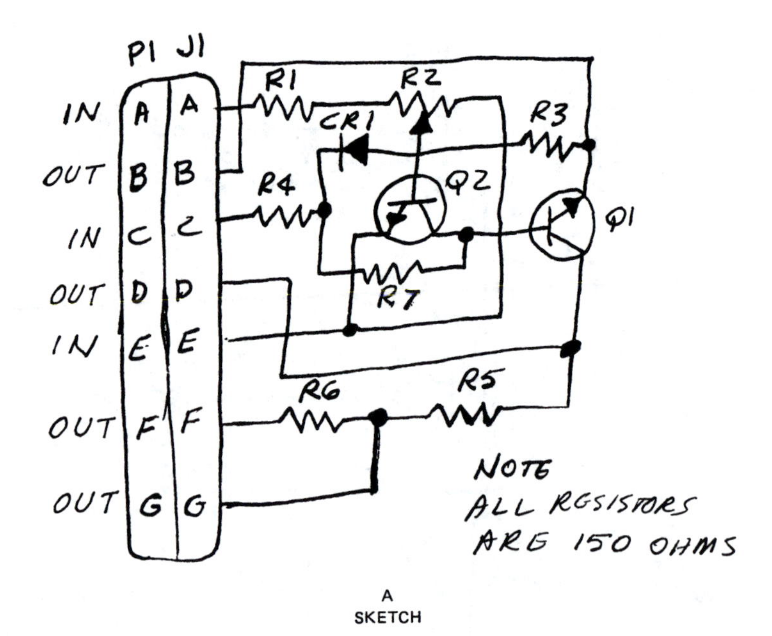

A
SKETCH

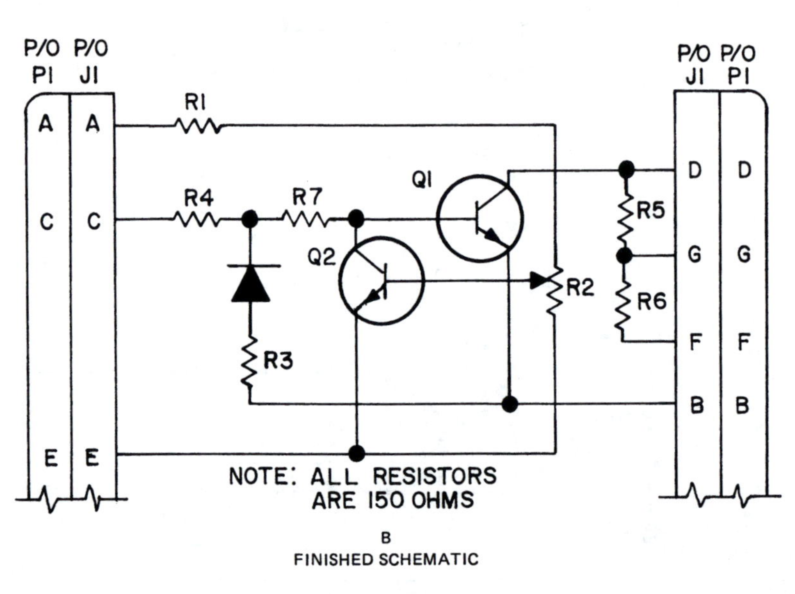

B
FINISHED SCHEMATIC

Figure 5–18 Example 4

Figure 5–18A. Looking at the connectors J1 and P1 on the finished schematic, note that A, C, and E are on the left-hand side, and D, G, F, B are on the right side. None of them is in order. Notice also that the value for the resistors is not given on each symbol, but that a note has been placed on the diagram showing the value of all the resistors.

SUMMARY

The ten basic rules for drawing schematic diagrams are:

Rule 1 – Inputs left, outputs right
Rule 2 – Break connectors and jumble pins
Rule 3 – Point grounds downward
Rule 4 – Point curved side of capacitor symbol to ground
Rule 5 – Arrange relay symbols
Rule 6 – Remove doglegs

Rule 7 – Avoid crossovers
Rule 8 – Observe minimum spacing
Rule 9 – Group leads
Rule 10 – Eliminate 4-way ties

These rules are guidelines in many cases, and may be modified as necessary.

Two other principles to remember are:

- Number symbols from left to right, and from top to bottom
- Leave adequate space for lettering

EXERCISE 5-1

Lay out correctly and draw a schematic diagram from the sketch shown. Use the rules for schematic drawing described in this chapter. Use a schematic template to draw resistors, diodes, and capacitors. Put the drawing lengthwise on an $8\frac{1}{2}'' \times 11''$ sheet. Terminals are open. Terminals and tie dots are drawn with a $\frac{1}{8}''$ diameter circle. Tie dots are blackened in.

This diagram cannot be greatly changed. Place inputs on the left and outputs on the right. Eliminate any unnecessary bends in lines.

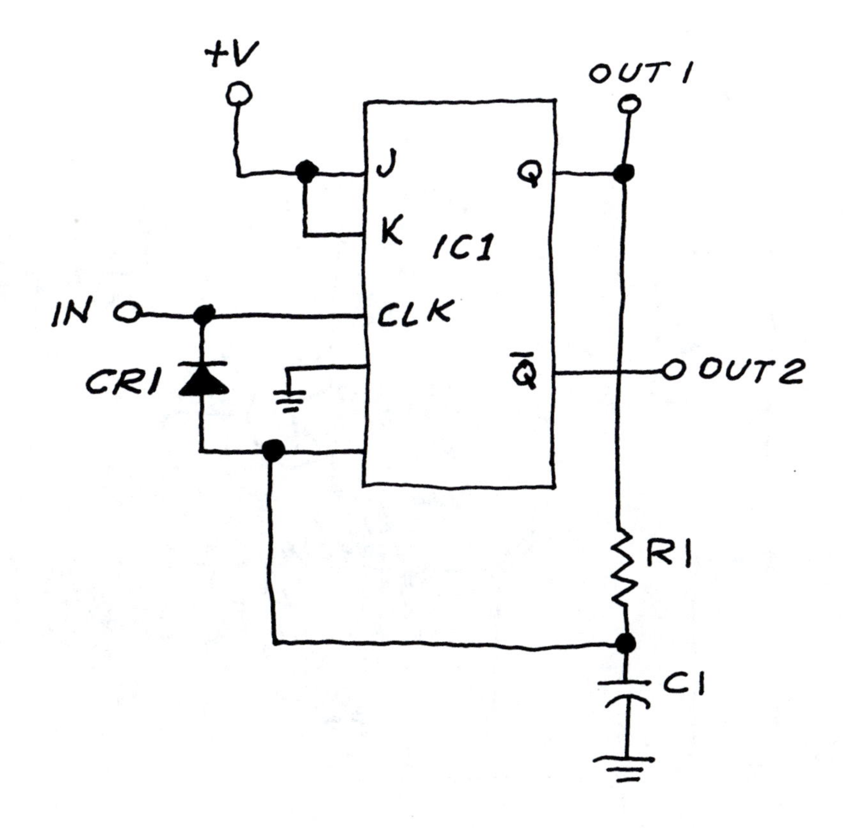

EXERCISE 5-2

Lay out correctly and draw a schematic diagram from the sketch shown. Use the rules for schematic drawing described in this chapter. Use a schematic template to draw resistors, diodes, and capacitors. If you do not have a good transistor symbol on your template, trace the ones in Chapter 3. Put the drawing lengthwise on an $8\frac{1}{2}''$ × $11''$ sheet. Terminals are open. Terminals and tie dots are drawn with a $\frac{1}{8}''$ diameter circle. Tie dots are blackened in. You may use the computer to redraw this sketch if you like.

This diagram cannot be greatly changed. The output may be placed on the right-hand side. The jog in the line just above R5 can be straightened. The crossovers on this diagram cannot be eliminated without a great deal of difficulty, so it is not necessary to do so. You may be able to tie all the grounds together if this looks like a reasonable way to lay out the drawing. If all the grounds are tied together, tie them to pin 5, the ground pin. Pin 5 can be input or output. It does not make any difference on which side of the schematic pin 5 is placed.

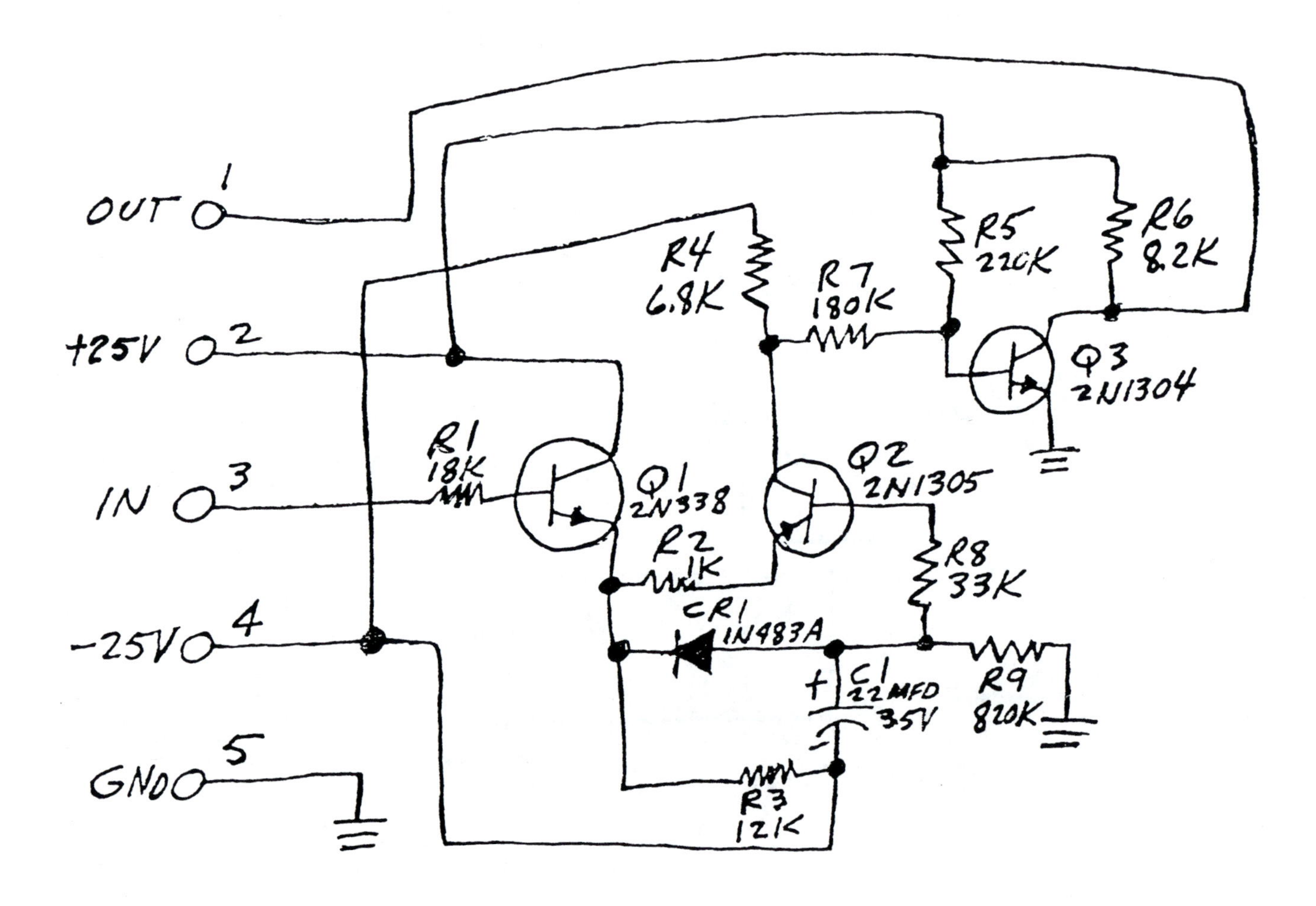

EXERCISE 5–3

Correctly lay out and draw this schematic on an 11″ × 17″ sheet. Use correct symbols (see Chapter 4). Place inputs on the left, outputs on the right. Remove doglegs. Hold crossovers to a minimum. Assign reference designators. Number symbols from left to right, and from top to bottom.

This exercise is a little more difficult than the previous one. Reference designators should be assigned after the schematic has been drawn. Do not assign reference designators on the sketch. Some components are identified with the reference designators. R7, R8, Q3, R9, R10, and Q4. This is done so that the components on the sketch can be identified for these instructions. Your reference designators may be entirely different from those on the sketch.

Three input lines are on the sketch. Line 1 is labeled IN and is at the top of the sketch. Everything in this schematic should fit between lines 1 and 3. Place pins 4 and 5, the two outputs, on the right side of the drawings. The components, R7, R8, Q3, R9, R10, and Q4 can be placed on the right side of the schematic between lines 1 and 3. For example, R10 can be tied on the far right side of line 3. To do this, extend line 3 to add R10. Tie one side of R10 at that point and extend the other side of R10 up into the emitter of Q4 (the leg with the arrow). Place a tie dot between R10 and Q4 and extend it out to pin 4. The other leg of Q4 is called the collector. Connect the collector of Q4 to line 1. Notice, if you follow the line on the sketch, the collector of Q4 weaves around to connect to line 1. All that is needed here is to extend line 1 a short distance to the right and draw a very short line from the collector of Q4 up to line 1. This is an example of how the rest of the numbered components should be arranged on the schematic. There will be a long crossover leading from R7 back to the second transistor on the sketch.

When the schematic diagram is completed it will be much clearer than the sketch if the components have been arranged between lines 1 and 3. This is a challenging schematic. If you have any trouble with it, see your instructor.

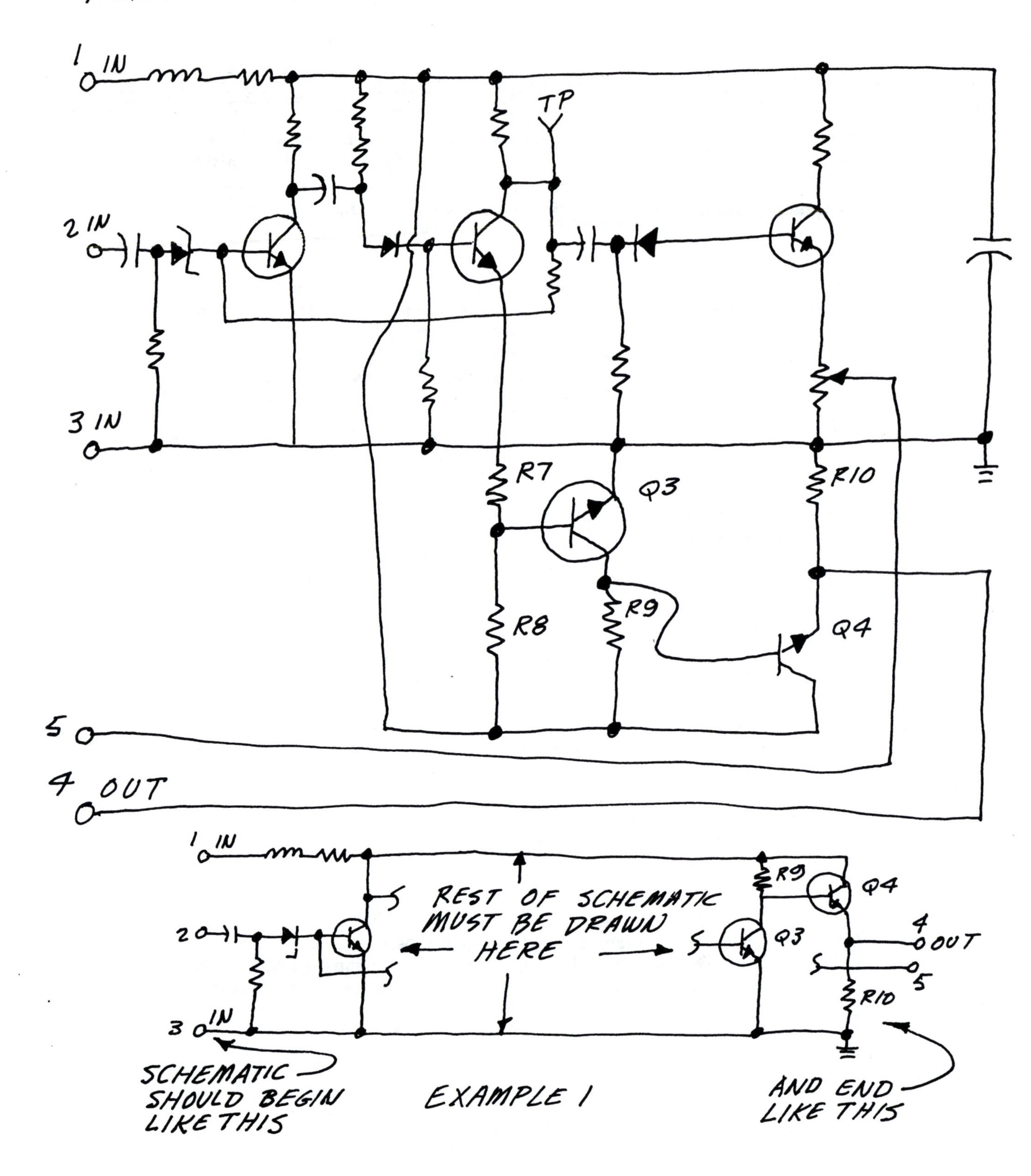

1 IN
2 IN
3 IN
TP
R7
Q3
R10
R8
R9
Q4
5
4 OUT
1 IN
2
3 IN
R9
Q4
REST OF SCHEMATIC
MUST BE DRAWN
HERE
Q3
4
OUT
5
R10
SCHEMATIC
SHOULD BEGIN
LIKE THIS
AND END
LIKE THIS

EXAMPLE 1

EXERCISE 5-4

Draw the wafer switch sketched here as a formal schematic diagram. Center it lengthwise on an $8\frac{1}{2}''$ × 11″ sheet. Remove doglegs and crossovers where possible. You cannot change the arrangement of numbered pins, so there is not much you can do to rearrange the symbols. Be sure to leave adequate room for $\frac{1}{8}''$ high lettering. Make all lines vertical or horizontal (no diagonal lines). Notice that the arrows on pins 2, 6, and 10 are longer than the others. They must be drawn exactly that way. Point all arrows toward the center of the symbol.

A wafer switch is a detail that is not encountered often. Two different wafers are shown on the same switch. The main thing you need to be concerned about is that you leave enough room for the lettering, and that all the lines are vertical or horizontal except for the arrows which should point toward the center. They should not extend any farther than they do on the sketch, but they should point toward the symbols on it. Use a circle template to trace the circles shown in this exercise. All of the pins 1 thru 12 will be on the same centerline around the outside of the wafer. The order of the pins and their positions cannot be changed. Everything must basically remain the way it has been arranged on the sketch. You may be able to move the wafer on the left up a little to eliminate some doglegs, but you should end up with a diagram that looks basically like the sketch. You may be able to eliminate a few doglegs and crossovers, but there is not a great deal that you can do about them.

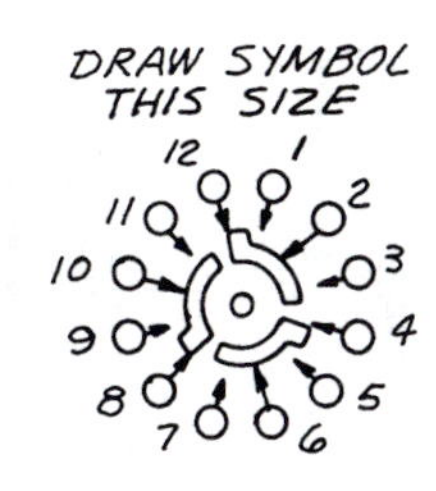

DO NOT CROWD LETTERING, AND DO NOT ALLOW ANY LETTERING TO TOUCH LINES OR SYMBOLS

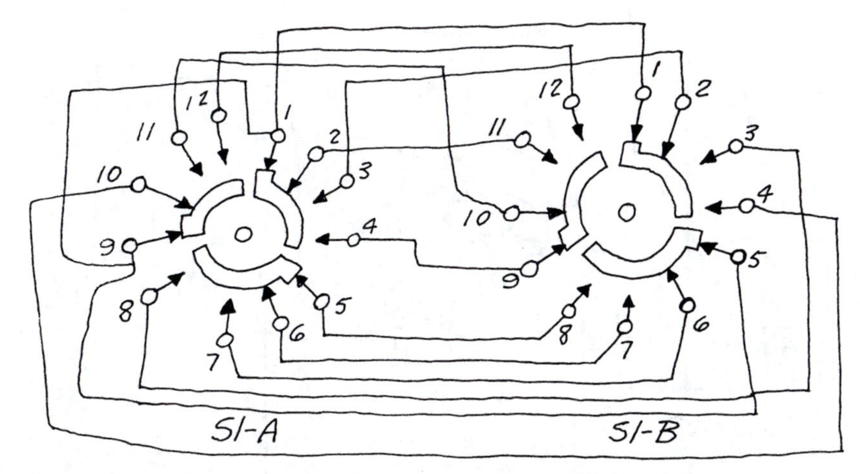

EXERCISE 5-5

Redraw the following sketch as a formal schematic. You may rearrange pins on the connectors, and you may stretch out the connectors and have unequal space between pins to eliminate doglegs. You may also put the pins of the relay on the left or right of the dashed line. Be sure the dashed line goes through the wipers of the switch only. Allow adequate space for lettering, and do not touch any lines with lettering. Label contacts of the switches as normally closed (NC), normally open (NO) or common (COM).

In this exercise, you can turn over the wiper from one side to the other. You can also turn over the contacts from one side to the other. The coil is the lower part of the relay symbol. You can put pin 7 on one side and pin 8 on the other, or both of them on one side. You will probably want one pin (8) on one side and the other pin (7) on the opposite side. You may jumble the pins on the connectors. You can extend the connectors so that the amount of space between the pins is unequal. With this much leeway, you should be able to produce a well-drawn schematic.

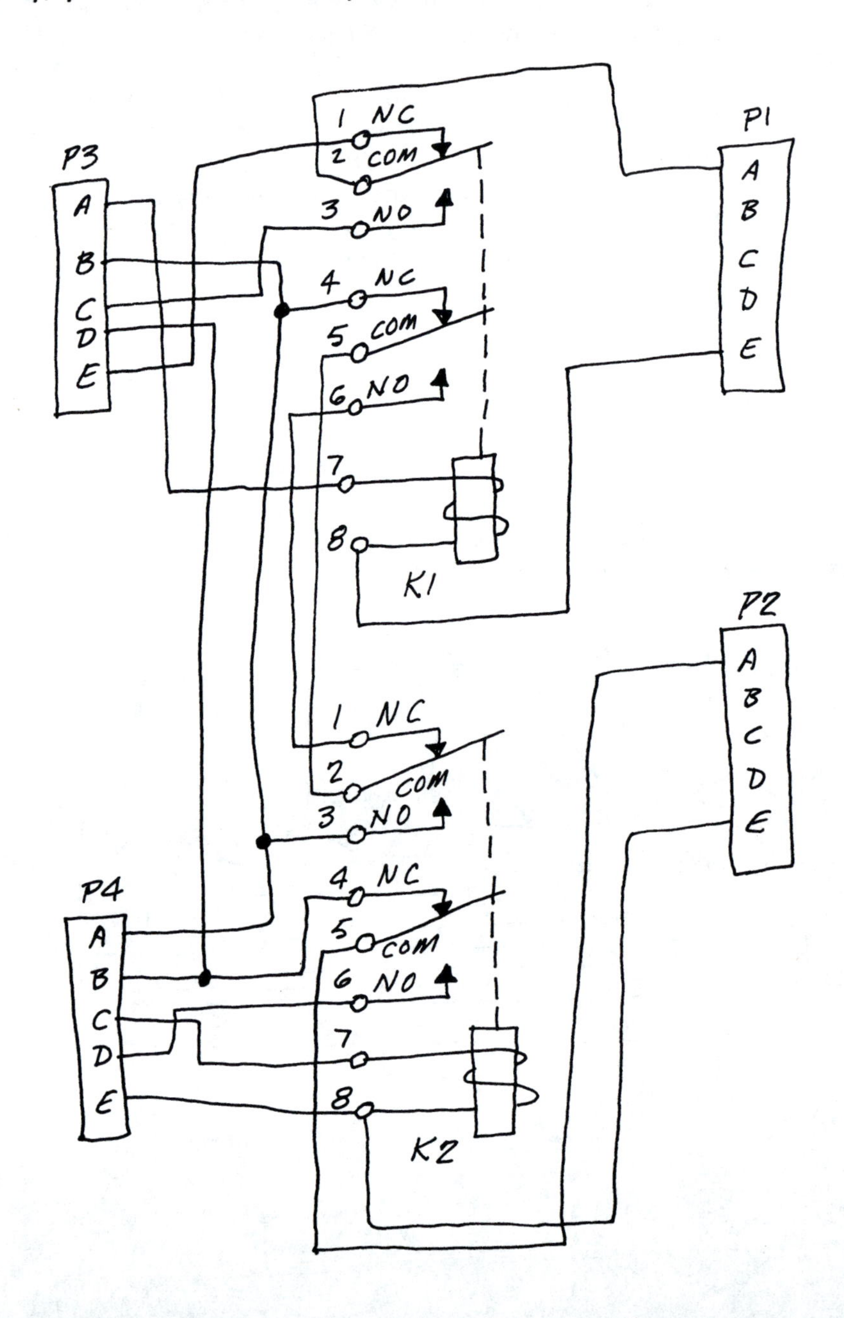

EXERCISE 5-6

Redraw the following sketch as a formal schematic. Pins on these ICs may *not* be rearranged. Draw the symbols as they are shown except that the boxes containing U20 and U30 may be drawn larger to accommodate the lettering. Try to eliminate doglegs and crossovers, if possible. It will not be possible to eliminate all of them. This drawing will fit on an $8\frac{1}{2}'' \times 11''$ sheet.
This schematic cannot be changed greatly. You will be able to eliminate a few doglegs and crossovers, and clean up things a little, but the positions of the pins must stay exactly as they are on the sketch. This is because engineers and other people using this diagram are accustomed to seeing these pins in these positions. For example, pin 11 is usually located across from pin 9.

The drafter should ask if the pins of a symbol may be moved, particularly if the symbol is new or unfamiliar. In this example, the part (also called a device) is LS112. Since the pins of an LS112 may not be rearranged, the layout cannot be as free of doglegs and crossovers as you would like. Follow all the rules for good layout, lettering, and line weights.

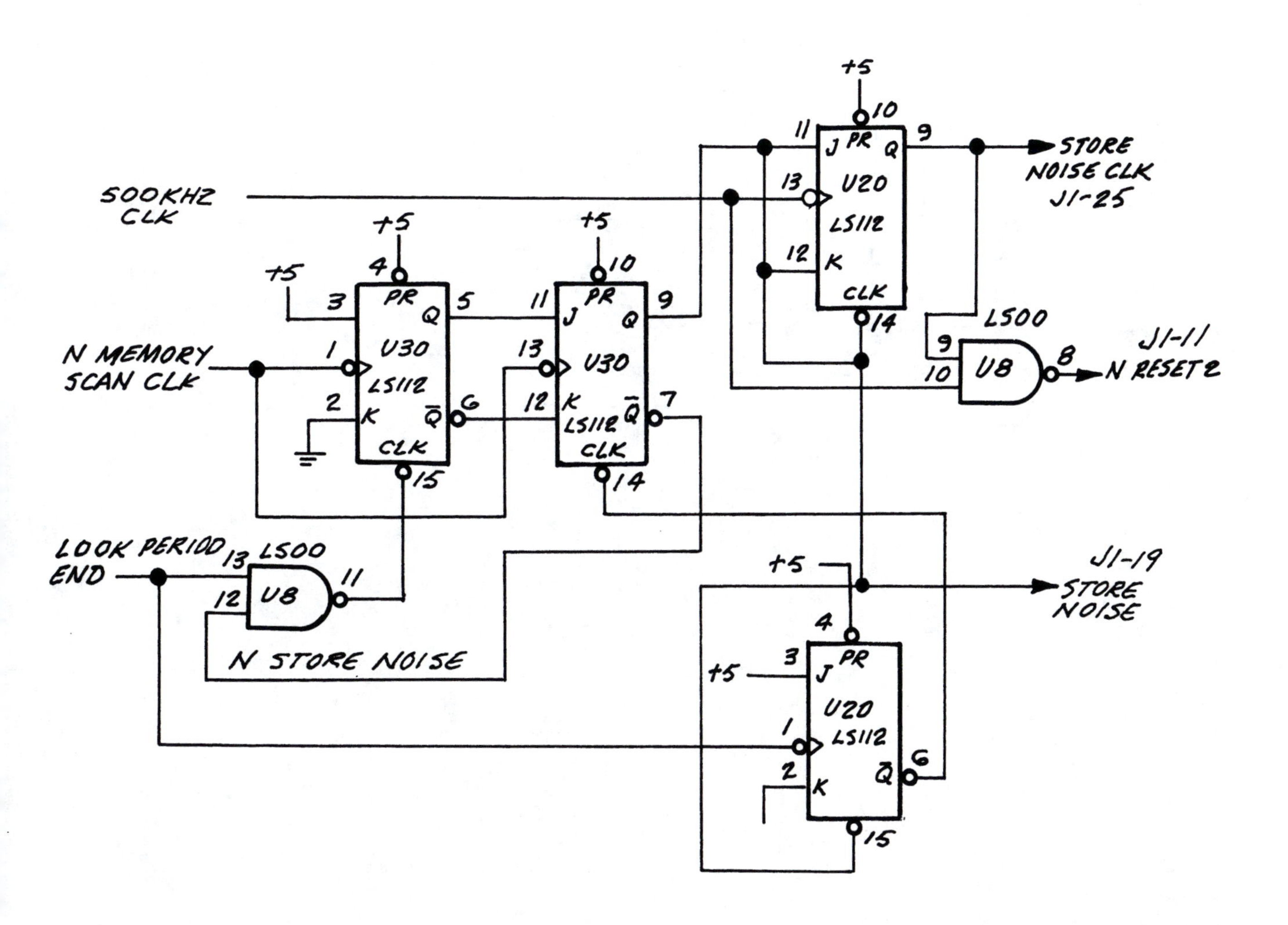

EXERCISE 5-7

Lay out correctly and draw a formal schematic diagram from the following sketch. Use the rules for good schematic layout described in this chapter. Pins on the DC/DC CONVERTER may be jumbled and placed on any side of the rectangle if that will improve the layout. Use a schematic template to draw the capacitors, the pins of the connector, and the diode. The voltage regulator (VR1) and the DC/DC CONVERTER should be drawn as they are shown in the sketch (neater, but as a circle and a rectangle). Arrange the diagram lengthwise on an 8½" × 11" sheet.

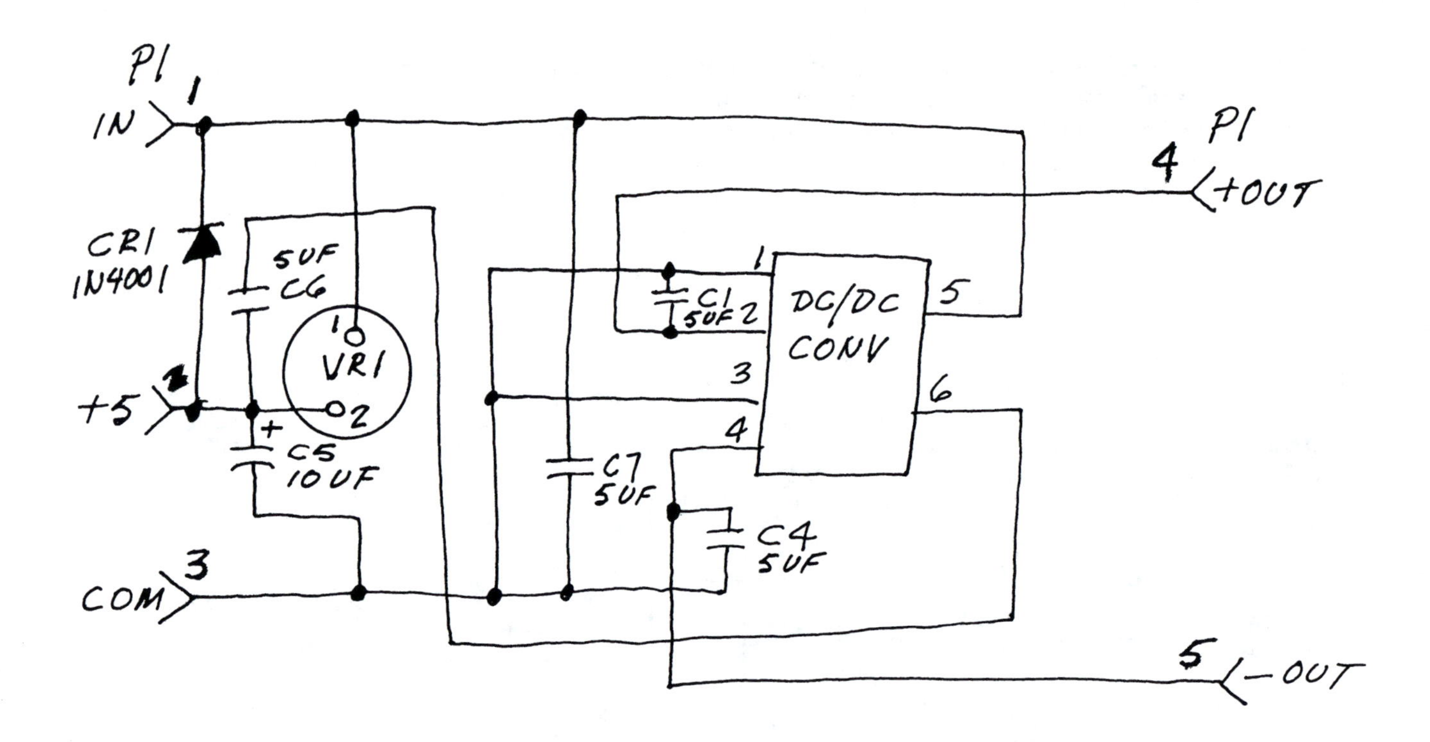

EXERCISE 5-8

Redraw the following sketch as a formal schematic diagram. Use the correct *symbol* for each component. Use good layout principles to remove doglegs and crossovers. Assign reference designators. Consult the Appendix for any unfamiliar symbols, such as the sensor and the earphone. Be consistent in arranging reference designators and component values. Center the drawing lengthwise on an $8\frac{1}{2}''$ × 11″ sheet.

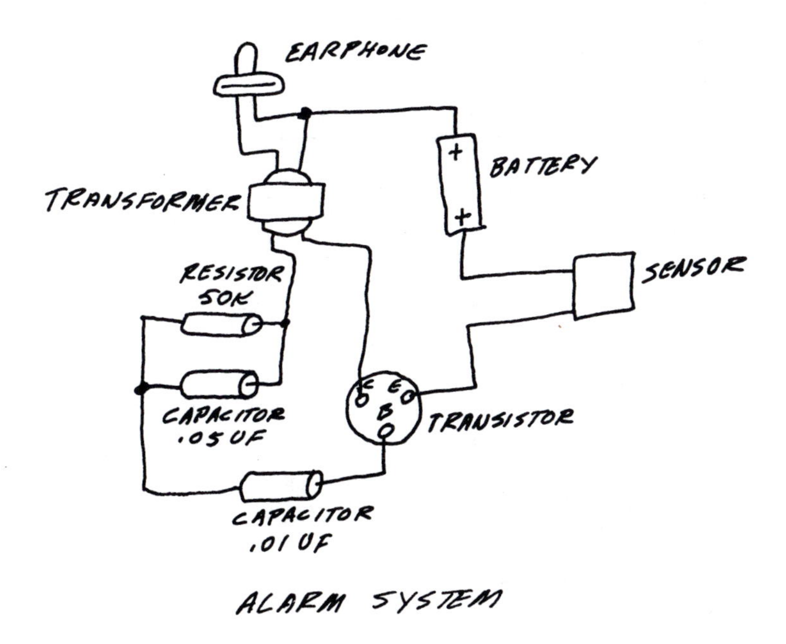

EXERCISE 5-9

Follow the same instructions as for Exercise 5-8, using the following sketch.

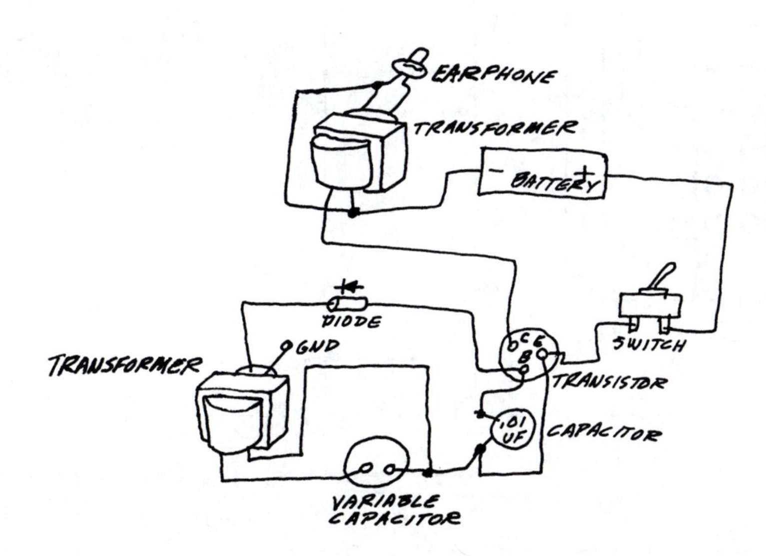

EXERCISE 5–10

Follow the same instructions as for Exercise 5–8 using the following design layout of a printed circuit board.

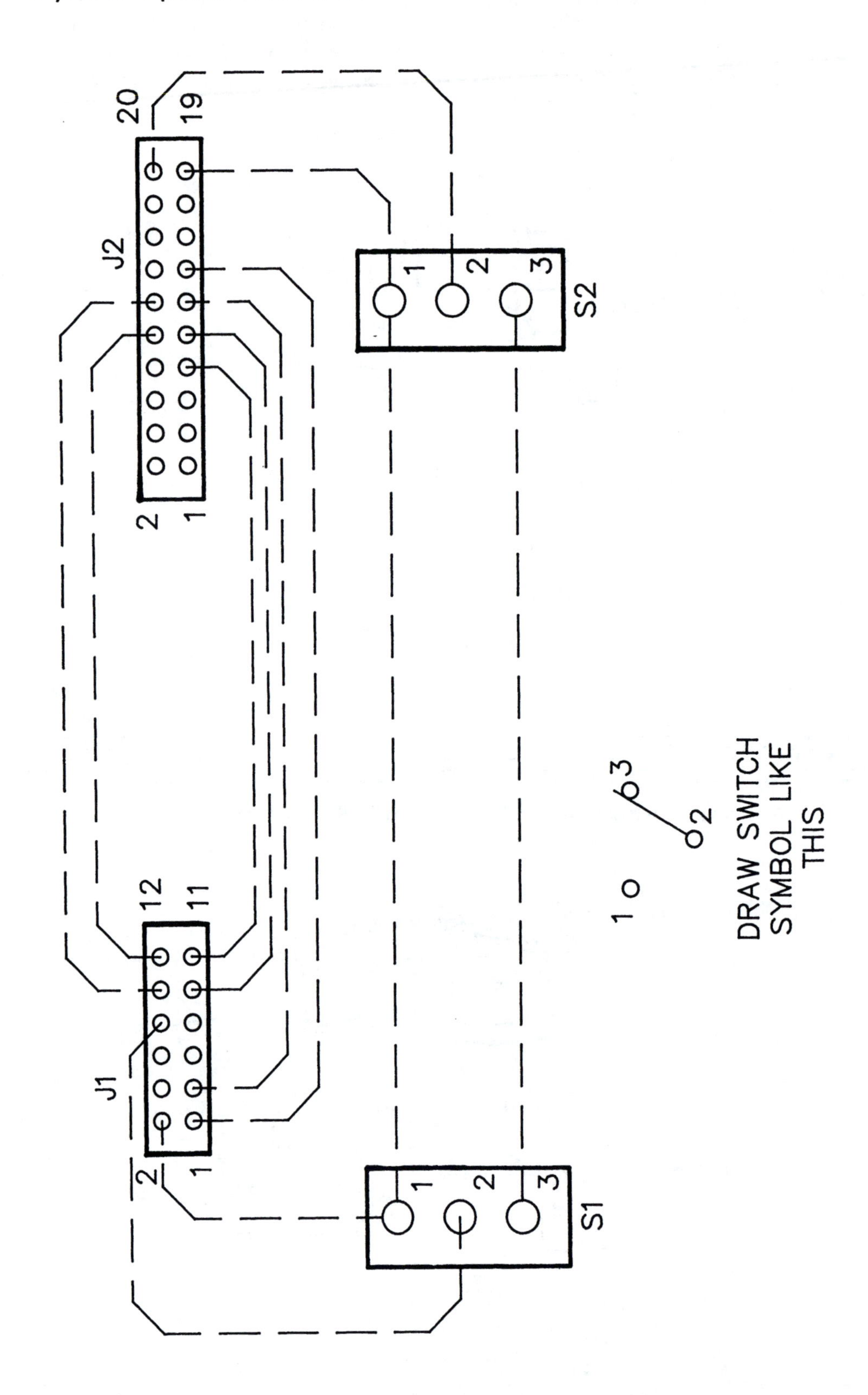

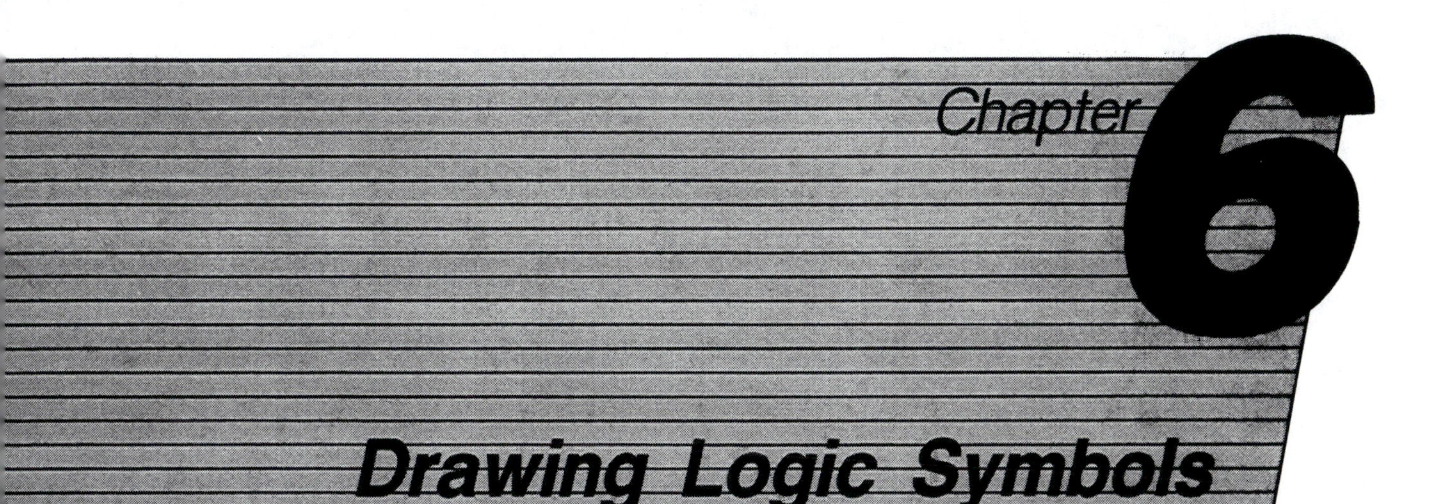

Chapter 6
Drawing Logic Symbols and Logic Diagrams

OBJECTIVES

After completing this chapter, you will be able to

- ✓ draw a logic diagram to acceptable industrial standards from an engineering sketch.
- ✓ apply the correct symbols for logic gates and the rules for correct layout of symbols and conducting lines.

GATES

Figure 6–1 is a logic diagram containing four integrated circuits, each of which is a combination of several gates. A *gate* is a miniaturized electronic circuit which performs a certain function within a larger circuit. Gates are

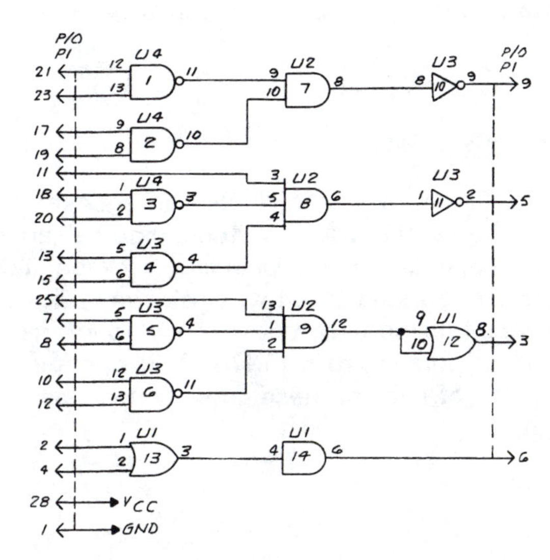

Figure 6–1 Logic diagram containing four integrated circuits, each having several gates

ACTUAL SIZE
AND SHAPE

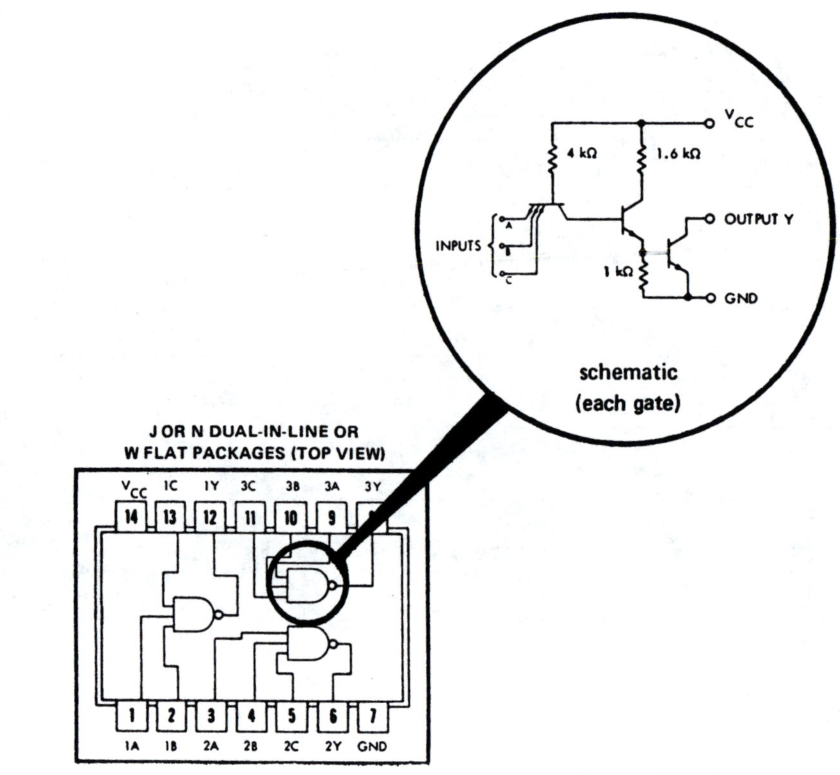

Courtesy of Texas Instruments, Inc.

Figure 6–2 Schematic for a gate

identified by their shape and the presence or absence of small circles on some part of the symbol. Each gate is a circuit in itself.

Usually, gates are not connected together in an integrated circuit package so that they may be used to form many different larger circuits. These larger circuits are used extensively in equipment requiring the use of switching circuits. Switching circuits are used in computers and other types of equipment which operate in a manner similar to a computer. The schematic for a particular gate is shown in the detailed enlargement in Figure 6–2.

INTEGRATED CIRCUITS

A number of different integrated circuit packages are shown in Figure 6–3. Figure 6–3A shows the very common external appearance of an integrated circuit package called a dual-in-line package (DIP). The seven different illustrations of integrated circuit types shown in Figure 6–3B indicate what gates are included inside each one of these packages. For example, the one with the arrow pointing to it (SN74LS26) has four different NAND gates in it. The function of each of these gates is described in greater detail in a later paragraph.

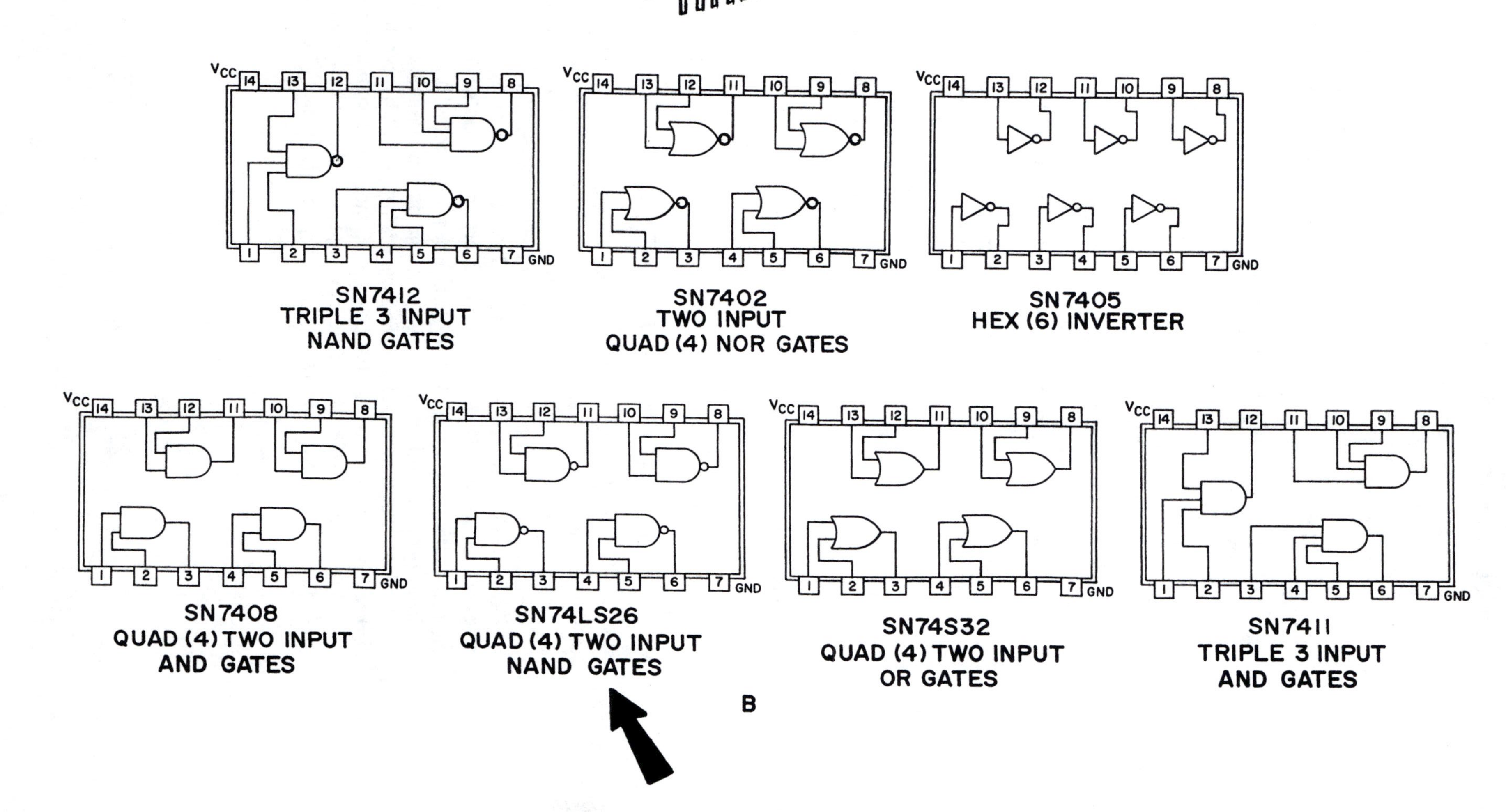

Figure 6–3 Types of integrated circuit packages

Inside of an Integrated Circuit

If one were to take apart an integrated circuit package, what would be seen is an integrated circuit chip similar to the one shown in Figure 6–4. *Chips* are the components and conducting lines which form the circuits represented by the schematic shown in Figure 6–3B. The circuits for each of the gates are physically separated from one another inside of the chip. The integrated circuit chip of Figure 6–4 has four NAND gates inside of it. The printed circuit board designer may use any number of these gates, along with other integrated circuits, to design a printed circuit board that will perform a specific function.

Printed Circuit Board Connector

Figure 6–5, a detail taken from the logic circuit in Figure 6–1, is a connector labeled P1. As shown in Figure 6–5A, the connector is actually part of a printed circuit (PC) board. The pins of the connector are called *fingers,* and they are etched onto the board. These fingers plug into a connector which is used to connect that printed circuit board with another printed circuit board or some other electronic or electrical part. Each of these fingers is given a number that is placed on the schematic diagram next to a wedgelike symbol which represents the finger, Figure 6–5B, or as the arrowlike symbols shown as P1 on Figure 6–1. A dashed line connects together all of the pins or fingers to show that they are all part of the same connector. The pins are connected to integrated circuits or other components by way of conducting lines etched into the printed circuit board. All of these connections are shown by a logic diagram such as the one in Figure 6–1.

LOGIC DIAGRAM EXPLAINED

Having discussed the details of the individual parts of the logic diagram, the way in which these parts fit together is described next.

The logic diagram for a printed circuit board, as given in Figure 6–1, now shows the final PC board itself, Figure 6–6. Gates 1, 2, 3, and 4 on the logic diagram are NAND gates. They are located in an integrated circuit package labeled U4. Gates 5 and 6 are in U3. Gates 10 and 11 are also in the integrated circuit U3. Although gate symbols 10 and 11 look different from gates 5 and 6, one gate function may sometimes be substituted for another through a process called *gate combining.* Chapter 10 discusses this process further. Do not be concerned that these gates are different from the other two. They should be drawn just as they are shown in Figure 6–6. Gates 12, 13, and 14 are in U1 and gates 7, 8, and 9 are in U2.

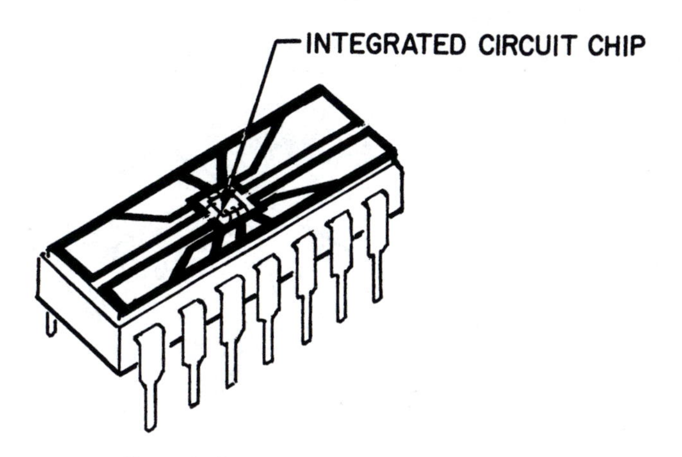

Figure 6–4 Integrated circuit chip (inside)

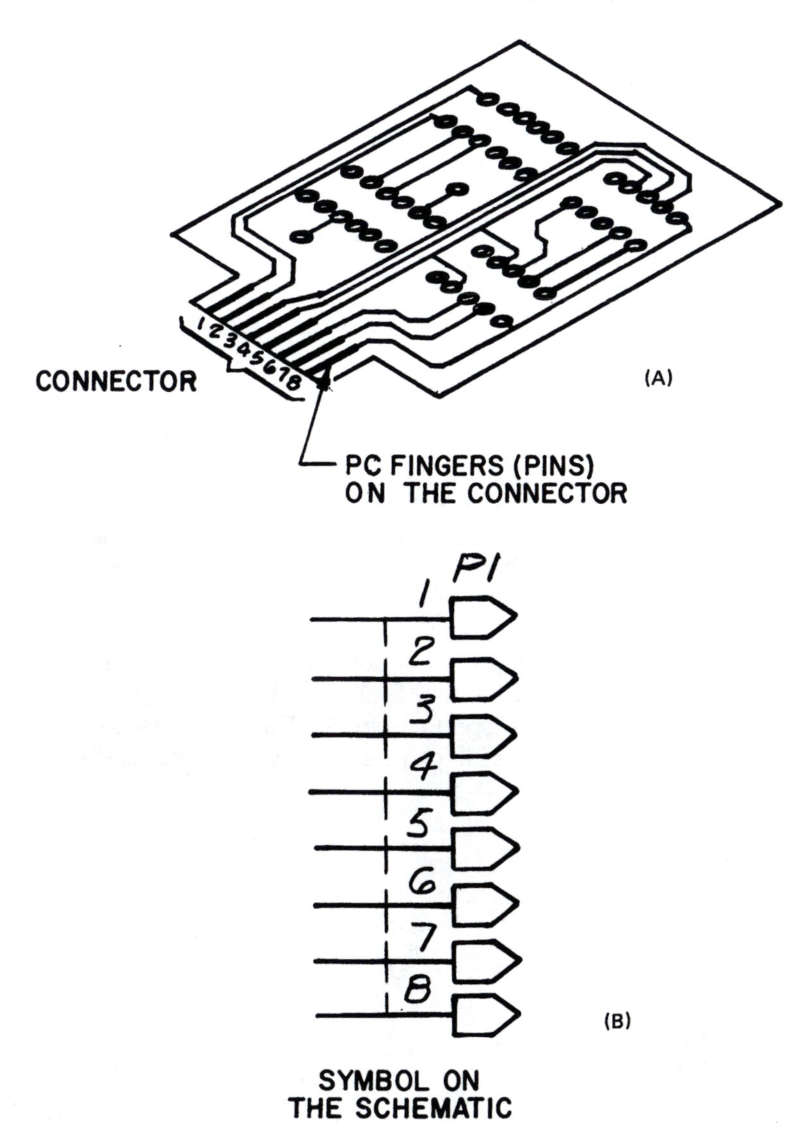

Figure 6–5 Printed circuit board connector and its symbol

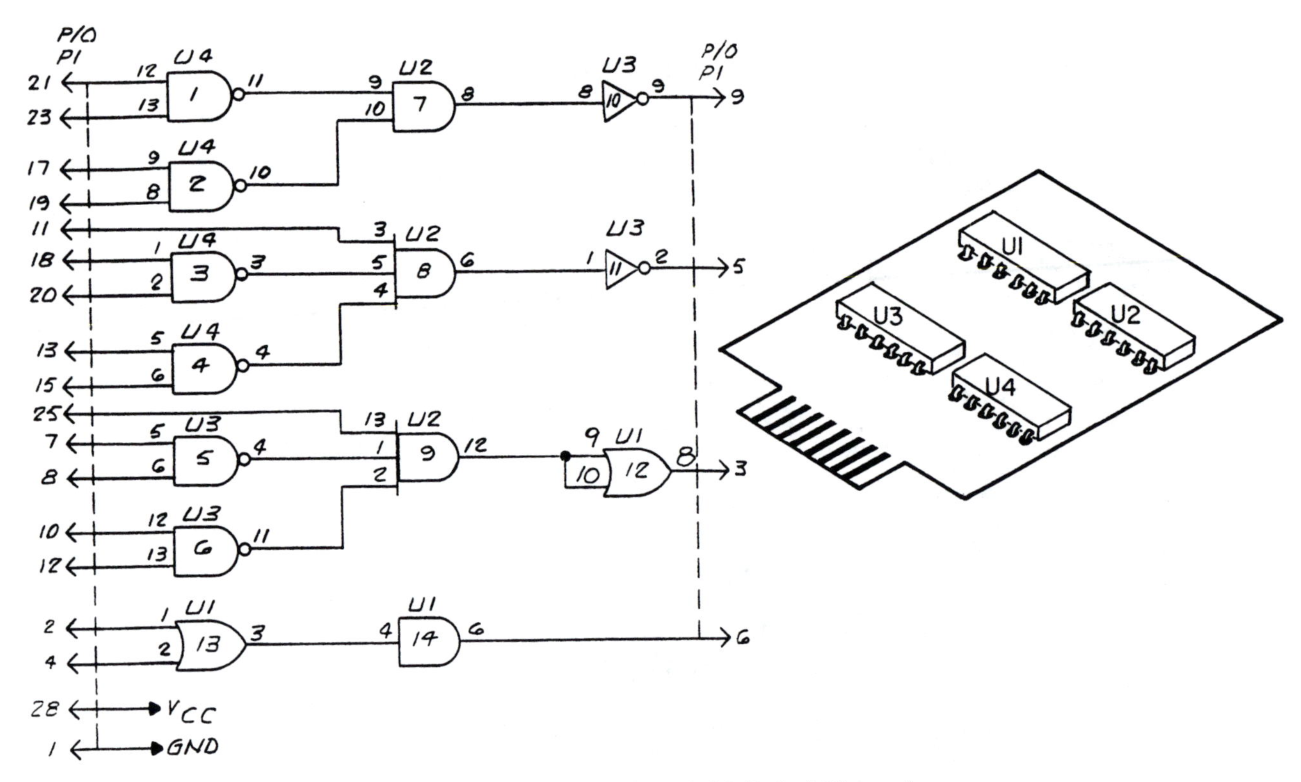

Figure 6–6 Logic diagram repeated (with final PC board)

Pin Numbers on the Integrated Circuit

Figure 6–7 shows a part of the logic diagram labeled gate 12. This is a two-input OR gate. This means that there are two lines leading into the left side of the OR symbol. These lines are numbered 4 and 5, Figure 6–7A. The output line of the symbol is numbered 6. These are the pin numbers as they would appear on an integrated circuit package. These pins are the leads coming out of the integrated circuit, and they are shown actual size and shape in Figure 6–7B. Figure 6–7C shows the inputs and outputs for all three gates inside of the integrated circuit. Gate 12 has been emphasized in this diagram just so that it can be described more easily.

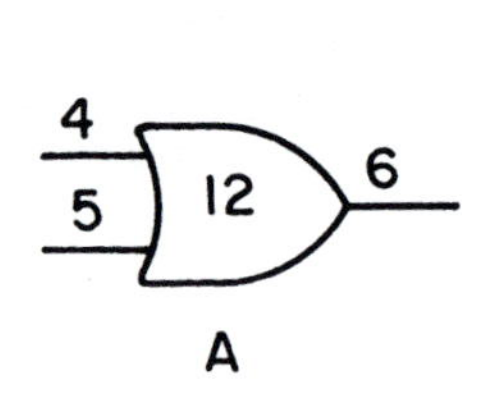

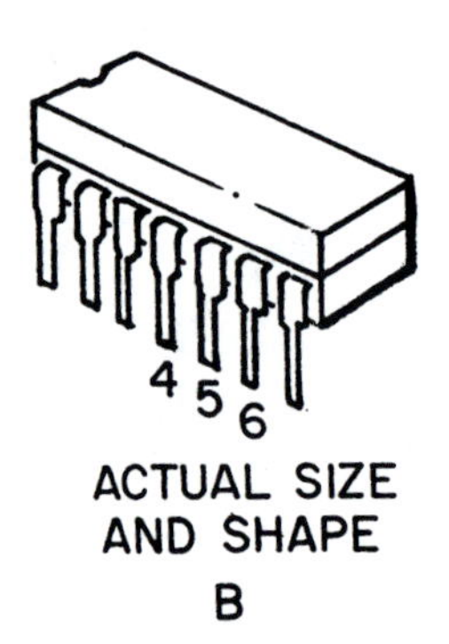

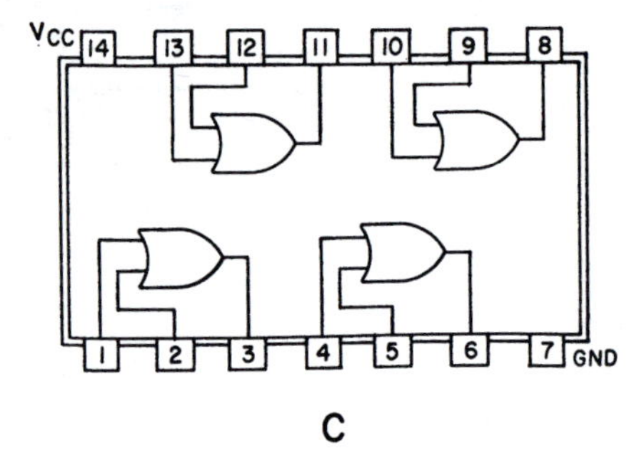

Figure 6–7 Two-input OR gate

Labeling Gates

Integrated circuit packages are given a reference designator of U, IC or, occasionally, other letters, Figure 6–8A. Each gate on the logic diagram, Figure 6–8C, is labeled with the reference designator of the integrated circuit of which it is physically a part. The IC type (74LS08) and the pin numbers of the inputs and outputs are also shown on each gate.

The printed circuit board itself is shown in Figure 6–8B. Rows of ICs are labeled TC and TB. Columns are labeled 1, 2, and 3. ICs, then, are labeled with the row and column number, such as TC1, TC2, TC3, and so forth.

V_{CC} and Ground

Notice in Figure 6–6 that there are two pins on connector P1 which do not seem to connect to anything. These pins are numbered 1 and 28. Pin 28 is labeled V_{CC}. This is the supply voltage used to run each of the integrated circuits of the PC board. It is usually not connected on the logic diagram because designers know that each IC must have V_{CC}. Pin 1 on the logic diagram is labeled ground. It is not connected on the logic diagram for the same reason. Notice that each IC in Figure 6–3 has pin 14 labeled V_{CC} and pin 7 labeled ground. On the PC board all pins 7 on the ICs will be connected to pin 1 on the PC board connector. All pins 14 on the ICs will be connected to pin 28 on the PC board connector.

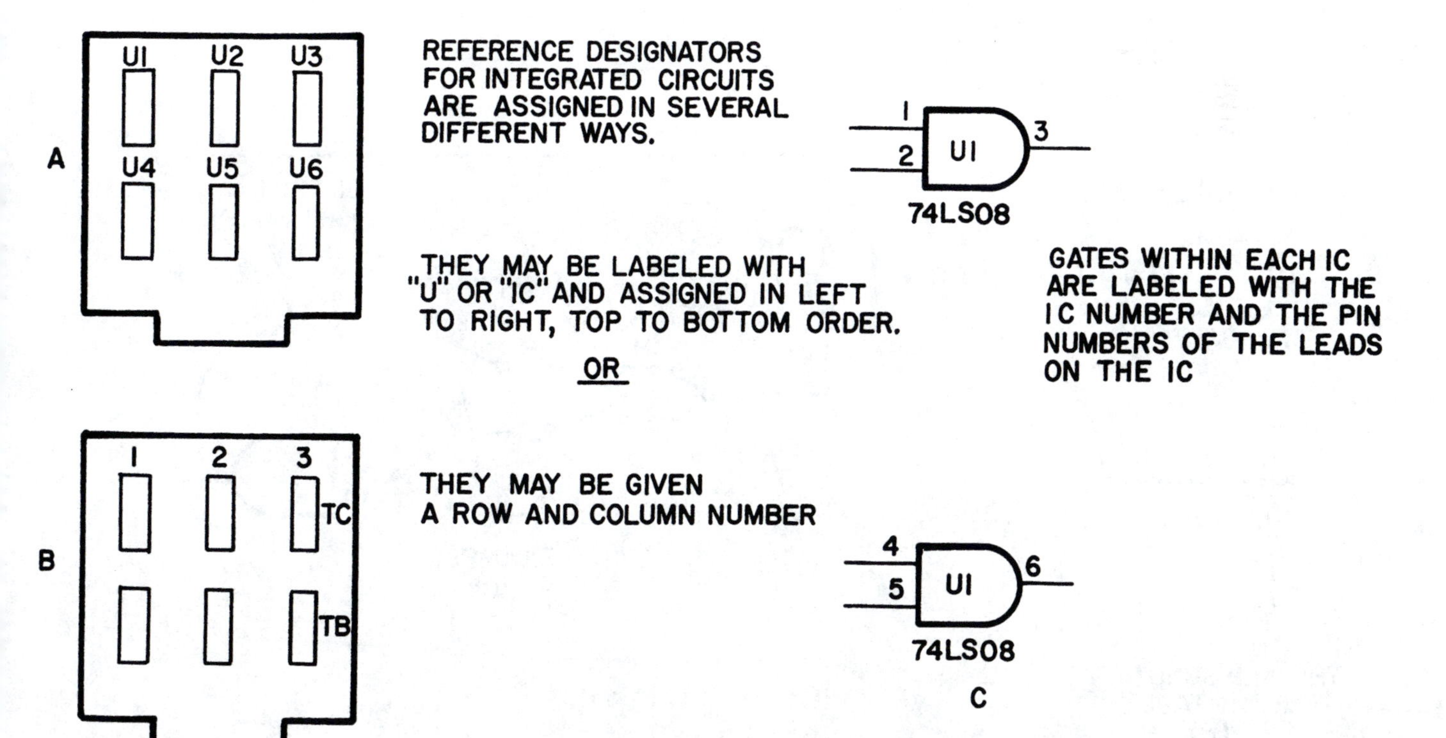

Figure 6–8 Labeling ICs and their gates

AND Gate

Figure 6–9 shows an AND gate. This gate has 2 inputs and 1 output. The table shown in this figure is called a *truth table.* A truth table describes how the gate functions under all possible conditions. The column labeled A in the truth table refers to the first input on the gate. Column B refers to the second input, and column C refers to the output of the gate.

The AND gate is aptly named. The truth table indicates that if there is an input at A *and* an input at B, there will be an output at C. Under no other conditions will there be an output at this gate. Inputs and outputs are shown as the digit one (1). No input or no output is shown as the digit zero (0). The first row of the table shows no input at A, no input at B, and no output at C. The second row shows no input at A, an input at B, and no output. The third row shows an input at A, no input at B, and no output. The fourth row shows an input at A *and* an input at B resulting in an output at C. The example of the switching circuits shown in Figure 6–9C illustrates the truth table AND function. The drawing shows that when both switches are on (inputs at A and B) the light will light. If either of these switches is off, the lights will not come on.

OR Gate

The OR gate, Figure 6–10, is constructed so that when an input occurs at either A *or* B *or* both places, there will be an output at C. The OR function truth table describes this gate well. The first row indicates no input at A and no input at B, resulting in no output at C. The second row indicates no input at A and an input at B, producing an output at C. The third row indicates an input at A, no input at B, and an output at C. The fourth row indicates an input at A and an input at B, resulting in an output at C. The example in Figure 6–10C shows that if either of the switches is on (*or* both), the bulb will light.

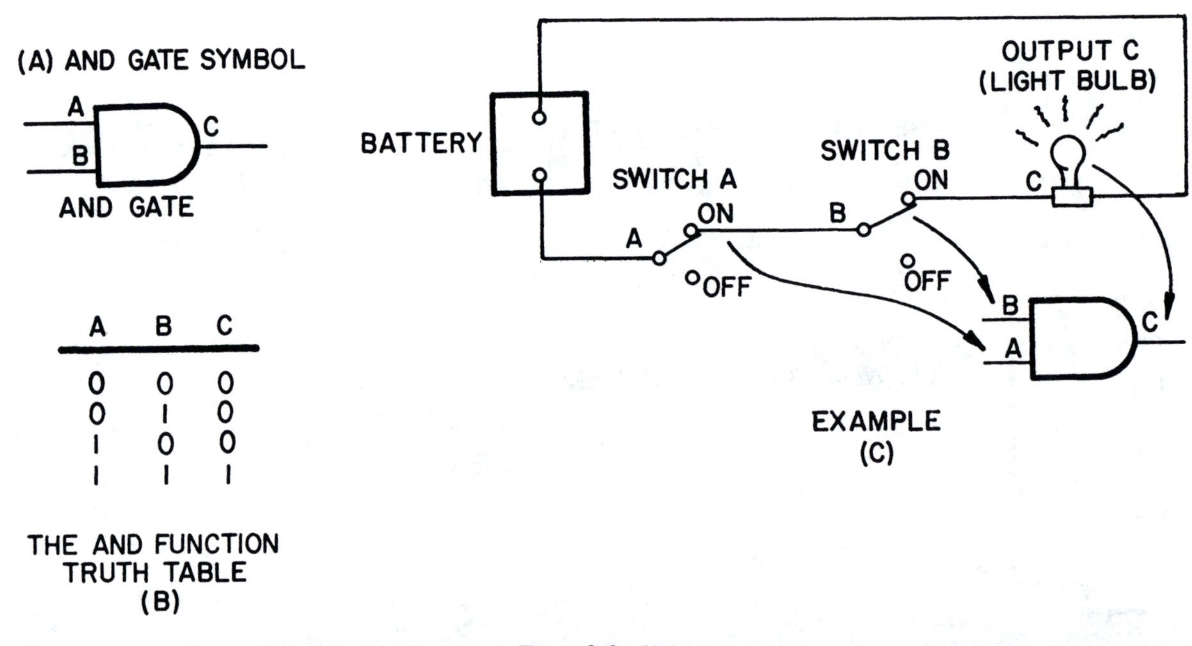

Figure 6–9 AND gate

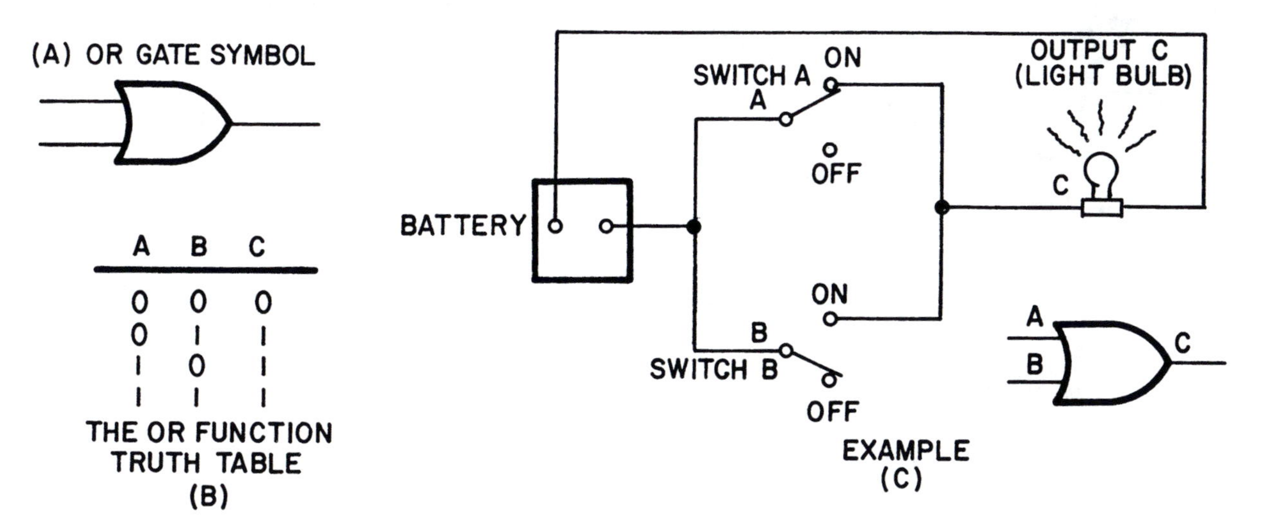

Figure 6–10 OR gate

EXCLUSIVE OR Gate

In the EXCLUSIVE OR gate, Figure 6–11, there must be an input at *either* A or B, but *not both* to get an output at C. The truth table shows the arrangement of inputs needed to get an output. The EXCLUSIVE OR function means that there cannot be inputs at both A and B for an output to exist at C. An example showing a switching arrangement was omitted from this figure because its complexity would serve no purpose for this chapter.

NAND Gate

The NAND gate, Figure 6–12, looks just like an AND gate with a small circle on the output. The truth table shows that the NAND gate is just exactly the opposite of an AND gate. It is called a *negative* AND gate. The truth table for the NAND gate shows that the outputs and inputs are the reverse of the AND gate.

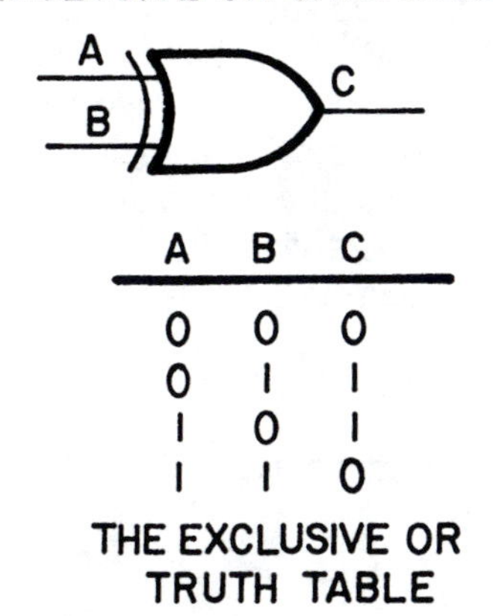

Figure 6–11 EXCLUSIVE OR gate

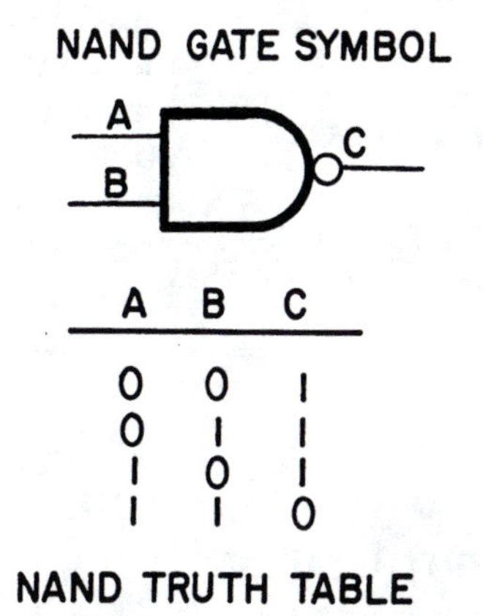

Figure 6–12 NAND gate

NOR GATE SYMBOL

A	B	C
0	0	1
0	1	0
1	0	0
1	1	0

NOR TRUTH TABLE

Figure 6–13 NOR gate

INVERTER SYMBOL

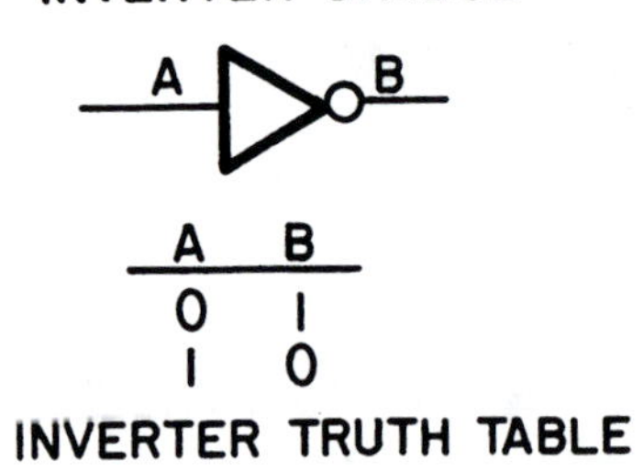

A	B
0	1
1	0

INVERTER TRUTH TABLE

Figure 6–14 INVERTER

NOR Gate

The NOR gate, Figure 6–13, has inputs and outputs that are exactly the opposite of the OR gate. When comparing the truth table of OR and NOR gates, it can be seen that a 1 output on the OR table is a 0 (negative) output on the NOR table. The symbol for a NOR gate looks just like an OR symbol with a small circle on its output. This small circle is referred to as negative logic, meaning that the symbol is the opposite of what was originally shown.

INVERTER

The INVERTER is shown as a small triangle with a small circle on the output, Figure 6–14. The truth table shows that if there is a 1 input into an INVERTER, there will be a 0 output. If there is a 0 input, there will be a 1 output. This truth table is much simpler than it is for any of the gates previously described.

BUFFER

The BUFFER symbol, Figure 6–15, is a small triangle without a circle on its output. The signal going in is the same as it is coming out. If it is a 1 input, it will be a 1 output; 0 in, 0 out.

Symbols

Figure 6–16 shows the symbols in their correct sizes and shapes for the exercises assigned in this text. Exercise 6–1 in this chapter requires that you use a small logic symbol template; or, you may trace the symbols shown in Figure 6–16 using a circle template and a straight edge. If you use a computer to make the drawings for this chapter, you will probably find these symbols included in your software drawing files. If not, you will have to construct them one time and place them in a library or file directory for future use.

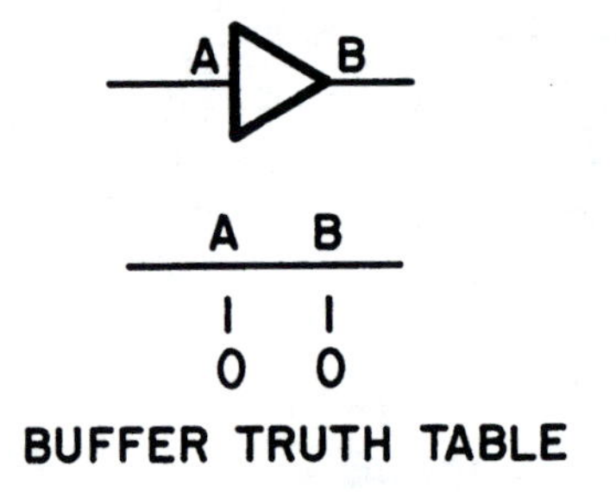

A	B
1	1
0	0

BUFFER TRUTH TABLE

Figure 6–15 BUFFER

Figure 6–16 Symbols

EXERCISE 6–1

Redraw the following sketch using the correct logic symbol for each of the gates. Use good layout principles.

Lettering should be $\frac{1}{8}$″ high. Do not letter the type of gate as was done on the sketch. Your drawing should resemble Figure 6–1. Use a logic symbol template for the gates if you have one, or trace the correct symbols from Figure 6–16 using a circle template and a triangle. Show device pin numbers outside of the gate and the reference designator inside of each gate. Position the drawing lengthwise on an $8\frac{1}{8}$″ × 11″ sheet.

This is a relatively simple diagram which may be laid out in the same manner as the sketch.

Be sure to use good lettering. Lettering and lines should be very dark and sharp. You may provide some contrast in the drawing by making the gates about twice as thick as the connecting lines.

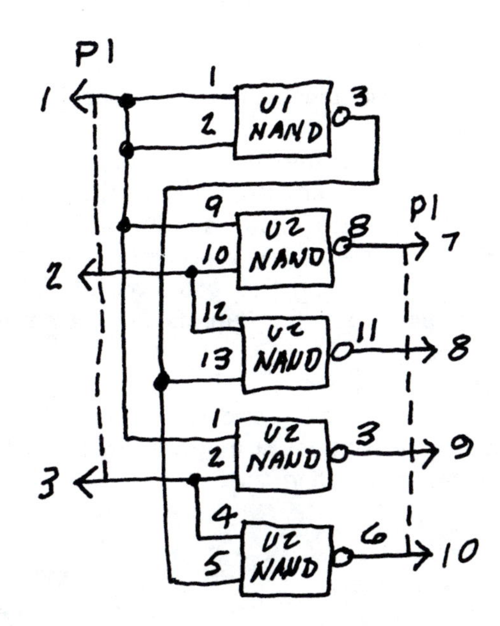

EXERCISE 6–2

Redraw the following sketch using the correct logic symbol for each of the gates. Use good layout principles.

Lettering should be $\frac{1}{8}''$ high. Do not letter the type of gate as was done on the sketch. Your drawing should resemble Figure 6–1. Use a logic symbol template for the gates if you have one, or trace the correct symbols from Figure 6–16 using a circle template and a triangle. Show device pin numbers outside of the gate and the device number inside of each gate. Position the drawing lengthwise on an $8\frac{1}{8}'' \times 11''$ sheet. Be sure to correctly label pins on the fourth gate of U1.

This is a relatively simple diagram which may be laid out in the same manner as the sketch. Substitute the correct symbols for the square shapes shown. For example, U1 is a NAND gate. Use the NAND symbol from Figure 6–16. Use good layout principles to eliminate any unnecessary doglegs or crossovers. Do not rearrange the drawing completely just to eliminate a dogleg or crossover, but concentrate on removing those that will leave the symbols in about the same position as they are on the sketch. Notice the note that asks you to place unused pin numbers on this gate. You will find these numbers on IC type SN74LS26 shown in Figure 6–3. This integrated circuit package has four NAND gates in it. They all have two inputs.

Refer to the gates labeled U1 in the sketch. The pins have already been labeled on the three U1 gates shown on the left side of the sketch. These numbers are exactly the same as the numbers shown on SN74LS26 of Figure 6–3. Refer to this IC and determine which of the three gates on the sketch of Exercise 6–1 have already been used in the IC. Then see which gate was not used and assign the unused pin numbers from SN74LS26 to the unlabeled gate on the sketch. Label all gates as U1, U2, U3, U4, and U5. Place pin numbers at all inputs and outputs on the gates. Also place pin numbers on the connector.

Be sure to use good lettering. Lettering and lines should be very dark and sharp. You may provide some contrast in the drawing by making the gates about twice as thick as the connecting lines.

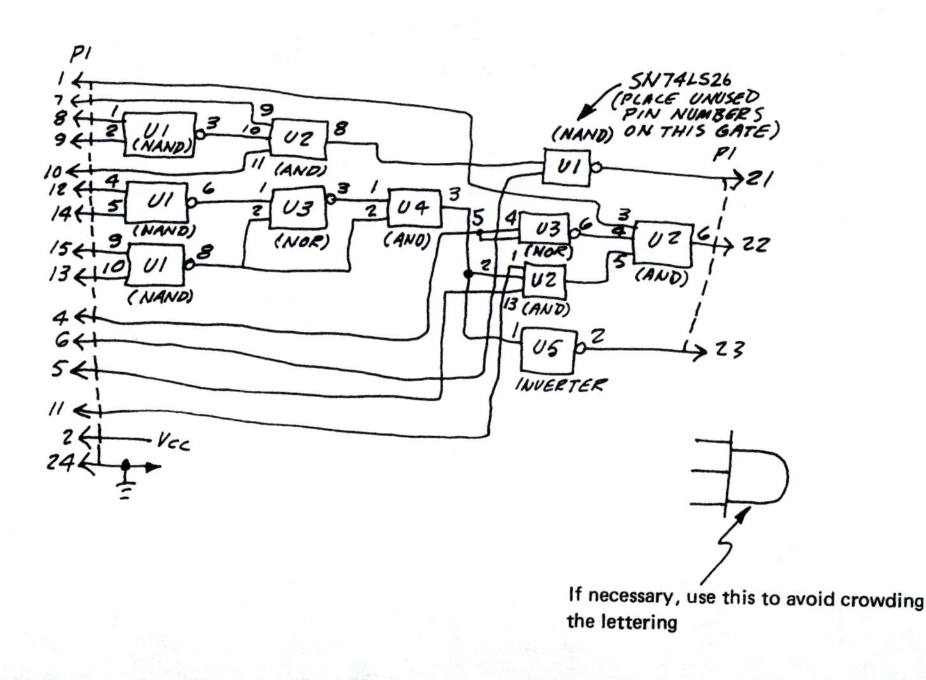

EXERCISE 6–3

Redraw the following sketch as a formal logic diagram. Remove doglegs and crossovers wherever possible. Be consistent in arranging reference designators and device types. The resistor pack should be drawn so that all of the resistors are aligned in the same direction. Consult Figure 6–3 to determine the pin numbers that should be placed on all gate inputs and outputs of IC1 and IC2. Place inputs on the left, outputs on the right. Center the drawing lengthwise on an $8\frac{1}{2}''$ × 11″ sheet.

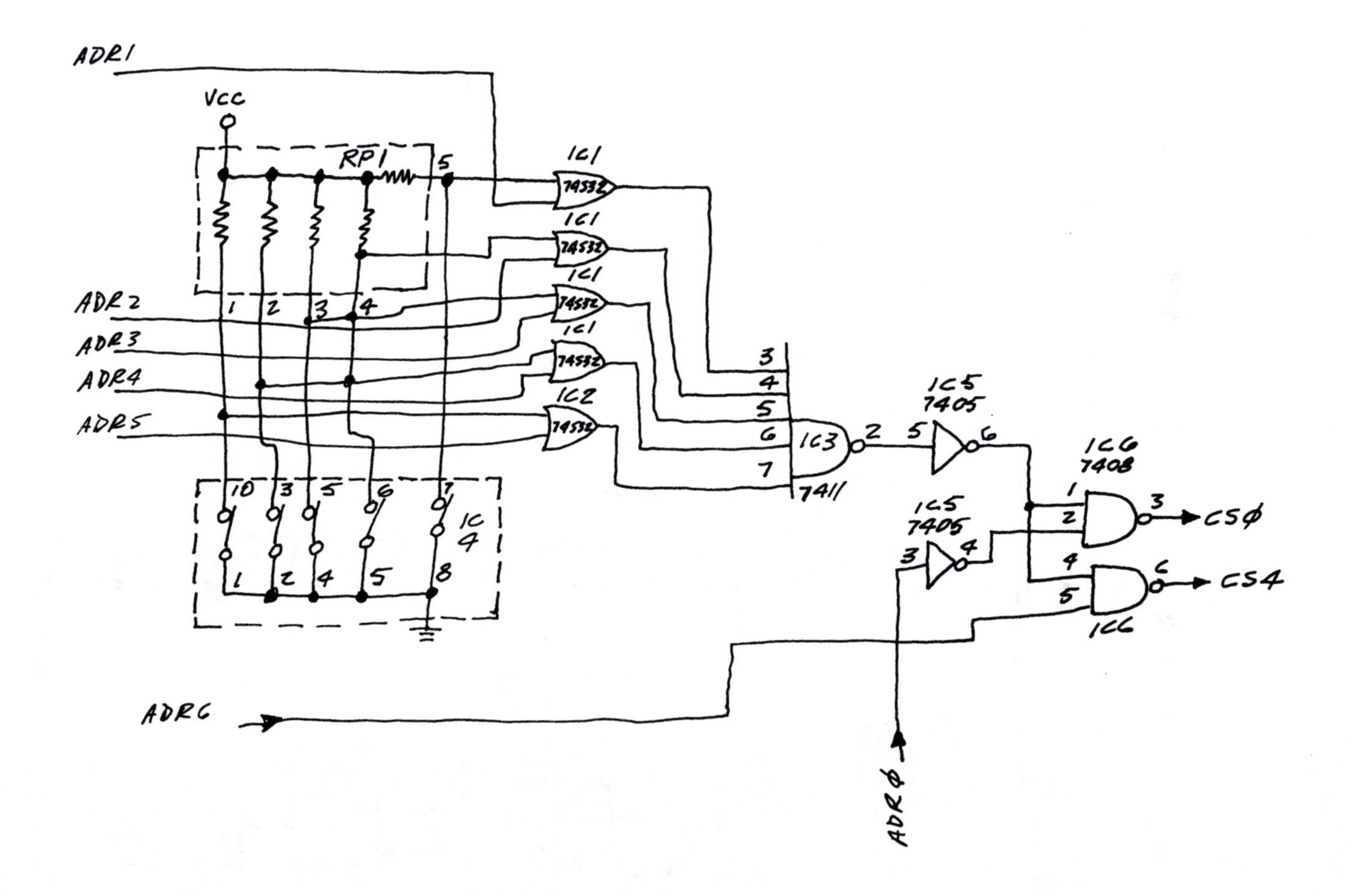

EXERCISE 6–4

Redraw the following sketch as a formal logic diagram. Use the correct logic symbol for all gates shown as IC21 and IC41. Consult Figure 6–3 to determine the correct symbols and pin numbers to be placed on the inputs and outputs of all gate symbols. Place a dashed line around the switches shown in IC52 and IC9 (similar to the one shown as IC4 on Exercise 6–3). Be consistent in the arrangement of reference designators and their values or device types. (The device type must be shown by each gate of the IC.) IC2 and IC50 should be drawn as rectangles as shown on the sketch. Center the drawing lengthwise on an 11″ × 17″ sheet.

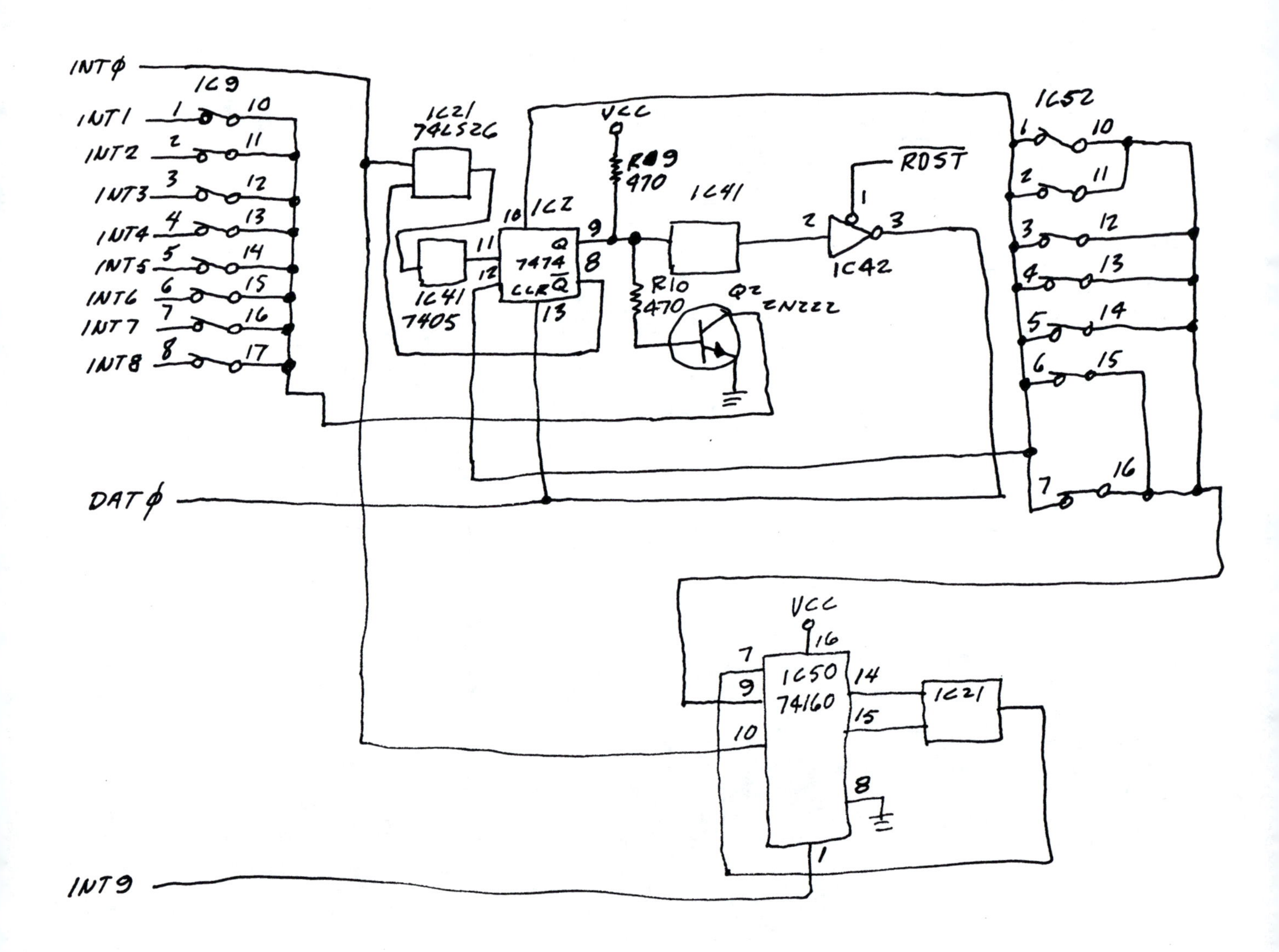

EXERCISE 6–5

Redraw the following sketch as a formal logic diagram. Use the correct symbols for all components and logic gates. Be consistent in the placement of reference designators and values or device types. Consult Figure 6–3 to determine the correct gate symbol to use on all gates labeled IC53. Place pin numbers on all gate symbols. Notice that pin numbers for two of the IC53 gates have already been assigned. Do not duplicate these numbers. Be sure to use good layout principles for logic and schematic diagrams. Center the drawing lengthwise on an $8\frac{1}{2}'' \times 11''$ sheet.

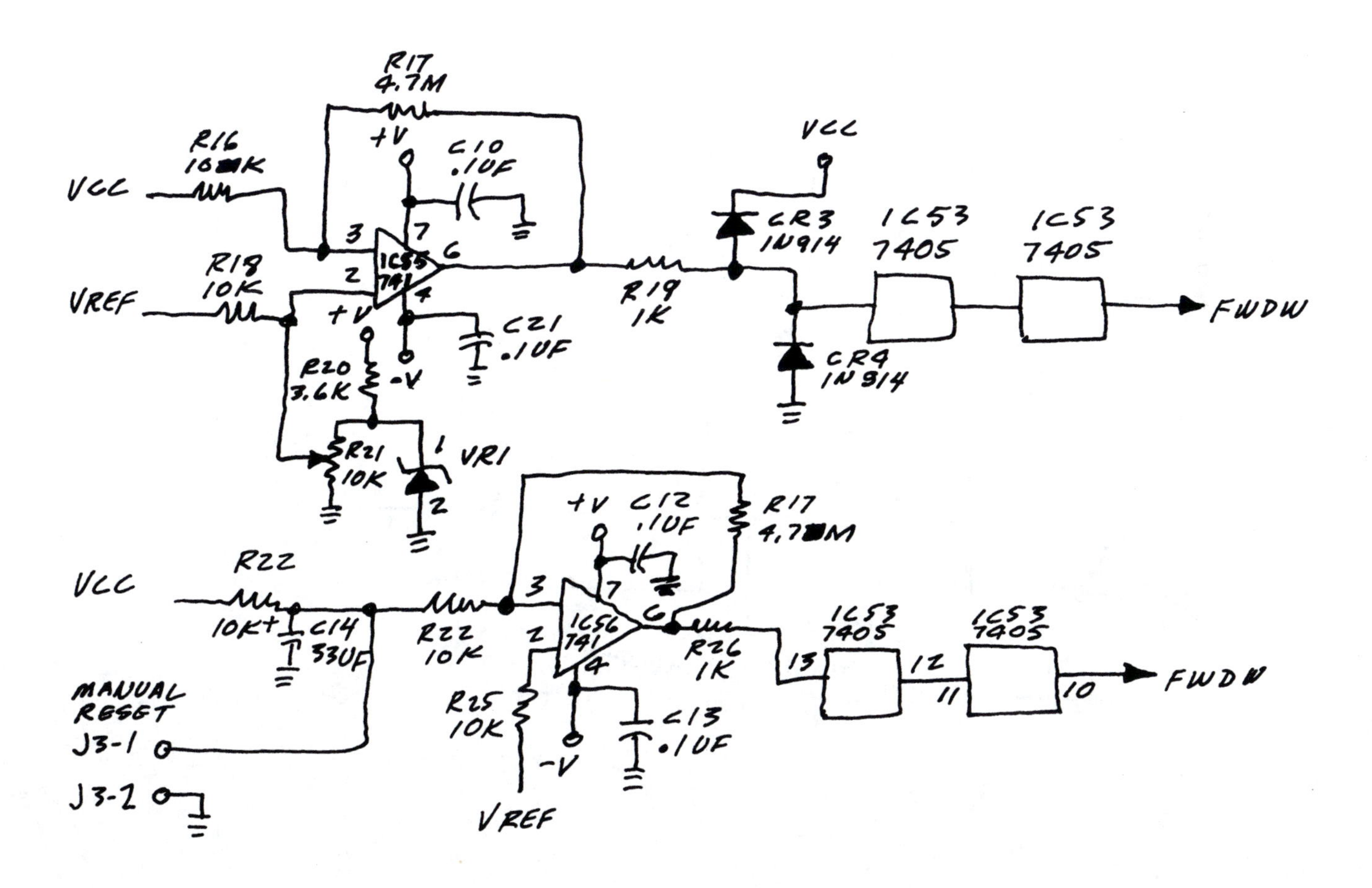

EXERCISE 6-6

Redraw the following sketch as a formal logic diagram. Use the correct symbols for all components and logic gates. Be consistent in the placement of reference designators and values or device types. Use "U" as the reference designator for all integrated circuits. Place pin numbers on all gate symbols. Be sure to use good layout principles for logic and schematic diagrams. Center the drawing lengthwise on an 11″ × 17″ sheet.

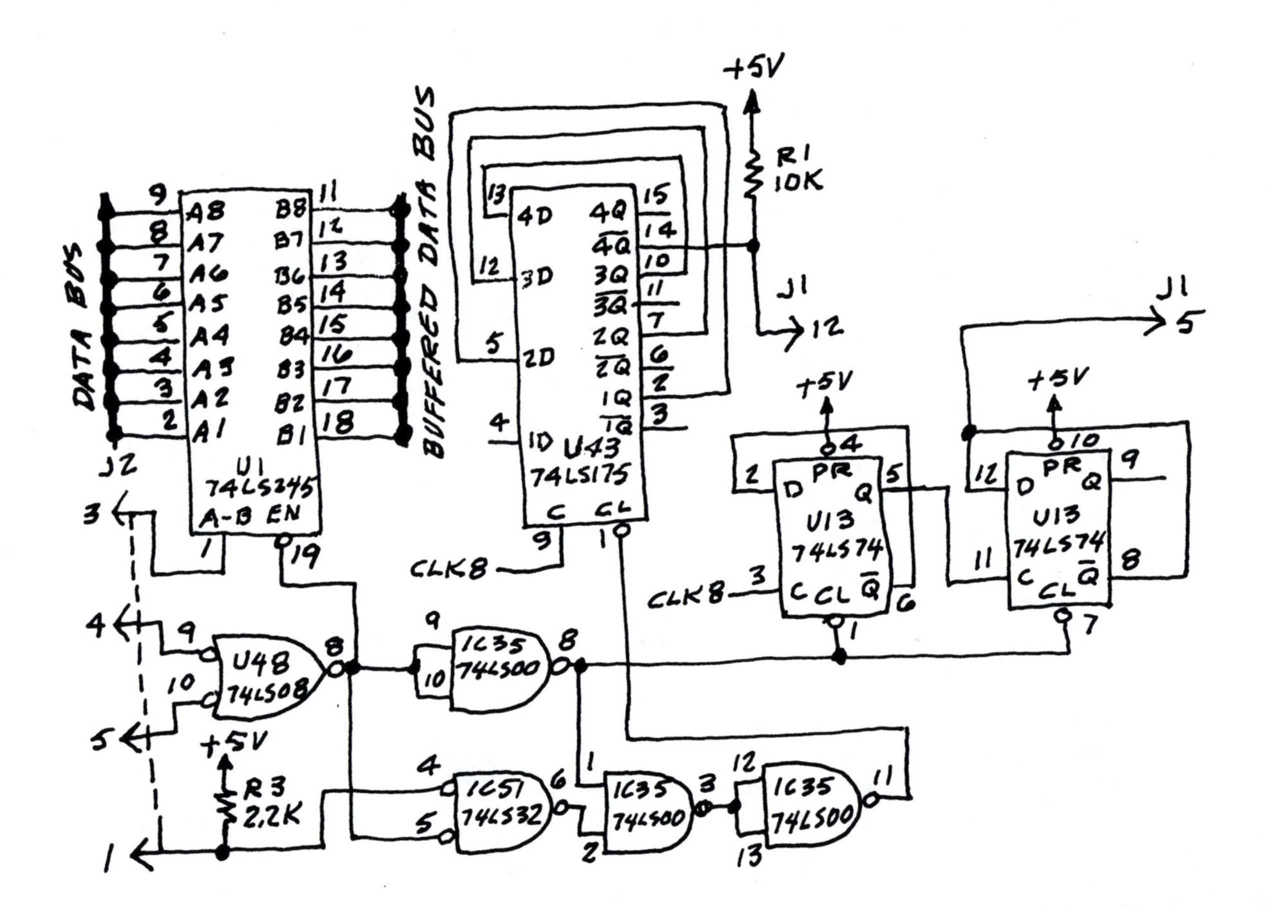

Chapter 7

Drawing Interconnecting Diagrams

OBJECTIVES

After completing this chapter, you will be able to

- ✓ draw to acceptable industrial standards a highway diagram from a point-to-point wiring sketch.
- ✓ list three ways a point-to-point interconnecting diagram is different from a highway diagram.
- ✓ describe the use of a wiring harness drawing.

INTRODUCTION

Interconnecting diagrams are drawings that describe how one electronic unit connects to another. Such connections are often made using a wiring harness or cable, Figure 7–1A. These connections are usually described on a point-to-point wiring list or a point-to-point drawing, Figure 7–1B. Because interconnecting diagrams often become very complex, they are sometimes simplified by the use of a highway diagram, Figure 7–1C. The first of these drawings to be described is the wiring harness and its accompanying drawing, the stake board.

WIRING HARNESS AND STAKE BOARD

A typical wiring harness is illustrated in Figure 7–2. A *wiring harness* is a bundle of wires of specific lengths and colors, usually having connectors that accept one or more wires. Wiring harnesses are assembled before they are connected to the electronic or electrical units, as shown in Figure 7–3.

A wiring harness is often built on a stake board or similar device. A *stake board* is usually a plywood sheet to which the full-size drawing of the wiring harness is attached, Figure 7–4. The drawing is covered with a clear film to allow it to be used many times. Special nails and springs are attached to the board so that the wiring harness can be built on the stake board. Wires are cut to the length indicated on the stake board, and connectors are attached to the wires according to the stake board drawing. Wire ties are used to hold the wires together. Their locations are shown on the stake board so that the ties can be placed in the same location on every harness.

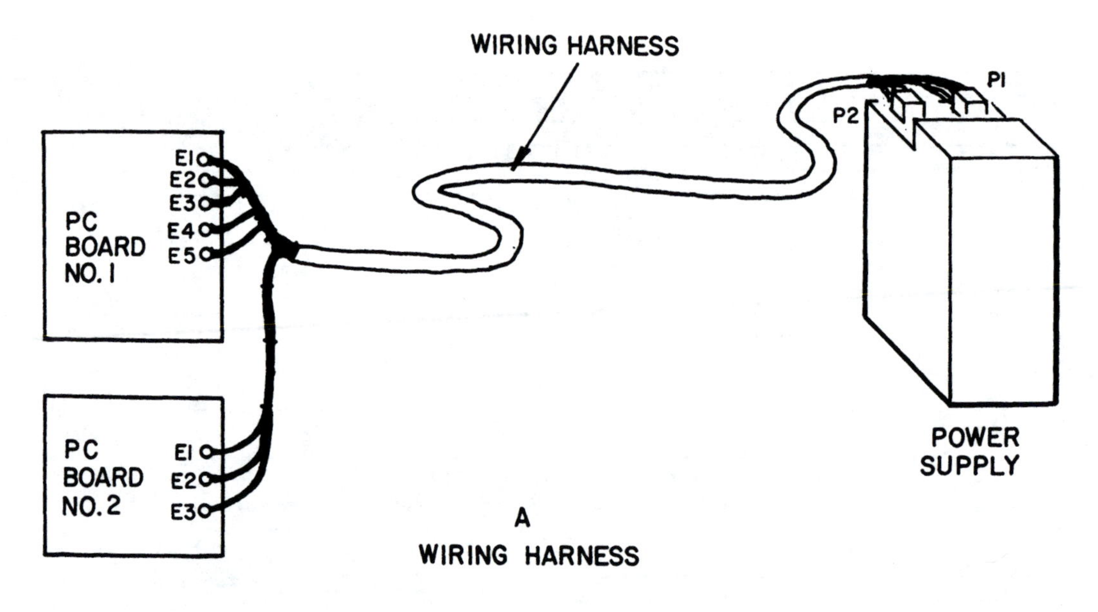

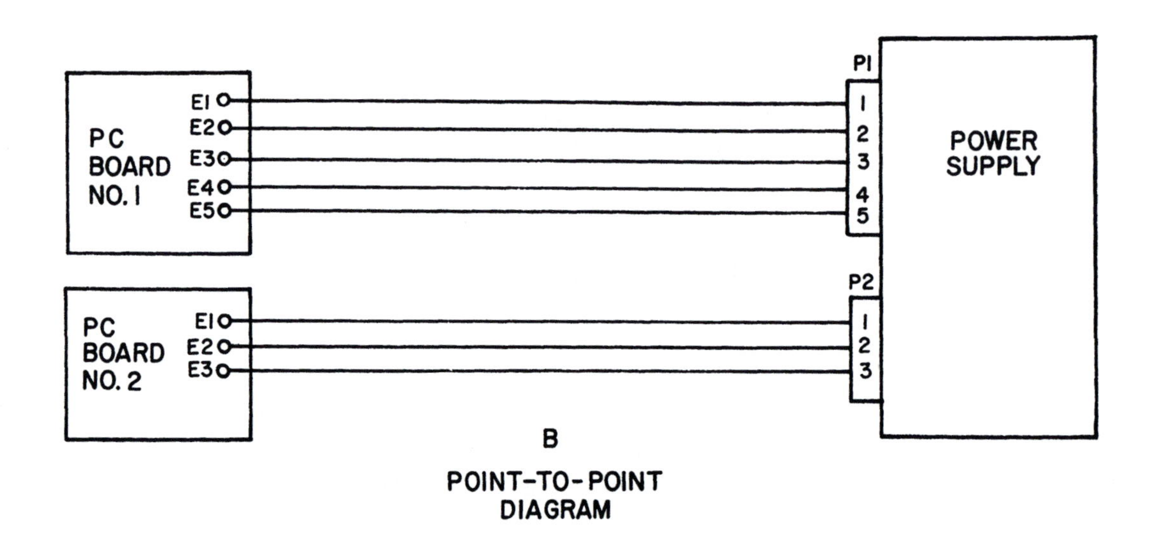

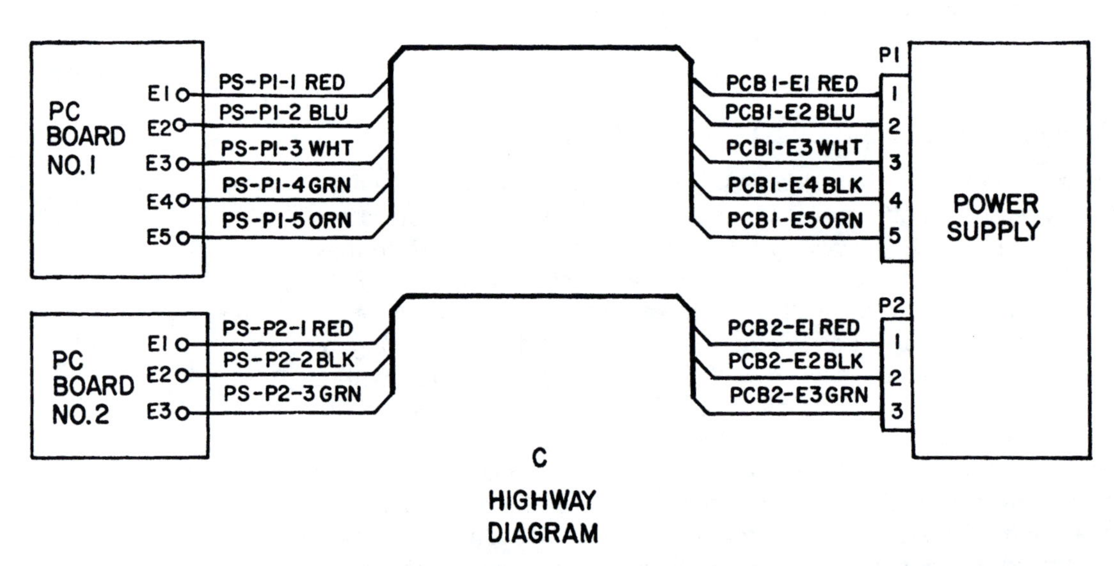

Figure 7–1 Interconnecting diagrams

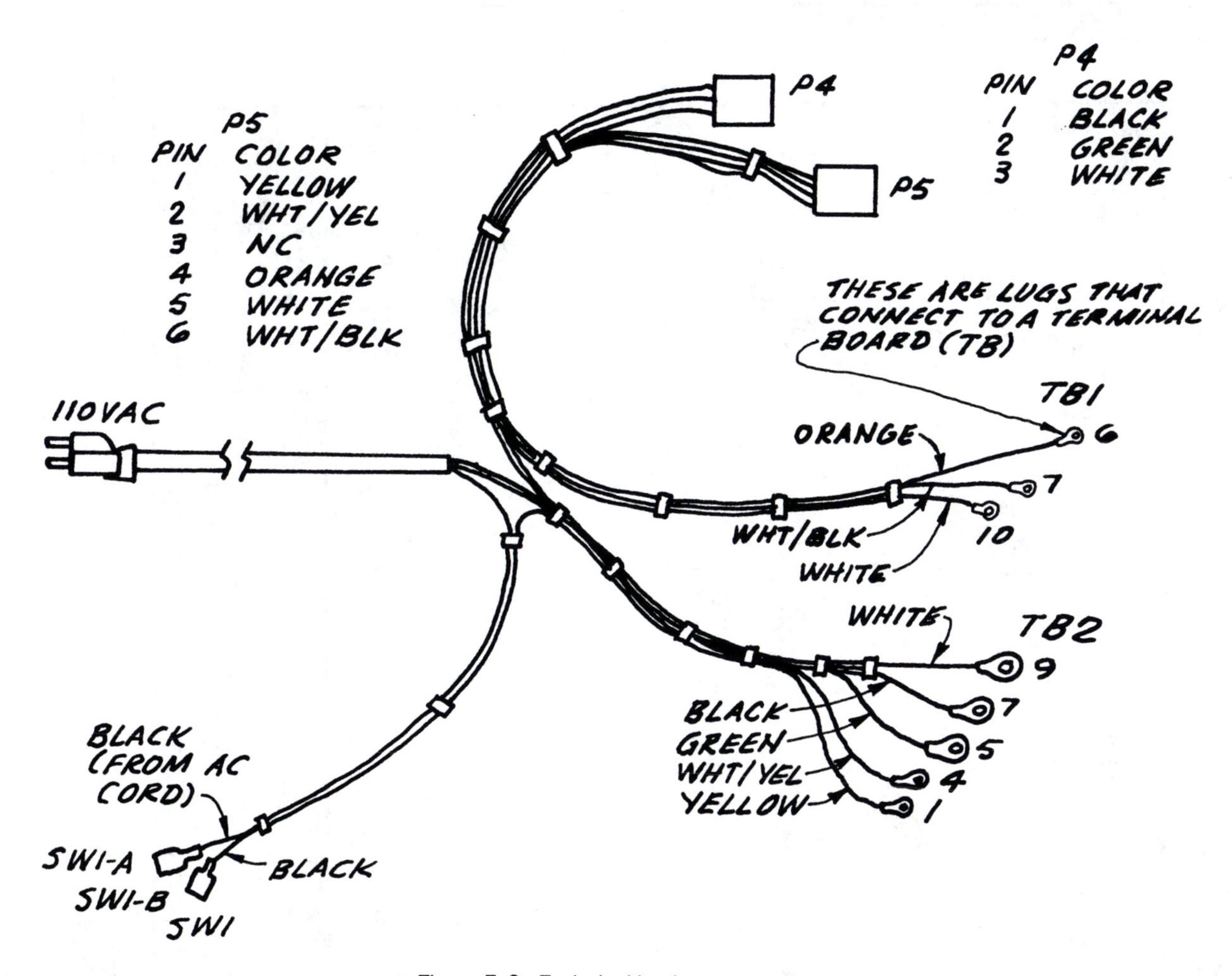

Figure 7–2 Typical wiring harness

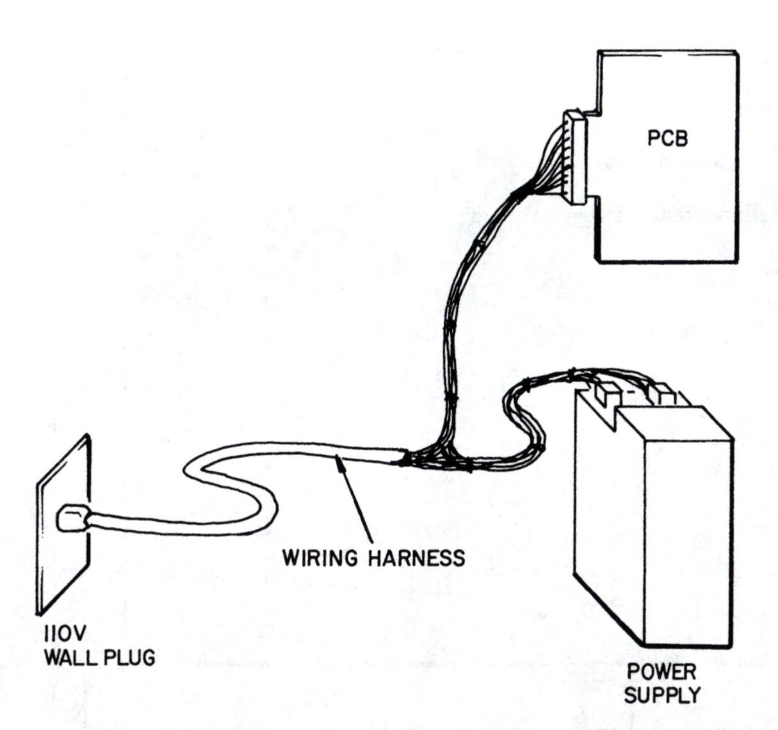

Figure 7–3 Wiring harness used to connect units together

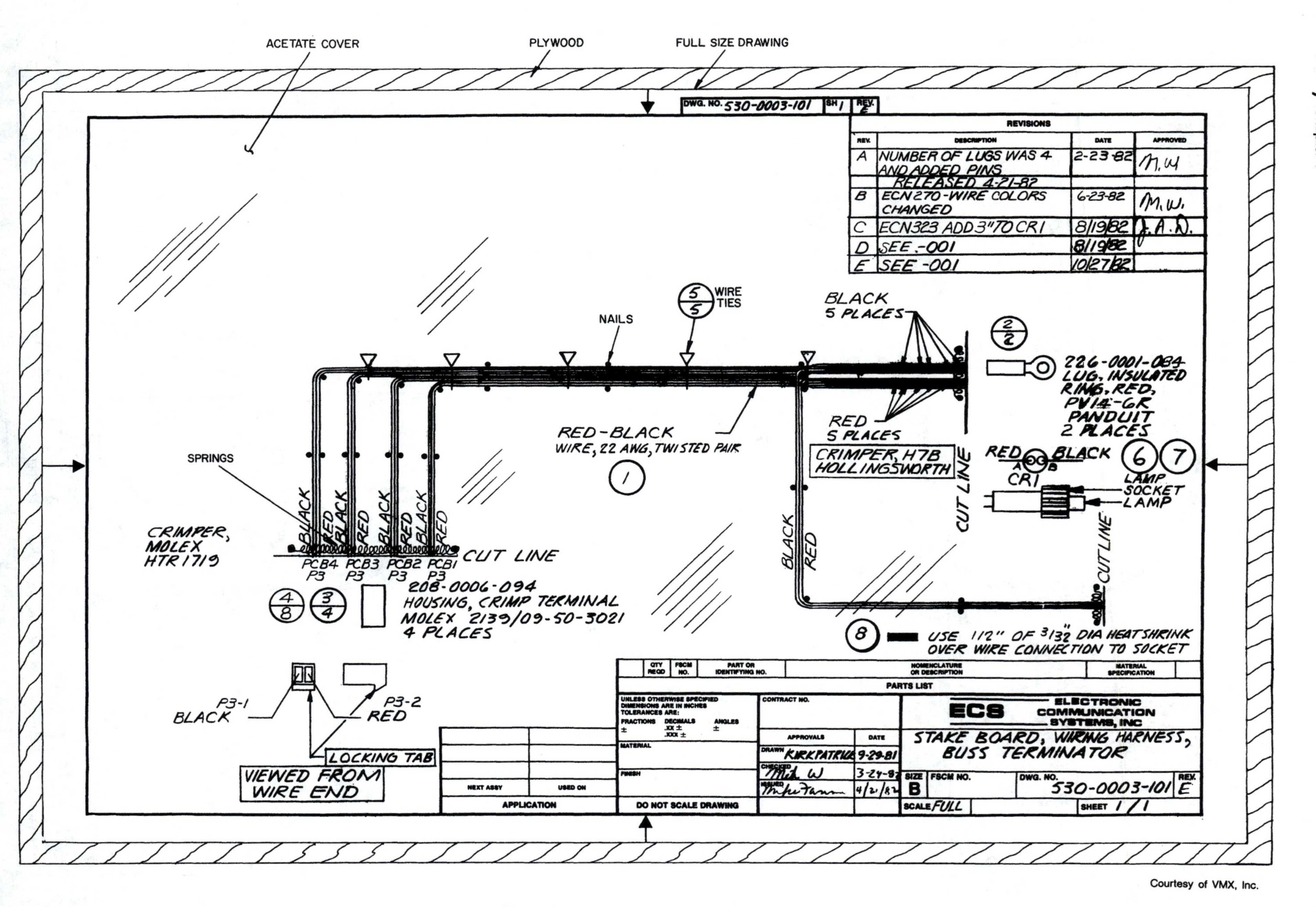

Figure 7–4 Example of a stake board

Items Identified on Stake Boards

The stake board drawing and the wiring harness drawing identify every wire, every connector, and every pin on the connectors. It also identifies all other items that are used to make the wiring harness, such as wire ties and insulating or other protective coverings. Many of these items are shown in Figure 7–4. Notice that the connector identified as P3 is illustrated on the stake board itself. The connector types and part numbers are shown and every pin has a line pointing to it to identify the color of wire that attaches to that particular pin of the connector. All other connectors are shown, such as the lugs on the black and red wires. The complete length of very long wires is sometimes not shown on the stake board because to show it would make the stake board unnecessarily long. Instead, a note on the stake board indicates the actual length, and the wire is rolled up out of the way.

Uses of Wiring Harness and Stake Board Drawings

Both the wiring harness drawing and the stake board drawing are usually needed. The small wiring harness drawing is used by purchasing and assembly people to order parts and to identify and assemble the wiring harness on the electronic equipment. The stake board drawing is used to build the wiring harness itself. After the many cables for connecting the major parts of complex electronic equipment are built, a document is required showing which cables are needed and how they are connected. These documents called interconnecting lists or diagrams, are used by assemblers, technicians, and servicing personnel. One type of interconnecting diagram is the point-to-point wiring diagram.

POINT-TO-POINT WIRING DIAGRAM

The *point-to-point wiring diagram* is a diagram showing the connections among all of the electronic units and each of the wires connecting those units, Figure 7–5.

For example, TRANSPONDER NO. 1 of this figure has three connectors: J109, J110, and J111. P109 plugs into J109 and goes to a signal source called TRIGGER IN. This signal source is often common to several units and is not always shown. The cable going to TRIGGER IN is numbered 510–0002, and it is colored red. P110 plugs into J110 and goes to an ANTENNA by way of a cable numbered 510–0001. P110 has only one pin on it; the wire connecting to that pin is colored red. P111 plugs into J111 and connects TRANSPONDER NO. 1 to TRANSPONDER NO. 2, the CONTROL PANEL, and the PRIMARY POWER source which is a distribution box or a power supply. The cable which connects all of these units together is NUMBER 510–0003. It has three connectors on it: P111 of TRANSPONDER NO. 1, P111 of TRANSPONDER NO. 2, and P114 of the CONTROL PANEL. The color of each wire is shown, and both ends of each wire are shown connected. Notice that the wire connected to Pin 1 of P111 on TRANSPONDER NO. 2 is spliced into the wire going from Pin 1 to P111 on TRANSPONDER NO. 1; one going to Pin 1 of P111 on TRANSPONDER NO. 2. All other connections on this point-to-point diagram are shown in the same manner as those just described.

A wire list, a variation of the point-to-point diagram, is sometimes used instead of a diagram. A wire list is usually compiled for every cable used to connect the electronic units.

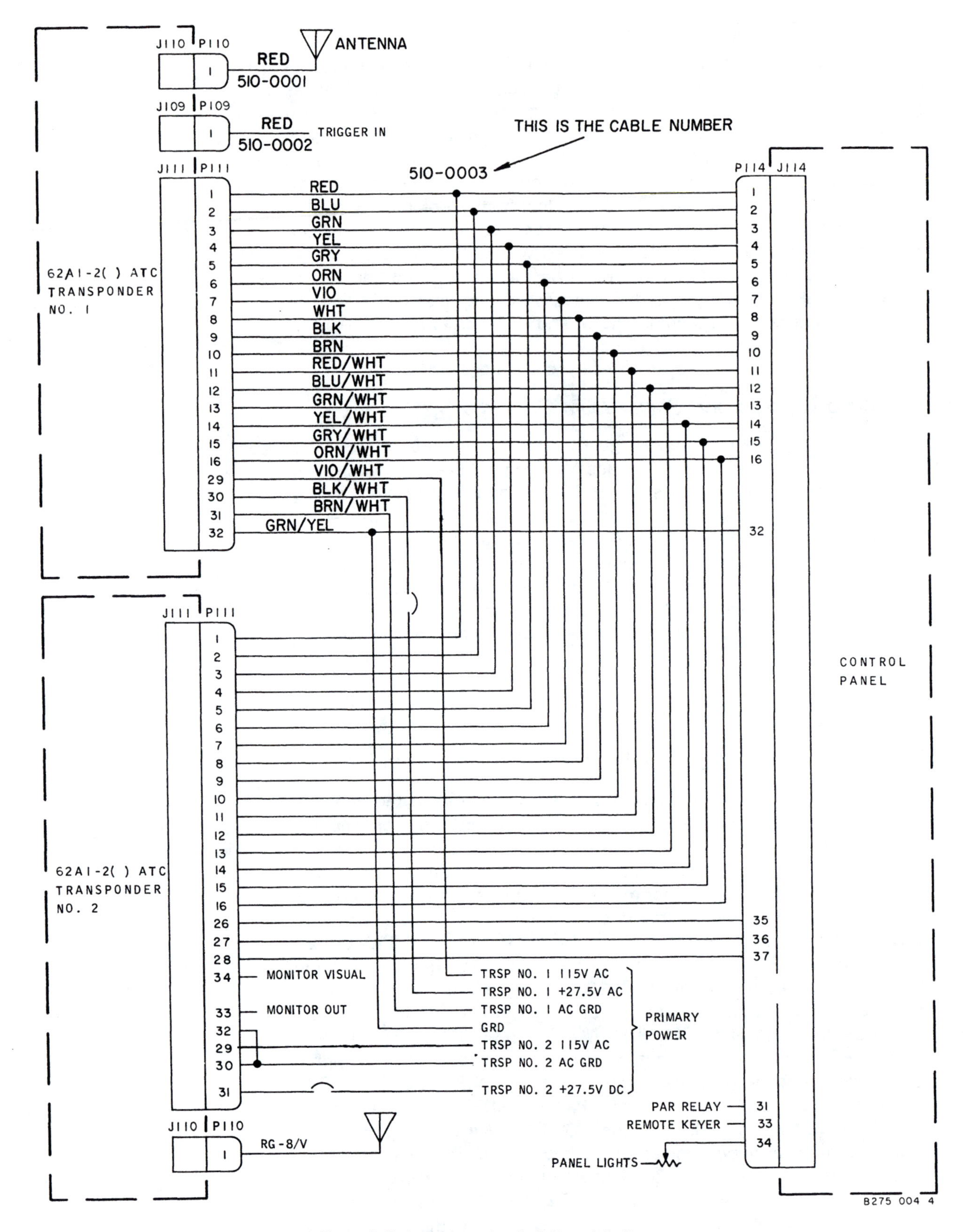

Figure 7-5 Interconnecting (point-to-point) diagram

Figure 7–6 is another type of point-to-point wiring diagram which does not show the wires connecting, but instead tells where each wire goes (its destination).

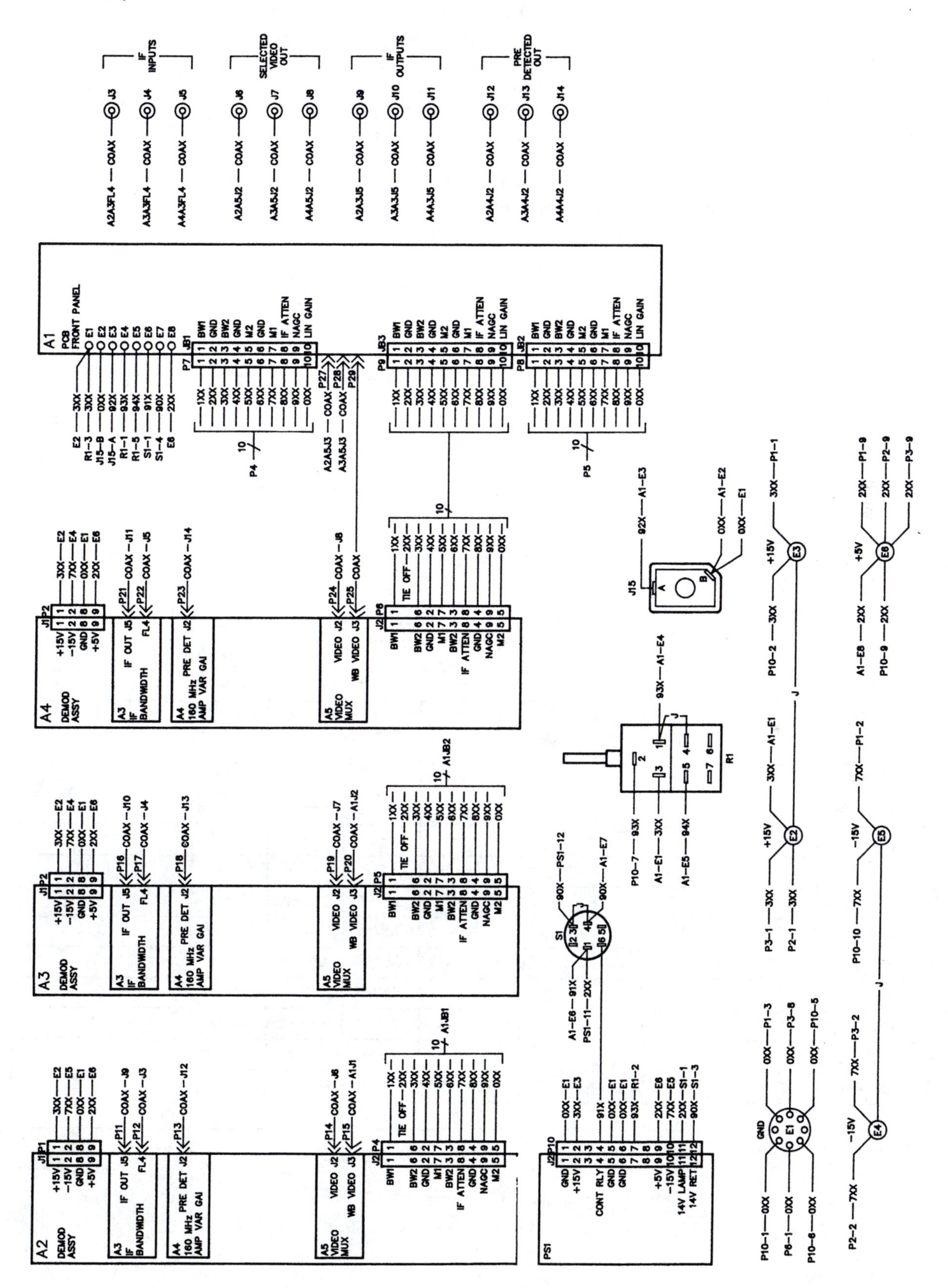

Figure 7–6 Point-to-Point Wiring Diagram

WIRE LIST

The wire list shown in Figure 7–7 contains the same information as a point-to-point wiring diagram. In many cases, a wiring list is used instead of a point-to-point drawing. As can be seen in the figure, every wire in the wiring harness is shown on the wire list. Each wire is described in detail. Details include the wire number, its gauge (diameter), its color, its length, where it comes from and where it goes to, and whether it has a connector attached to the end. (Wire gauge sizes are shown in the appendix; the smaller the number, the thicker the wire.)

In most drawing systems, all wiring harnesses have a wire list and many complex harnesses also have an interconnecting diagram (either point-to-point or highway diagram) to support the wire list. Often, an interconnecting diagram is used to show several wiring harnesses and cables connecting units. These diagrams are often simplified into a drawing called a highway diagram.

WIRE LIST

ECS — ELECTRONIC COMMUNICATION SYSTEMS, INC

REV LTR: K | DATE:

WIRE LIST NO. 000-0002-001 (MASTER)
FOR CABLE PART NO. 700-0002-001 TITLE AC DISTRIBUTION BOX
ENGINEER ______ DATE 8-27-81 SHEET 1 OF 2
RELEASE ______ DATE ______

WIRE NO.	DESCRIPTION	P/N OR STYLE	LENGTH	FROM		TO		REMARKS
				TERM P/N	REF DES	TERM P/N	REF DES	
1	6AWG GRN	UL-1028-8 133/29	6 1/2"	—	J1-GND	—	TB1-14	
2	6 AWG GRN	UL-1028-8 133/29	7 1/2"	—	TB1-16	PN8-10R	TB2	
3	8AWG GRN	"	6"	—	TB1-15	—	P1-GND	
4	8AWG GRN	"	10"	—	TB1-13	—	P2-GND	
5	6 AWG BLK	"	6¼"	—	J1-X	PN6-10R	CB2-1	
6	6 AWG BLK	UL-1028-8 133/29	9"	—	TB1-24	PN6-10R	CB2-3	
7	8 AWG BLK	"	4⅞"	—	TB1-26	PN8-10R	CB1-1	
8	8 AWG BLK	"	4¼"	—	TB1-2	PN8-10R	CB1-3	
9	8AWG BLK	UL1028-8 133/29	5⅝"	—	P1-X	—	TB1-25	
10	8AWG - BLK	"	9½"	—	P2-X	—	TB1-23	
11	6AWG WHT	UL	5"	—	J1-W	PV6-10R	CB2-2	
12	6 AWG, WHT	UL	6"	—	TB1-20	"	CB2-4	
13	8 AWG, WHT	UL1028-8 133/29	3¾"	—	TB1-22	PN8-10R	CB1-2	
14	8 AWG, WHT	"	3"	—	TB1-8	"	CB1-4	
15	8 AWG, WHT	"	5 1/4"	—	P1-W	—	TB1-21	
16	8 AWG WHT	UL1028-8 133/29	9"	—	P2-W	—	TB1-19	
17	16 AWG BLK	UL1429-16 19/29	5"	DV14-250F	CB3-1	PN14-8SLF	P3-1	

Courtesy of VMX, Inc.

Figure 7–7 Wire list

HIGHWAY DIAGRAM

A *highway diagram* shows the same information as a point-to-point diagram, but the complete length of each wire is not shown. As can be seen in Figure 7–8, the units where each wire or cable begins and ends are shown, as well as the connectors of the units.

A highway diagram may be used to connect various electronic units or components within a single electronic unit. When the diagram is used to connect components within a single unit, the connector, such as the connector P1, is shown with all of its pins. A line leads from each pin of P1 and stops next to the information which identifies the wire and describes where it is going. For example, the wire leading from P1–1 on the receiver shown in Figure 7–8 goes to Pin 10 of a connector labeled P6. The wire is 24 gauge (quite thin) and it is white. The angle leading into the highway shows that the reader of the drawing must follow the highway in the direction of the angle to find the wire destination on P6. Connector P6 pin 10 then

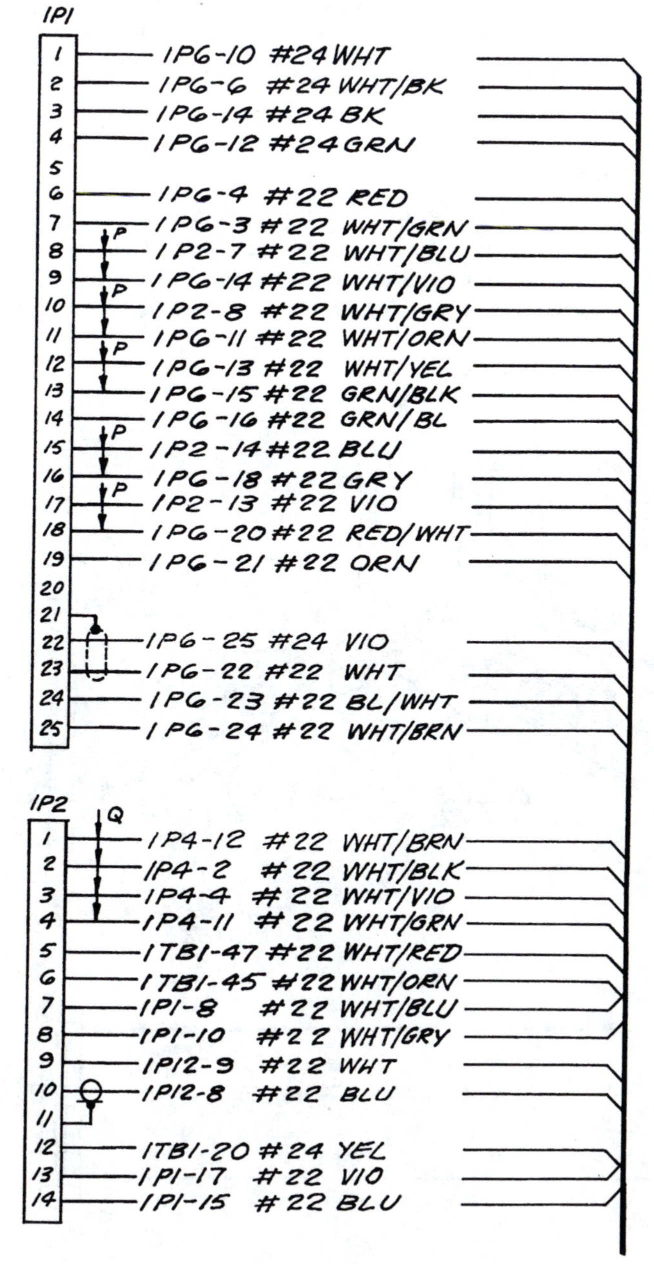

Figure 7–8 Highway diagram

would have a line showing its destination as pin 1 of P1. It would also show 24 gauge wire and the same color as shown on the other end of the wire.

Highway Diagram Details

Other details shown on Figure 7–8 need to be explained. The short lines labeled P indicate by their arrows that two wires are twisted together. This arrangement is called a *twisted pair.* In Figure 7–8, lines 8 and 9 are twisted together, as are 10 and 11, 12 and 13, 15 and 16, and 17 and 18.

The oblong-shaped dashed lines represent a shield. The wires coming out of pins 22 and 23 have a wire mesh or similar material covering them as a shield to prevent the electrical characteristics of the current flowing through these two wires from interfering with the electrical functions of other components or vice versa. In this example, the shield is connected to pin 21 of P1. Wires leading from pins 1, 2, 3, and 4 of P2 are also twisted together. This is a twisted quadruple and is labeled Q. Pins 10 and 11 are connected together to form a coaxial cable which is represented by the symbol shown around the line coming out of pin 11.

RIBBON CABLES

Ribbon cables are manufactured by molding several separate conductors (wires) into a thin flat band. These conductors are then attached to a connector such as the one shown in Figure 7–9. Two of the most common sizes of ribbon cable are 50 conductor and 25 conductor.

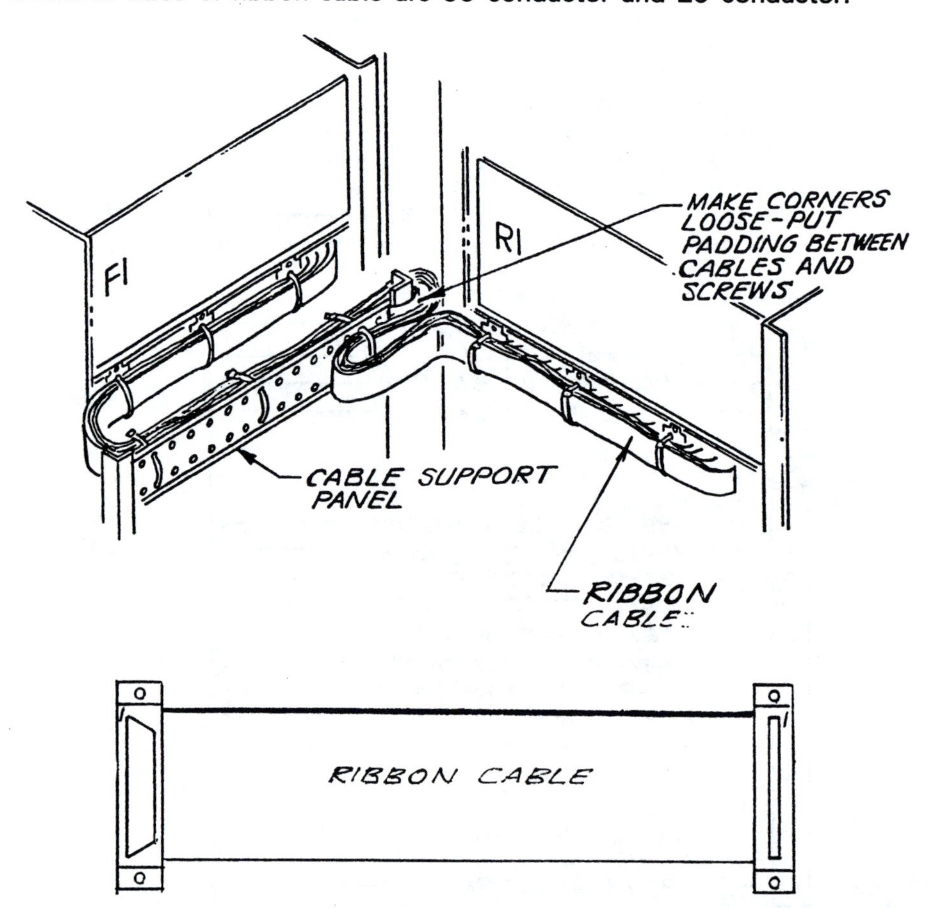

Courtesy of VMX, Inc.

Figure 7–9 Ribbon cables

In many cases, use of ribbon cables makes connections from printed circuit boards and other components within a system much simpler than with other methods. The connectors are easily attached to the cable and often one ribbon cable may have several connectors on it. Mating connectors for connecting the ribbon cables to other parts of the system are often permanently attached to printed circuit boards or sheet metal brackets.

Examples of the manner in which ribbon cables are used are shown in Figure 7–10.

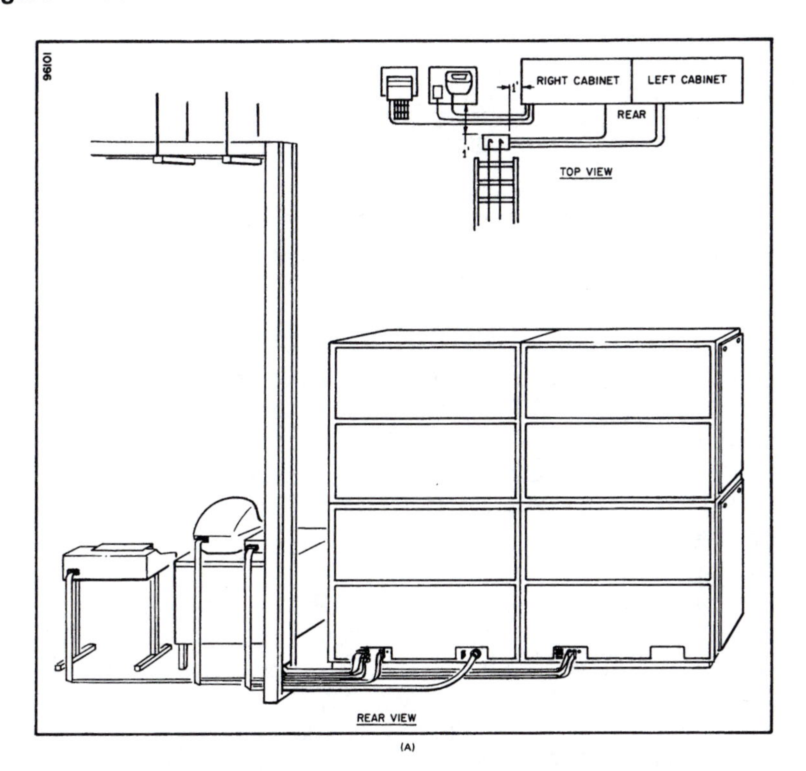

(A)

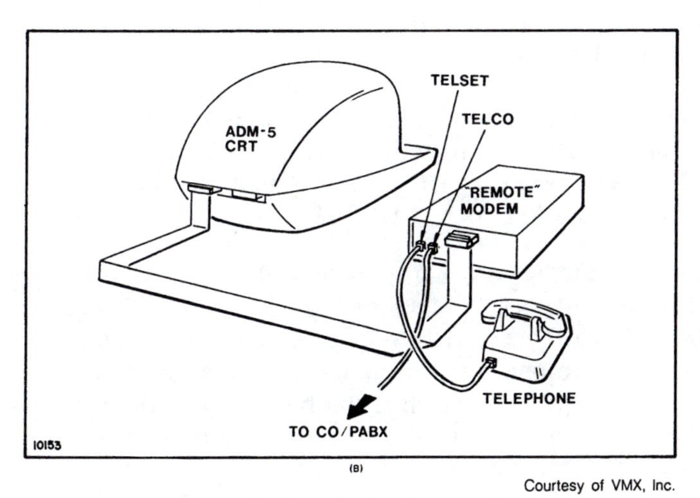

(B)

Courtesy of VMX, Inc.

Figure 7–10 Uses for ribbon cables

EXERCISE 7-1

On a separate piece of paper,

1. list three ways in which a point-to-point diagram differs from a highway diagram.
2. describe the major use of a wiring harness diagram.

EXERCISE 7-2

Redraw the point-to-point diagram shown on the following page. Use three separate highways. Make boxes enclosing units twice as thick as connecting lines, and the highway itself twice as thick as the boxes. Study the EXAMPLE and the layout shown beneath the point-to-point diagram and use these illustrations as guides.

Destinations should include *unit, connector number, pin number,* and *wire color.* Make the lead into the highway at a 45° angle. Point it in the direction that the user of the drawing must follow on the highway to find the destination. Put the drawing on a 17″ × 22″ sheet. Make the boxes and connectors as large as needed to allow all lettering to fit comfortably without crowding. Make all lettering $\frac{1}{10}$″ high.

This is an exercise in changing a point-to-point diagram to a highway diagram. The highway diagram should resemble the EXAMPLE shown. The destinations, which are shown by the lettering from the middle of each one of these lines, should include the unit, the connector number, and the wire color. Look at UNIT 1 in the example. For example, the wire leading out of pin 1 of that connector goes to UNIT 2, CONNECTOR 1, which is also labeled P1. It is connected to pin 1 on that connector and the wire is blue. Now look at pin 4 on UNIT 1. Pin 4 on plug 1 goes to UNIT 3, CONNECTOR P1, pin 1, and the wire color is blue and white. Notice that the wire as it enters the highway shows the direction that you must follow on the highway to find the destination. Looking at UNIT 2 on the EXAMPLE, you will see that Pin 1 goes to CONNECTOR P1, pin 1 on UNIT 1. The angle points to the left, so follow the highway to the left to find UNIT 1.

The only differences from the EXAMPLE are some individual components shown inside the UNITS. For example, UNIT 4 has R1 and R2 in it. The destination coming from UNIT 2 to R1 on UNIT 4 should say 4–R1 and then the wire color red. Remember to use three separate highways for this diagram. You could do it with one highway (which is the best way to draw this diagram). A single highway, however, does not give you the problems for practice that you will have with three highways. This is a simple example for this type of diagram, and it will give you the experience you will need to draw a much larger diagram of the same type.

Once again, draw dark lines, following the directions for the sizes of lines for the boxes and interconnecting lines. Also, follow the directions for the UNIT labels and the other lettering. If you use a computer for this drawing use one layer for the boxes or units and a different layer for text and interconnecting lines. You may then plot the drawing with a thick pen for the "units" layer and a thin pen for the text and interconnecting lines. The highway should be even thicker than the unit outlines.

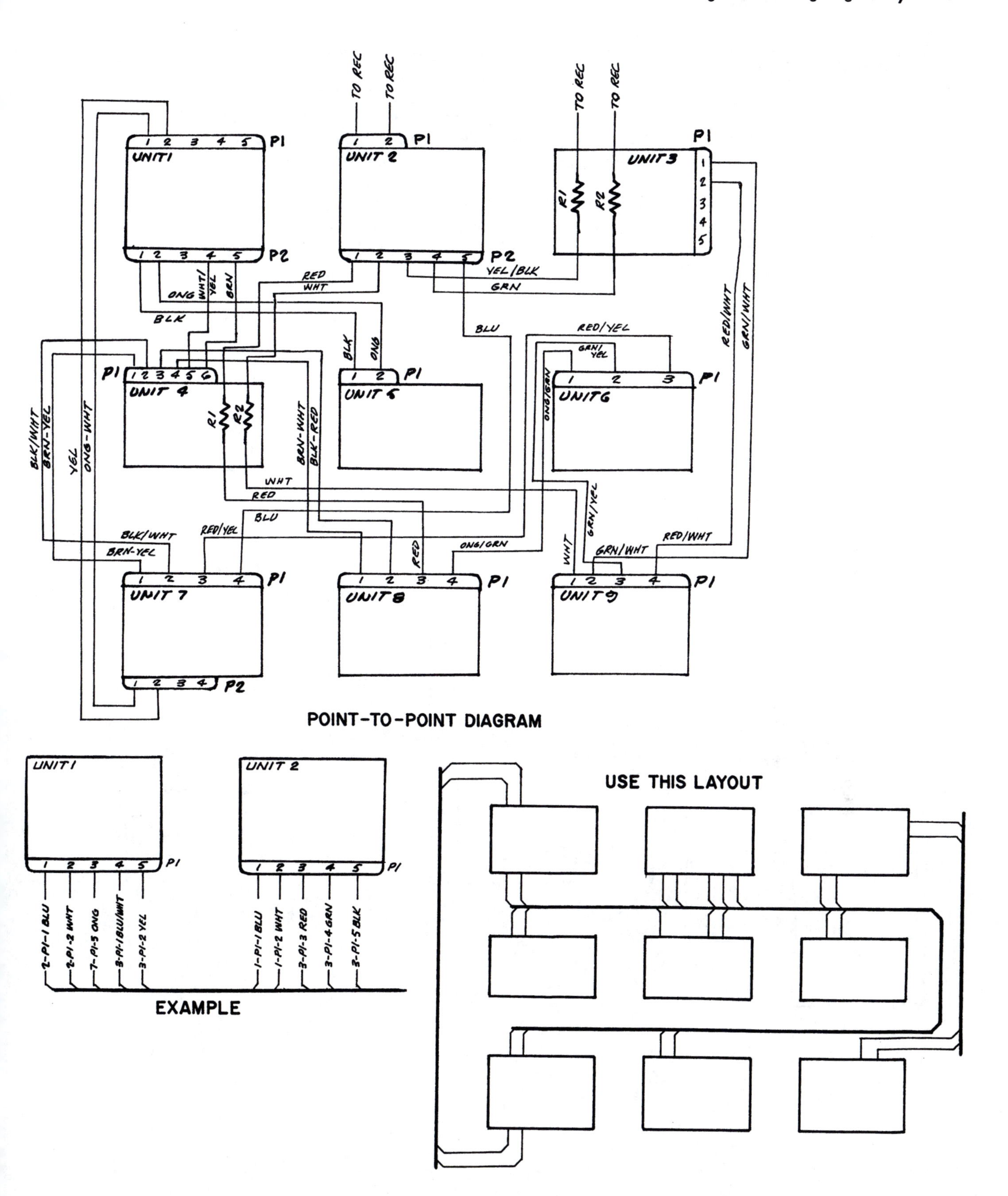
TO REC
TO REC
TO REC
TO REC
UNIT 1
P1
P2
UNIT 2
UNIT 3
R1
R2
YEL/BLK
GRN
RED
WHT
ONG
WHT/YEL
BRN
BLK
BLU
RED/YEL
RED/WHT
GRN/WHT
GRN/YEL
UNIT 4
UNIT 5
UNIT 6
ONG/GRN
BLK/WHT
BRN-YEL
YEL
ONG-WHT
BRN-WHT
BLK-RED
UNIT 7
UNIT 8
UNIT 9
POINT-TO-POINT DIAGRAM
2-P1-1 BLU
2-P1-2 WHT
7-P1-5 ONG
3-P1-1 BLU/WHT
3-P1-2 YEL
1-P1-1 BLU
1-P1-2 WHT
3-P1-3 RED
3-P1-4 GRN
3-P1-5 BLK
EXAMPLE
USE THIS LAYOUT

EXERCISE 7–3

Redraw the following sketch as a wiring diagram similar to the example shown as Figure 7–6. Use no highways. Make boxes enclosing units twice as thick as connecting lines. Study Figure 7–6 and use it as a guide.

Destinations should include *unit* (A1, A2, PS1, FL1, CB1, S1), *connector number, pin number,* and *wire color.* Put the drawing on a 17″ × 11″ sheet. Make the boxes and connectors as large as needed to allow all lettering to fit comfortably without crowding. Make all lettering $\frac{1}{8}$″ high.

Place the connectors A1/J2 and A1/J1 inside of A1 similar to the connectors shown on PS1.

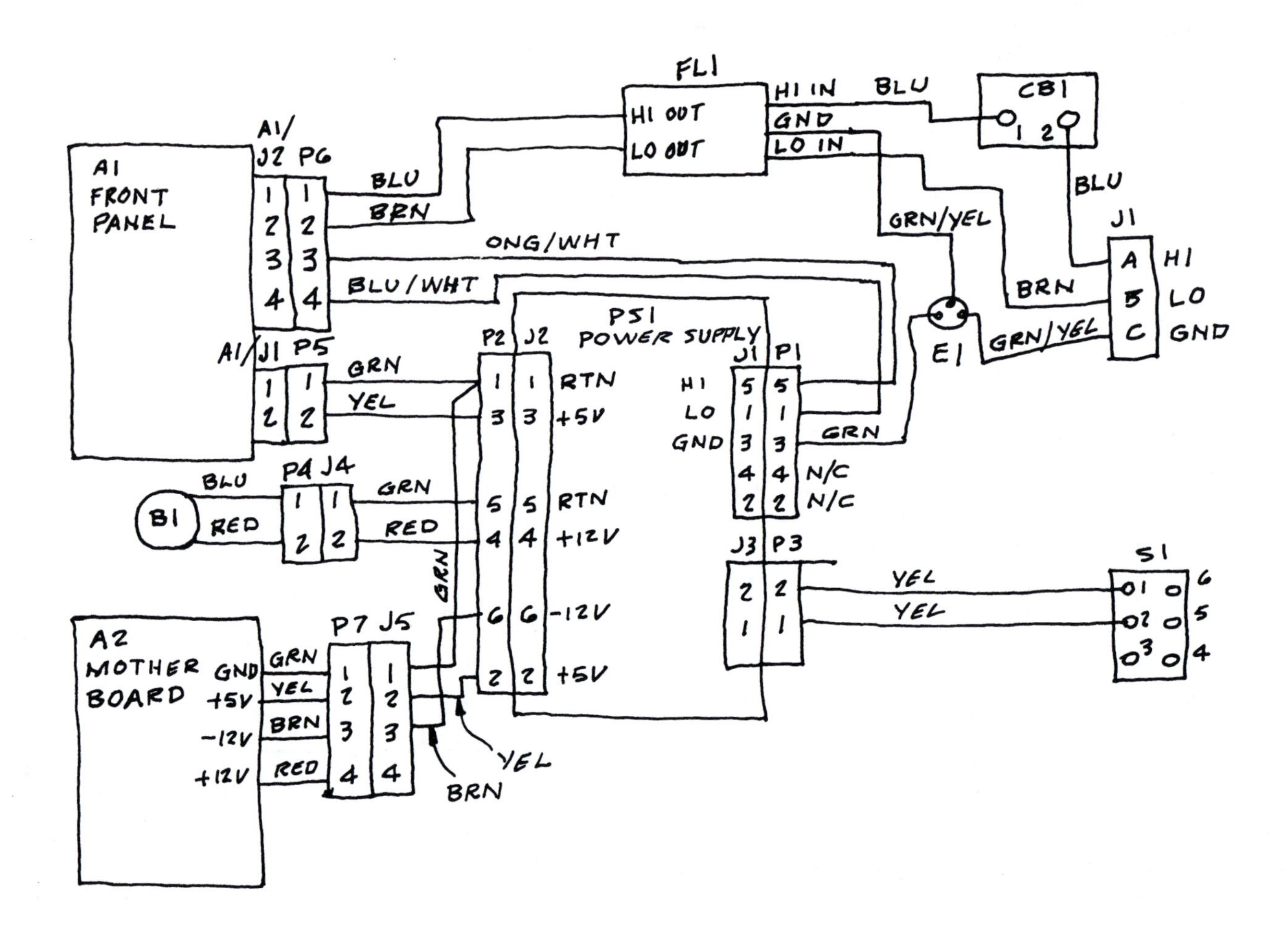

EXERCISE 7–4

Redraw the following sketch as a formal wiring diagram similar to Figure 7–6. Use no highways. Be sure to show the wire color on both ends of each wire (you will have to search for it in some cases). Make all boxes twice as thick as connecting lines. Destinations should include unit, connector number, pin number, and wire color. Make the boxes and connectors as large as needed to allow all lettering to fit easily without crowding. Put the drawing on an 8½" × 11" sheet lengthwise.

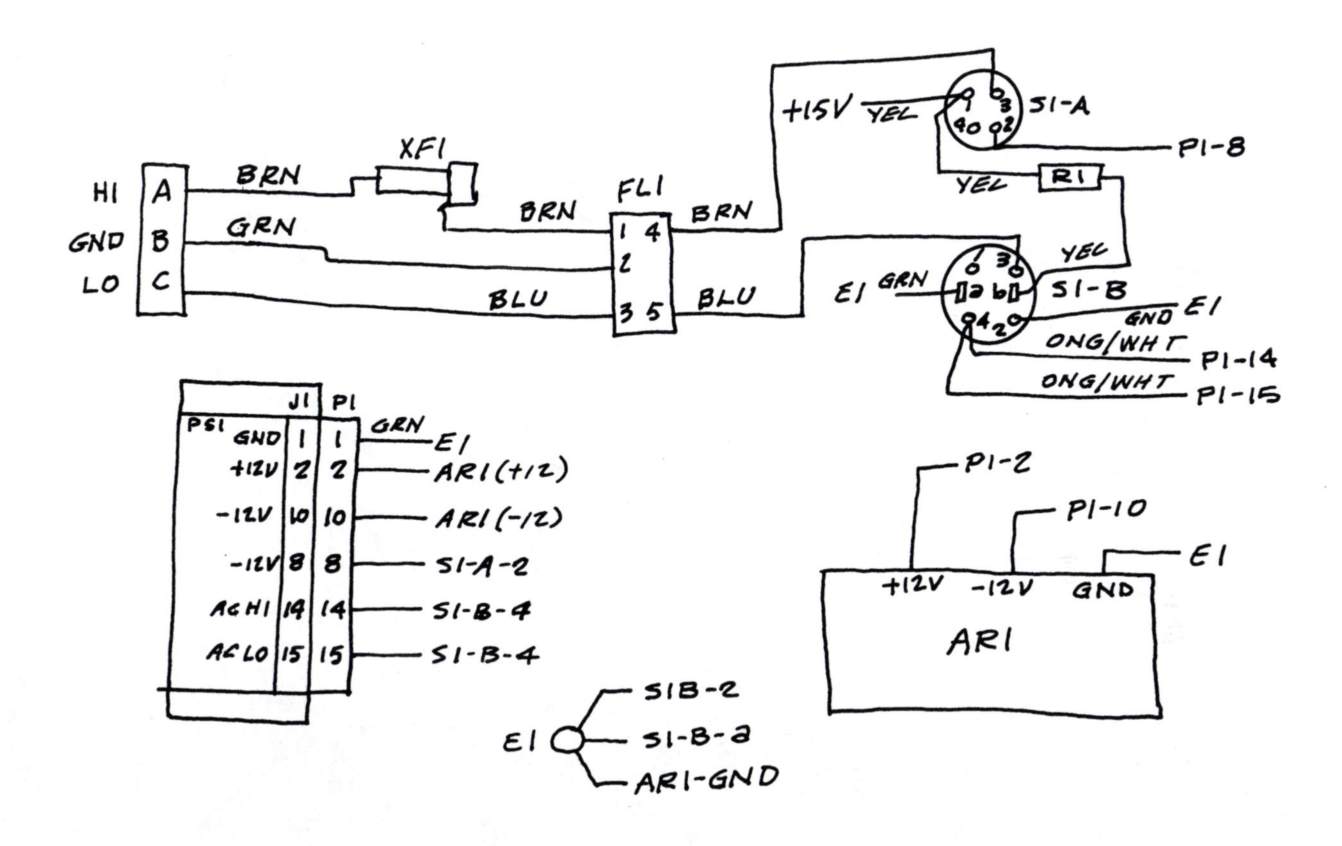

Chapter 8

Introduction to Printed Circuit Boards

OBJECTIVES

After completing this chapter, you will be able to

- ✓ describe the basic function of a printed circuit board.
- ✓ name three types of printed circuit boards.
- ✓ list three types of drawings which must be used to build a printed circuit board.
- ✓ list five steps in manufacturing a printed circuit board.

INTRODUCTION

In this chapter you will learn what a printed circuit board (PCB) is used for, the three basic types of PCBs, the types of drawings needed to build a printed circuit board, and how they are manufactured. This explanation begins with a simple illustration of just what a printed circuit board is.

FUNCTION OF A PRINTED CIRCUIT BOARD (PCB)

A printed circuit board (PCB) functions to connect components in an electronic circuit. Figure 8–1 shows the evolution of a simple, single-sided printed circuit board. The schematic diagram at (A) is where every PCB begins. An engineer or technician designs the circuit using this drawing form. The drafter or PCB designer lays out the board as a component assembly (B), and then the board is manufactured as shown in (C). Notice that the connections in the component assembly drawing are exactly the same as those of the schematic diagram. The symbols have been replaced with drawings of real components the shape of those actually placed on the manufactured board.

The board shown in Figure 8–1 is a single-sided board, meaning that the circuitry is on only one side of the board. Notice the circle placed at each end of all resistors and the capacitor. These circles represent holes where the leads or wires from the components pass through the board. These leads are then soldered to the circuit on the other side of the board.

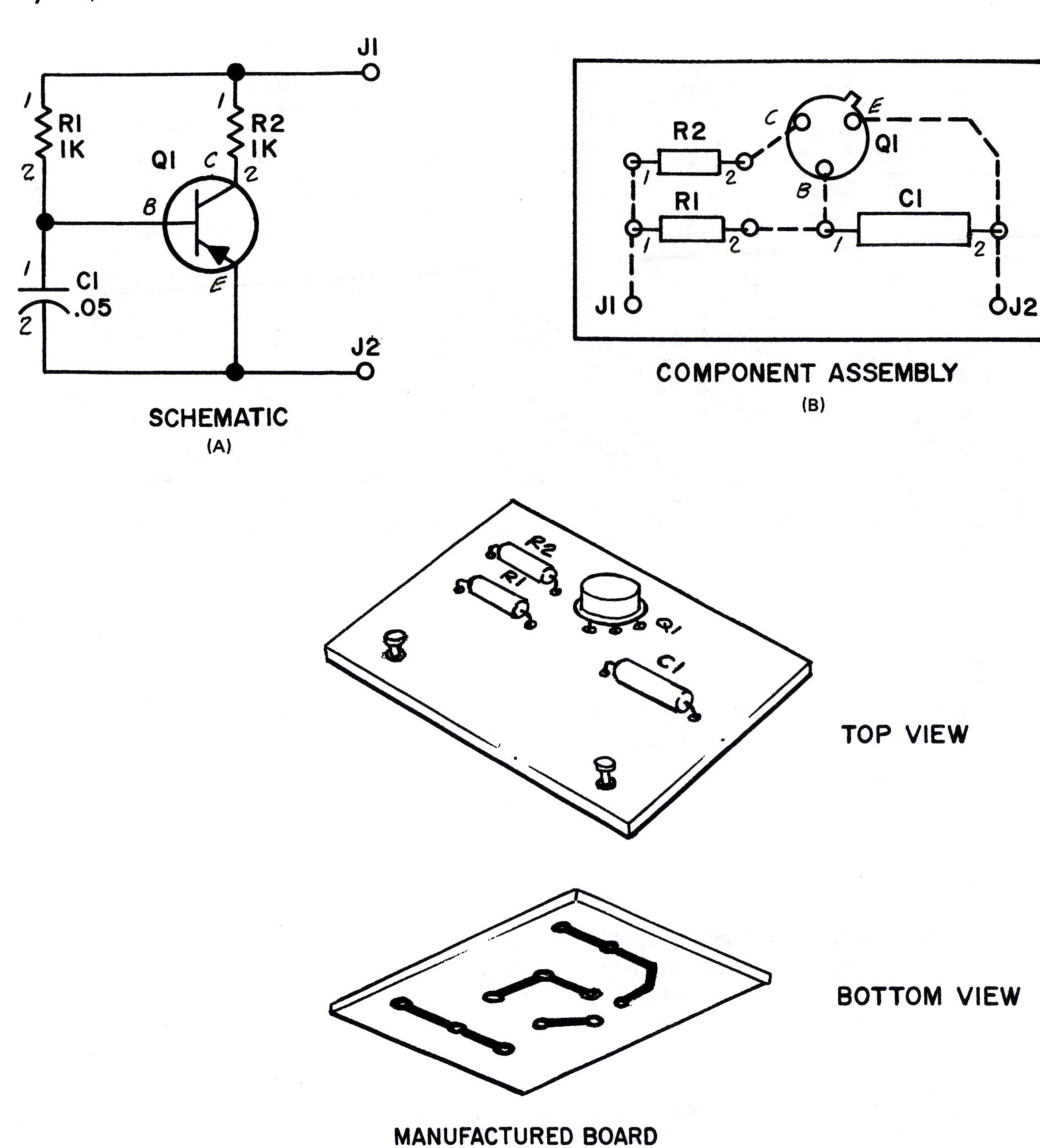

Figure 8–1 A single-sided circuit board

The transistor Q1 has three circles on it for its three leads to pass to the other side of the board. To lay out this board accurately, the designer must have a parts list and specifications which describe the exact size of each component and the size of the manufactured board. In summary, Figure 8–1 shows the schematic diagram created by the engineer, the component assembly created by the PC designer, and the top and bottom view of the manufactured board.

PCB TYPES

Figure 8–2 illustrates the three major types of printed circuit boards. Figure 8–2A is the single-sided board described in Figure 8–1. Components

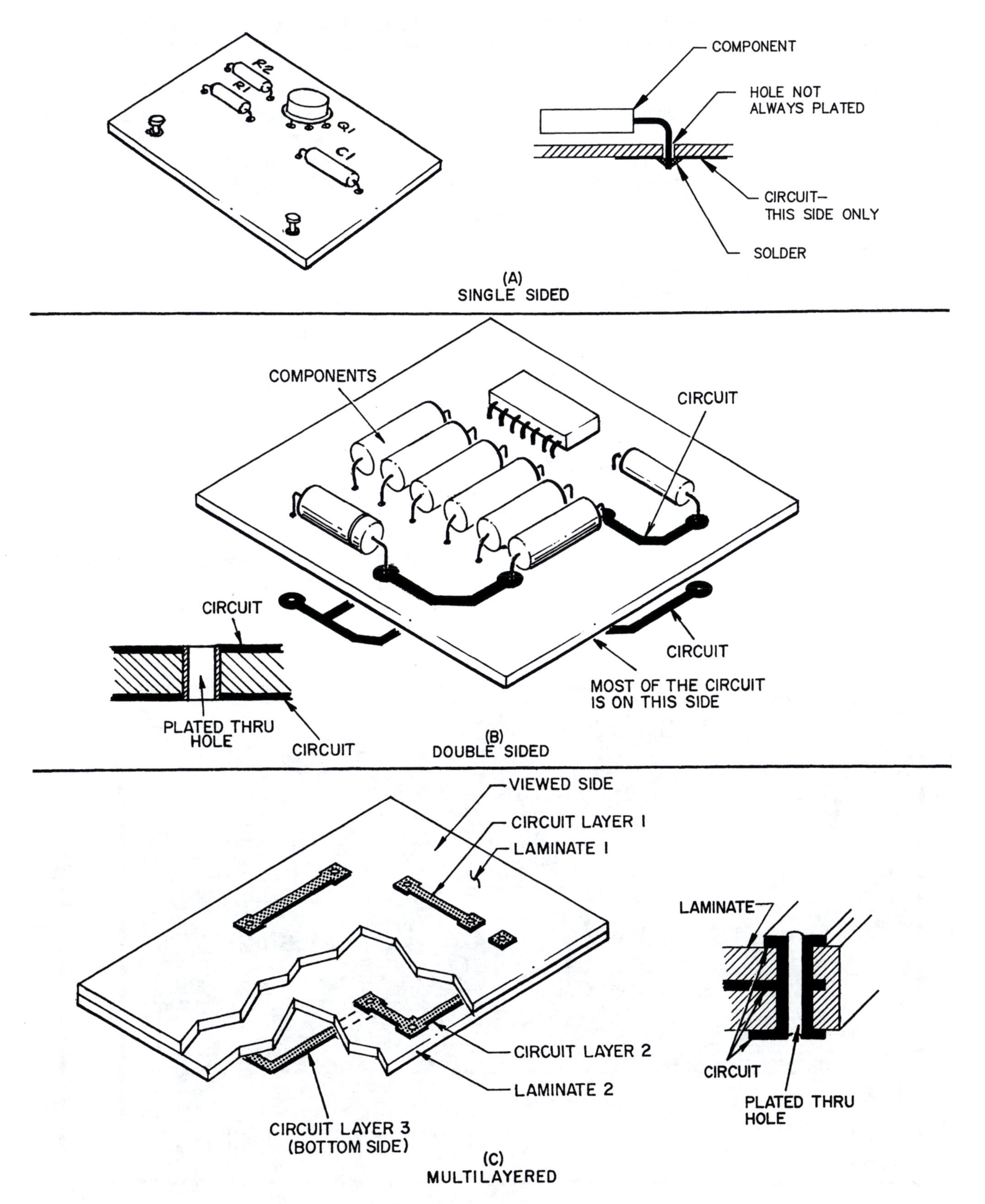

Figure 8–2 Printed circuit board (PCB) types

are mounted on one side and the circuit is etched on the other side. Figure 8–2B is a double-sided printed circuit board showing circuits on both sides of the board. Components are mounted on only one side. Figure 8–2C shows a multilayered board. This is a PCB made of two or more boards which have circuits on one or both sides. These boards are carefully aligned so that the holes in all of them match, and then they are permanently assembled together. Components are not often mounted on both sides of a board, even the multilayered type. Printed circuit connections may be made on any of the several circuit layers. Notice the small sectional views showing plated-thru holes. These views show that electrical contacts are extended from one layer to another using the plating material which conducts electricity.

PCB DRAWINGS

Three major types of drawings are needed to build a PCB. They are the component assembly drawing, artwork, and the drill plan, also called the fabrication drawing.

Component Assembly Drawing

Figure 8–3 shows a component assembly drawing. This type of drawing is produced by breaking the circuit lines into small dots that screen the circuit so that the components may be more easily seen. The components are then drawn in solid or produced photographically from another drawing. All components are then labeled with reference designators to show the assembler where to place the parts on the board which has had holes drilled in the center of each pad. The outline of the board edge is usually shown

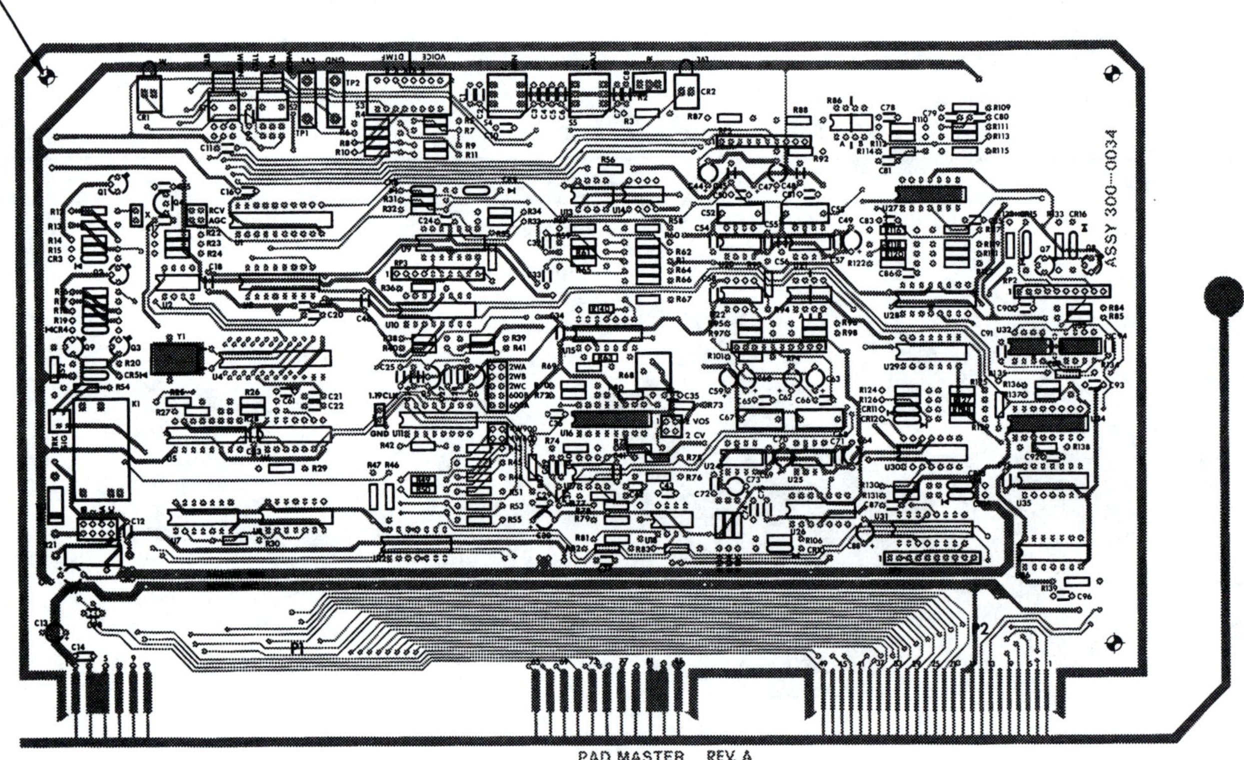

Courtesy of VMX, Inc.

Figure 8–3 Component assembly drawing

with tape identifying the corners of the board. Register targets, such as the three shown on this drawing, are used to make sure that all drawings used to build the board are precisely aligned with one another.

In practice, component assembly drawings, such as those shown in Figure 8–3 and 8–4, are usually the final drawings produced. A preliminary drawing, called a design layout, is similar in appearance to the component assembly. This is the worksheet from which the other drawings are produced to build the board. The first drawing resulting from the design layout usually is the artwork.

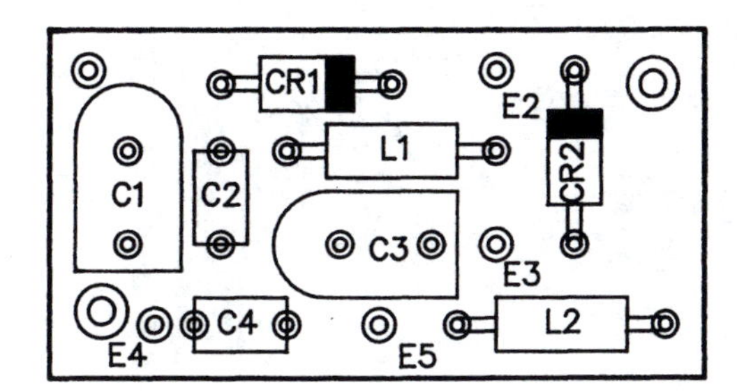

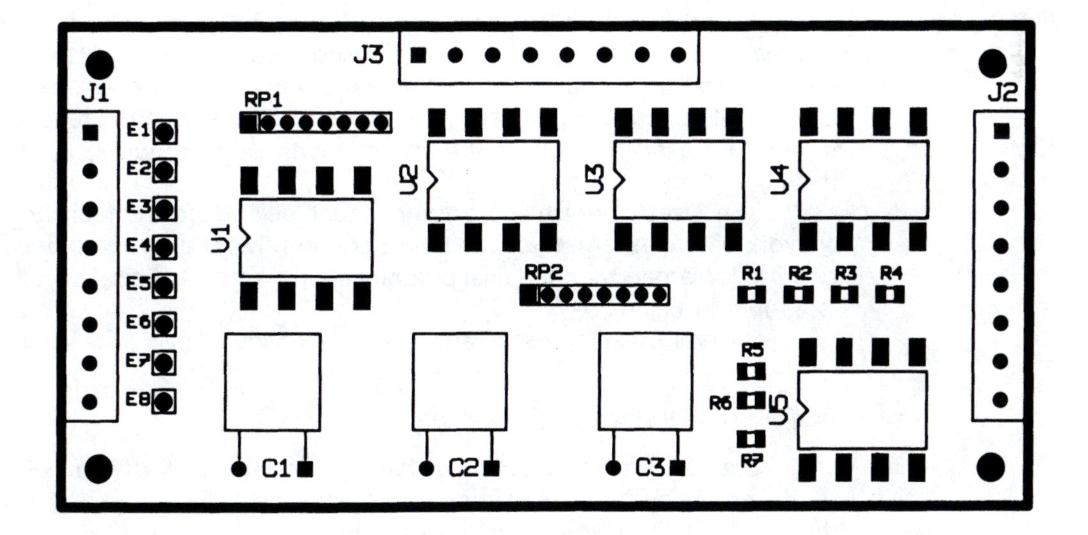

Figure 8–4 Typical assemblies

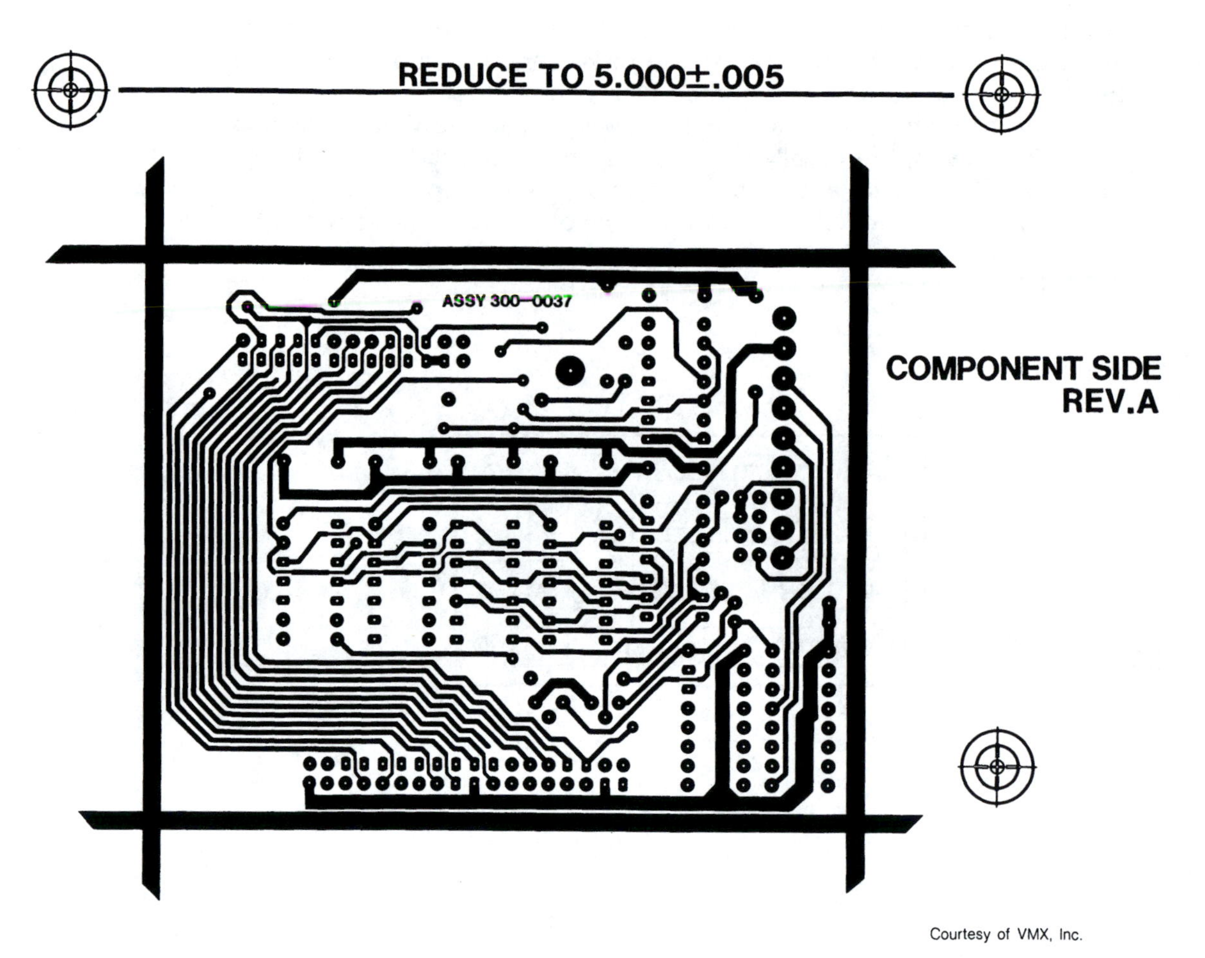

Courtesy of VMX, Inc.

Figure 8–5 Artwork

Artwork

An artwork is a drawing made from a plotter or with tape (with the increasing use of computers, tape is used less and less). Artwork usually shows just the circuit to be etched onto the board. Occasionally, it also shows reference designators. The corners of the board are shown and targets are used to register the artwork with other artwork, if any, Figure 8–5.

The size to which the artwork must be reduced is indicated on the artwork between targets or the board corners. Most artwork is drawn twice as large (or larger) than the final printed circuit board. A typical circuit artwork is shown in Figure 8–6.

Several other types of artwork are described in a later chapter.

Drill Plan (Fabrication Drawing)

The drill plan is a drawing that shows where all of the holes are to be drilled in the board and the size of each hole. Figure 8–7 is a drawing that explains the elements of a drill plan.

A point on the board, such as a hole center, is established as an origin (0–0) from which each hole is located. A letter is placed alongside each hole to identify all holes of the same size. This letter is then placed in a chart called a *hole schedule.* As stated before, boards are often priced on the basis of the number of holes to be drilled, so that the hole schedule is very important from a cost standpoint as well as in assembly. Holes are

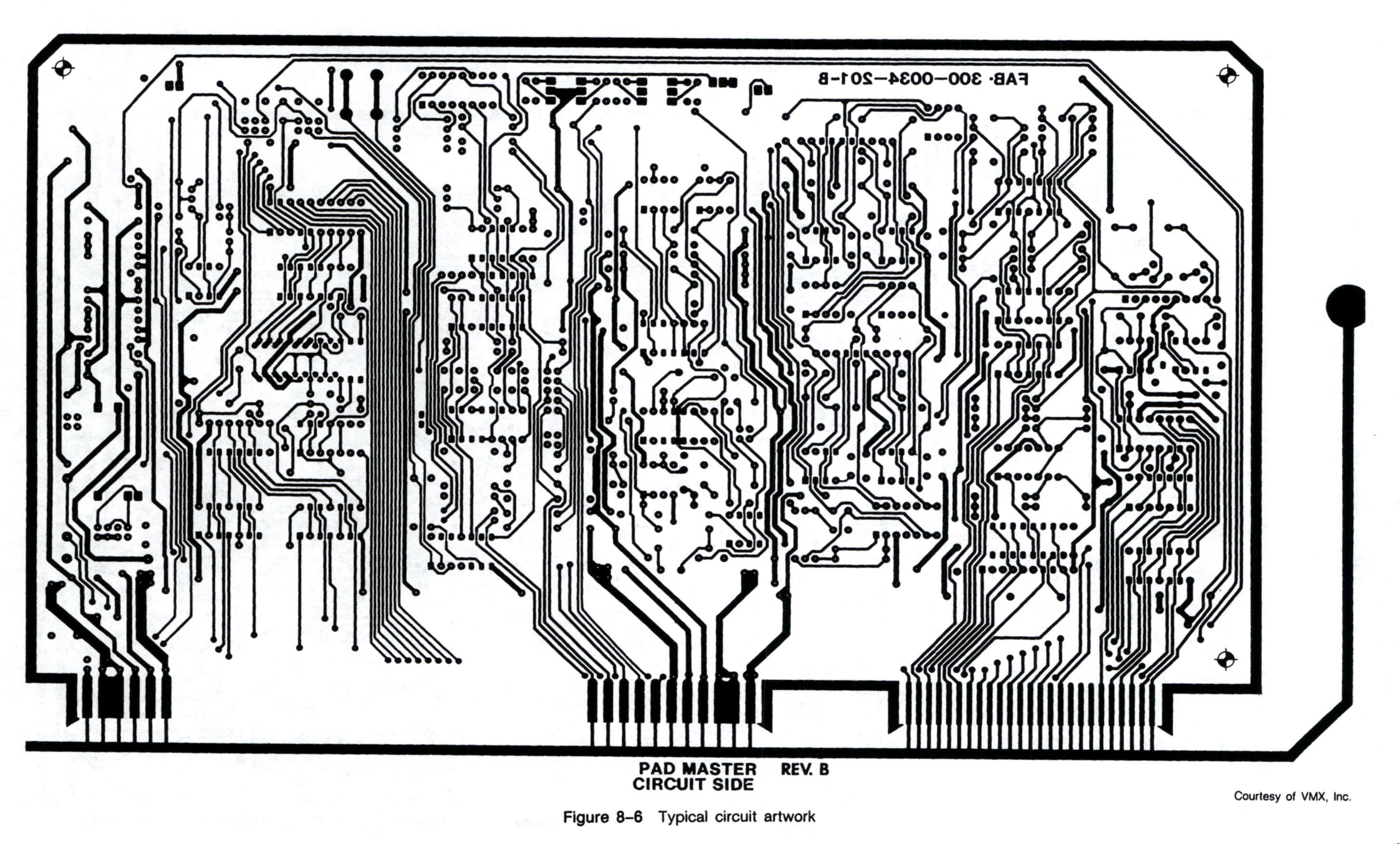

Courtesy of VMX, Inc.

Figure 8–6 Typical circuit artwork

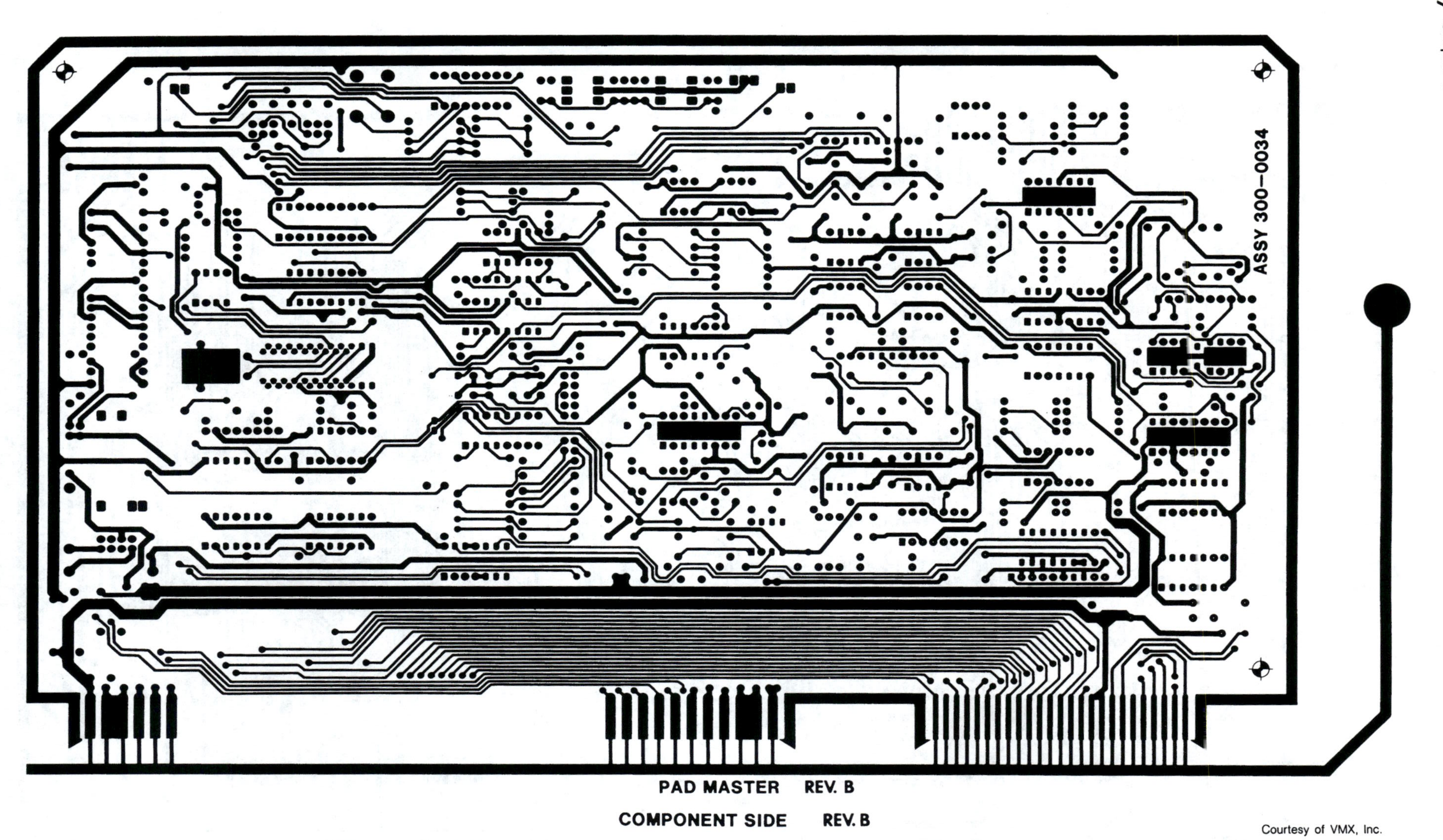

Figure 8-6 (continued)

Courtesy of VMX, Inc.

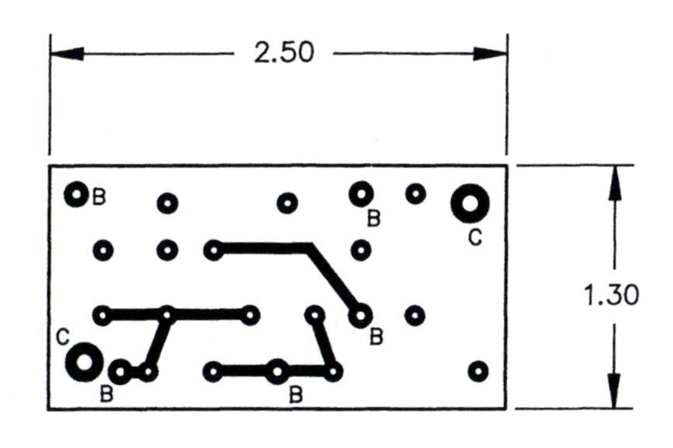

HOLE CHART

LTR	QTY	DESCRIPTION
UNMARKED	X	.032 DIA ± .003
B	9	.042 DIA ± .003
C	2	.094 DIA ±.005

3. HOLE SIZES ARE AFTER PLATING.
2. FRONT TO BACK REGISTRATION: ±.005
1. ALL HOLES ARE PLATED THRU.

NOTES:

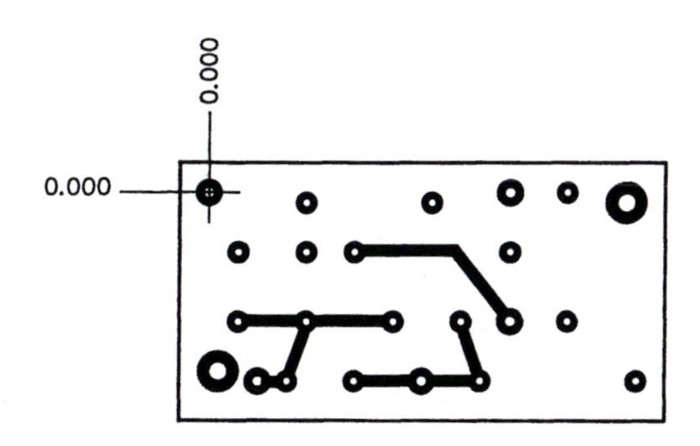

Figure 8–7 Elements of a drill plan

sized according to the lead diameters of the components to be mounted on the board. If the hole is too small, the lead will not go through it. If the hole is too large, the lead will not be firmly bonded to the board.

Figure 8–8 shows a type of drill plan commonly used in industry. The artwork is photographically copied onto another sheet. The board size is shown, and all holes are identified. A hole schedule and notes relating to the board specifications are placed on the drill plan. A drill tape is produced by locating the center of each hole from a full-size picture of the circuit artwork. *NOTICE that most of the holes are not located with dimensions.* They will be located using an etched board so that the hole will be placed in the exact center of the pad (the area surrounding each connection point). Drill plans that you create should resemble Figure 8–7.

PRODUCING THE CIRCUIT

After full-size circuit artwork has been reproduced by photography, the process of manufacturing begins.

Etching the Board

The board, which is made of a special type of plastic or fiberglas and layered with a sheet of copper or silver, is coated with a material called

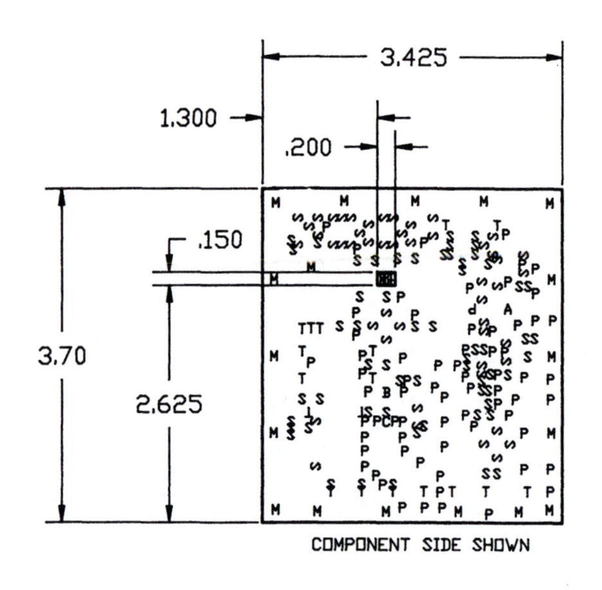

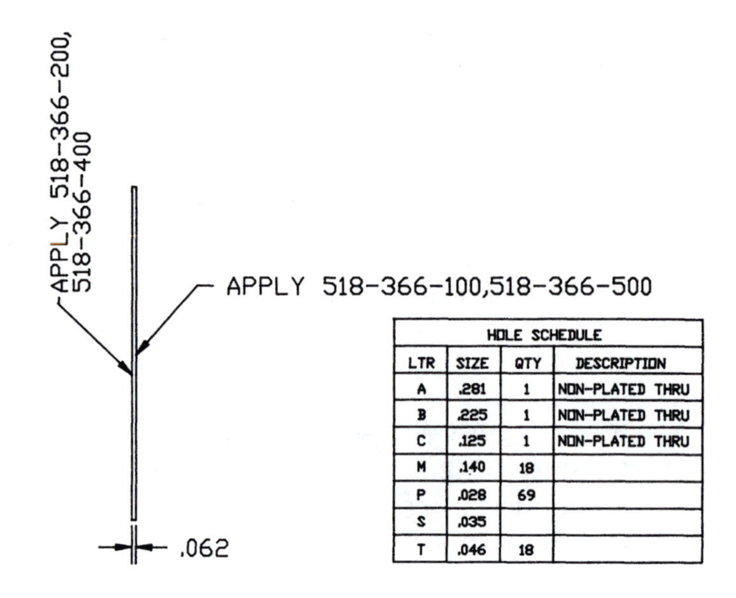

NOTES;UNLESS OTHERWISE SPECIFIED.
1. ALL HOLES TO BE PLATED-THRU.
2. FRONT TO BACK REGISTRATION TO BE WITHIN .005.
3. HOLE SIZES ARE AFTER PLATING.
4. APPLY FAR-SIDE;518-366-100.518-366-500
5. APPLY NEAR-SIDE;518-366-200,518-366-400.
6. WRAP AROUND GND PLANE TO BE USED.

HOLE SCHEDULE			
SYM	DESCRIPTION	QTY	COMMENTS
NONE	.031±.003 DIA AFTER PLT	114	SEE NOTE 3
B	.067±.003 DIA AFTER PLT	16	SEE NOTE 3
C	.125±.005 DIA	3	

NOTES:

1. MATERIAL: EPOXY GLASS LAMINATE FOIL CLAD TYPE FLG.062 Cl/1 PER MIL-P 13949.
2. FINISH: CIRCUITRY-ELECTROPLATED TIN-LEAD PER QQ-S-571 AND REFLOW 500 MICROINCHES MINIMUM THICKNESS AFTER REFLOW.
3. MINIMUM TOTAL WALL THICKNESS TO BE .001 AFTER REFLOW.
4. NO PLATING ALLOWABLE IN HOLES.
5. SYMBOLIZATION TO BE SCREEN PRINTED USING BLACK EPOXY INK AND ITEM 2, COMPONENT SIDE ONLY.

Figure 8–8 Example of an industrial drill plan

photo resist, Figure 8–9. This coating is photographically exposed in a darkroom using a full-size positive reproduction of the circuit artwork. The board is then placed in a solution which etches away the unwanted copper leaving only the copper circuit. Pads on the copper circuit surround areas

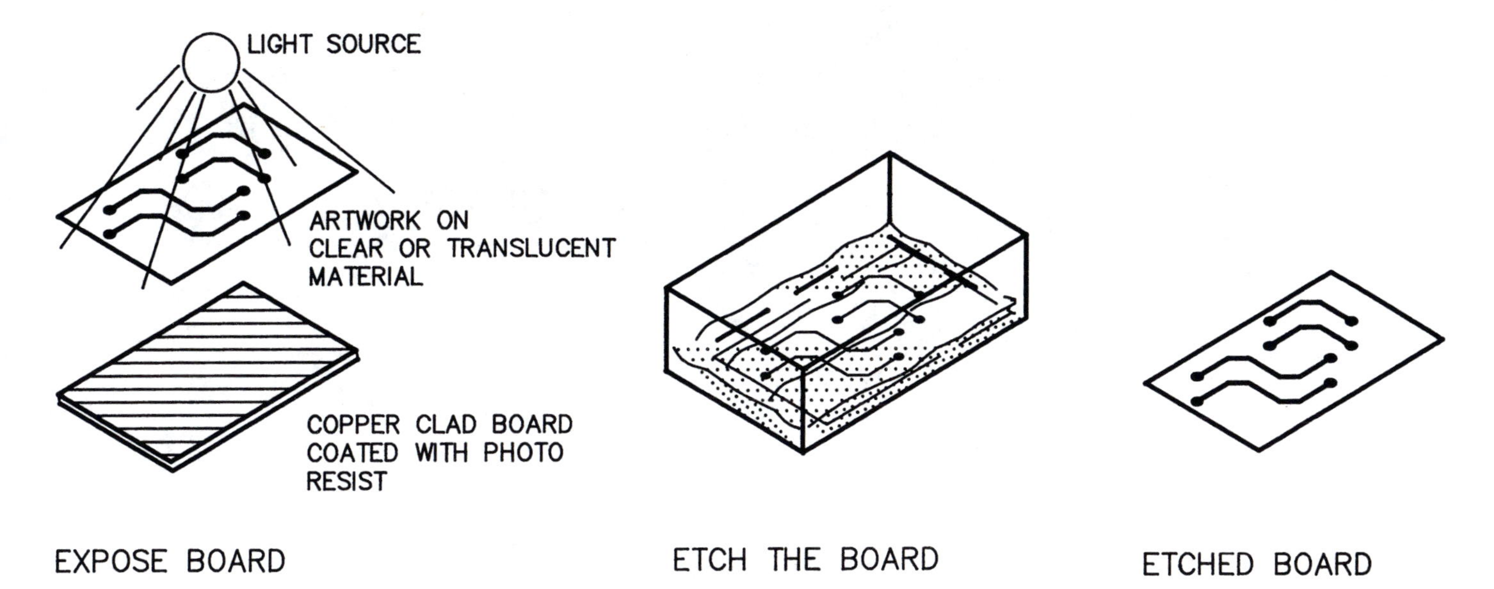

Figure 8–9 Etching board

which must be drilled out to allow component leads to be soldered to the pads.

Drilling the Holes

Using a drill plan drawing and the etched board, a numerically controlled drill is programmed to drill holes in a PCB, Figure 8–10. This procedure allows carefully sized holes to be precisely drilled in the correct locations. These holes are drilled through the center of each pad. After holes have been drilled, the board is ready for the insertion of components (such as resistors, capacitors, connectors, and integrated circuits) into the holes.

Loading Components

Loading components on a PCB, or *stuffing the board* as it is sometimes called, is the process of inserting components into the holes in the board, Figure 8–11. This loading can be done by hand or by machine, according

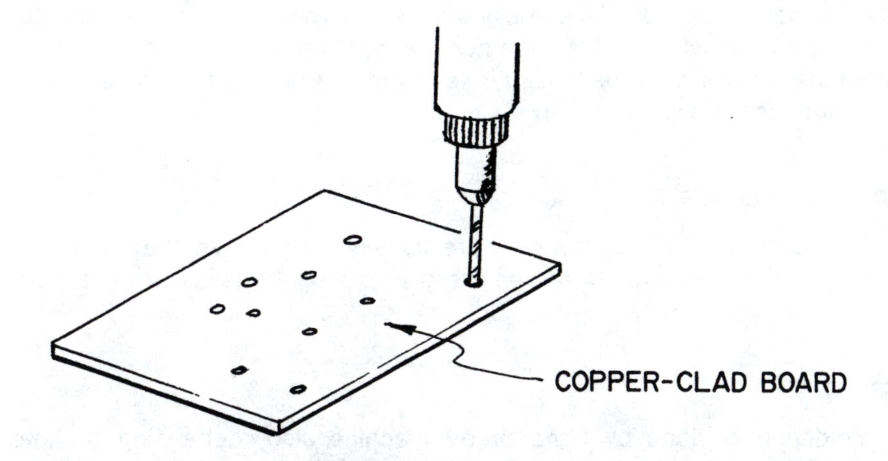

Figure 8–10 Drilling holes in a PCB

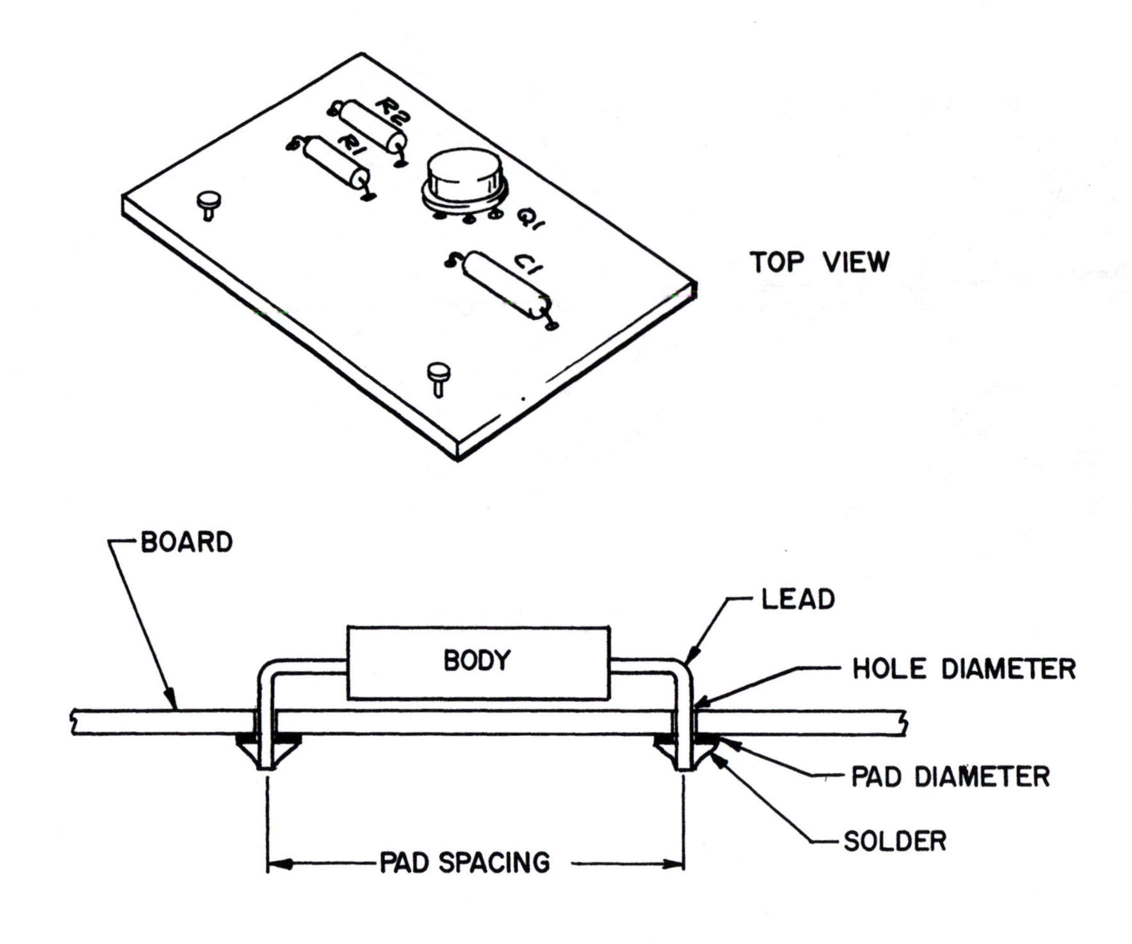

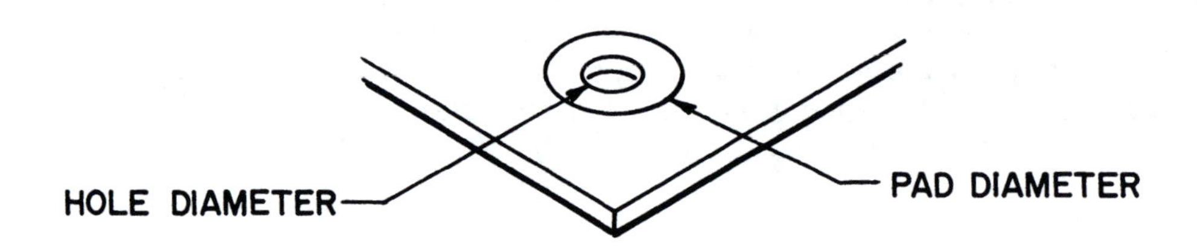

Figure 8-11 Loading components on a PCB

to the number of boards. In some instances, a machine can be used to insert components where there are a number of boards of the same pattern. If the number of boards to be loaded is few, the components can be inserted by hand into the holes.

Trimming the Leads

The leads on the components are usually longer than they need to be, so, after the components are soldered in place, the leads are trimmed off.

Soldering

Soldering is done by hand or by machine also, depending on the number of boards to be soldered. The diameter of the hole must be carefully controlled to allow the component lead to fit into it correctly. The diameter of the pad surrounding the hole must be large enough to allow the lead to be soldered firmly in place.

The drafter is often required to make a pattern called a *solder mask*. A solder mask is used to prevent stray solder from causing electrical shorts. This pattern is an artwork showing only the pads. The pads are enlarged (often photographically) to form the solder mask. The mask includes the holes which surround only the pads thus produced. The mask becomes part of the manufactured board before components are placed on it, and soldering begins. The mask prevents solder from being splashed onto areas where it could create a problem. After soldering is finished, the leads are trimmed and the board is complete.

Surface Mount Components

A relatively new form of component (such as integrated circuits, transistors, resistors, capacitors, and diodes) allows components to be mounted directly on the circuit surface as shown in Figure 8–12.

The advantages of this type of component are:

- The size of the printed circuit board may often be reduced appreciably because:
- The component itself is often smaller.
- Allowances for bending of leads to go through holes in the board is eliminated.
- Manufacturing cost is reduced because holes no longer have to be drilled in the board.
- Surface mount components are more suitable for automated assembly and testing.
- Circuit paths are often much shorter resulting in better circuit functioning.

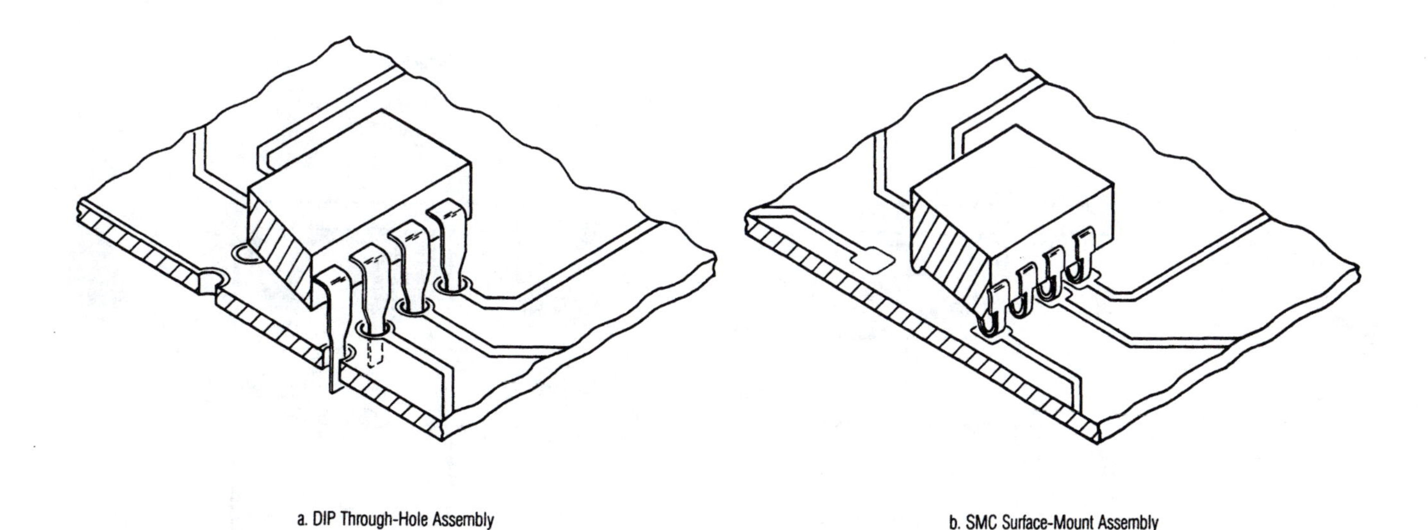

a. DIP Through-Hole Assembly

b. SMC Surface-Mount Assembly

Courtesy of Texas Instruments Inc.

Figure 8–12 Through the board vs surface mount components

The Surface Mounting Process

Surface mounting is the process by which a component is mounted on the surface of a printed circuit board instead of inserting the leads into holes in the board, Figure 8–13. The top and bottom views of a surface mounted integrated circuit are shown in a and b of this figure. View c shows the footprint of the component. The footprint is the metal pattern on the PCB to which the component is soldered. View d shows the process of applying a solder paste to the footprint in preparation for soldering the component to the footprint. Figure 8–13e shows the component placed and 8–13f shows what the component looks like after it has been soldered to the board.

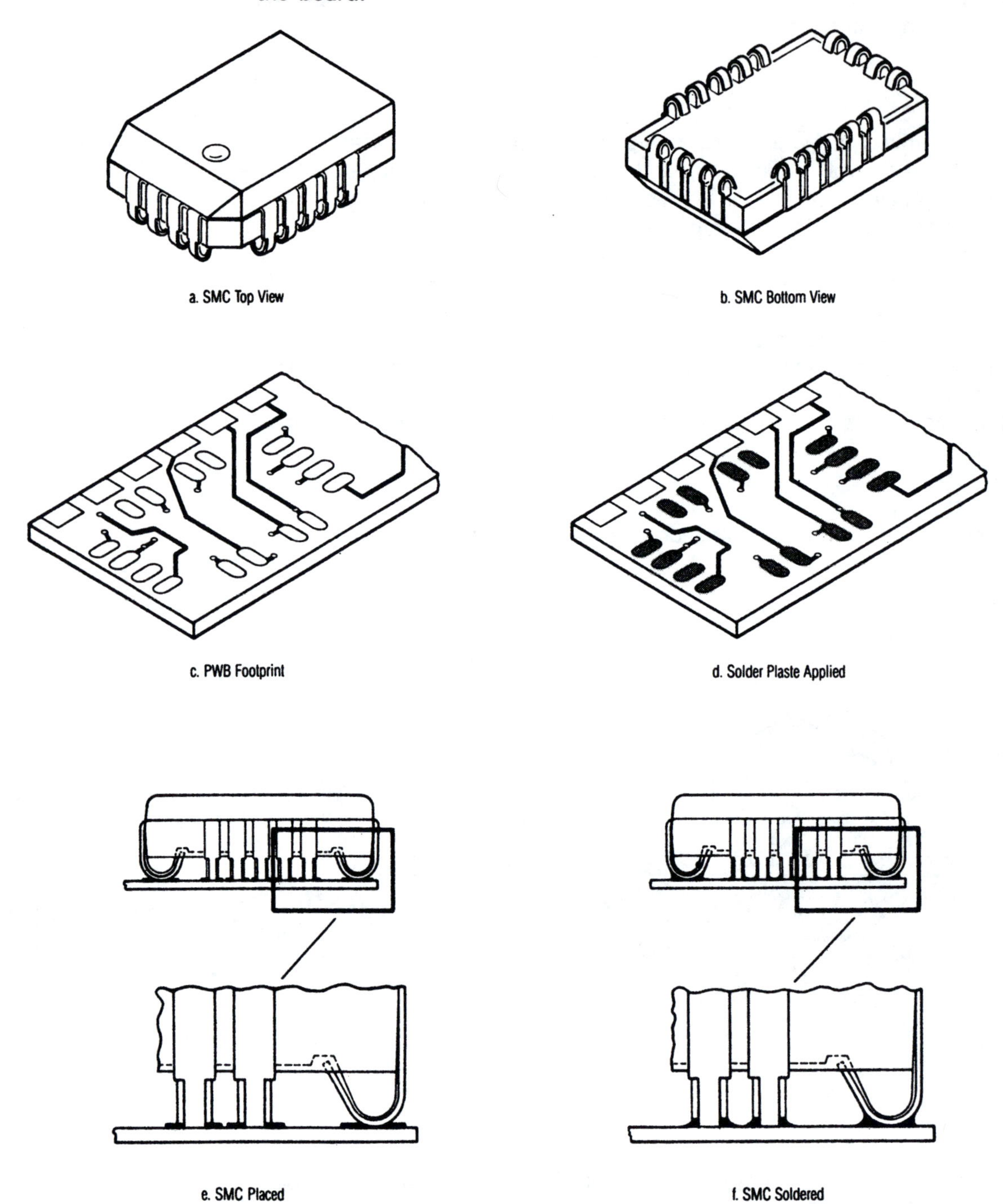

Figure 8–13 The surface mounting process

EXERCISE 8–1

Write your answers to the following on a separate sheet of paper.

1. Describe the function of all printed wiring boards.
2. List three types of printed wiring boards.
3. List the three types of drawings used to manufacture a printed wiring board.
4. List the four basic steps in manufacturing a printed wiring board.
5. List four advantages of surface mounted components.

Chapter 9

Design of Discrete Component Single-Sided PC Boards

OBJECTIVES

After completing this chapter, you will be able to
- ✓ make a design layout drawing of a discrete component single-sided PC board from a schematic diagram.
- ✓ draw artwork from design layout drawings.
- ✓ produce a drill plan from printed circuit artwork.

INTRODUCTION

A single-sided PC board has components mounted on one side of it, with the circuit on the other side. *Discrete components* are single components, such as a resistor or a capacitor. These components are mounted individually on the printed circuit board, as opposed to the integrated circuit which contains a predetermined combination of components.

The illustrations in Chapter 8 described the terms *layout, artwork,* and *drill plan.* Review Chapter 8 at this time to be thoroughly familiar with these terms.

DESIGN LAYOUT DRAWING

A design layout drawing is used to develop all of the other drawings used to produce a printed circuit board. With simple boards, this layout can be done with one color, but for complex boards the layout is often done with three or more colors. The design layout drawing in Figure 9–1 is shown here in screens of one color. Turn to the inside of the front cover for a three-color reproduction of this figure.

To produce a design layout, the designer needs the schematic diagram of the board and information pertaining to the physical features of all components, such as size, hole spacing, and lead diameter. Any other characteristics, such as weight, heat, and electrical characteristics which would affect the placement of components, must also be known. The board size and shape is also needed before the layout can be produced. To begin the layout, the hole spacing and hole size of all components must first be determined.

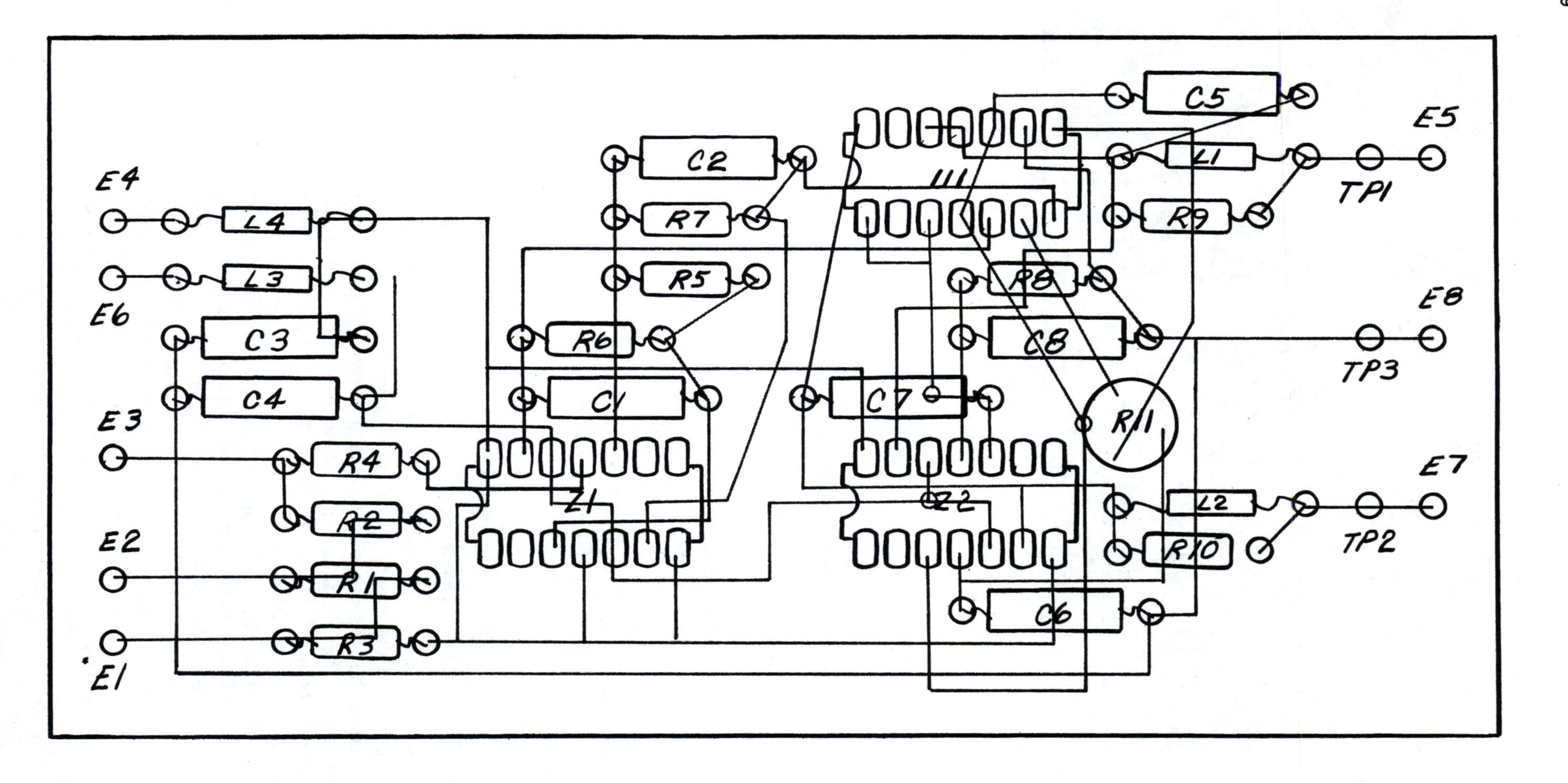

Figure 9–1 Design layout drawing

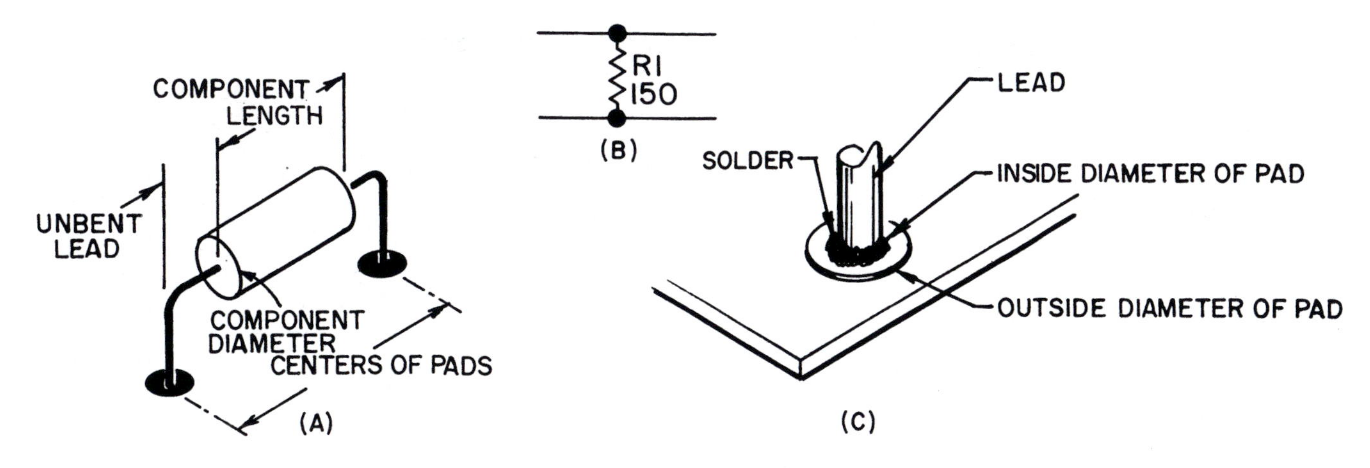

Figure 9–2 Determining hole spacing and hole size

Determining Hole Spacing and Hole Size

Figure 9–2 shows the elements necessary to determine hole spacing for a typical part.

In Figure 9–2A, the length of the component, plus the length of the unbent lead on both ends is used to approximate the distance between holes. The precise measurement is a result of determining the distance from the center of one lead to the center of the other, and the nearness of this distance to the center of a grid on a layout sheet. Most layouts are done on gridded sheets so that drilling and other physical features of the board can be measured accurately. For example, if the center-to-center distance between leads is .895″ and the drawing is done on a $\frac{1}{10}$″ grid, the measurement is rounded off to $\frac{9}{10}$″ to allow holes to be drilled in the center of grid lines. Component templates are commonly used. These templates are designed to include component size, lead length, and lead bend, which always comes out on $\frac{1}{10}$″ grid.

The component symbol R1 is seen in Figure 9–2B and would appear that way on the schematic diagram. Figure 9–2C shows the component lead passing through the PC board and the center of a pad. A *pad* is an etched area on the end of a trace. A *trace* is the etched circuit path. The pad must be larger than the lead so that the lead can be firmly soldered to the pad.

Schematic to Design Layout

Figure 9–3 expands on the previous figure. Two resistors are shown in schematic form in Figure 9–3A. Figure 9–3B shows these two resistors in a design layout. The resistors are drawn actual size or larger so that the pads may be accurately located, and for drilling holes through the centers of the pads. A pad is in place wherever a lead must pass through the board.

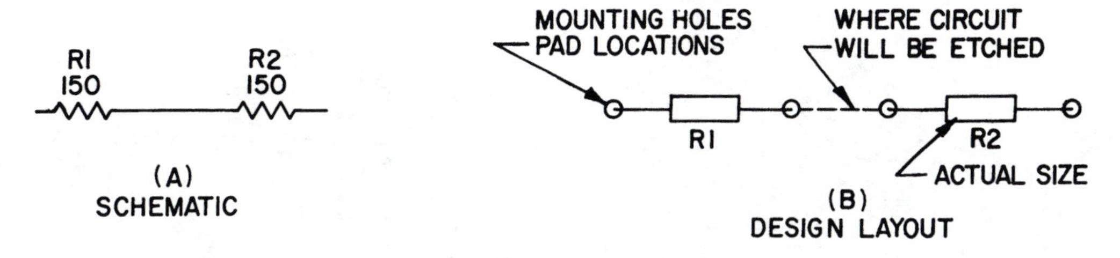

Figure 9–3 Schematic to design layout

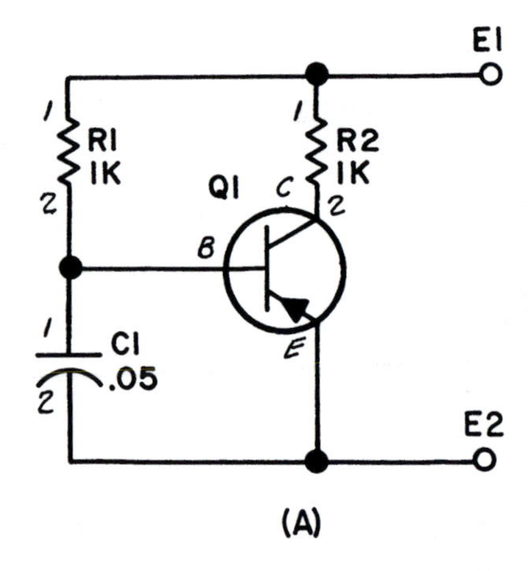

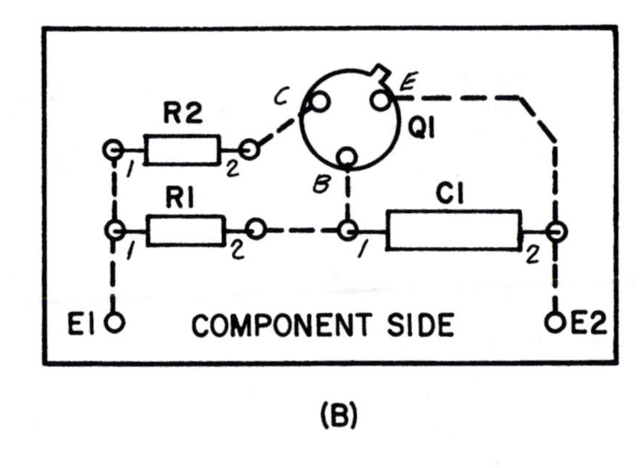

Figure 9–4 Layout of four components

Layout of Four Components

A schematic consisting of four components is shown in Figure 9–4A. The two resistors and the capacitor have two leads each that must pass through the board in the same manner as the resistors in Figure 9–3. The transistor, Q1, has three leads and must have three pads with holes drilled through them to allow the transistor leads to pass to the circuit side of the board, Figure 9–4B. The circuit is shown in this figure with dashed lines. The dashed lines show that the circuit is on the side of the board opposite the side with the components mounted on it. The side of a printed circuit board that has the components mounted on it is called the *component side.* The opposite side is called the *circuit side* or *solder side.*

Connectors

Before discussing some design principles of layout, the termination of traces should be identified. Figure 9–5 shows some standard means of terminating traces on a board. Figure 9–5A shows etched fingers on the edge of a PCB. These fingers comprise a connector which plugs into another connector allowing the PCB to perform its function. A similar arrangement, a PCB connector, is shown in Figure 9–5B. Figure 9–5C shows a terminal

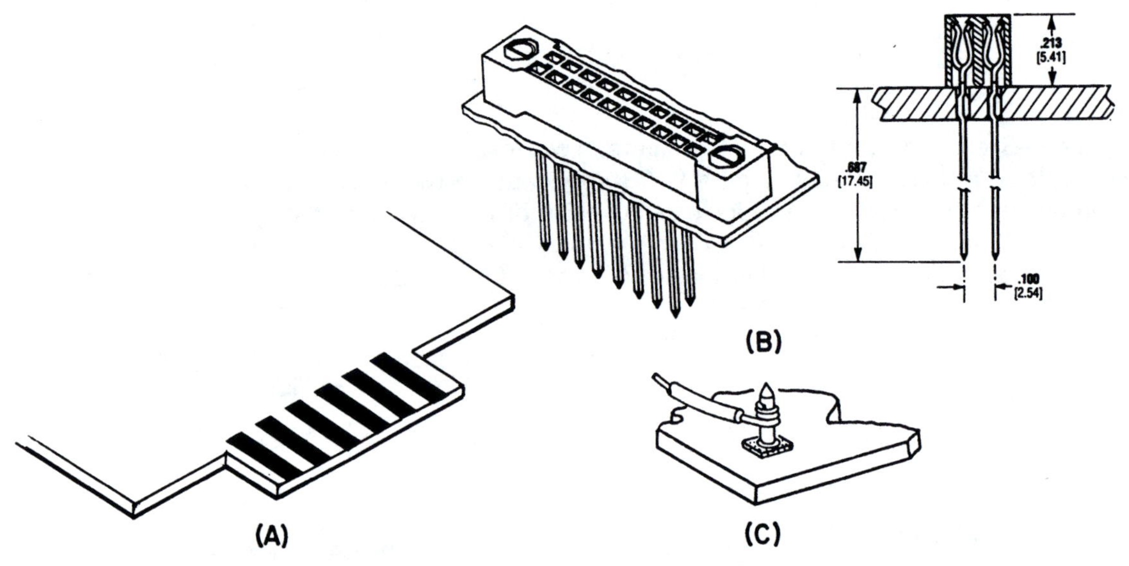

Figure 9–5 Connectors

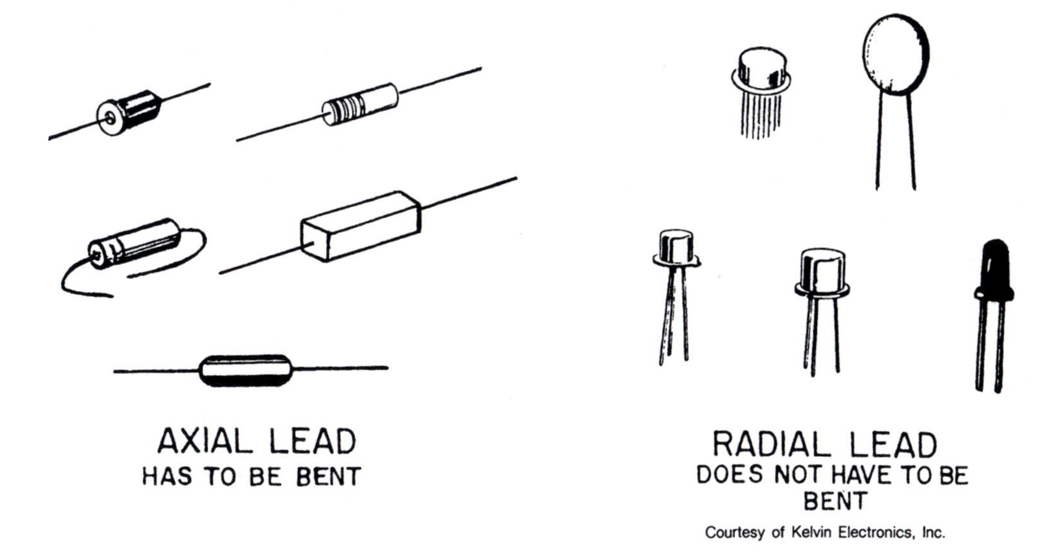

Figure 9–6 Radial lead and axial lead components

to which a wire may be attached to connect one part of the circuit to another.

Radial and Axial Lead Components

Figure 9–6 shows the two major types of components: axial lead and radial lead.

Axial lead components have leads that must be bent to pass through the PC board. Resistors, capacitors, and diodes are examples of components having axial leads.

Radial lead components have leads that do not have to be bent to pass through the board. The hole pattern for such radial lead components as transistors, connectors, ICs, LEDs, and sockets can be taken directly from the base of the component. These leads are often spread slightly to allow them to fit into holes which are not the exact pattern of the leads as they protrude from the base of the component.

Component Size Features

Figure 9–7 shows two components having axial leads: a resistor and a capacitor. To accurately lay out a PCB, the designer must know three size features of each component: its length, its diameter, and the diameter of its lead.

Although the difference in the length of the two components illustrated is .061, an adjustment in the length of the straight lead before the bend would allow both components to have the same hole spacing on a grid. Common hole spacing on a grid can simplify drilling and save money.

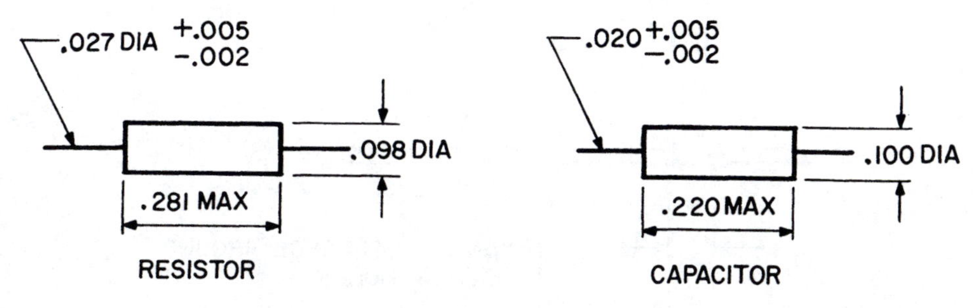

Figure 9–7 Components having axial leads

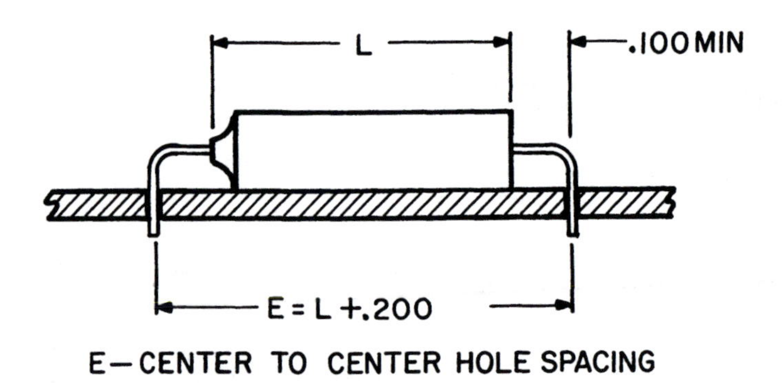

Figure 9–8 Component lead length

However, the holes for the leads of these two components must be sized differently because one lead is .027 in diameter and the other is .020. These holes must have a certain minimum clearance between the hole and the lead to allow solder to be properly applied. (The chart in Figure 9–24 shows the correct size of hole for each lead size.) The diameters of the two components (.098 and 1.00) are so nearly the same that they can be ignored in the design layout.

Component Lead Length

Figure 9–8 shows an illustration commonly used in determining hole spacing. The length of the component includes any feature of the body that protrudes from it before the bare lead is exposed. The .100 maximum straight lead length could be lengthened somewhat to allow common hole spacing for the two components shown in Figure 9–7. Often the leads of axial lead components are bent and inserted into the board by machine. Such a machine is called automatic insertion equipment.

Automatic Insertion • Automatic insertion equipment allows components to be placed on boards much more quickly than they can be placed by hand. For example, a large quantity of resistors can be fed into the machine which is programmed to bend the leads to a specific length and then to drop the resistors into holes drilled into the board. Because the board moves in a straight line through the insertion equipment, the more resistors that are turned the same direction, the fewer times the board must pass through the machine and the faster the process of insertion is accomplished.

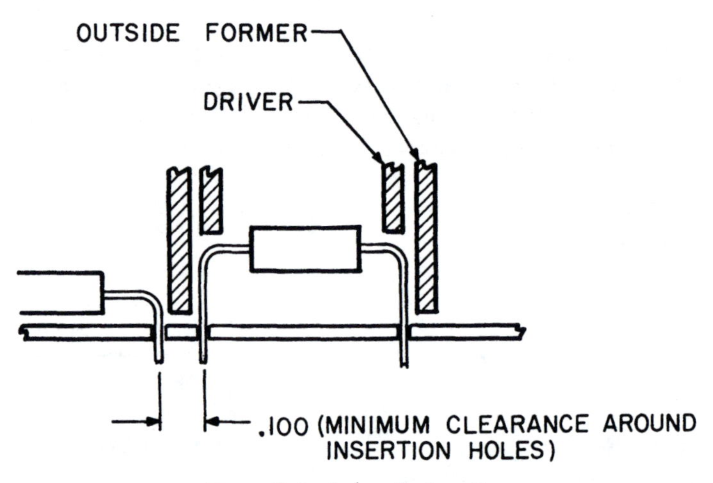

Figure 9–9 Automatic insertion

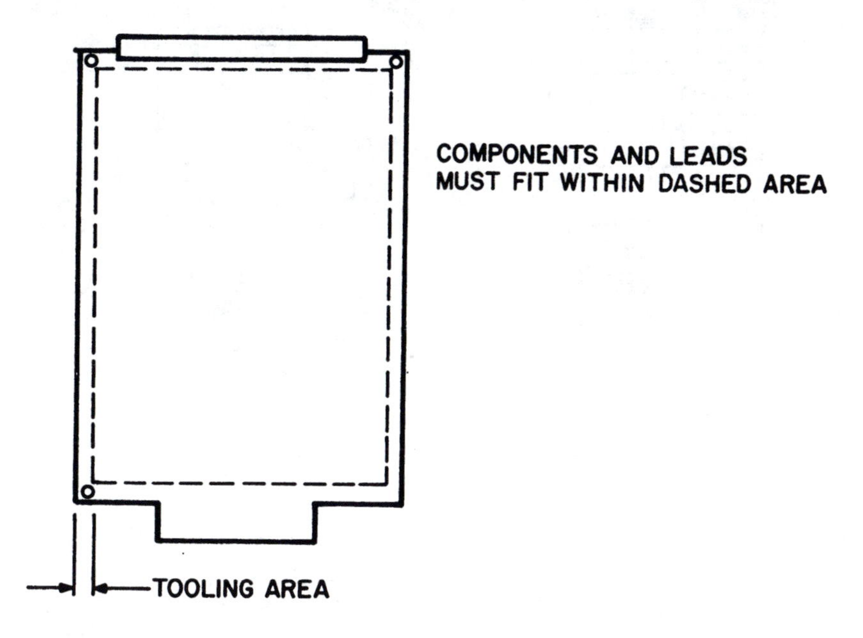

Figure 9–10 Component area and tooling area

When planning to use automatic insertion, provisions must be made to allow the tools to move properly. Figure 9–9 shows the *outside former* which determines the outside edges of the leads, and the *driver* which pushes the component into the holes in the board. Notice that this particular equipment requires that component leads be not closer than .100, or the outside former will deform or break the lead of the component already in place. While components are being inserted into the board, the board must be held accurately in place.

Component Area and Tooling Area • Holes are provided in PC boards to allow the board to be accurately held in a fixture while the operations of component insertion, soldering, and lead trimming take place, Figure 9–10. These holes are placed in an area around the edges of the board. This area is called a *tooling area,* and no components or their leads can be placed in that area.

Heavy Component Arrangement

Figure 9–11 shows where to place heavy components which generate heat, such as transformers. Locating heavy components on the edges of the board, perpendicular to the connector, allows the board to be handled

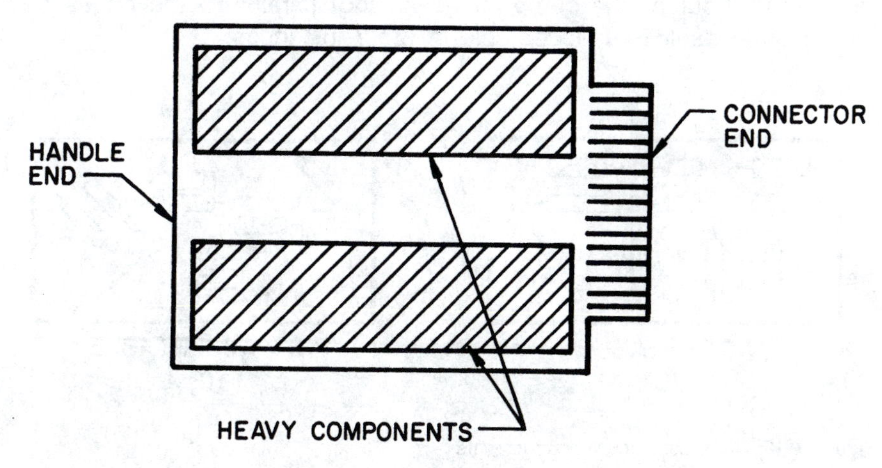

Figure 9–11 Heavy component arrangement

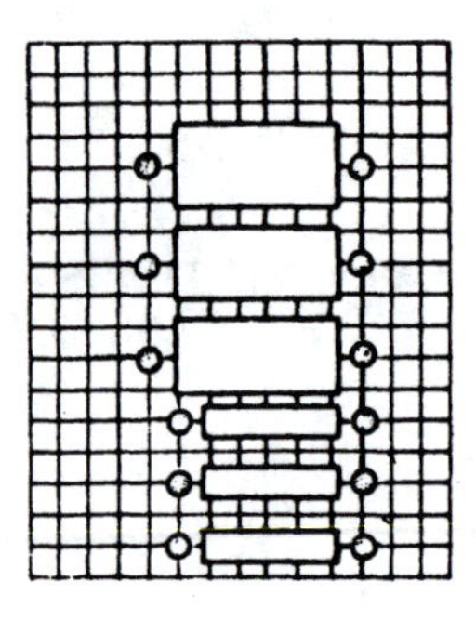

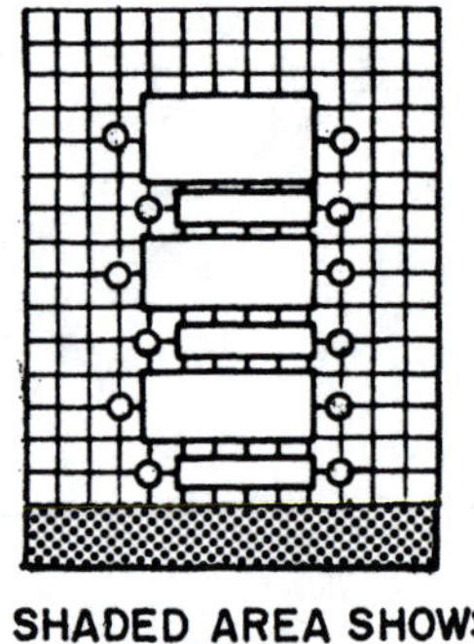

Figure 9–12 Alternating large and small components

more easily when inserting it or removing it from the system. This arrangement is also useful in that it allows air to pass over the heavy components to cool them. If heavy components were placed in the center of the board surrounded by smaller components, airflow would be restricted and cooling would be hindered.

Alternating Large and Small Components

Because holes must be located at the intersection of grid lines, it is often an advantage to alternate large and small components, Figure 9–12. This results in a saving of space on the board. The space on a PC board is very valuable, so any saving is usually important.

Axial Component Arrangements

Figure 9–13 illustrates three different arrangements of axial components. The preferred arrangement, Figure 9–13A, allows for economical drilling of the board and insertion of components by either automatic or manual means. In addition, turning the + side of all polarized components in the same direction reduces the possibility of error in placing these components. Figure 9–13B shows components turned in two 90° directions, but still well organized and capable of being manually or automatically inserted. The final arrangement, which is unacceptable, Figure 9–13C, shows components randomly placed. Four of the components are not parallel to the edges of the board. This in itself is unacceptable in most instances.

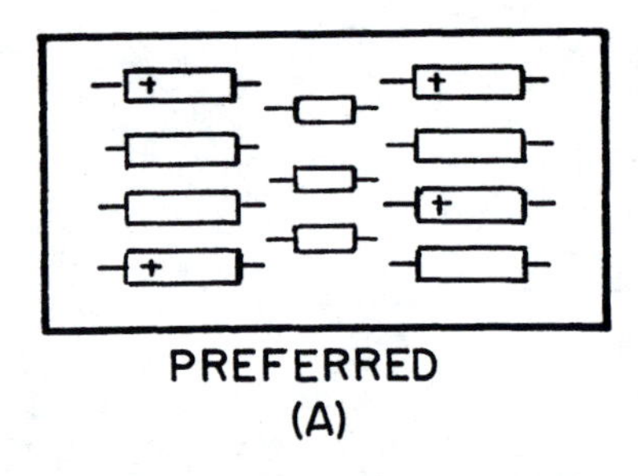

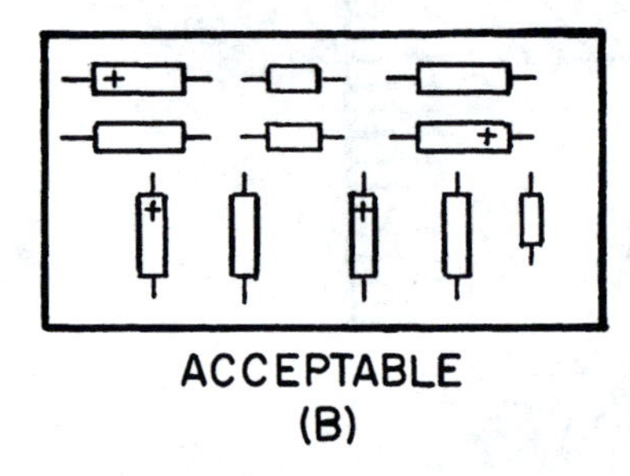

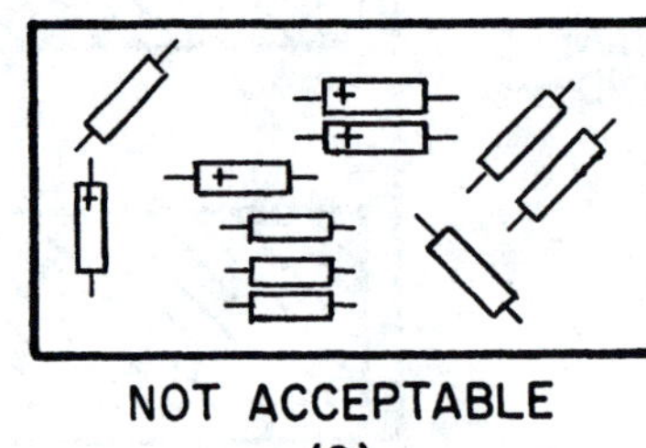

Figure 9–13 Axial component arrangements

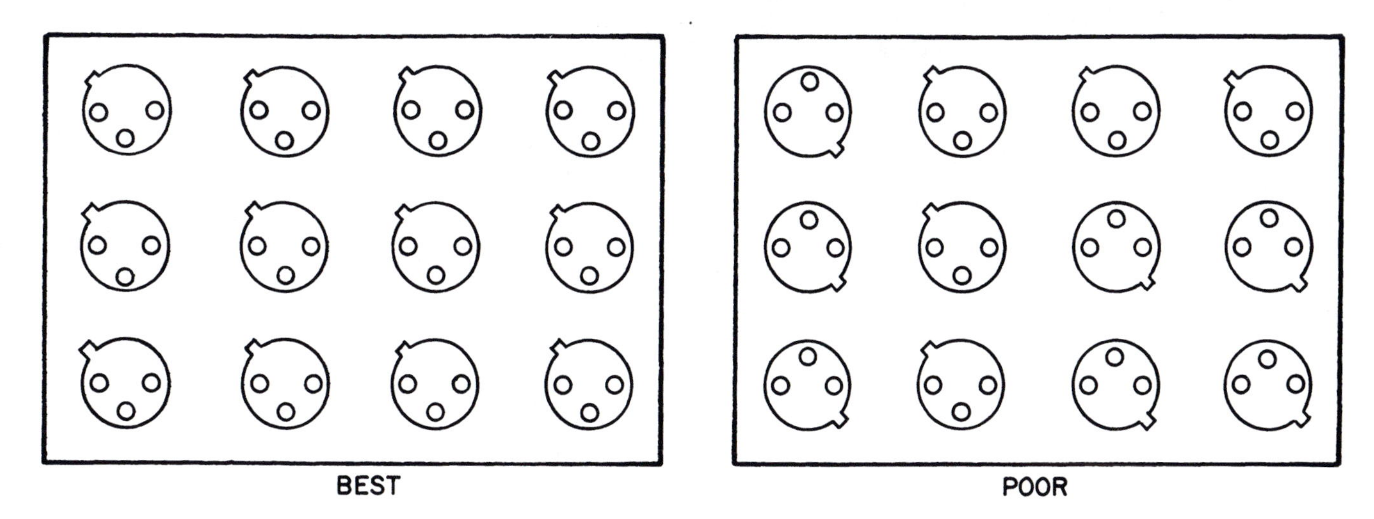

Figure 9–14 Radial arrangements

Radial Component Arrangements

In Figure 9–14, all of the transistors of the BEST arrangement are positioned in exactly the same manner. The tab indicates which is the emitter lead. This arrangement reduces the possibility of error in component insertion, and allows for more economical hole drilling and inspection of the finished product.

Board Layout

A schematic from which a PCB is designed is shown in Figure 9–15A. An organized approach to board layout with components placed parallel to one side of the board is shown in Figure 9–15B, labeled GOOD LAYOUT. This layout combines several of the best design principles, as follows: all components are parallel to one edge of the board, polarized components are turned in the same direction, large and small components are alternated to save space, the board shape is simple, and the size of the board is conveniently small.

The POOR LAYOUT of Figure 9–15C requires a larger board with an irregular shape which would be more expensive than it needs to be. Also, drilling and insertion become more expensive (C2 must be manually inserted), and inspection of the final product becomes more difficult.

The SCHEMATIC for both poor and good layout is shown in Figure 9–15C.

Figure 9–16 shows additional examples of good and poor board layouts. The SCHEMATIC in Figure 9–16A shows the fingers of a connector which is etched on an edge of the board. Inputs and outputs are grouped on the actual board itself and are located on a tongue which will slide into a mating connector in a card cage (a rack to hold several PC boards). In the POOR LAYOUT example, Figure 9–16B, the board is too large, it is irregularly shaped, and its components are randomly arranged. In the GOOD LAYOUT example, Figure 9–16C, all axial components are in one column, the transistors are oriented with the emitter in the same location, and the potentiometer (R2) is located so that traces to it are as short as possible. The board size is small and its outline is simple.

After the design layout is complete, the other drawings in the PCB package may be made. The first of these is the artwork.

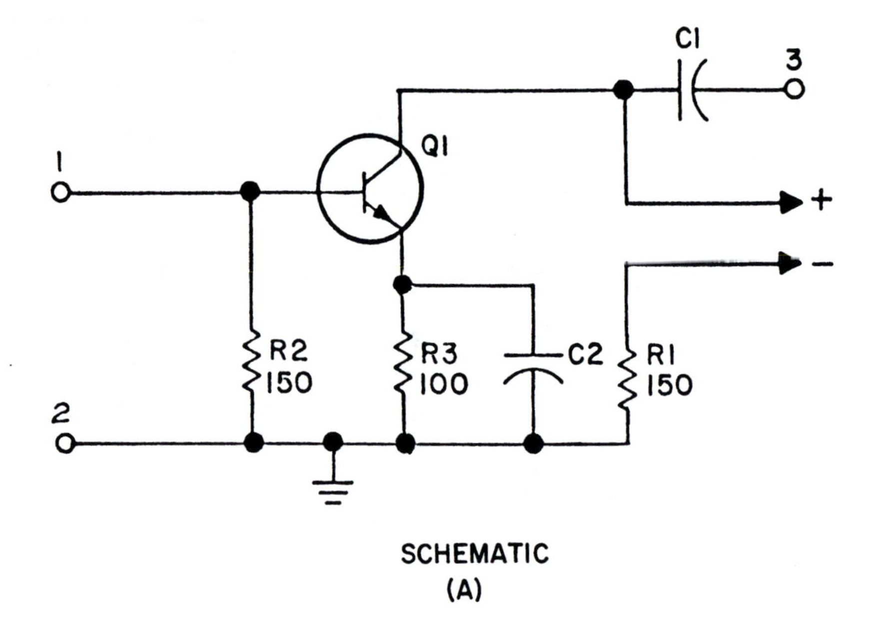

SCHEMATIC
(A)

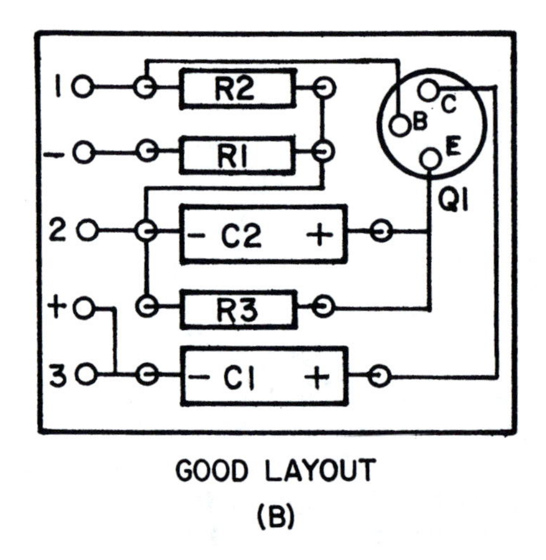

GOOD LAYOUT
(B)

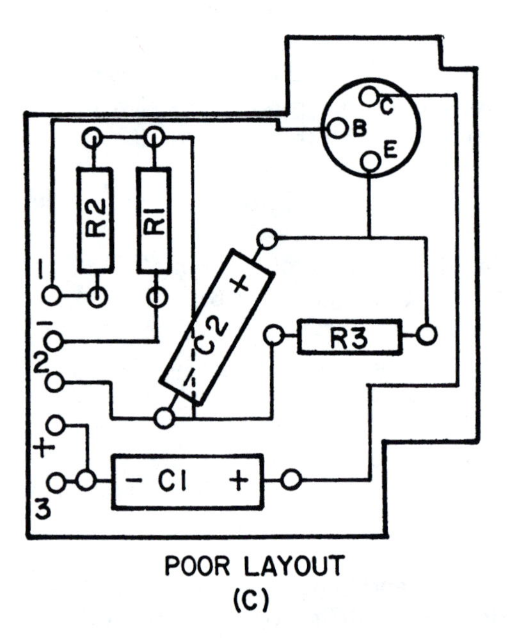

POOR LAYOUT
(C)

Figure 9–15 Board layout comparisons

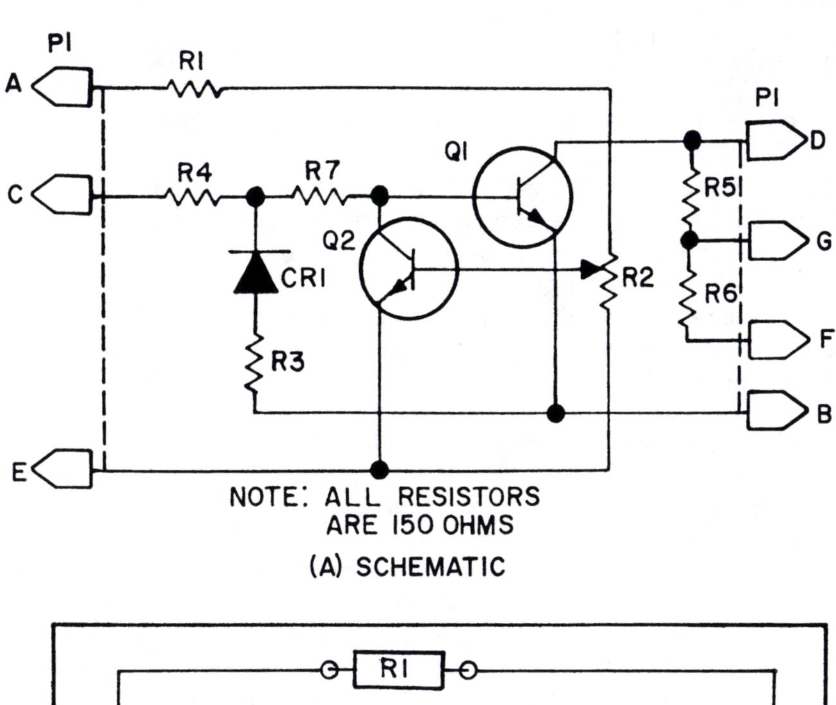

(A) SCHEMATIC

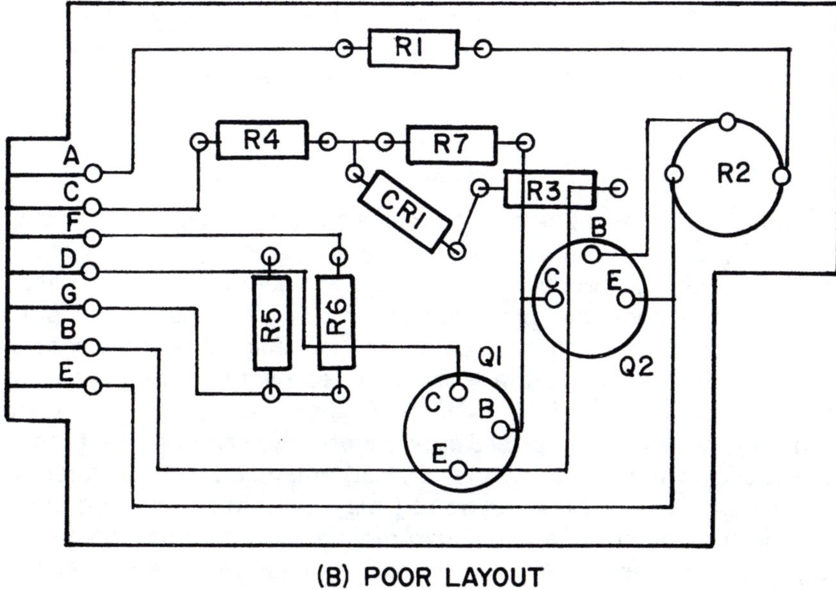

(B) POOR LAYOUT

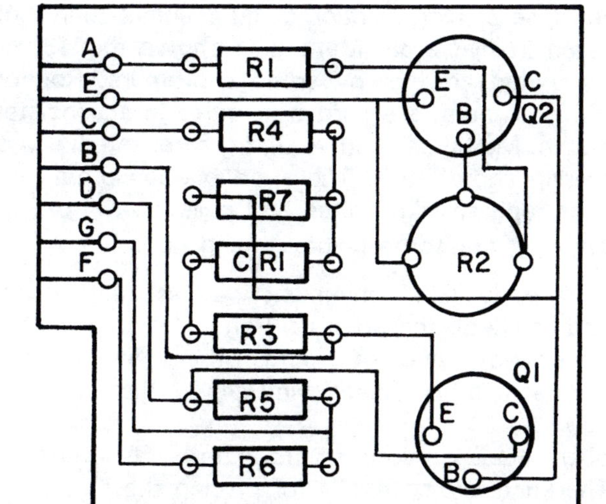

(C) GOOD LAYOUT

Figure 9–16 Additional comparisons of board layouts

ARTWORK

Because a one-sided PC board has circuitry on one side only, Figure 9–17A, only one artwork is needed for circuitry. Some customers require that reference designators and the outline of major components be shown with a paintlike ink on the component side of the board to aid assemblers in placing components on the board. A silk screen artwork, Figure 9–17B, is often made for this purpose. This artwork is produced by a photographic process wherein registration targets are placed in areas outside the board, where they will not reproduce photographically, but assuring that artwork on both sides of the board are in the proper positions. Artwork for circuitry is generally required to be extremely accurate, so it often is made two or four times as large as the size of the finished board.

There can be no breaks in the traces or in the areas where traces meet pads. If the artwork is made manually with tape, beginners often do not allow enough tape to overlap onto the pad. Tape tends to shrink a little and, if the overlap is too small, it will draw away from the pad. This can create an electrical short in the board, and result in considerable expense if a board has to be scrapped because of the short. Notice in Figure 9–17A that the traces in the artwork are as short as possible and have few 90° corners. Avoiding 90° corners prevents cracking which can occur in those areas.

Good Taping Methods for Artworks

Figure 9–18 describes several contrasting desirable and undersirable methods of routing the circuit for an artwork. Although many companies no longer use tape to produce artwork, these methods of routing apply to both manual and computer produced artwork. In the figure, the approved method A shows that a short trace can be run diagonally instead of using method B. Note that C has the trace centered on the pads instead of on the pad edges as shown in D. Method E shows a 45° corner around an isolated pad and the trace, providing a uniform appearance to the artwork. Method G shows that obtuse angles and 90° intersections are preferable to the acute angles in H. The situation at I may be a little confusing until it is understood that the schematic diagram requires only that the three pads be tied together. Therefore, the simplest and shortest route is the best. Both of the examples at J have longer traces, and the acute angles shown are undesirable because etching solution could accumulate in those pockets and weaken the etch in that area. Method O shows the 45° corners that are preferred to rounded corners by some customers. Rounded corners, as compared to 45° corners, are harder to make in a consistent pattern, and require more time. Method Q shows that greater overlap is preferred to the very short overlap shown in R. The short overlap often results in an open circuit when the tape shrinks a little and pulls away from the pad.

The following are some helpful taping hints.

- Arrange the pads in a convenient way so that you can place them quickly and easily on the artwork. Some designers put the pads on the hand they are not using for taping. Others devise a way to line them up on a scale or other object. Find a way that is suitable for you and use it.
- Keep a sharp blade on your X-acto knife©. This makes cutting the tape easier. The knife is also useful for placing the pads.
- For taping a corner, cut about three-fourths of the way through the tape and bend it to the angle you need, Figure 9–19. That way you have no ragged edges or broken places in the tape.
- Overlap the tape onto the pads for a clean, solid connection.

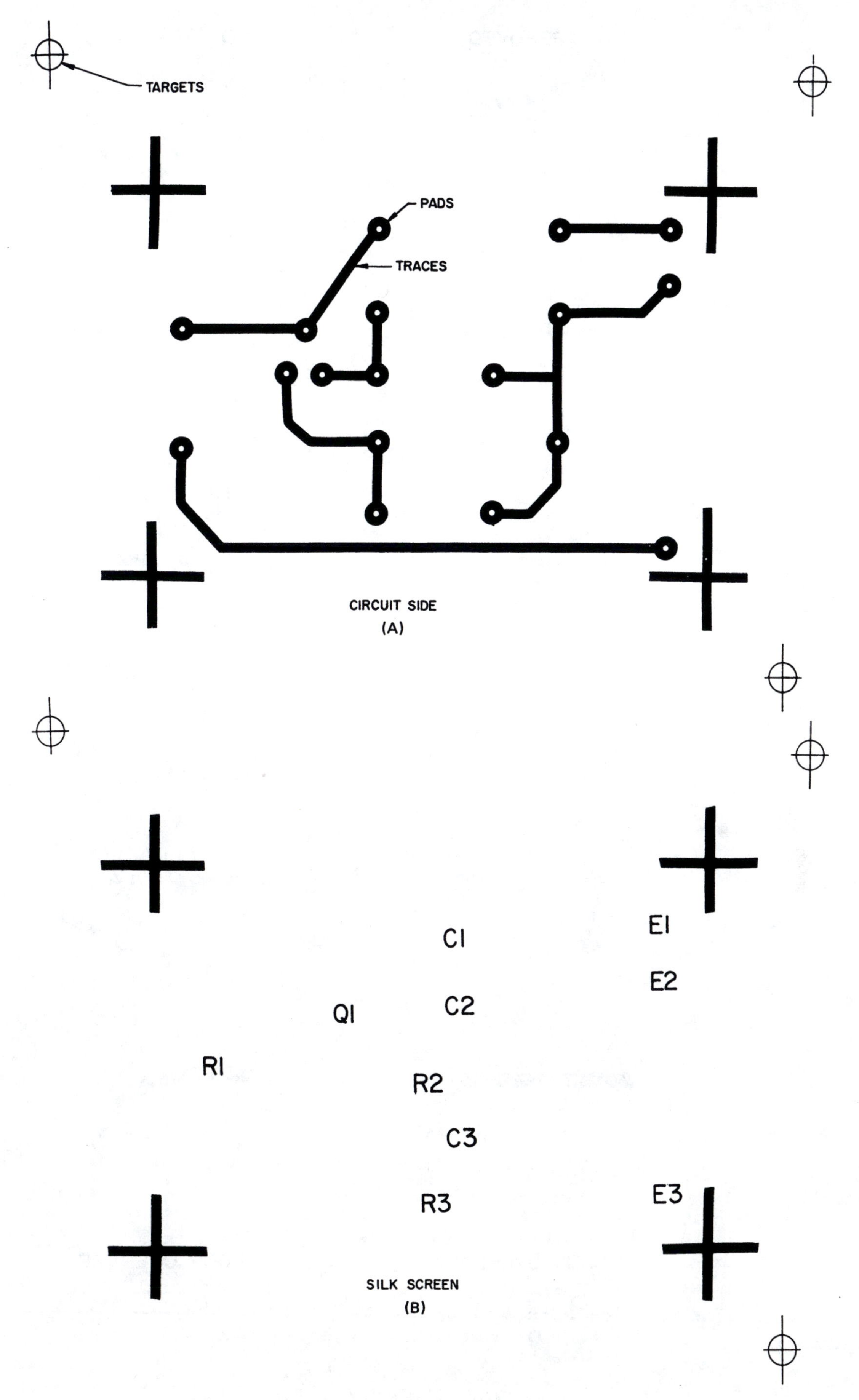

Figure 9–17 Artwork for a one-sided PC board

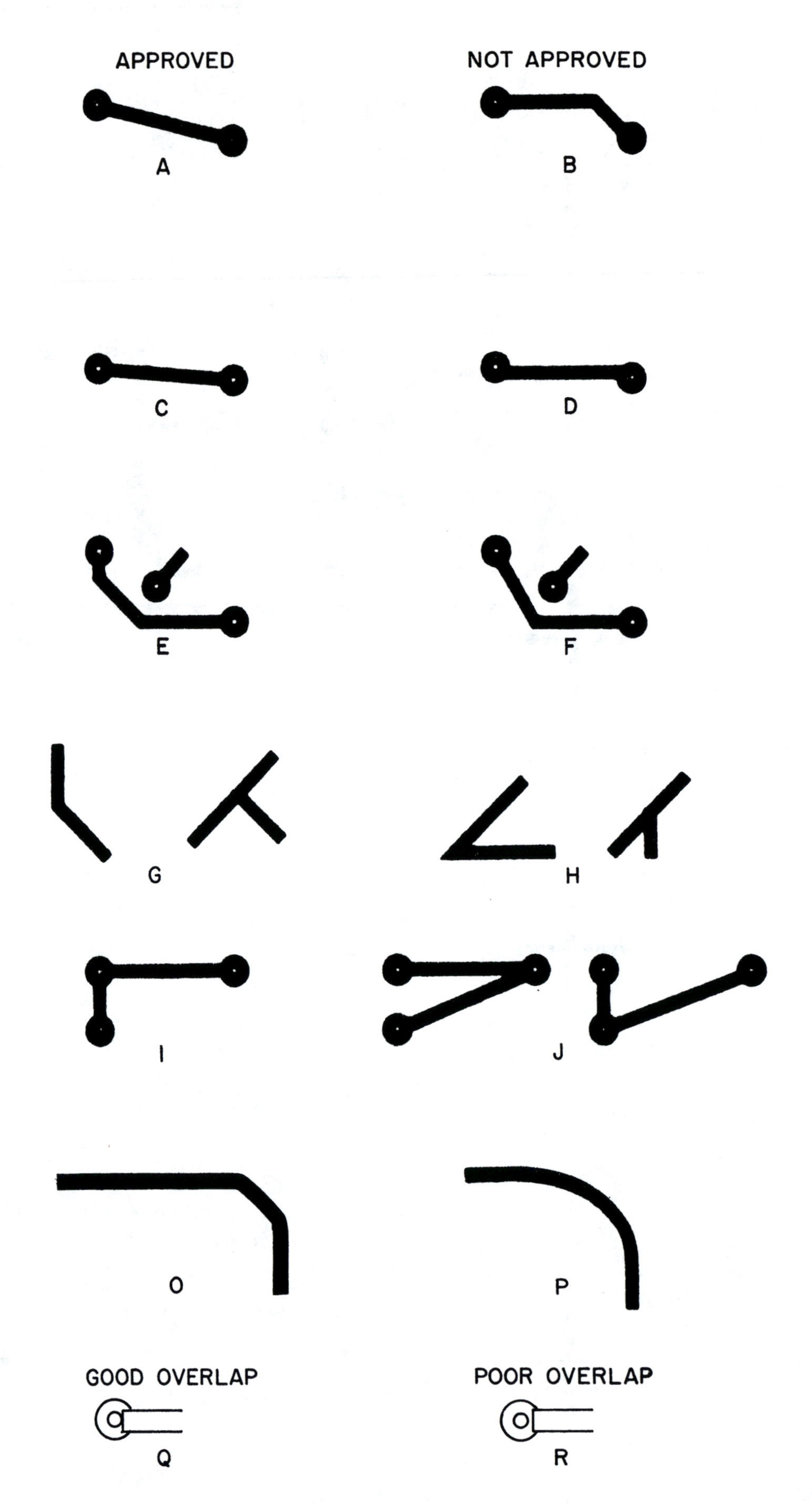

Figure 9-18 Taping methods

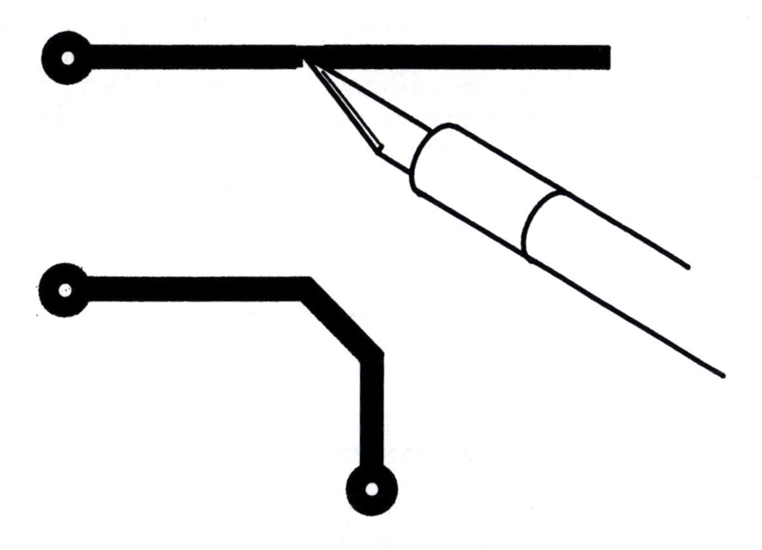

Figure 9–19 Cutting and bending the tape

Silk Screen Artwork

The artwork used to produce a silk screen mask is shown in Figure 9–20. This mask is used like a stencil to allow a paintlike ink to be applied to the component side of the PCB. This artwork must be carefully *registered* (aligned with the targets) so that the component outlines and their reference designators are in the correct location when the component leads are pushed into their holes on the board.

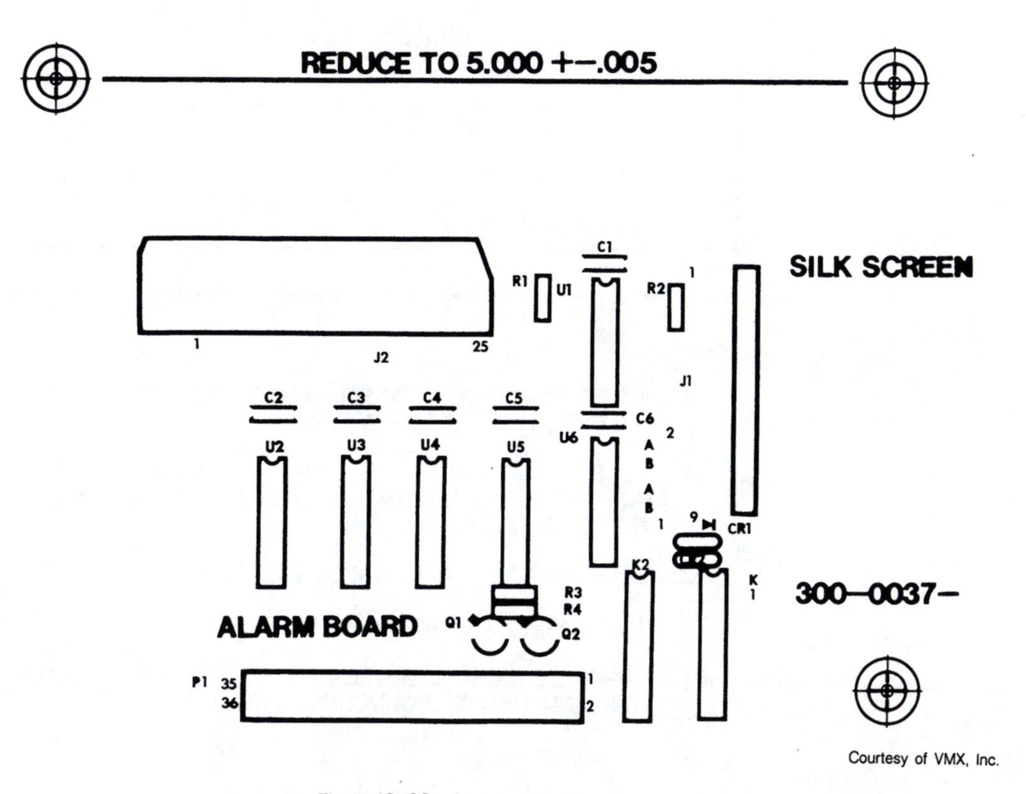

Courtesy of VMX, Inc.

Figure 9–20 Silk screen artwork

DRILL PLAN (FABRICATION DRAWING)

A drill plan (fabrication drawing) is another document needed to produce a PC board. The drill plan, Figure 9–21, is produced by using a reproduction of the artwork (sepia copy or photographic copy) spliced into a format sheet. All holes are identified by a letter (or the lack of one). The largest number of holes of the same size are unmarked to save time. All other holes have a letter placed next to them or a dashed box enclosing them with the letter in the box.

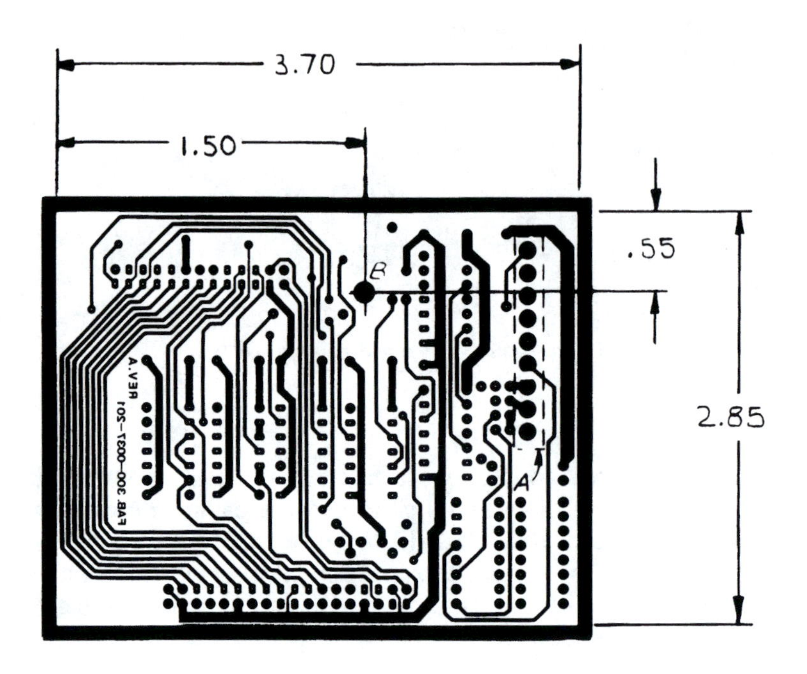

HOLE SCHEDULE			
SYM	DESCRIPTION	QTY	COMMENTS
NONE	.031±.003 DIA AFTER PLT	114	SEE NOTE 3
A	.067±.003 DIA AFTER PLT	9	SEE NOTE 3
B	.125±.005 DIA	1	

NOTES:

1. MATERIAL: EPOXY GLASS LAMINATE FOIL CLAD TYPE FLG.062 CI/1 PER MIL–P 13949.
2. FINISH: CIRCUITRY–ELECTROPLATED TIN–LEAD PER QQ–S–571 AND REFLOW 500 MICROINCHES MINIMUM THICKNESS AFTER REFLOW.
3. MINIMUM TOTAL WALL THICKNESS TO BE .001 AFTER REFLOW.
4. NO PLATING ALLOWABLE IN HOLES.
5. SYMBOLIZATION TO BE SCREEN PRINTED USING BLACK EPOXY INK AND ITEM 2, COMPONENT SIDE ONLY.

Courtesy of VMX, Inc.

Figure 9–21 Drill plan

The drill plan also shows the board outline dimensions and any other notes necessary to produce the board to the exact specifications. The hole schedule shows the size and quantity of holes, and any other information needed to drill and plate the holes.

Some companies produce PC boards on automated equipment. If the drilling is to be done on a numerically controlled machine, that machine may be controlled by a drill tape. When numerically controlled drills are used in conjunction with computer-aided drafting and design, the drill tape is produced by the computer.

COMPONENT ASSEMBLY

The final drawing required to produce a PCB is a component assembly drawing, Figure 9–22. This is often produced by combining circuit artwork and silk screen artwork into a single picture. The circuit artwork is subdued photographically by screening. *Screening* is a means of breaking a solid line or an area into small dots spaced very close together. This screening subdues the circuit artwork, allowing the component artwork or silk screen to stand out from the background. The components are then identified by an index number which corresponds to an item number on a parts list. Parts are purchased using the parts list, and they are inserted into the board by using both the assembly drawing and the parts list.

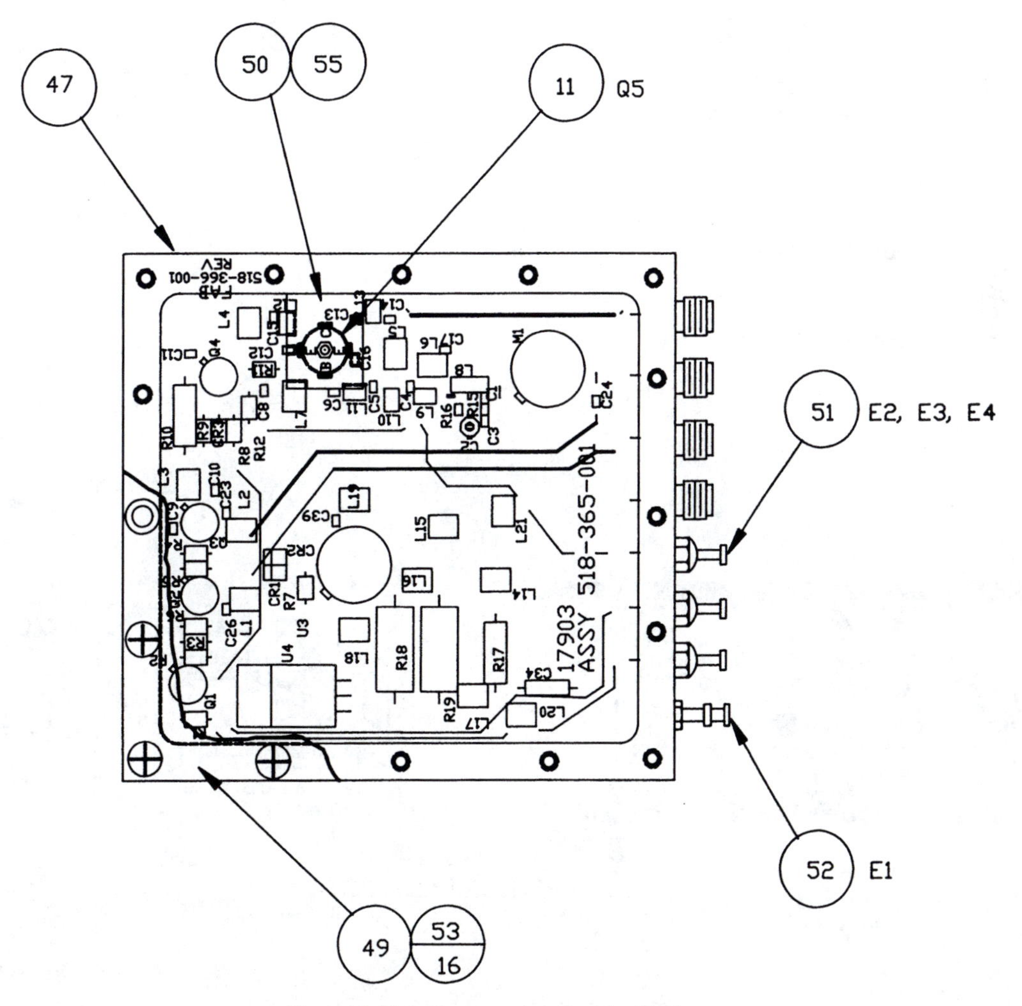

Figure 9–22 Component assembly drawing

SUMMARY (DRAWING PACKAGE)

The drawings needed to produce a single-sided PCB are shown in Figure 9–23.

- The design layout is the accurately drawn preliminary drawing from which all other drawings are made. It may be discarded after the other drawings have been prepared.
- The circuit artwork is the picture needed to produce the etched circuit.
- The silk screen is the artwork needed to make a silk screen mask for inking component outlines and reference designators on the board.
- The drill plan (fabrication drawing) is used to drill holes accurately and to cut the outline of the board to the correct size and shape.
- The component assembly drawing is used to assemble components onto the board and solder them in place.

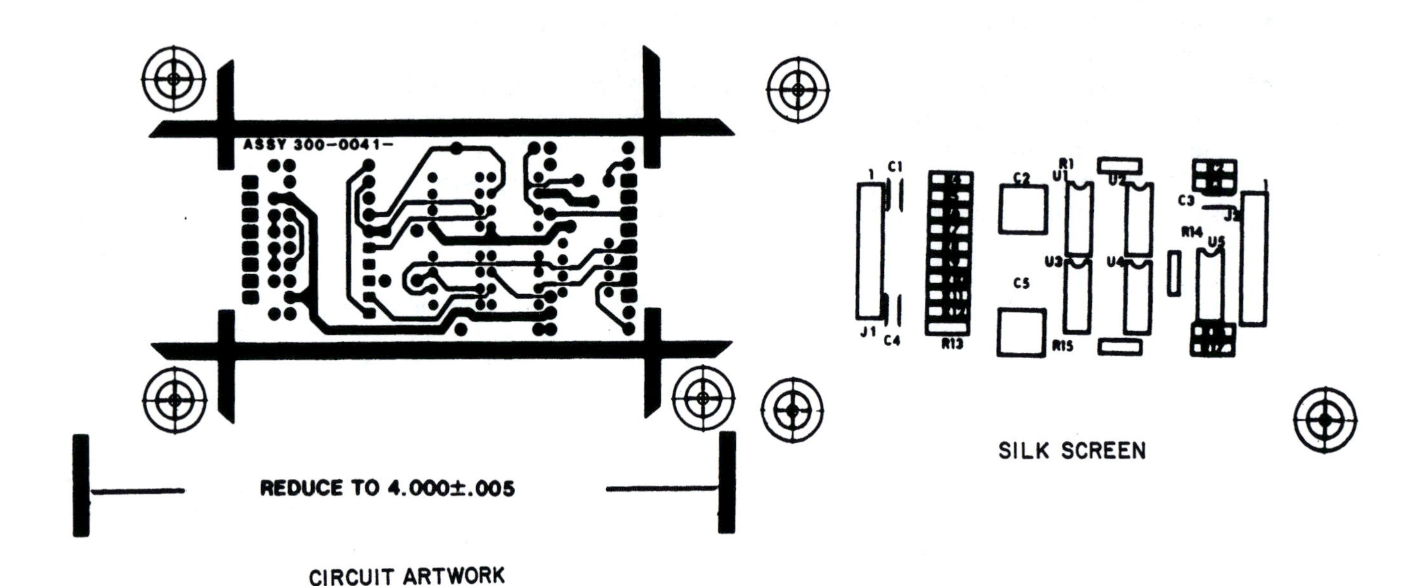

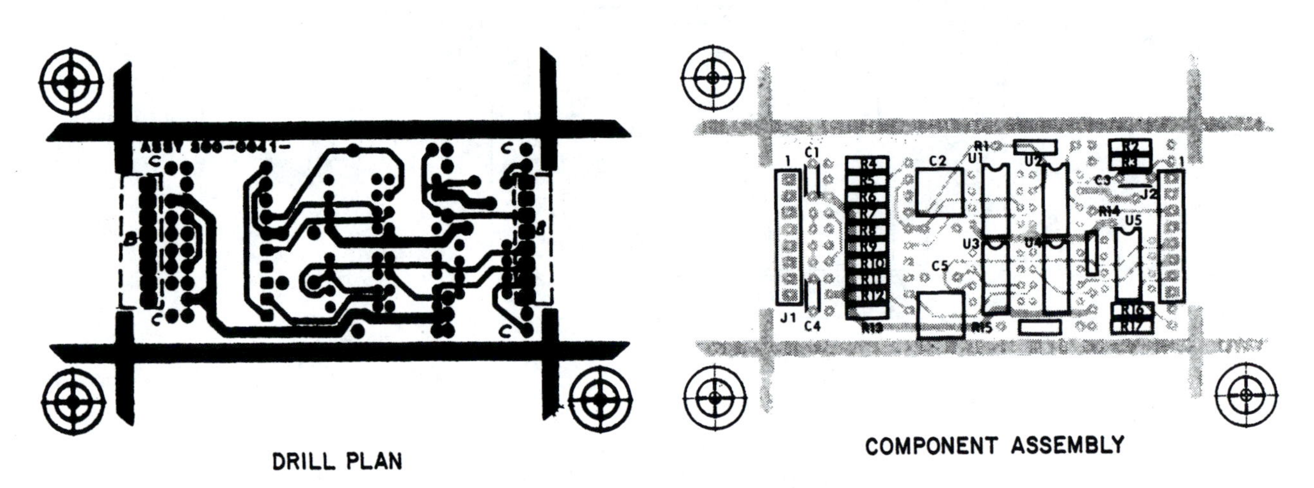

Courtesy of VMX, Inc.

Figure 9–23 Drawing package

WIRE GAUGE	LEAD DIAMETER	DRILL SIZE	HOLE SIZE DIAMETER
30–36	.005–.010	#76	.020
25–29	.011–.018	#71	.026
23 & 24	.019–.024	#68	.031
20–22	.025–.032	#60	.040
19	.033–.038	#56	.046
18	.039–.044	#55	.052
16 & 17	.045–.051	#53	.059
———	.052–.055	#52	.063

Figure 9–24 Hole sizes

HOLE TABLE

Figure 9–24 is a table showing the size hole needed for common sizes of leads. For example, a 26-gauge wire (a resistor lead) has a lead diameter in the .011 to .018 range. The size hole which must be drilled for that lead is .026 inches, and drilling it requires a #71 drill bit. Both the diameter of the hole and the drill bit size in parentheses are often shown in the hole schedule. This practice reduces the possibility of drilling the incorrect size hole.

DESIGNER'S CHECKLIST

The following are some questions that the designer should ask about a schematic when it is received.

1. Check to see if there are any special-size components. Are all the components standard?
   For example:
   - All resistors are 1/4W, 5%
   - All capacitors are ceramic disc
   - All transistors are 5018 case
   - All ICs are 7400
   - All diodes are 1N914
2. Are there any critical leads on the board? Are there special sizes for some leads?
3. Are there preassigned pins on the connectors?
4. Are there any variable resistors or capacitors, and where should they be placed on the board?
5. Should there be any test points? (Try to place all test points on the edge.)
6. Do any components require heat sinks? (Heat sinks are discussed in a later unit.)
7. By using the maximum size of all the components, will all the components fit on the board? [This is called component density. *Component density* is the area of components divided by the area of the board. For example, if you have 20 components, the total area of which (measured in the flat view of the component as it appears on the component assembly) is 2″ and the board measures 2″ × 2″, the component density is .50 or 50%.]

$$(2 \div 4) = .5 = 50\%$$

EXERCISE 9–1

Prepare a design layout drawing, a circuit artwork, and a drill plan from the schematic drawing below. Measure the component sizes from the parts shown. The sizes shown are twice the size. Your drawings should also be twice the size. The outer dimensions of the final full size board are 1.000″ x 2.000″. Your drawings will measure 2.000″ x 4.000″. Make all pads the same size and all leads $\frac{1}{10}$″ at 2 times size.

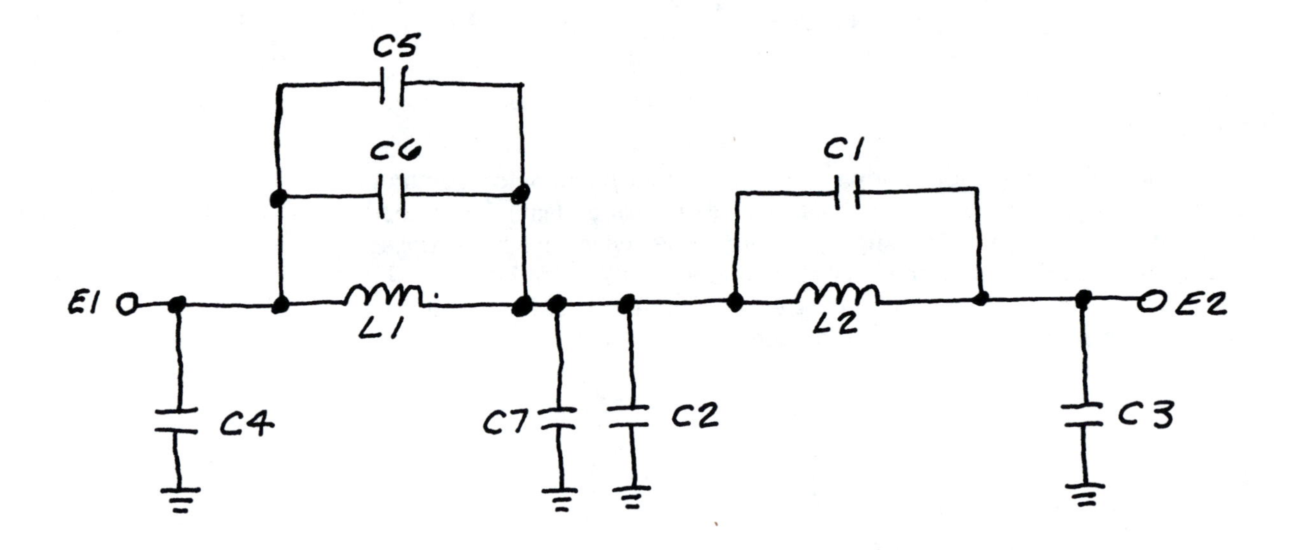

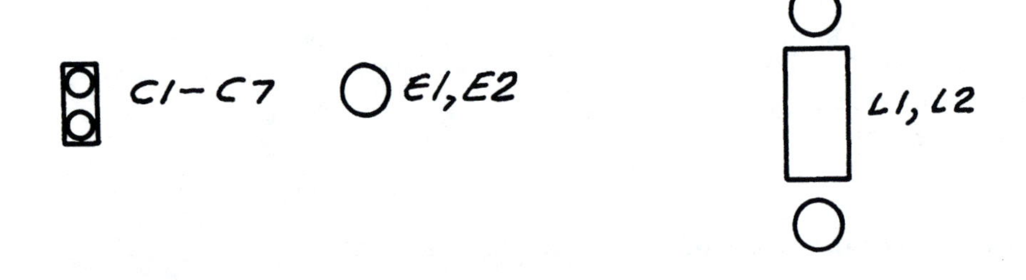

EXERCISE 9-2

Prepare a design layout drawing, a circuit artwork, and a drill plan from the schematic drawing below. Measure the component sizes from the parts shown. The sizes shown are twice the size. Your drawings should also be twice the size. The outer dimensions of the final full size board are 3.000" x 4.000". Make all pads the same size and all leads $\frac{1}{10}$" at 2 times size.

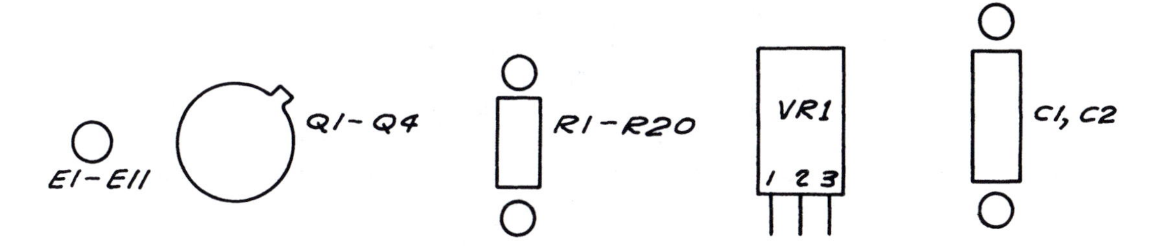

SPECIFICATIONS FOR EXERCISE 9-2

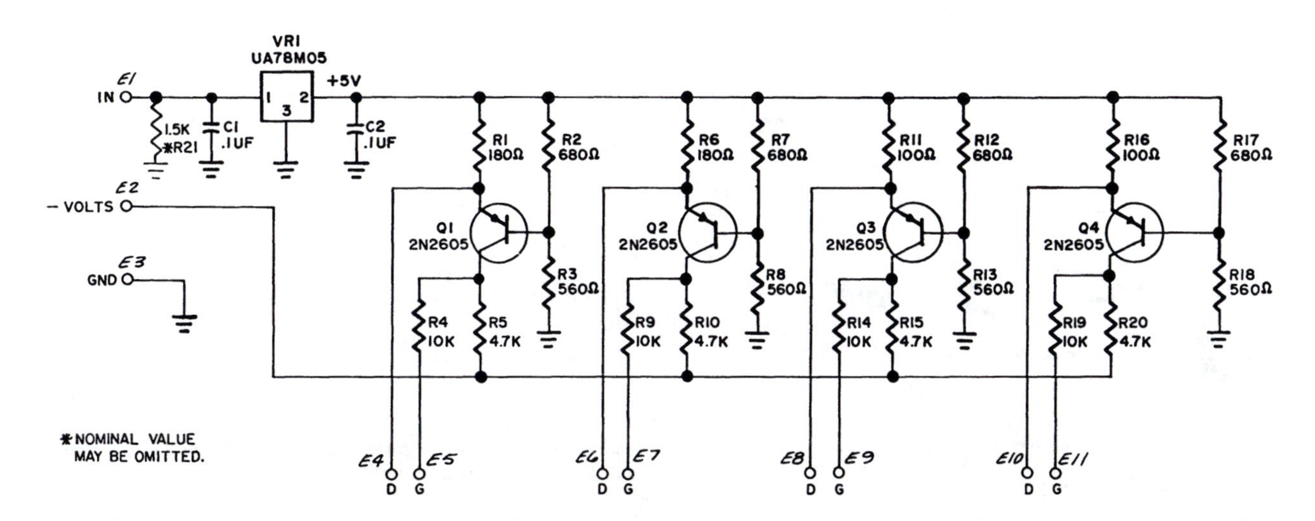

SCHEMATIC FOR EXERCISE 9-2

EXERCISE 9-3

Make a design layout drawing, a circuit artwork, and a drill plan from the schematic on the facing page. Use gridded vellum sheets (10 by 10 to 1 inch) for your drawings.

Component sizes are shown in Figure 9–25. Make the drawings twice full-size. The manufactured board should measure the size shown at the top of the schematic. Your drawings will be twice that size.

Tie all of the grounds together and connect them to the ground terminal, Pin 5. Make the ground traces .200 wide. Be sure to center pads at the intersection of grid lines and center connecting lines on a grid as shown in Figure 9–26. Use .10 tape and .250 outside diameter pads with .05 inside diameters on the circuit artwork. Prepare a drill plan by running a sepia copy of the artwork, splicing it into a format sheet, and adding board outline, dimensions, hole letters, and a hole schedule.

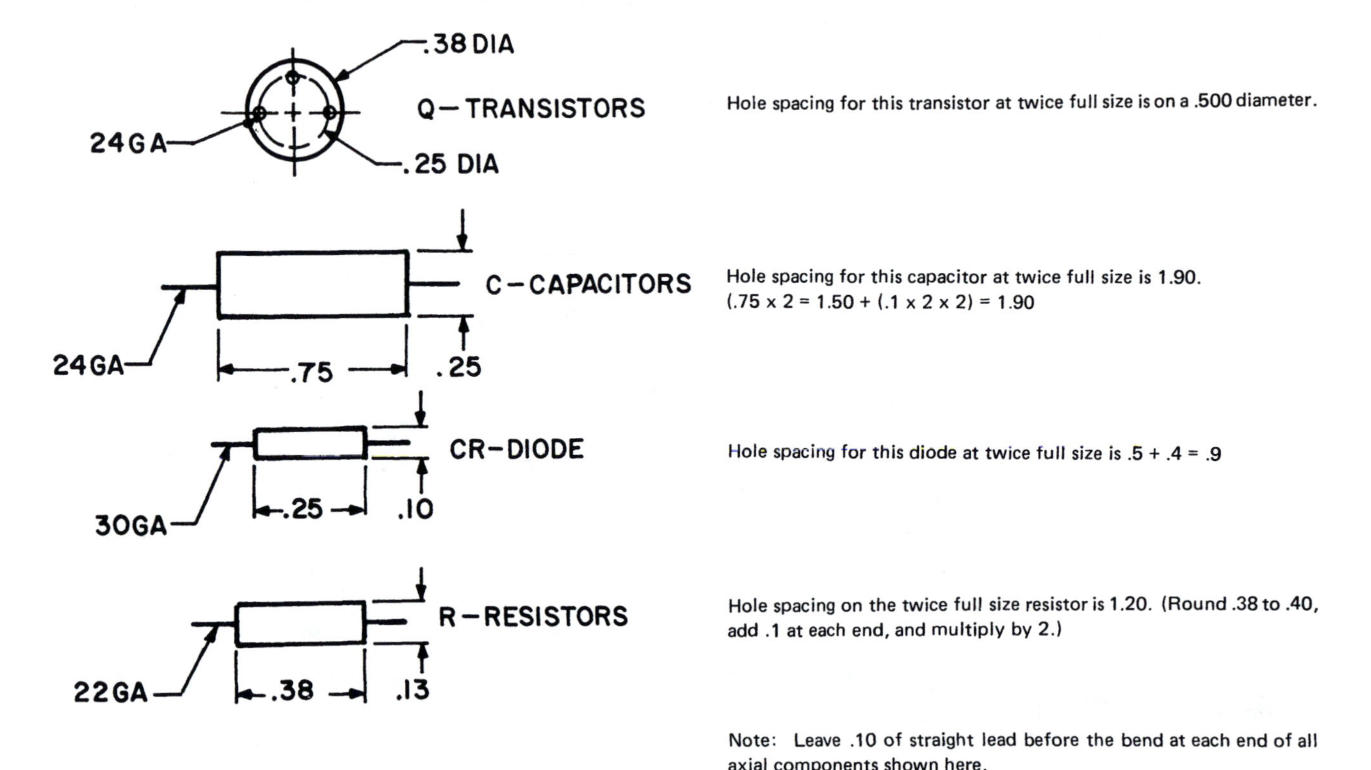

Figure 9–25 Component sizes

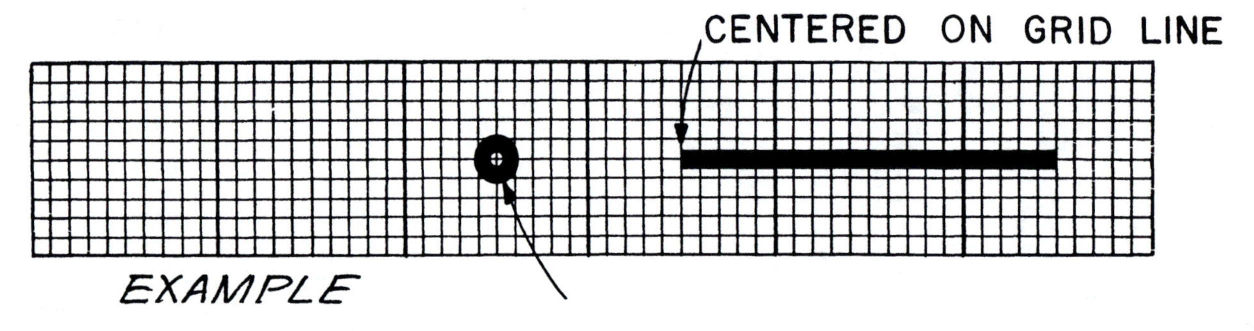

Figure 9–26 Centering tape and pads

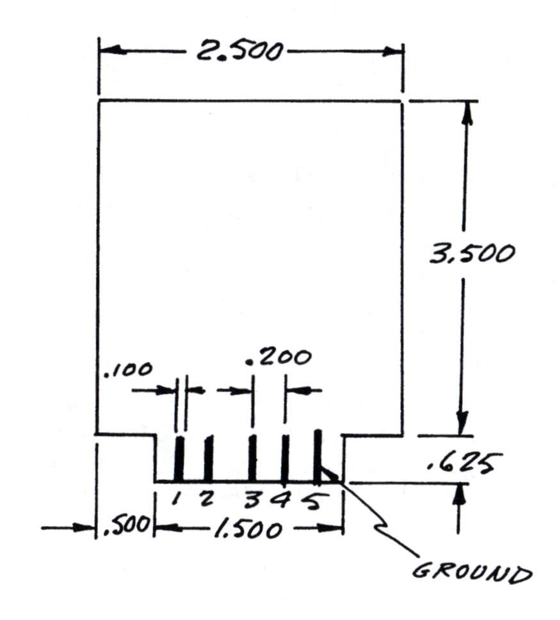
2.500
3.500
.100
.200
.625
1 2 3 4 5
.500
1.500
GROUND

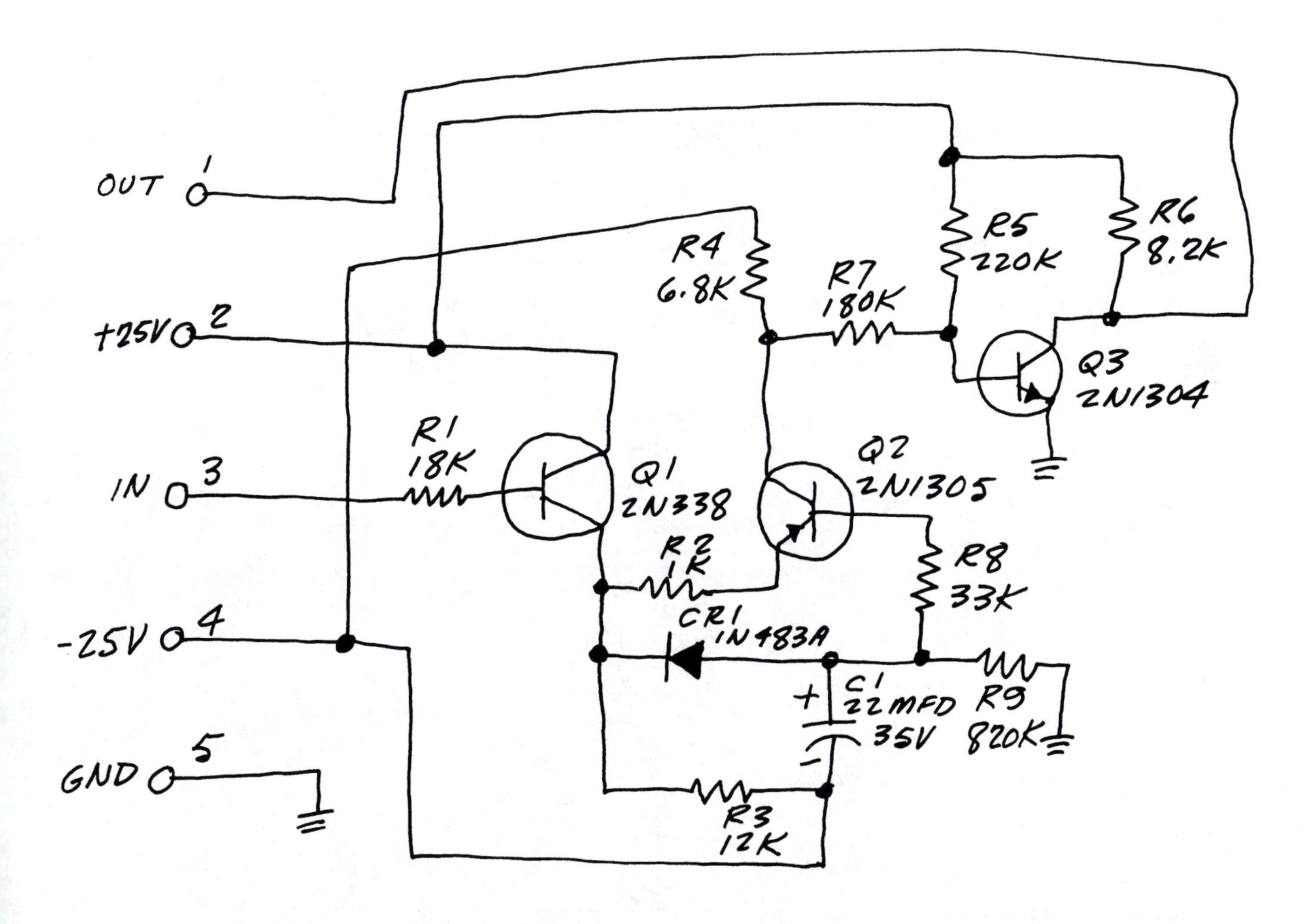
OUT 1
+25V 2
IN 3
-25V 4
GND 5
R1 18K
Q1 2N338
R2 1K
R3 12K
R4 6.8K
R7 180K
R5 220K
R6 8.2K
Q3 2N1304
Q2 2N1305
R8 33K
CR1 1N483A
C1 22MFD 35V
R9 820K

EXERCISE 9-4

Use the following schematic and parts specifications to design a double sided printed circuit board. Provide:

- A design layout
- A circuit artwork for the circuit side
- An artwork for the component side of the board
- A drill plan
- An artwork to be used for silk screening reference designators on the component side of the board.

The board size should be as small as practical without crowding components or standing them on end. Consult figures in this chapter if there is any doubt about how these four drawings should look. E1 thru E6 should be in order, centered on one edge of the board .375" from the edge of the board to the center of the terminal (E), and .375" from center-to-center of the terminals. Allow $\frac{1}{4}$" area around the outside edge of the board for tooling holes. Notice that there are two resistors labeled R10 on the schematic sketch. Renumber the second R10 as R11 and renumber all resistors from R11 to R26.

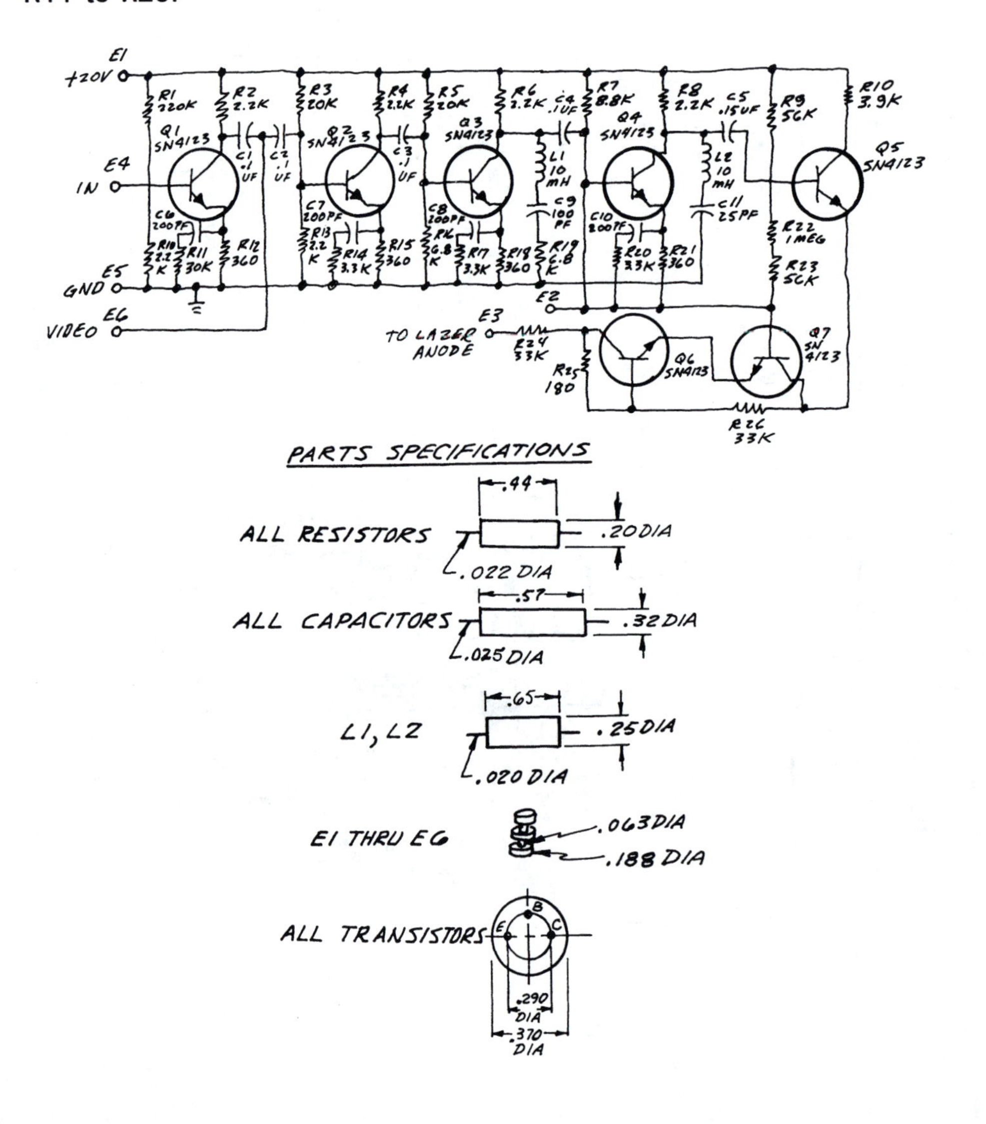

EXERCISE 9–5

Prepare a design layout, circuit side artwork, component side artwork, and a drill plan for a double-sided printed circuit board from the schematic that follows, and the specifications on the next page. The DC/DC converter must be mounted on one side of the board and the connector and all other components must be mounted on the other side of the board. There will be circuitry on both sides of the board.

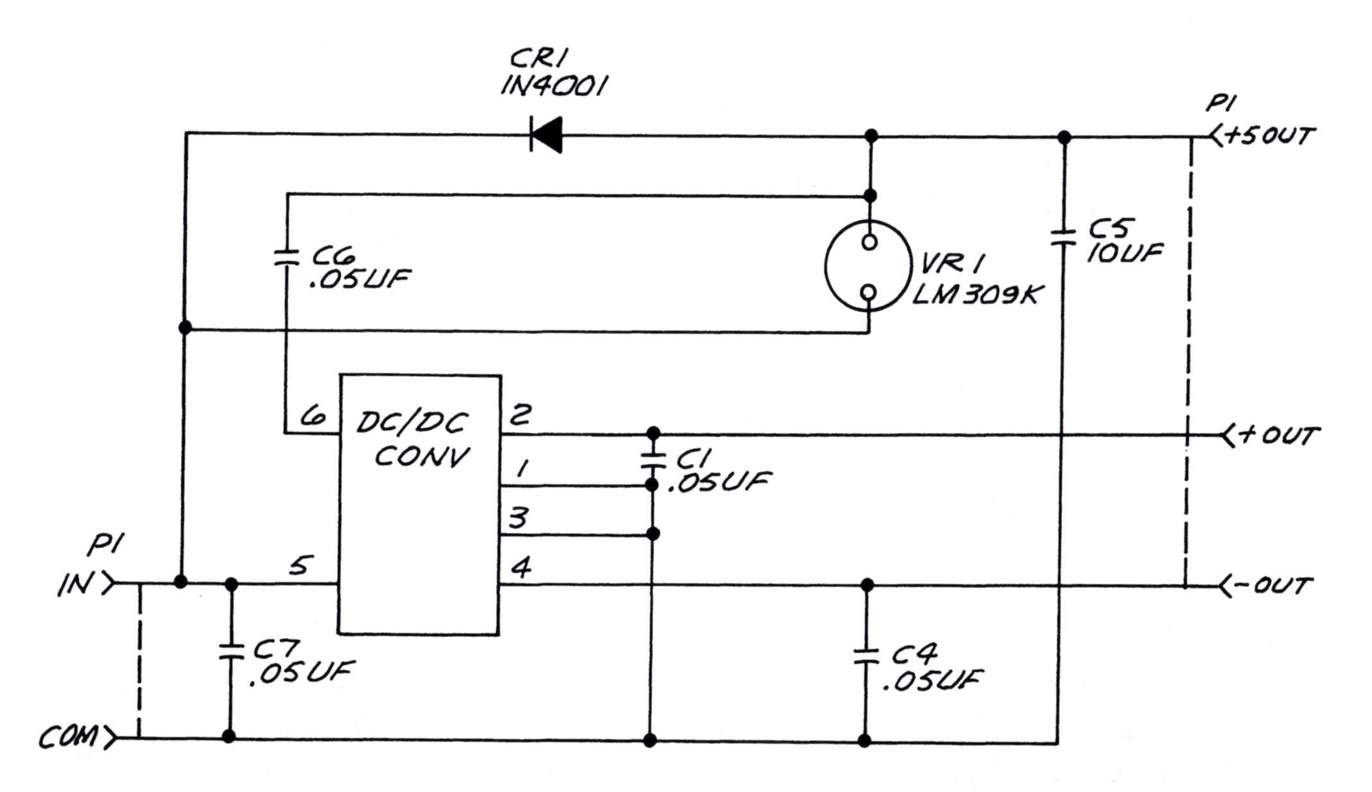

SCHEMATIC FOR EXERCISE 9–5

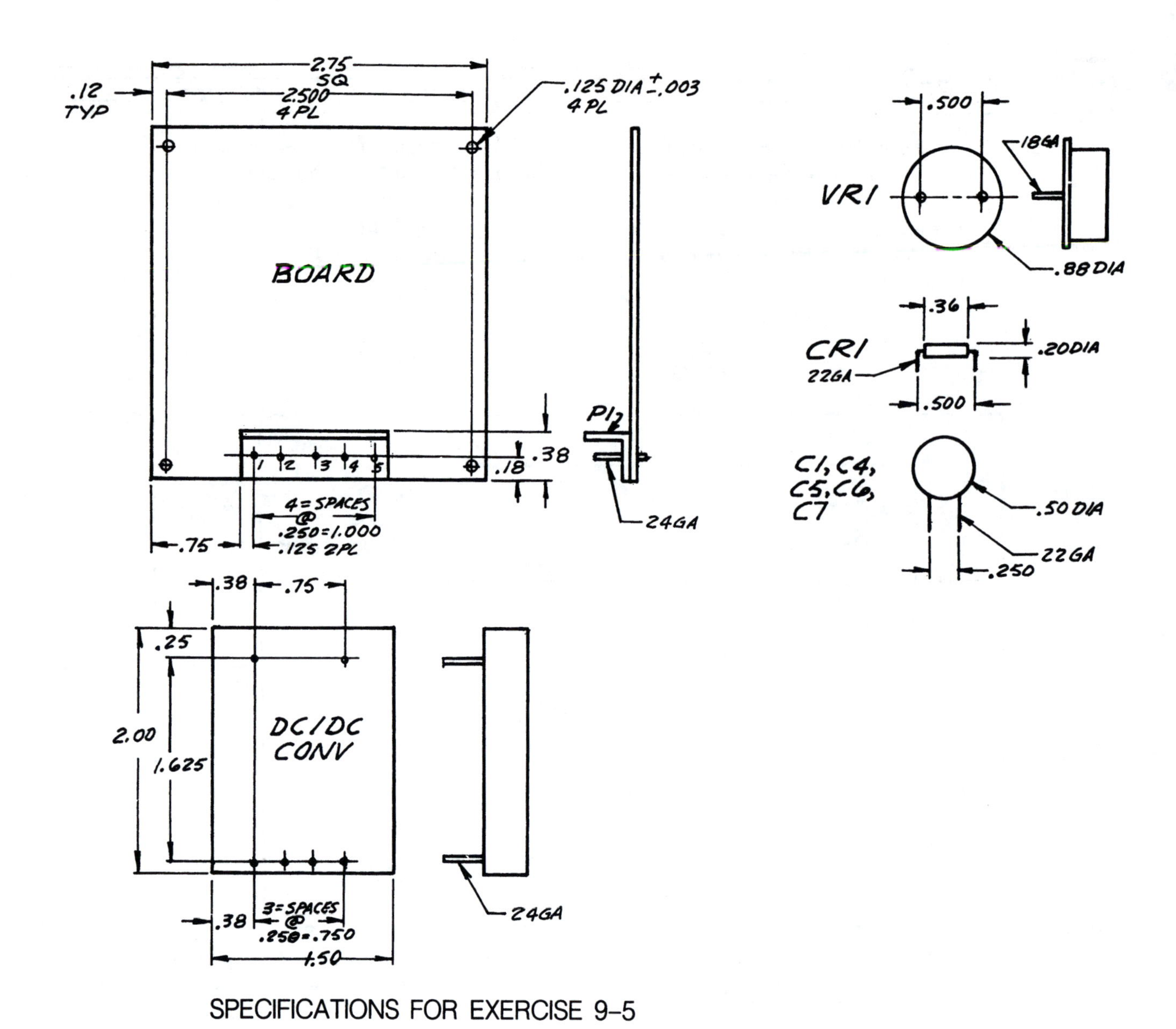

SPECIFICATIONS FOR EXERCISE 9–5

Chapter 10

Design of a Double-Sided PC Board with Integrated Circuits

OBJECTIVES

After completing this chapter, you will be able to

- ✓ group gates, using manufacturers' specifications for integrated circuit packages so that the least number of integrated circuit packages may be used for a given schematic.
- ✓ calculate the area on a PC board to be used for integrated circuit packages.
- ✓ make a chart of integrated circuit connections.
- ✓ produce a design layout drawing for a two-sided printed wiring board from a given schematic.
- ✓ assign pin numbers and reference designators for integrated circuit gates, and pin numbers for the connector on the given schematic.

LOGIC SYMBOL REVIEW

Review the logic symbols described in Chapter 6 and determine what you need to know about them to design a PC board from a schematic diagram. Figure 10–1 shows six of the commonly used gates. The AND gate, NAND gate, OR gate, and NOR gates all have two inputs and one output. The INVERTER and the BUFFER symbols have a single input and a

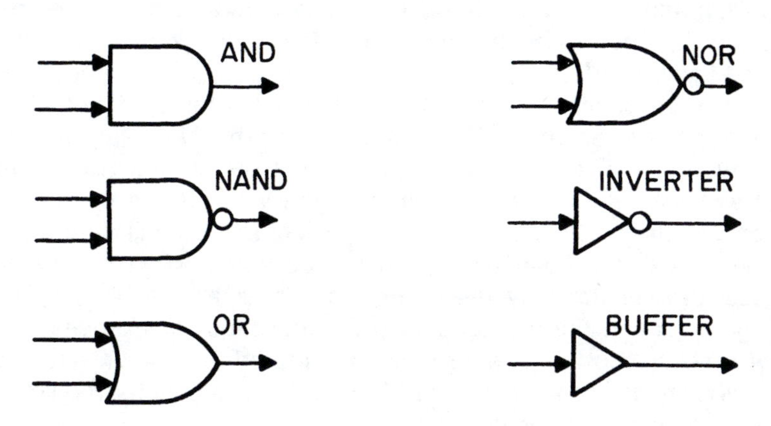

Figure 10–1 Commonly used gates

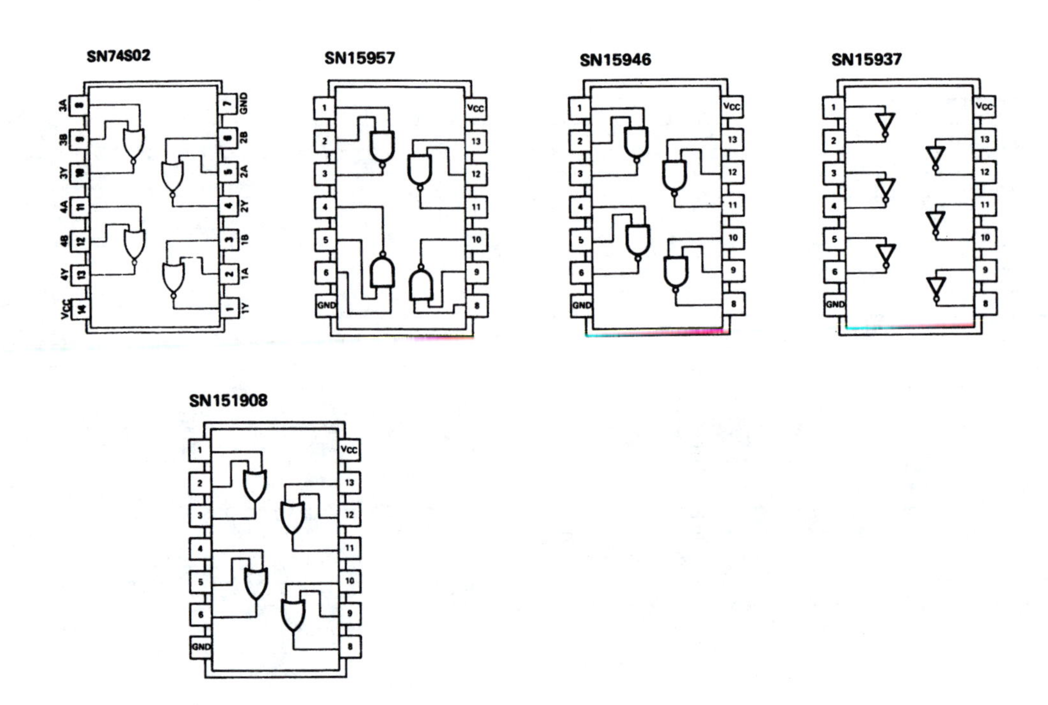

Courtesy of Texas Instruments, Inc.

Figure 10–2 Integrated circuit functions

single output. The functions of these gates are described in Chapter 6. While these functions should be of interest to you, a complete understanding of them is not needed to design a PC board using integrated circuits. All you really need to know is how to identify the symbols, and how to relate them to the integrated circuit (IC) packages.

Notice that the AND gate has no circle on the output side which distinguishes it from the NAND gate. The OR gate has no circle on the output which distinguishes it from the NOR gate. The BUFFER has no circle on its output, distinguishing it from the INVERTER. How these symbols relate to the IC packages is shown in Figure 10–2.

INTEGRATED CIRCUIT PACKAGES

The integrated circuits shown in Figure 10–2 are the same ones you will be using to design the drawings in Exercise 1 at the end of this chapter. Notice the illustration of the SN15946 chip. The inputs of one of the gates are Pins 1 and 2. The output of this gate is Pin 3. Pin 7 of all of the ICs shown in the figure is GROUND, and the power pin (V_{CC}) is Pin 14. All Pins 7 and 14 must be connected to pins on the connector of the PC board. Not all of the gates in each IC will be used every time. Be aware that each different IC type has different gates with different pin arrangements. Study these illustrations briefly and remember that you will refer to them later.

When you design PC boards with IC packages, you will often be able to lower the cost of the boards by using a process called *gate combining.* This process may be done by the drafter, but it usually requires approval of the design engineer. Further combining can be done by a process called *mixed logic* which allows gates to be substituted for other gates that ultimately perform the same functions. Mixing logic functions requires the services of an electrical engineer.

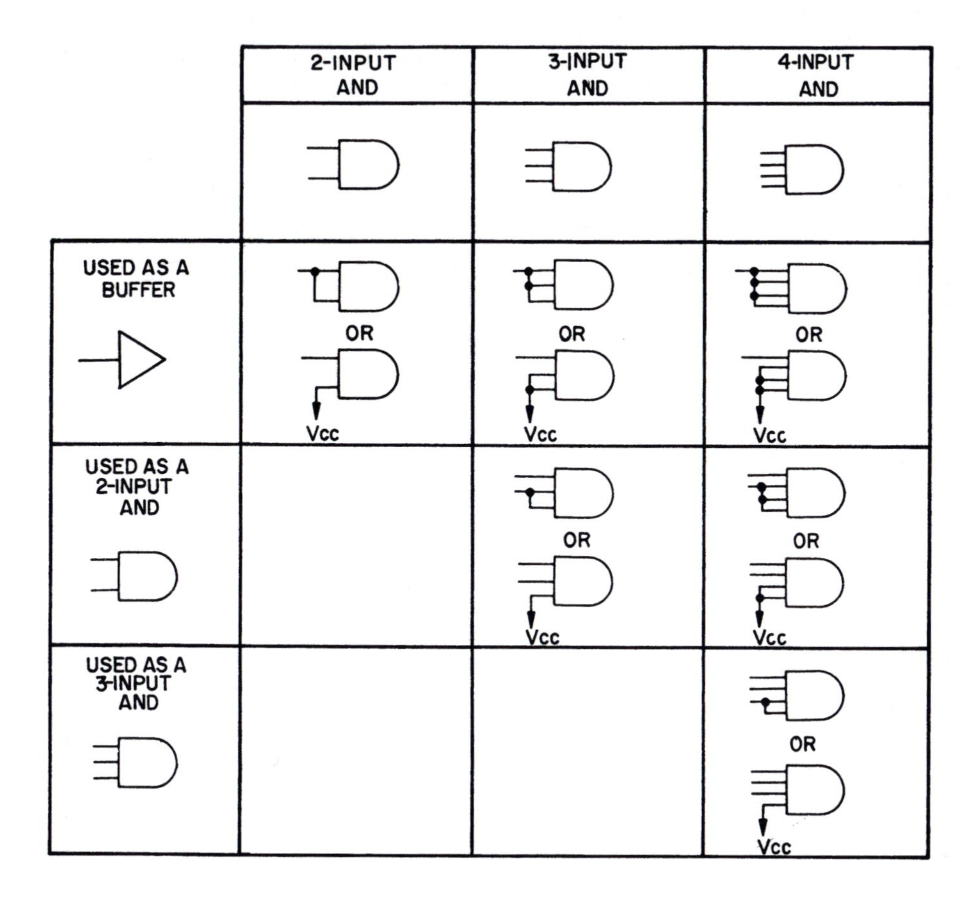

Figure 10–3 AND gate combining

AND Gate Combining

Figure 10–3 shows a chart which describes how the inputs of AND gates can be connected to have the gate perform functions other than the original function. For example, a two-input AND gate can be used as a BUFFER either by tying its two inputs together or by tying one of the inputs to V_{CC}. The three-input AND gate can be used as a BUFFER by tying all three inputs together or by tying two of its inputs to V_{CC}. The three-input AND gate can be used as a two-input AND gate by tying two of its inputs together or by tying one of its inputs to V_{CC}. The four-input AND gate can be used as a BUFFER, a two-input AND or a three-input AND in a similar manner. How gate combining is used to conserve costs of IC packages is described later in this chapter.

NAND Gate Combining

Figure 10–4 describes how the inputs of NAND gates can be connected to form INVERTERS or NAND gates with fewer inputs. All of the inputs of two-, three-, and four-input NAND gates must be connected together or all but one of the inputs must be connected to V_{CC} to form an INVERTER. To make a three-input NAND gate function as a two-input NAND gate, two of its inputs are tied together or one of the inputs is tied to V_{CC}. Other combining may be done in a similar manner by referring to other columns in this figure.

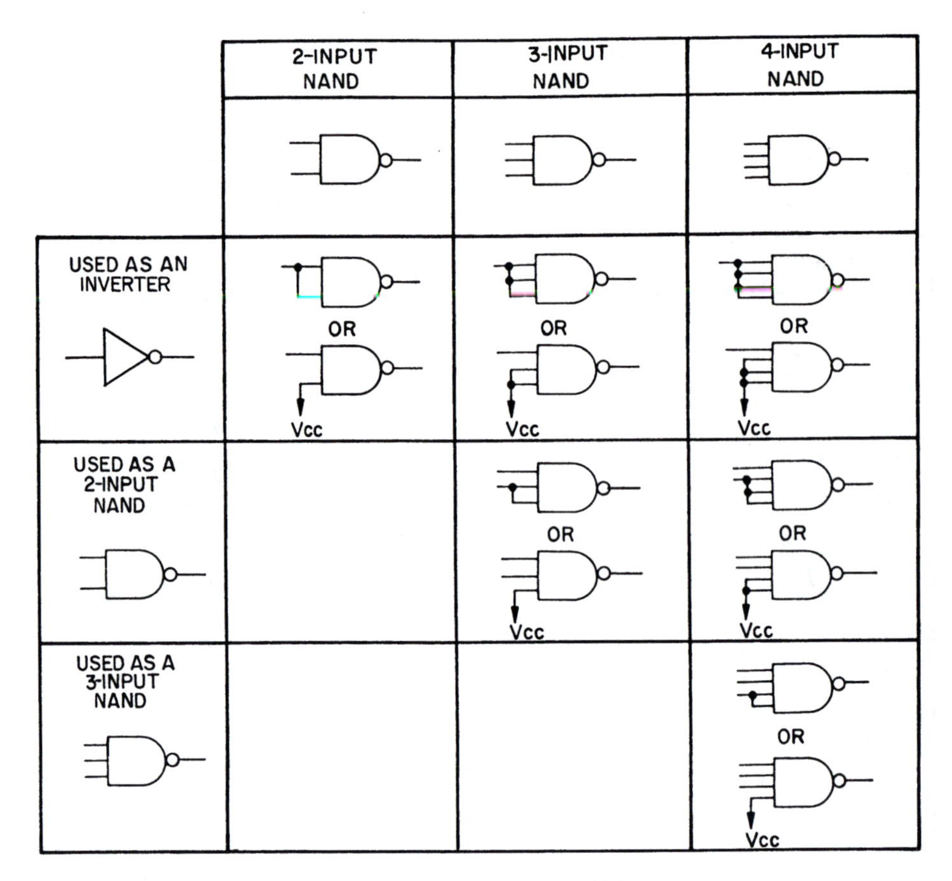

Figure 10–4 NAND gate combining

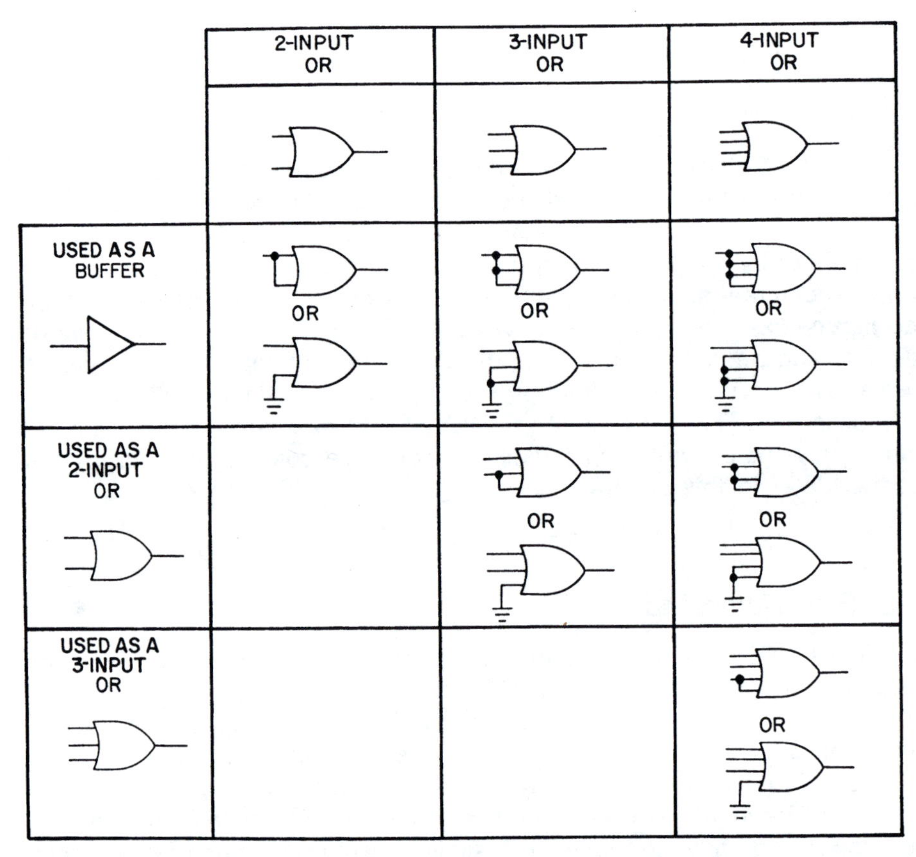

Figure 10–5 OR gate combining

	2-INPUT EXCLUSIVE OR	3-INPUT EXCLUSIVE OR	4-INPUT EXCLUSIVE OR
USED AS A BUFFER	ONLY THIS	ONLY THIS	ONLY THIS
USED AS A 2-INPUT EXCLUSIVE OR		OR	OR
USED AS A 3-INPUT EXCLUSIVE OR			OR
USED AS AN INVERTER	ONLY THIS Vcc	ONLY THIS Vcc	ONLY THIS Vcc

Figure 10–6 EXCLUSIVE OR gate combining

Other Gate Combining

OR gate combining is shown in Figure 10–5. EXCLUSIVE OR gates are shown in Figure 10–6. NOR gates are described in Figure 10–7, and EXCLUSIVE NOR gates are shown in Figure 10–8. The same principles apply to these figures as were used in other gate combining figures, with the exception that some inputs must be connected to GROUND instead of V_{CC}. For example, Figure 10–6 shows that EXCLUSIVE OR inputs must be tied to GROUND to form a BUFFER or to V_{CC} to form an INVERTER. All of the other charts show similar connections to form different gates or functions.

Example PC Board without Using Gate Combining

Figure 10–9 shows a schematic diagram which has been planned using six integrated circuit packages. For example, Gates 1, 2, 3, and 4 have been assigned to IC1. These are all two-input NAND gates. All of the gates in that chip have been used, so there are no spares. Gates 5 and 6 are also NAND gates, so another IC must be used. Because there are no other NAND gates in the schematic, the other two gates in IC2 become spare or unused gates. Gates 7 and 9 require another IC (IC3). Because there are no other two-input AND gates, two of the gates in IC3 also become

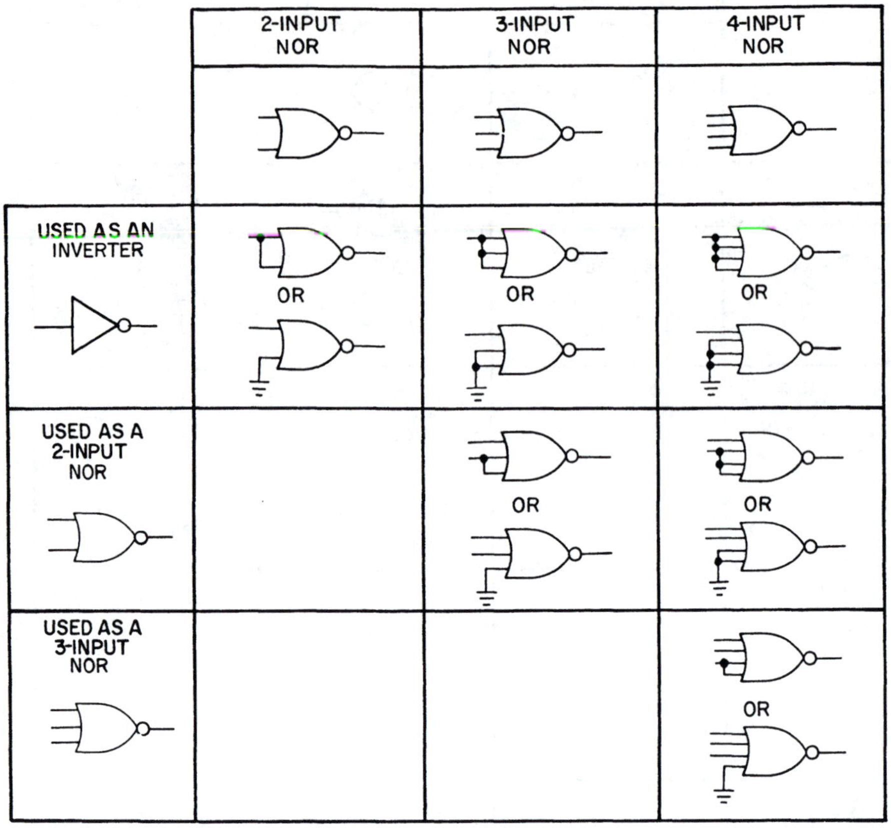

	2-INPUT NOR	3-INPUT NOR	4-INPUT NOR
USED AS AN INVERTER	OR	OR	OR
USED AS A 2-INPUT NOR		OR	OR
USED AS A 3-INPUT NOR			OR

Figure 10-7 NOR gate combining

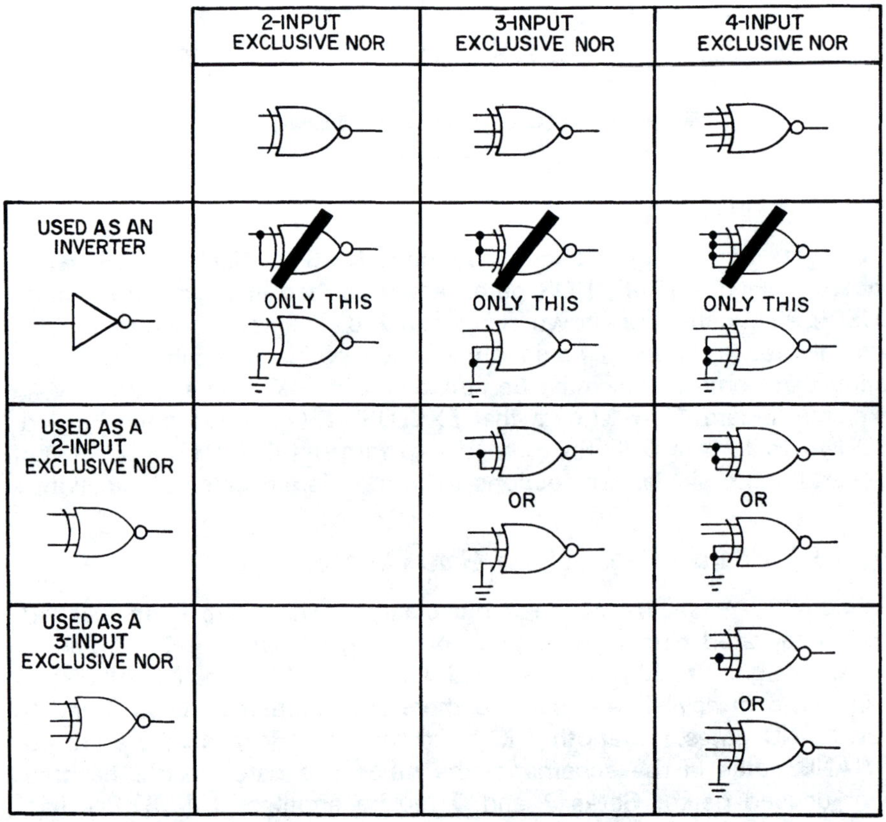

	2-INPUT EXCLUSIVE NOR	3-INPUT EXCLUSIVE NOR	4-INPUT EXCLUSIVE NOR
USED AS AN INVERTER	ONLY THIS	ONLY THIS	ONLY THIS
USED AS A 2-INPUT EXCLUSIVE NOR		OR	OR
USED AS A 3-INPUT EXCLUSIVE NOR			OR

Figure 10-8 EXCLUSIVE NOR gate combining

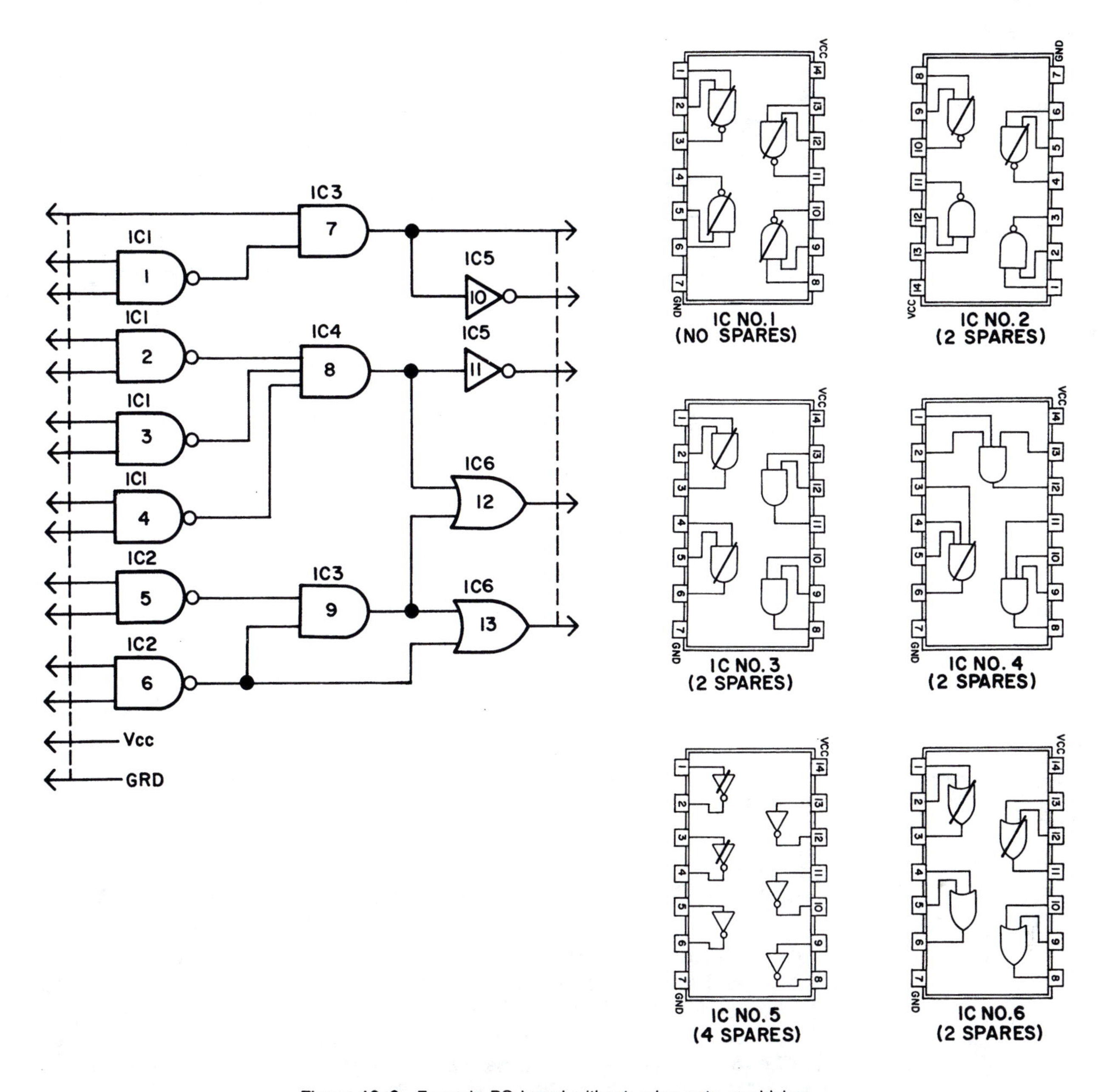

Figure 10–9 Example PC board without using gate combining

spares. Gate 8 is a three-input AND gate which leaves two spares in IC4. Gates 10 and 11 are inverters assigned to IC5 so that four spares are left in that IC. Finally, Gates 12 and 13 are taken from IC6 which also has two spares.

Without gate combining, six IC packages are needed to make a PC board that will perform the functions required of the schematic shown in Figure 10–9. A total of twelve unused gates is found in these six ICs.

Example PC Board Using Gate Combining

Figure 10–10 shows the same schematic diagram as is shown in Figure 10–9. Gates have been combined so that only four integrated circuits are needed to perform the functions required of the schematic. In the schematic of Figure 10–10, Gates 10 and 11 have been changed from

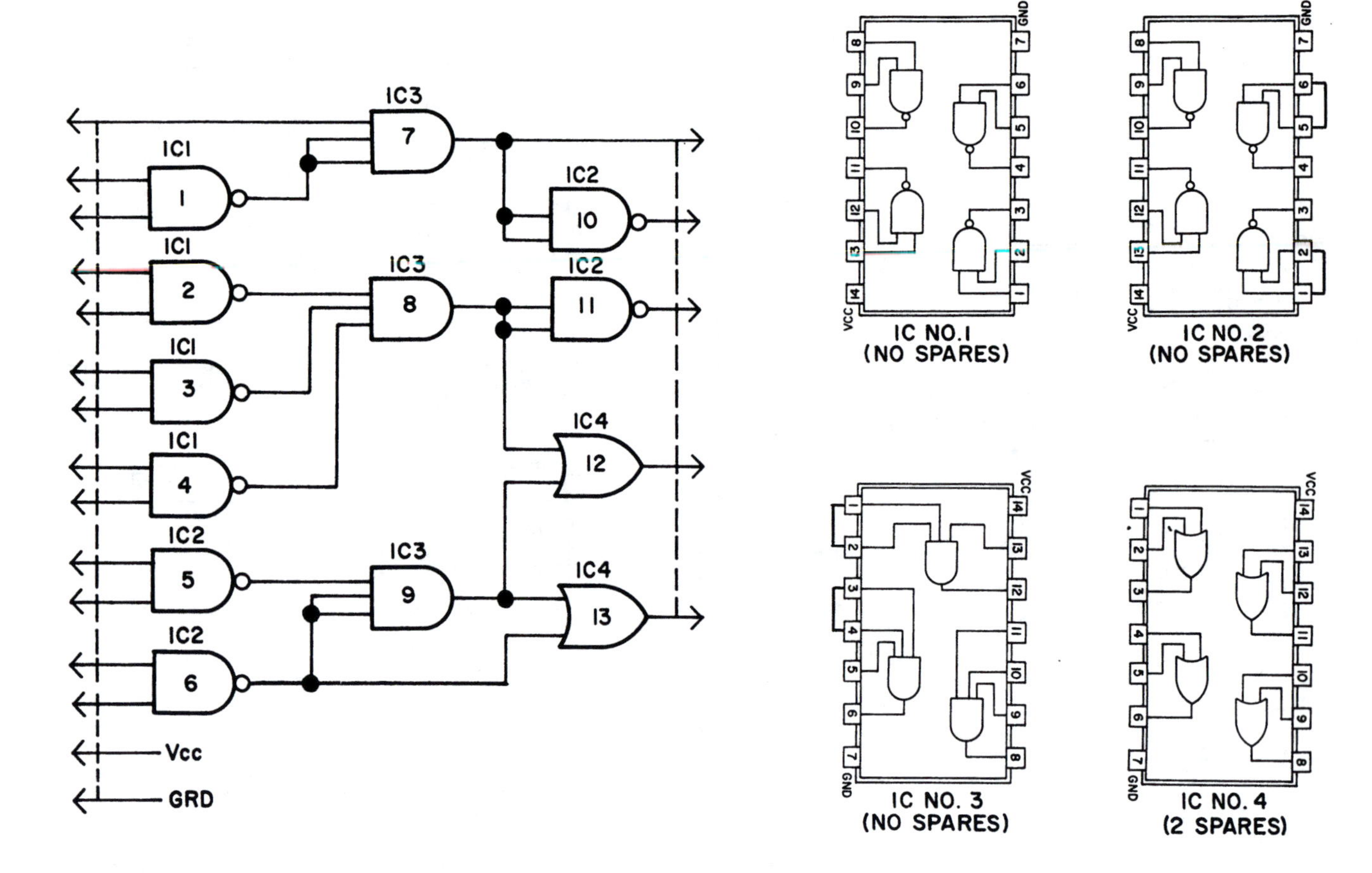

Note: This schematic was changed only for clarity in the text explanation. Do not change the schematic in practice or in Exercise 1 at the end of this chapter.

Figure 10–10 Example PC board using gate combining

NAND gates to INVERTERS by connecting their inputs together. As a result, two gates from IC2 can be used as INVERTERS and the other two gates can be used as NAND gates, 5 and 6. Gates 7 and 9 have been changed from three-input AND gates to two-input AND gates allowing all three gates of IC3 to be used. IC4 contains Gates 12 and 13 and has the only two spares in all of the four ICs.

Gate combining in this case has resulted in a saving of two ICs per PC board and has lowered the cost of the board considerably. Notice the note on Figure 10–10 which explains that gate combining does not require that the schematic be changed to the gates actually used. Now that the number of ICs has been determined, the board on which they are to be placed should be analyzed.

2.00"

2.50"

Figure 10–11 Board profile

Board Area

A board profile must be determined for the example schematic shown in Figure 10–10. This determination is made by measuring the space available for the board. The board profile is shown in Figure 10–11. The area available for integrated circuit packages is 2.00″ by 2.50″ = 5.00 square inches. To arrive at the space available for each IC, 5.00 square inches must be divided by the number of ICs (4): 5.00 ÷ 4 = 1.25 square inches. This is adequate

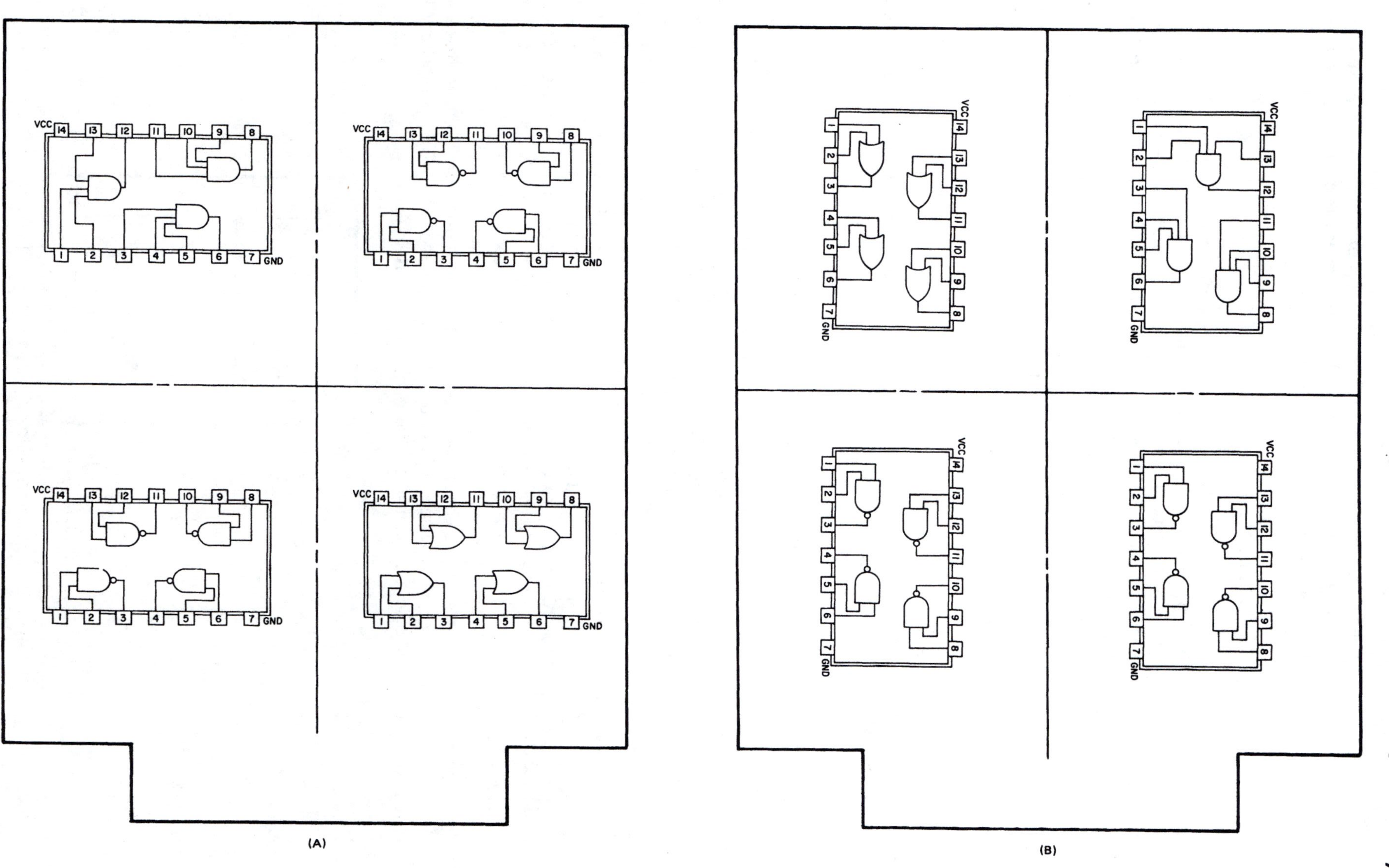

Figure 10–12 Two arrangements for IC placement on the board

space. In general, each IC will require approximately one square inch. If the available area results in less than one square inch per IC, the board may not lend itself to a single-board design.

IC Placement

Figure 10–12 shows two arrangements for IC placement on the board. Notice that all ICs have been turned so that Pin 1 is in the same position on each IC. Figure 10–12A has Pin 1 located in the lower left corner of each IC. Figure 10–12B has Pin 1 located in the upper left corner of each IC. Turning each IC in the same direction reduces the number of possible errors in hooking up V_{CC}, GROUND, and all other pins of the integrated circuits.

In general, the vertical arrangement of Figure 10–12B is the preferred layout because it allows more leads to be connected to the board fingers without passing over other ICs.

Connection Matrix

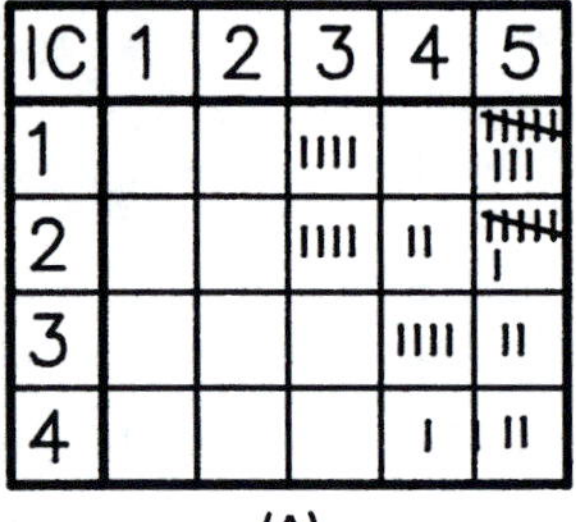

IC	1	2	3	4	5
1			IIII		~~IIII~~ III
2			IIII	II	~~IIII~~ I
3				IIII	II
4				I	II

(A)

IC	1	2	3	4	5
1			4		8
2			4	2	6
3				4	2
4				1	2

(B)

Figure 10–13 Connection matrix

After the general arrangement has been decided upon, some decisions must be made regarding the specific location of each IC. A connection matrix, such as the one shown in Figure 10–13, is often very helpful in making these decisions. The ICs are numbered vertically and horizontally, with one additional vertical (5) column for the connector. The first IC number referenced corresponds with the vertical axis of the matrix; the second IC number corresponds with the horizontal axis.

The schematic of Figure 10–10 was used to find the number of connections between ICs and between ICs and the board connector (P1). For example, in Figure 10–10, Gate 1 of IC1 has two connections to the board connector, and one connection to IC3. Therefore, one mark should go in the IC1–IC3 box (the IC3–IC1 box is not used because that would duplicate connections unnecessarily). Two marks should go in the IC1–P1 box, and so forth, Figure 10–13A. Study the matrix and compare it to Figure 10–10 carefully to determine how the numbers in 10–13B were calculated.

After the matrix has been formed, it is clear that ICs 1 and 2 should be placed nearest P1, the board connector. IC3 has four connections to IC1, four connections to IC2, and two connections to IC4. IC3 then could be placed on either the upper right or the upper left because there seems to be no advantage in one position over the other. The preliminary layout should look similar to the one in Figure 10–14.

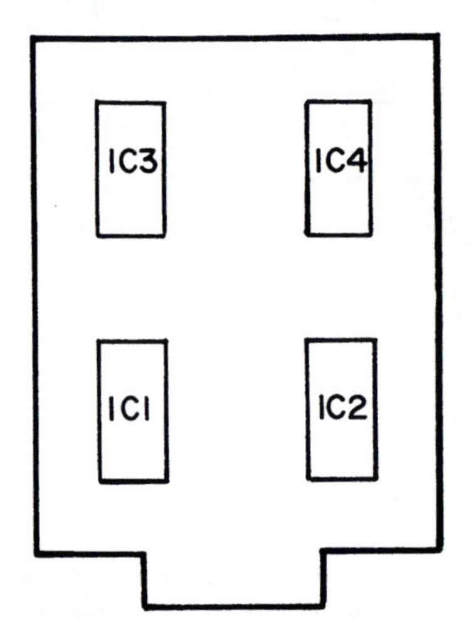

Figure 10–14 Preliminary layout

Board Connector Pin Numbers

The board connector for the example schematic of Figure 10–10 is shown on a blank board, Figure 10–15. The red numbers are odd numbers, and they are located on the front side of the board. The blue numbers are even numbers and they are located on the back side of the board. Notice that the schematic in Figure 10–10 has no pin numbers assigned to the connector. Those will be assigned as leads are drawn from the ICs to the pins on the connector.

DESIGN GUIDELINES

To begin the design of a two-sided board, a few guidelines are often helpful. These guidelines are:

- Keep most of the horizontal lines on one side of the board
- Keep most of the vertical lines on the other side of the board

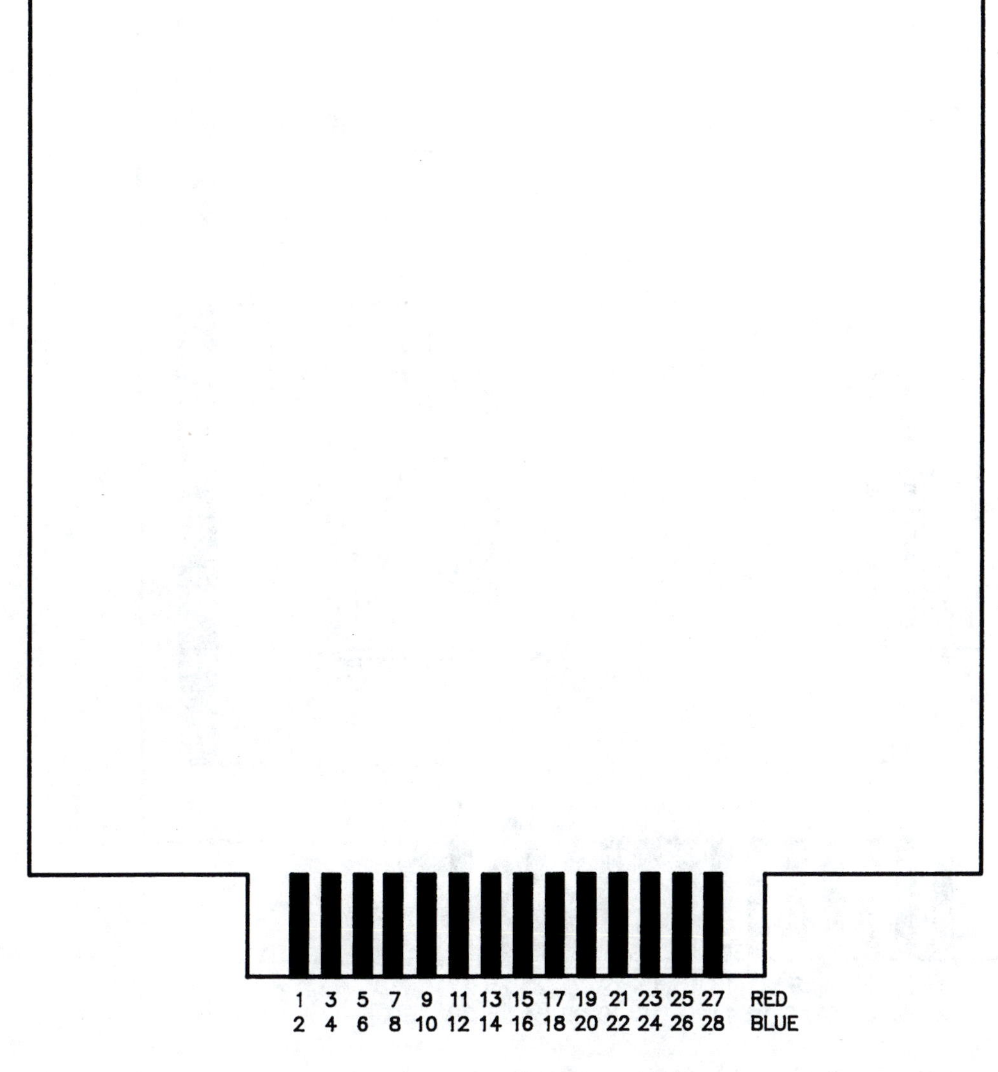

Figure 10–15 Pin numbers of the board connector

- Hook up V_{CC} and GROUND first
- Keep feed-thrus to a minimum
- Avoid diagonal traces

Connect V_{CC} To Begin Red/Blue Layout

All of the V_{CC} pins of each IC must be tied to a pin on the connector. In this example, the V_{CC} pins are Pin 14, Figure 10–16. All of these are connected to a wide strip on one edge of the board. This strip is then tied to a pin on the connector. In this case, V_{cc} was tied to Pin 27 which is on the front side of the board. The logic symbols shown inside of each IC are

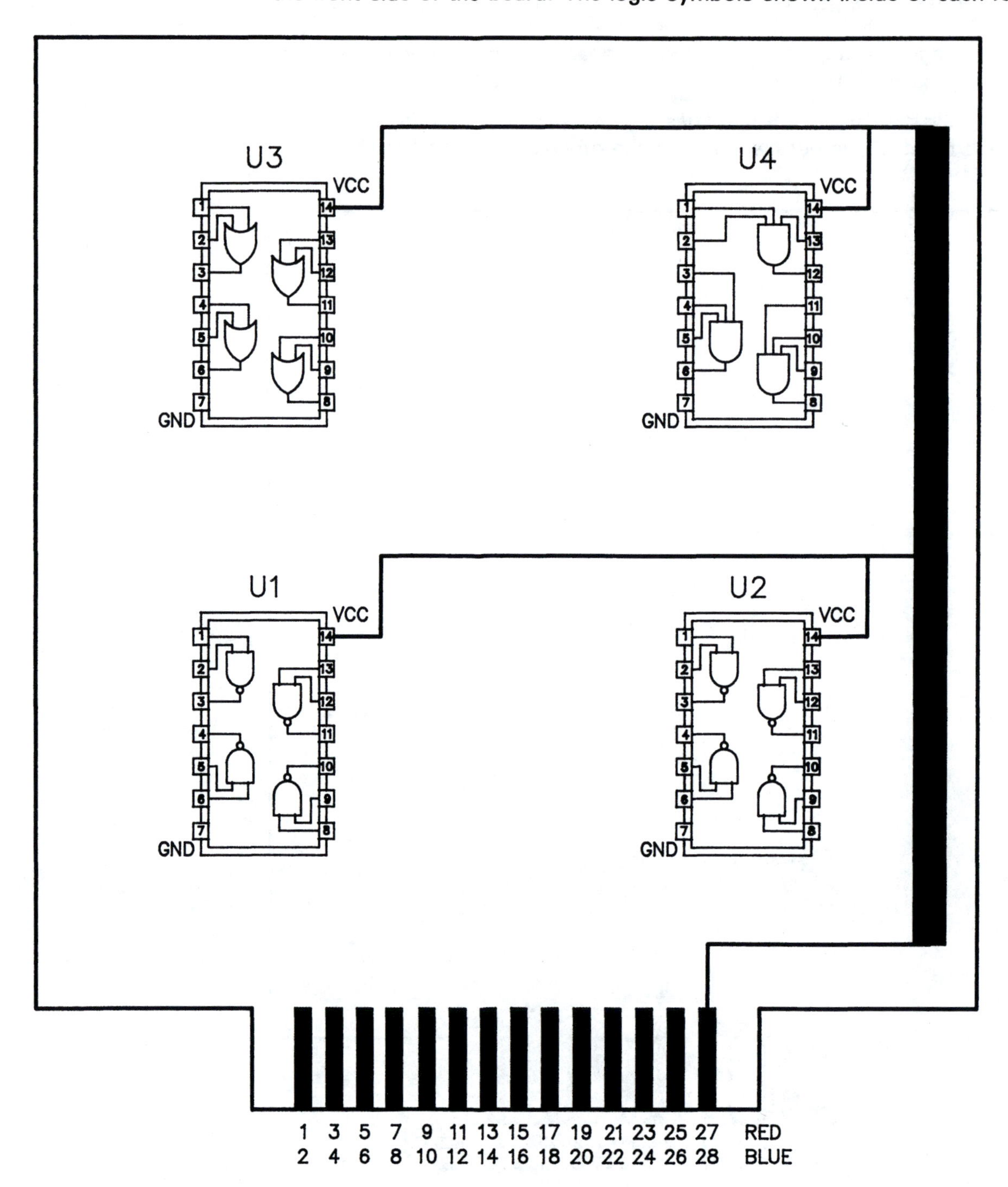

Figure 10–16 Connecting the V_{CC} pins

there for reference only. In the actual design of the board, only the pattern for the pads of each IC is needed. Lines are then accurately drawn in red pencil on a grid to show all connections on the front side of the board. This layout is then used to photo plot or manually tape an artwork from which the board is etched.

Connect Grounds

The next step is to connect all of the GROUND pins to a pin on the connector, Figure 10–17. In this case, all of the GROUND pins are Pin 7.

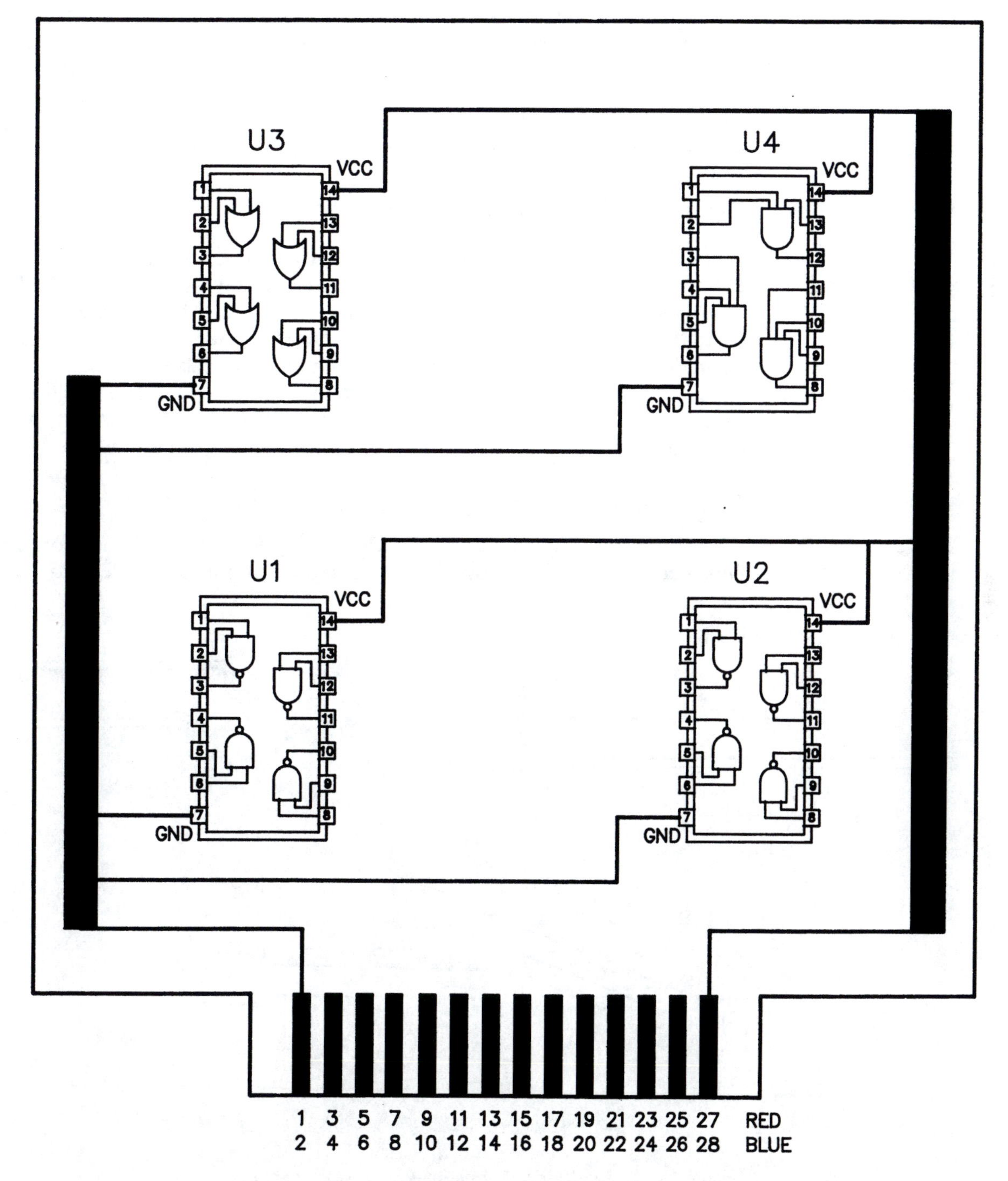

Figure 10–17 Connecting the GROUND pins

All of these GROUND pins are connected to Pin 1 on the front side of the board. They could just as easily have been connected to Pin 2 on the back side of the board, but the designer of this board decided to place both V_{CC} and GROUND connections on the front of the board.

Avoid Diagonal Lines

Before going on to connect all other leads of the ICs, it is necessary to discuss the reason to avoid using diagonal lines on a design layout, Figure 10–18. Many PC boards are very crowded and, as a result, must have leads placed as close together as possible. A minimum spacing allowance is always required for the distance from one conductor to another. If lines are laid out on a horizontal and vertical grid, the minimum spacing requirement is automatically achieved when the grid lines are selected. If diagonal lines are used or if pads are placed in a diagonal relationship, as in 10–18B and 10–18C, the minimum distance between conducting lines or pads must be carefully calculated. This results in more design time and can cause electrical shorts if the conductors are spaced too close together.

Drawing Lines Connecting ICs

An analysis of the matrix in Figure 10–13 and the schematic of 10–10 shows that IC1 and IC3 have four connections between the two ICs. Figure 10–19 shows how those connections were made. These ICs were placed so that one was directly above the other. This allows these connections to be made with the least number of feed-thrus (also called VIAs). VIA in this case means "by way of." All four connections were made on the same side of the board. This could change as the layout proceeds but, since a designer has to start somewhere, this was the beginning chosen.

Drawing Lines Connecting ICs to the PCB Connector

Next is the connecting of the ICs to the printed circuit board connector. These connections are shown here in a screen of one color in Figure 10–20.

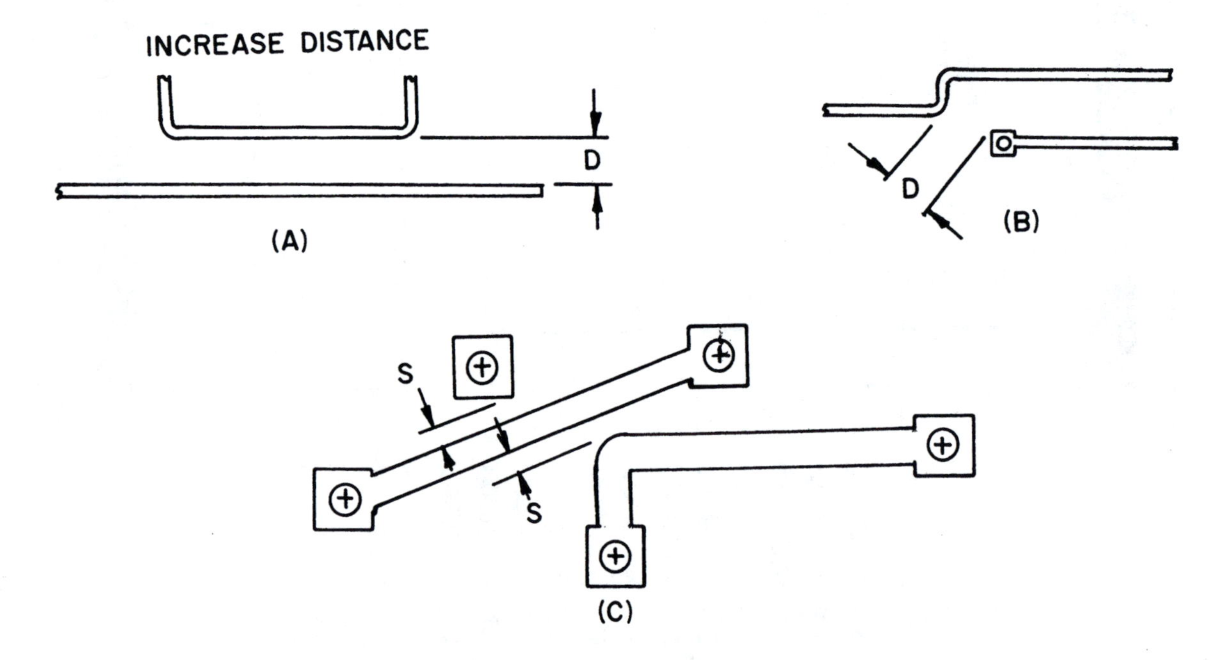

Figure 10-18 Avoid using diagonal lines on a design layout

When the design layout is complete, it will appear as shown in the three-color reproduction of Figure 10–24 on the inside back cover. The color rendition shows all of the leads of IC1 connected to the PCB connector with red lines representing lines on the front side of the PC board. These connections could have been made on the back (blue lines) side of the PC board. The front and the back side of the board each has pads for all pins

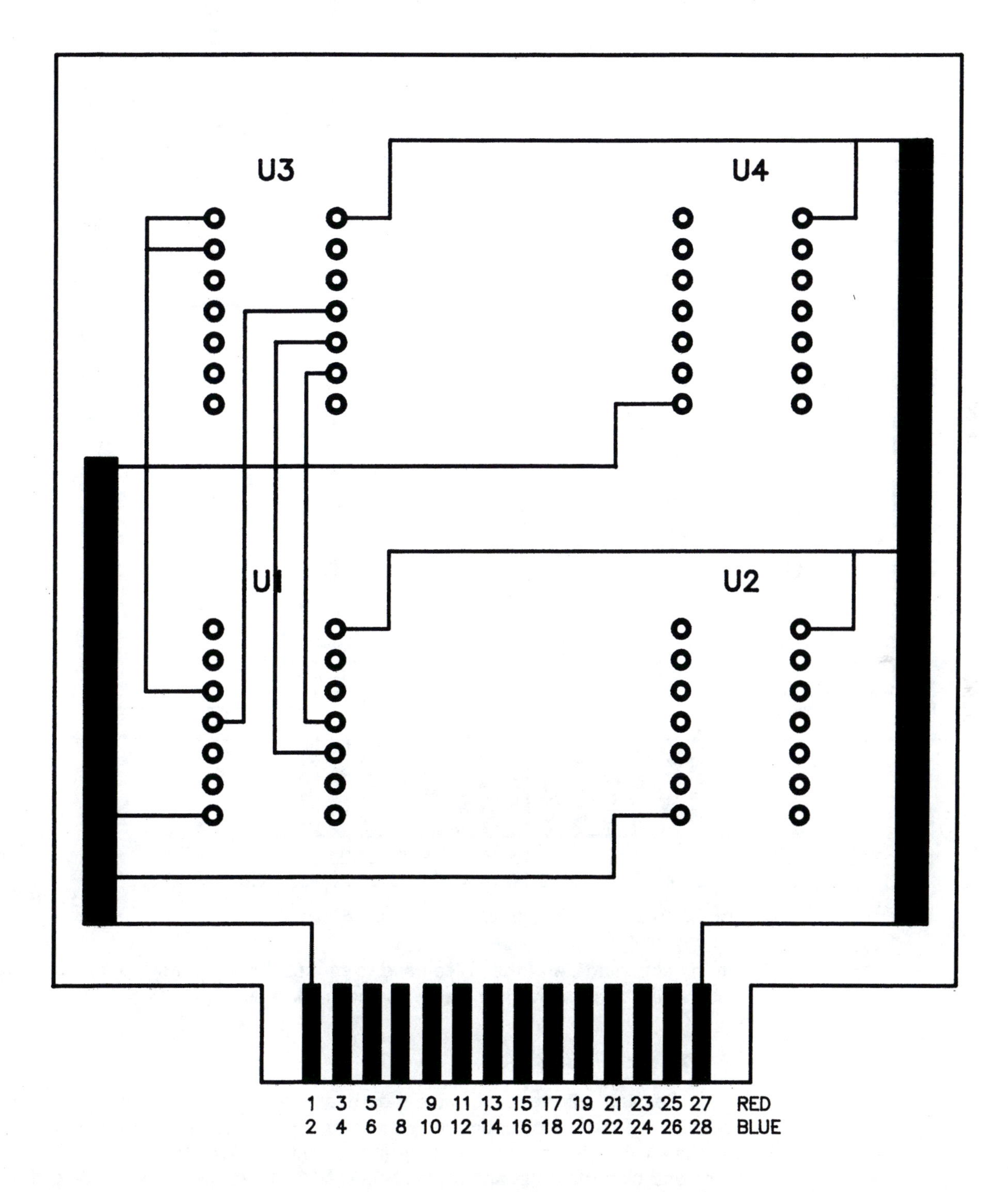

Figure 10–19 Connecting lines between ICs

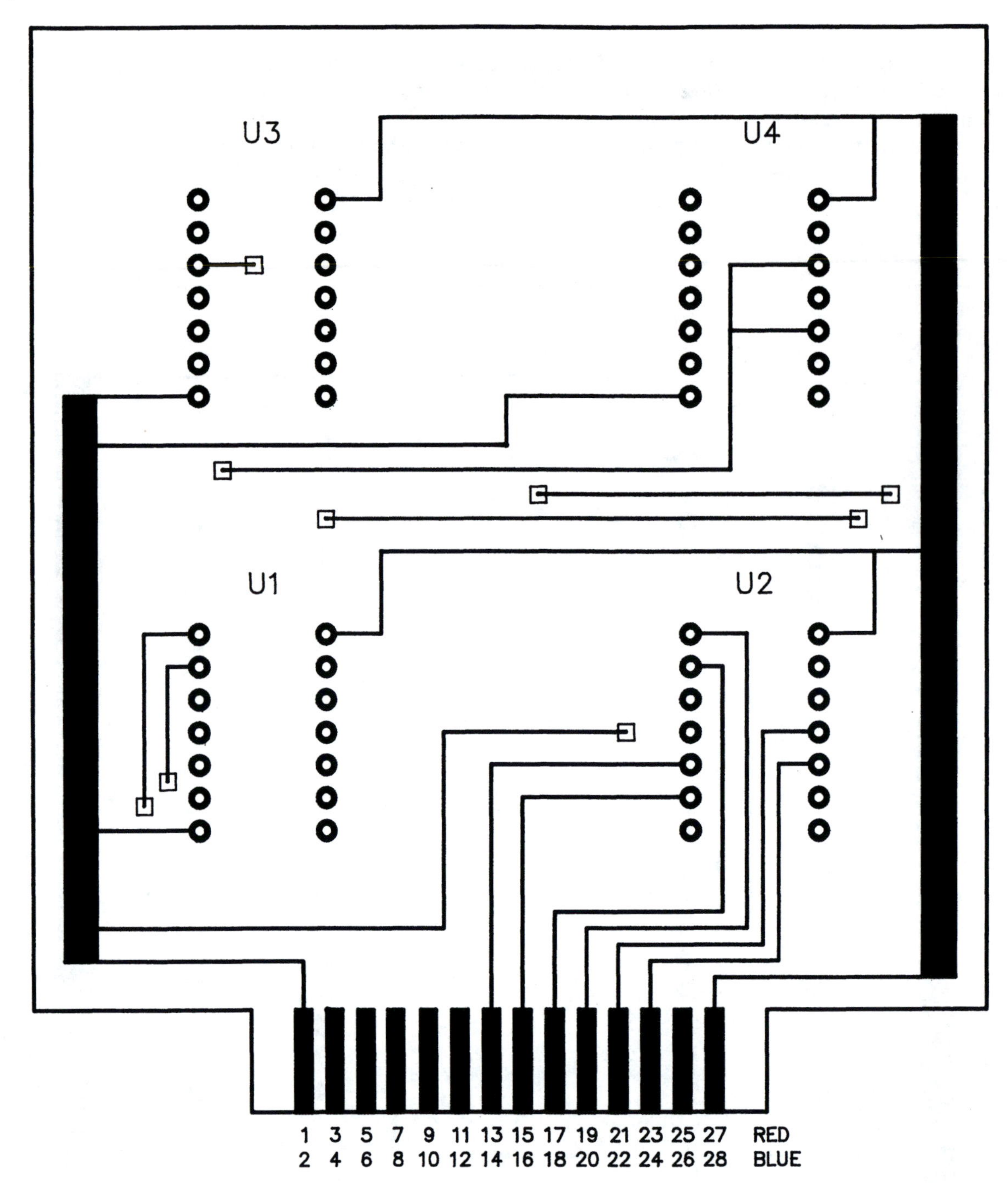

Figure 10–20 Connecting the ICs to the PC board connector

on the connector, and the holes used to mount ICs are plated thru to provide a connection from the front side to the back side.

Using Feed-Thrus (VIAs)

Because all traces on the board are the same electrically as bare wires, they must not touch one another. As a result, it often becomes necessary to use a plated-thru hole (a feed-thru or VIA) to pass a conductor from one side of the board to the other. VIAs have pads surrounding them on both sides of the board. Usually the pads for VIAs are shaped or sized differently from the other pads so that they can be easily identified. The size of plated-thru holes drilled through the center of these pads are usually

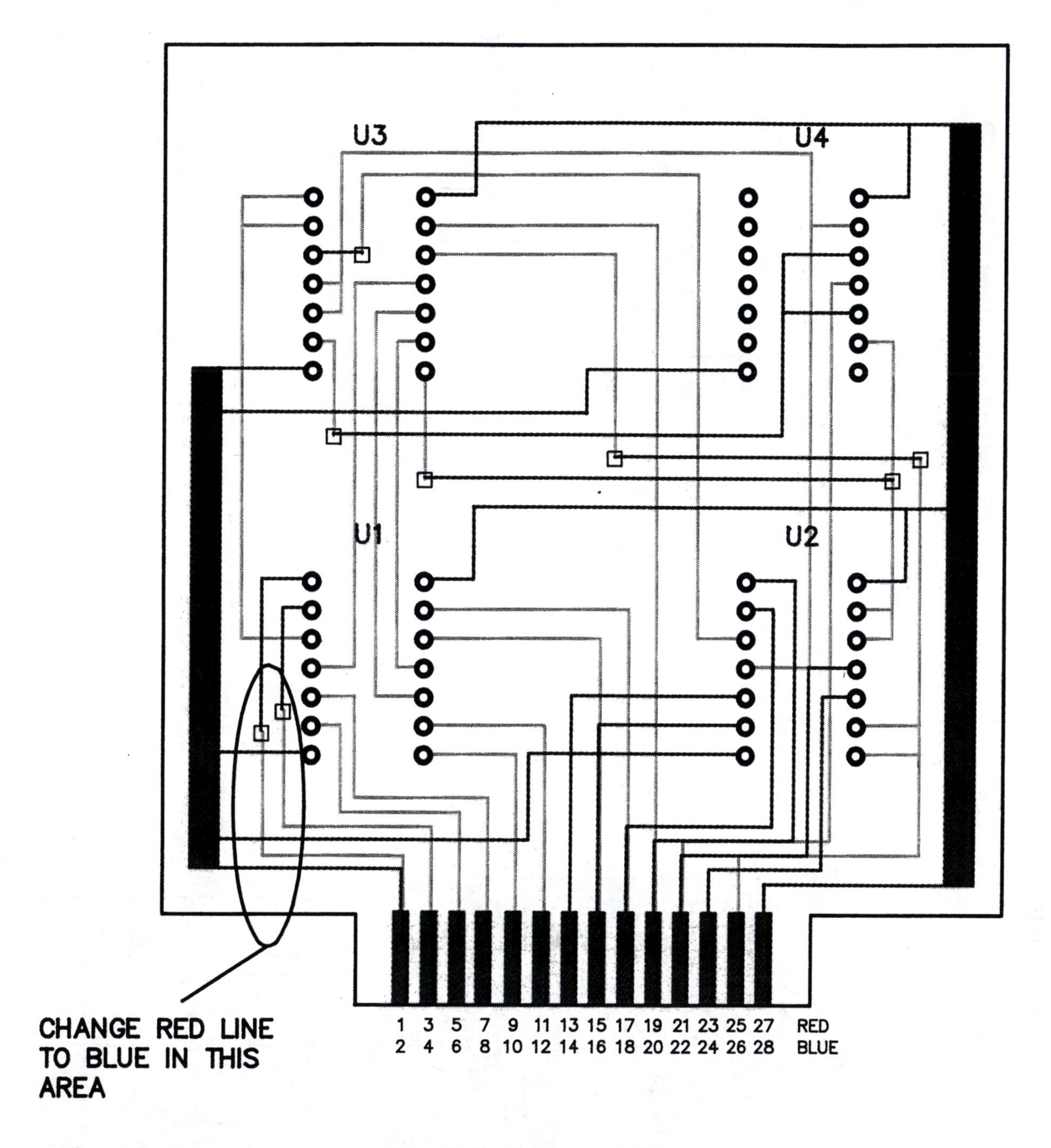

Figure 10–21 Typical use of a VIA

different from the size of holes for the IC leads. Figure 10–21 shows a typical use for a VIA where the lead from Pins 1 and 2 of IC1, which were originally red, were changed to blue in the circled area. The VIA allowed the trace to pass to the back side of the board so that the GROUND lead no longer touches the vertical leads passing from IC1 to the connector.

Labeling the Schematic

After lines are connected on the design layout, they must be labeled on the schematic diagram, Figure 10–22. This is the same schematic as the one shown in Figure 10–10. Now, however, the leads on the schematic from U3 and U1 are numbered with the pin numbers taken from the ICs

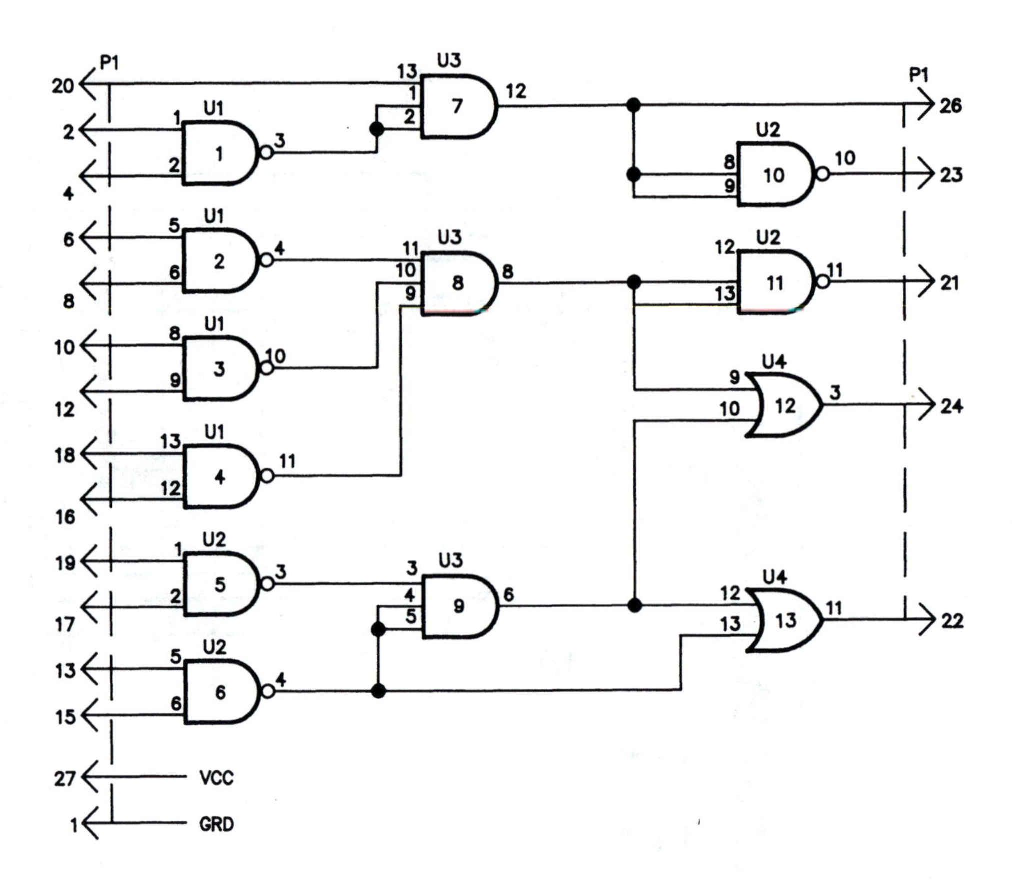

Figure 10-22 Labeling the connecting lines on the schematic

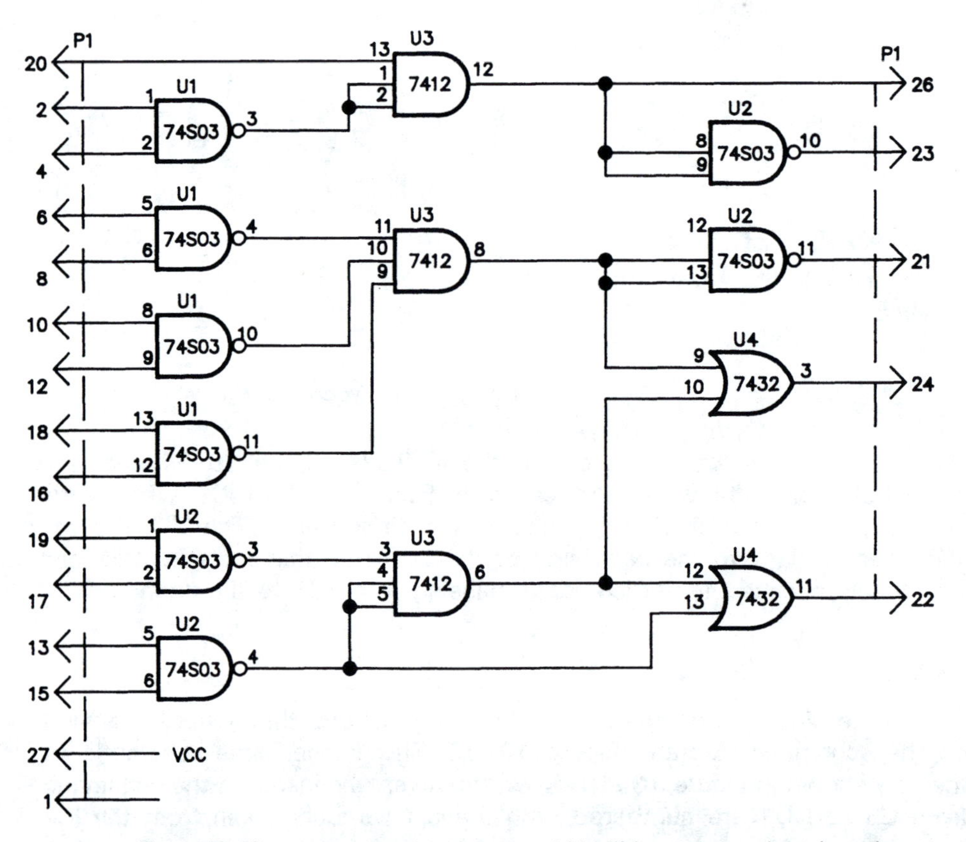

Figure 10-23 Labeling the pin numbers and identifying gates on the schematic

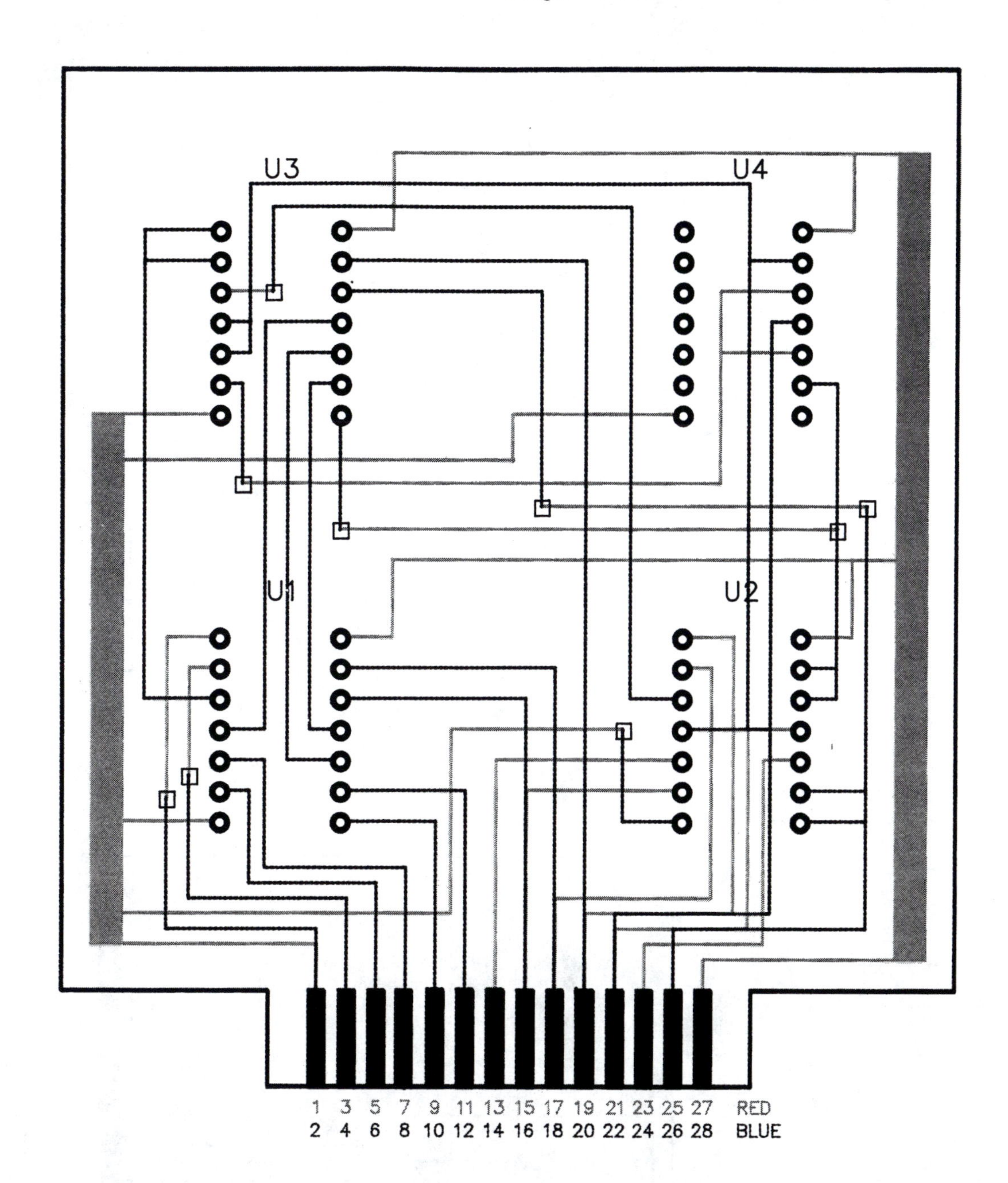

and the connector used on the complete design layout, Figure 10–24. This figure is shown here in a screen of one color. A similar layout appears in three colors on the inside back cover. Your layout for the exercises in this chapter should be drawn in red, blue, and black colored pencils just like the drawing on the inside back cover. All pin numbers on each gate and on the connector must be labeled on the schematic, Figure 10–23. Each gate must also be identified with the IC number and its type. Notice that the original IC numbers have been changed.

ETCHED BOARD ARTWORK

After the design layout has been completed, an artwork is made of each side of the board, Figure 10–25A and 10–25B. Notice that pads are shown for each pin on each IC on both the front (component side) and the back (circuit side) of the board. In addition, a silk screen artwork, Figure

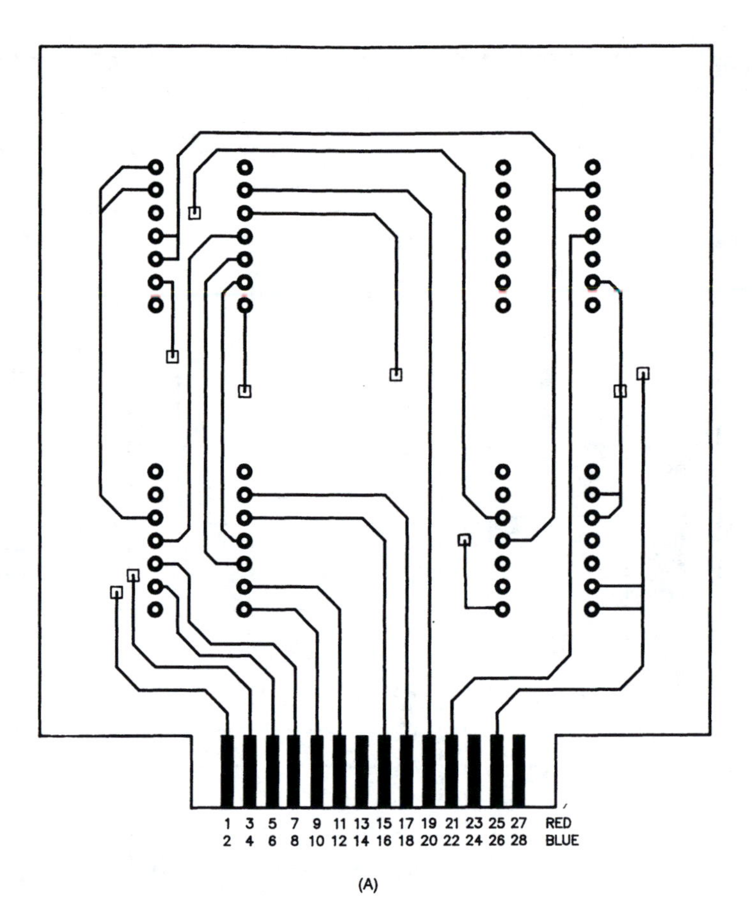

(A)

1 3 5 7 9 11 13 15 17 19 21 23 25 27 RED
2 4 6 8 10 12 14 16 18 20 22 24 26 28 BLUE

(B)

Figure 10–25 Artwork of each side of the board (targets not shown)

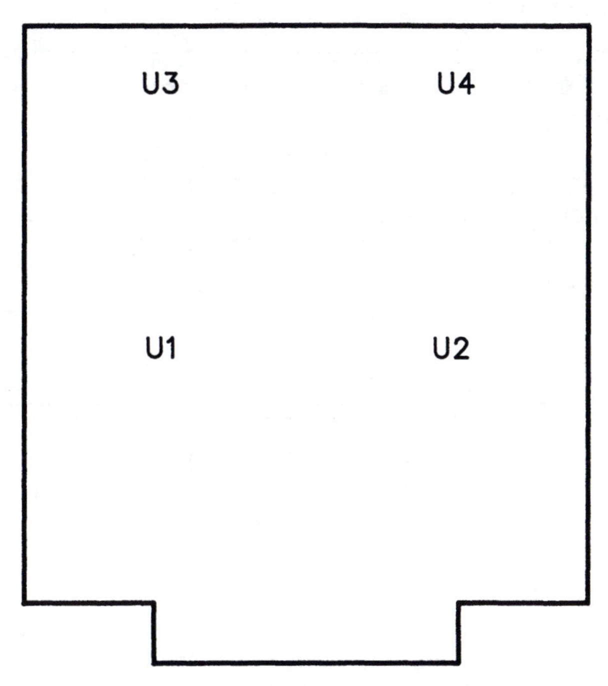

Figure 10–26 Silk screen artwork

10–26, is also made in nonconductive ink for reference designators, the part numbers of the board, and any other board markings which must be printed on the board in permanent, paintlike ink. All artwork must have targets placed outside of the board area so that it can be properly aligned (registered) with other parts when the board is assembled. Figure 10–27 shows the completed board with ICs mounted and soldered in place.

SUMMARY

This chapter describes the design of a two-sided printed circuit board which has four integrated circuits as components. The design begins with a schematic diagram that has logic symbols, but no pin numbers or integrated

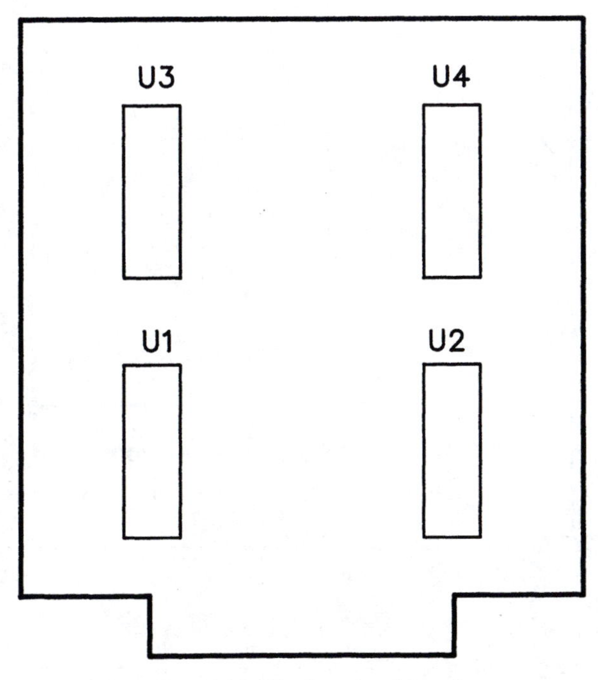

Figure 10–27 Completed board

circuit numbers are identified. The principles of gate combining allow the board to be designed using the least possible number of integrated circuits. Using five design guidelines, a complete red-blue design layout is drawn showing traces on the front and back of the board. The schematic is completed by labeling all IC and connector pin numbers and reference designators. Finally, the artworks for the component (front) side, circuit (back) side, and silk screen markings are shown along with a drawing of the completed board.

EXERCISE 10-1

Choose the schematic diagram in either Figure 10–28 or Figure 10–30. Make a layout on 10 × 10 = 1″ grid paper for a printed circuit board, and redraw the schematic on an $8\frac{1}{2}$″ × 11″ sheet. Your drawings should look similar to the ones in Figure 10–23 and Figure 10–24 (the three-color reproduction on the back cover. Combine gates so that the least number of ICs will be used.

As you make your layout, mark IC numbers, pin numbers, and connector numbers on the schematic diagram that you have redrawn. Assign pin numbers to a 28-pin connector. Make Pin 1 GROUND and Pin 28 V_{cc}. Use the dimensions shown on either Figure 10–28 or Figure 10–30 for the board profile, but double them for your design layout. Use the twice size patterns in Figure 10–29 for ICs and finger connector size.

Make the V_{cc} and GROUND planes .200 twice size. Use a $\frac{1}{8}$″ (twice full-size) square pad for feed-thrus. Allow a minimum of .10 space between conducting lines or pads (.050 full size). Allow a minimum of .050 full-size (.100 twice full-size) from the board edge to any trace or component. Show three .188 DIA (twice full-size) tooling holes located .25 (twice .125 full-size) from each corner of the board.

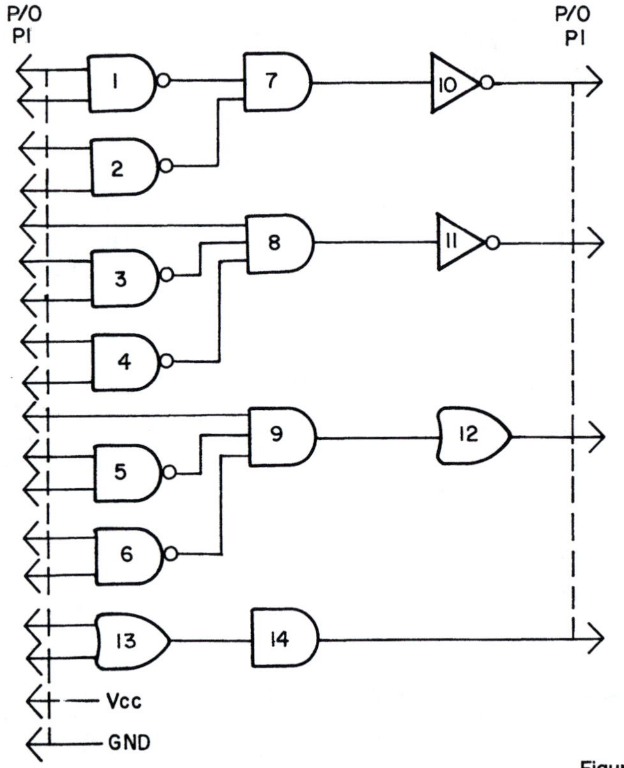

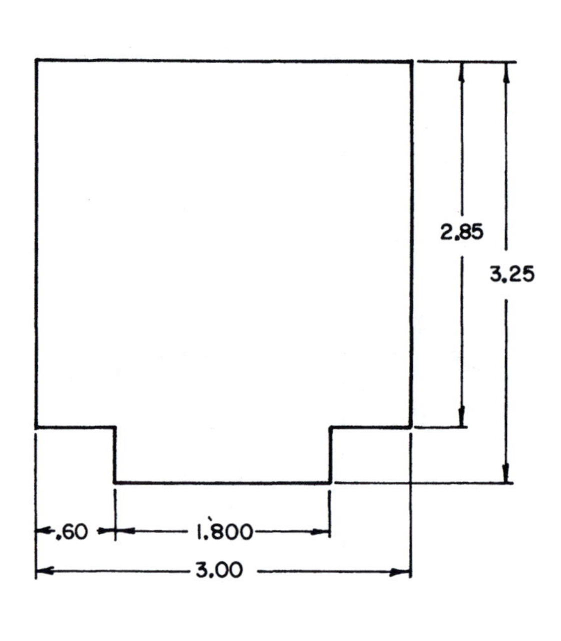

Figure 10–28

Use red lines for the front side traces, and blue lines for the back side traces. Make board outline, feed-thrus, fingers, and IC patterns in black.

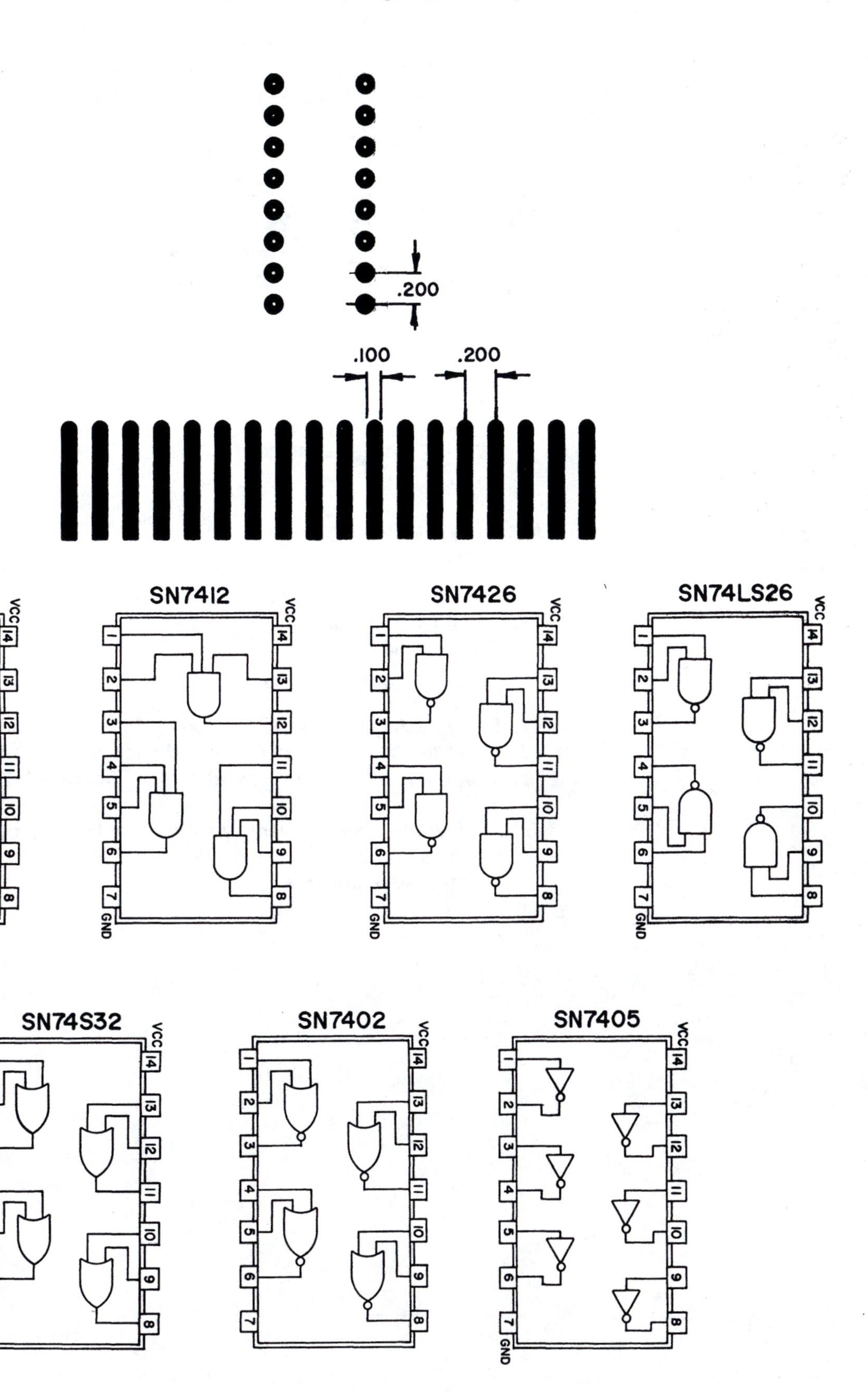

Figure 10–29

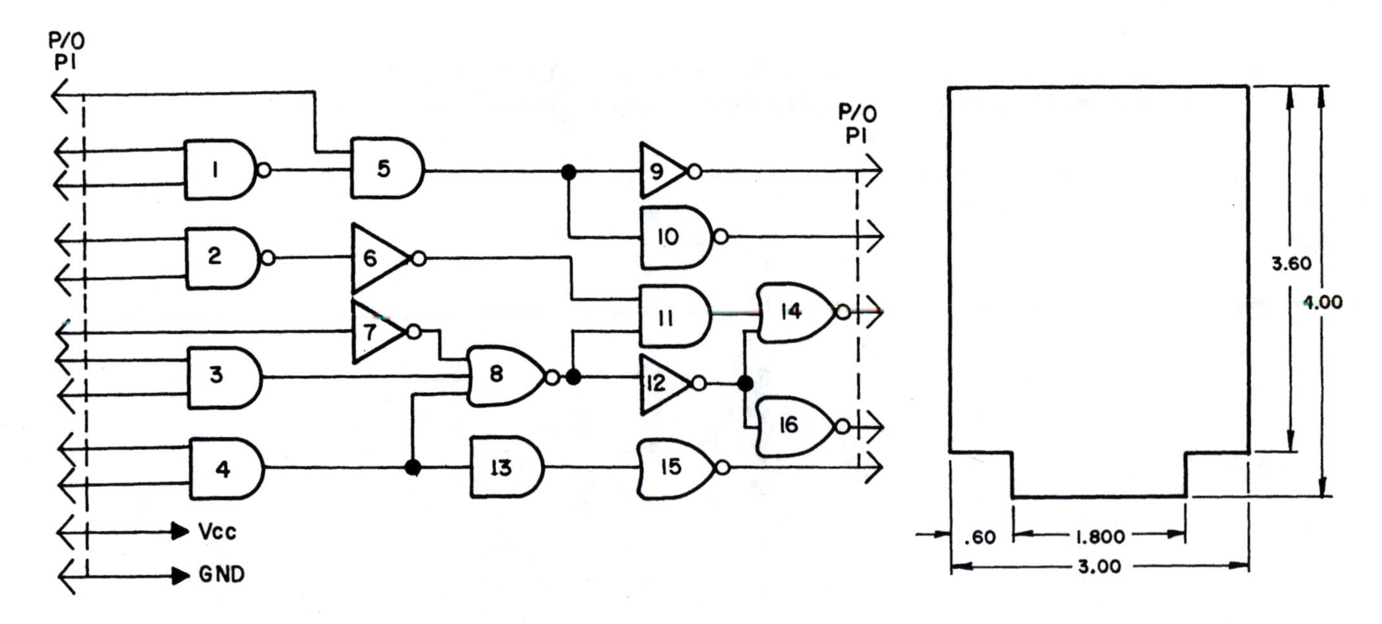

Figure 10–30

On the layout drawing:

1. Label each IC with reference designator (U1, U2, U3, U4) and type (SN741509). (Find each type from Figure 10–29).
2. Label pin numbers on the connector.
3. Black in Pin 1 of each IC.

On the schematic (logic) diagram:

1. Label each gate with IC reference designator and IC pin numbers.
2. Label the connector as P1 and label all pins with their correct number.
3. Label V_{cc} and GROUND as shown in Figure 10–23.

NOTE: If you chose to make your design layouts on the computer, use the same colors as described for manual layout keeping each color on a separate layer.

EXERCISE 10-2

Prepare a design layout, circuit side artwork, and silk screen artwork for the component side, and a drill plan for a single-sided printed circuit board from the following schematic and specifications. Notice that the connectors J1 and J2 are already positioned on the board. Both connectors are the same size. The hole spacing and the component sizes are shown for the ICs, capacitors, and resistors. Notice that the E points are also located on the board. These are holes in the board with a standard-size pad around them on the circuit side of the board. They are used for making minor modifications to the circuit after the board is built. It may be necessary to use thinner tape to pass between the E points on the way to the connector. Parts specifications are shown on page 204.

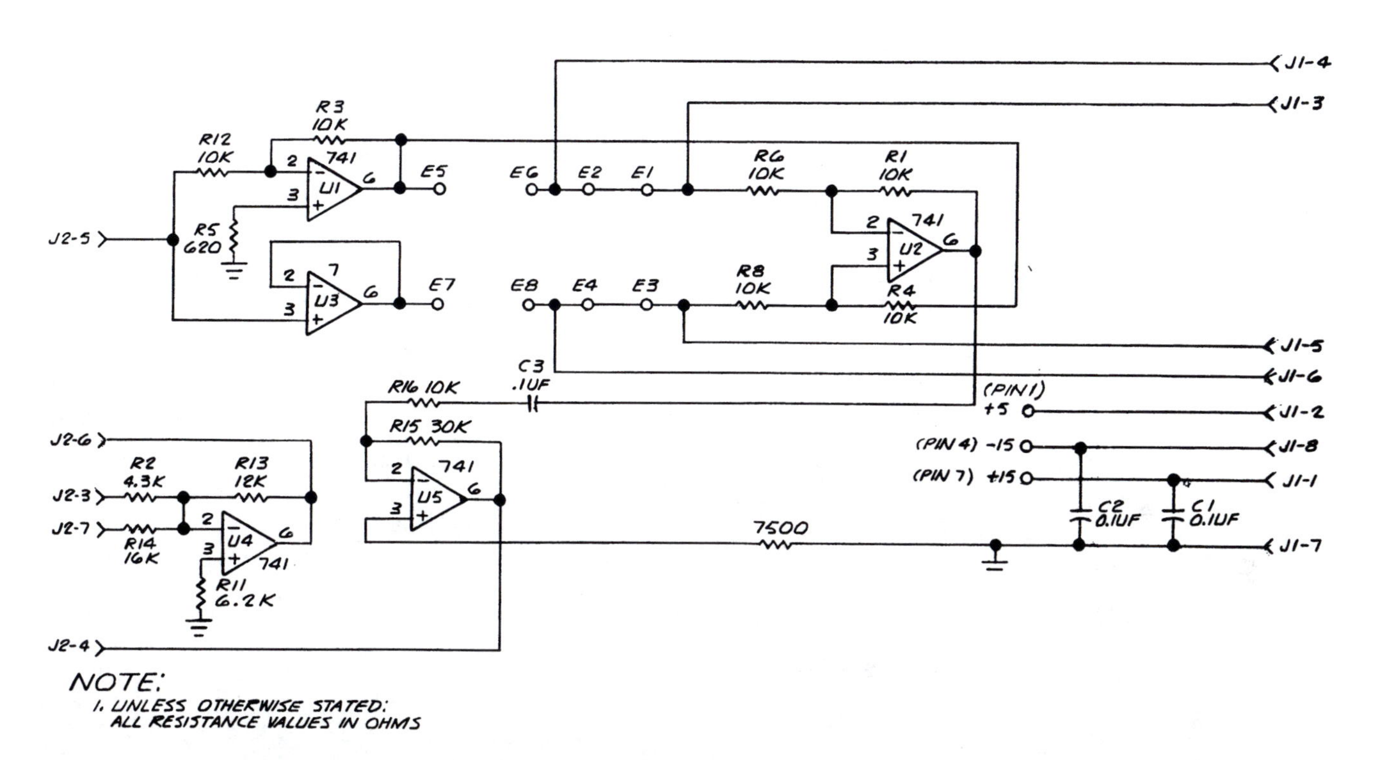

SCHEMATIC FOR EXERCISE 10-2

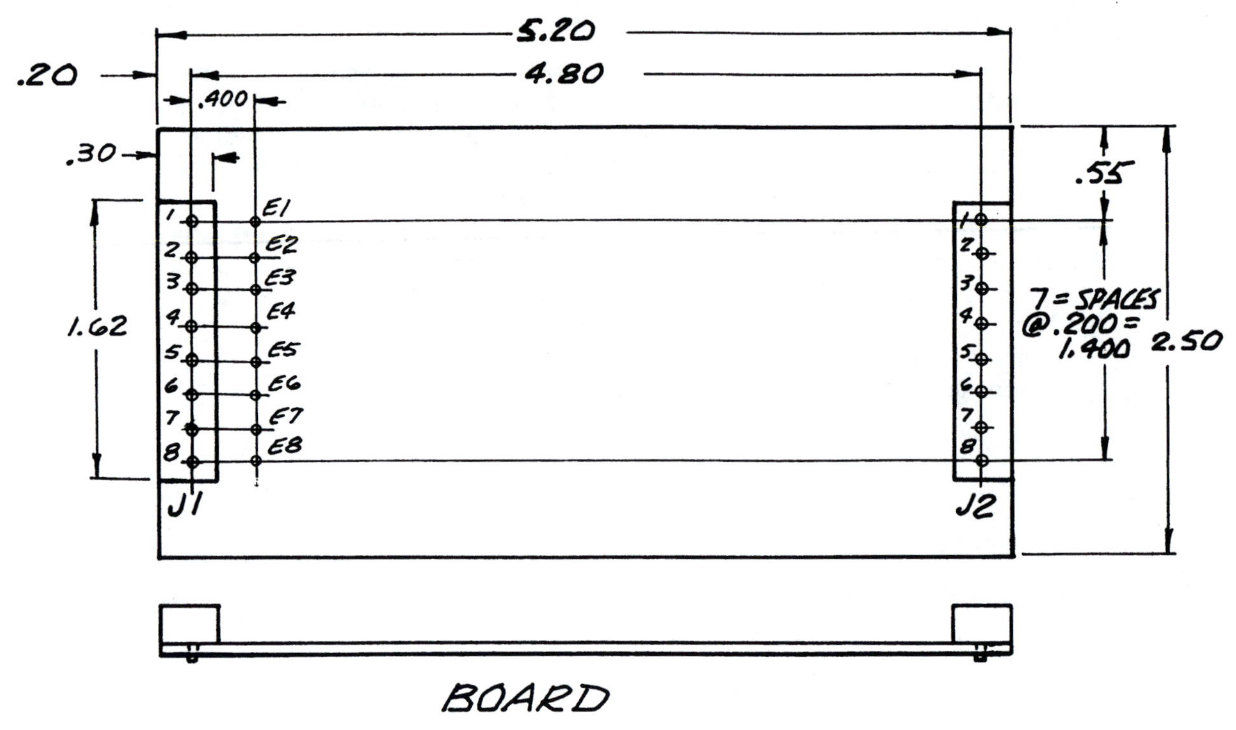
PARTS SPECIFICATIONS
5.20
4.80
.20
.400
.30
.55
E1
E2
E3
E4
E5
E6
E7
E8
1.62
7 = SPACES @.200 = 1.400
2.50
J1
J2
BOARD

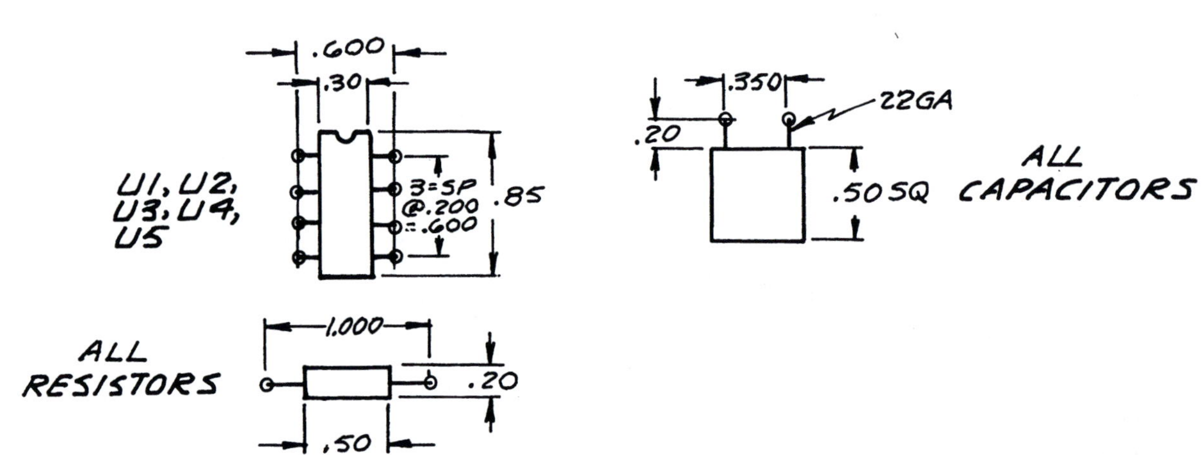
.600
.30
U1, U2, U3, U4, U5
3=SP @.200 =.600
.85
.350
22GA
.20
.50 SQ
ALL CAPACITORS
1.000
ALL RESISTORS
.20
.50

EXERCISE 10-3

Prepare a design layout, circuit side and component side artworks, silkscreen artwork, and a drill plan from the following schematic. Redesign the board shown in the SPECIFICATIONS part of the figure to fit onto a frame such as the one shown which has been changed in size. The new size is 1.500″ × 5.000″. Provide six mounting holes for the board instead of the present four. Take all component sizes and hole patterns from the full size drawing shown in the specifications. Make all drawings twice size.

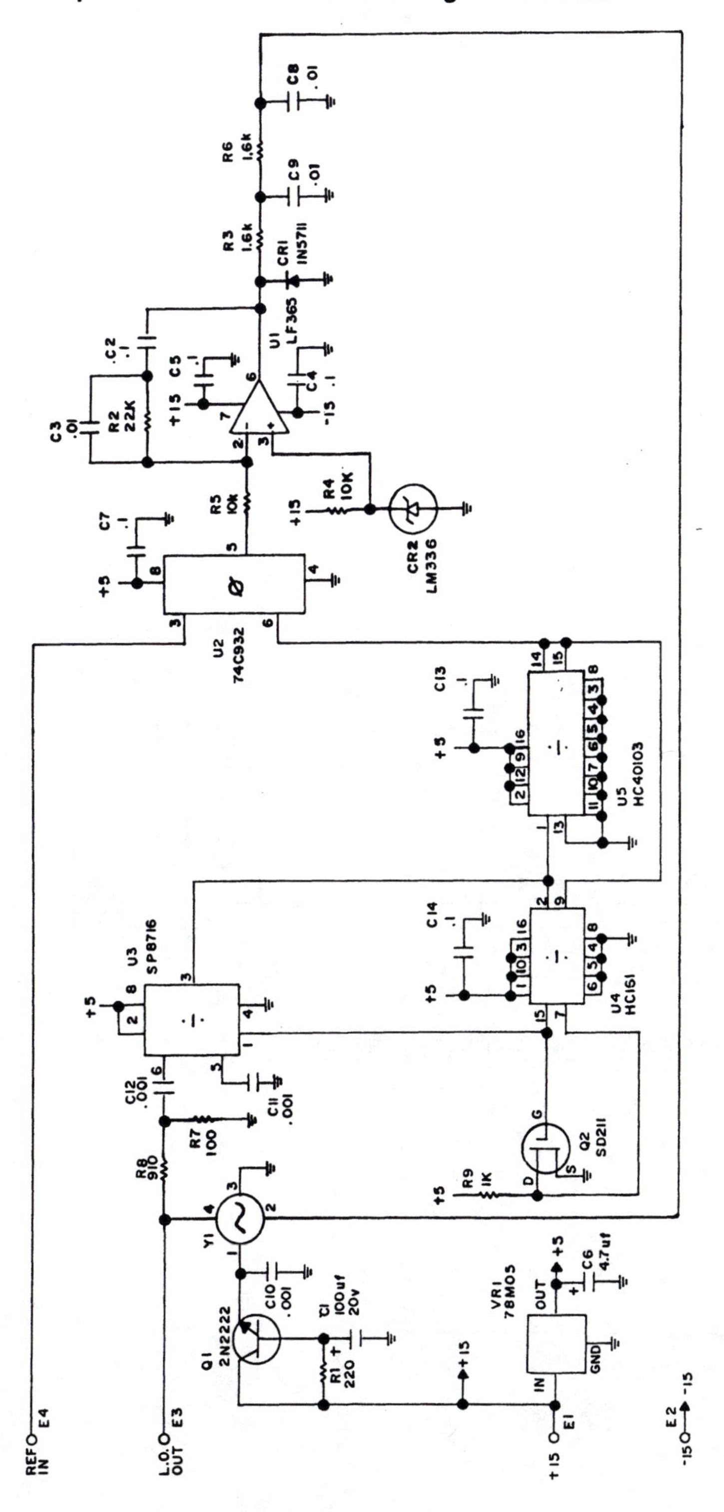

SCHEMATIC FOR EXERCISE 10-3

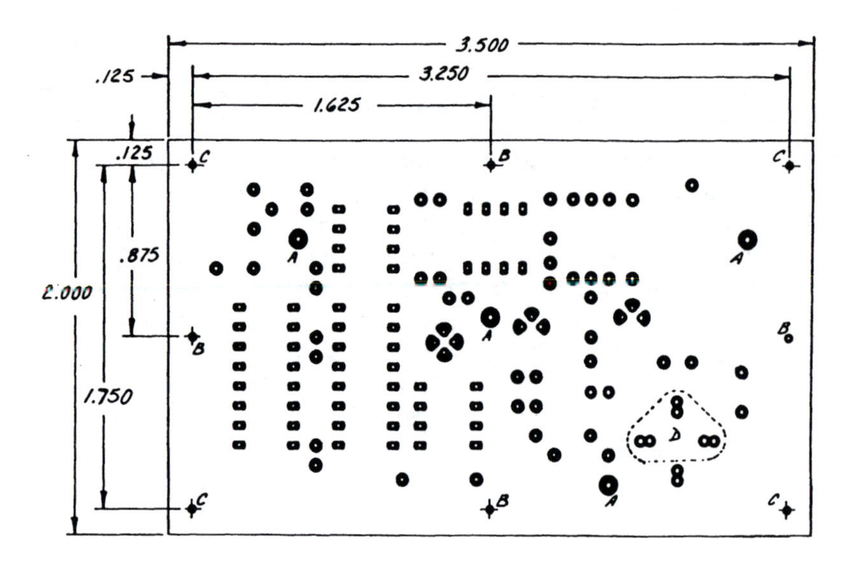

HOLE SCHEDULE			
LTR	SIZE	QTY	REMARKS
—	.032	—	ALL UNMARKED HOLES
A	.042	4	
B	.111	4	
C	.140	4	
D	.032	6	N.P.T.

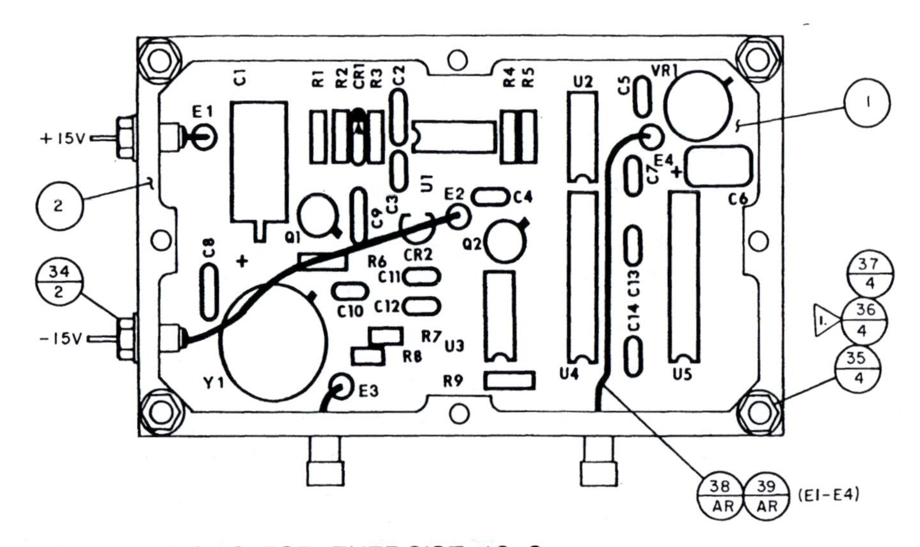

SPECIFICATIONS FOR EXERCISE 10-3

EXERCISE 10-4

Follow the same instructions for Exercise 10–3. Redesign the board to a size of 1.500″ × 5.000″. Make all drawings twice size.

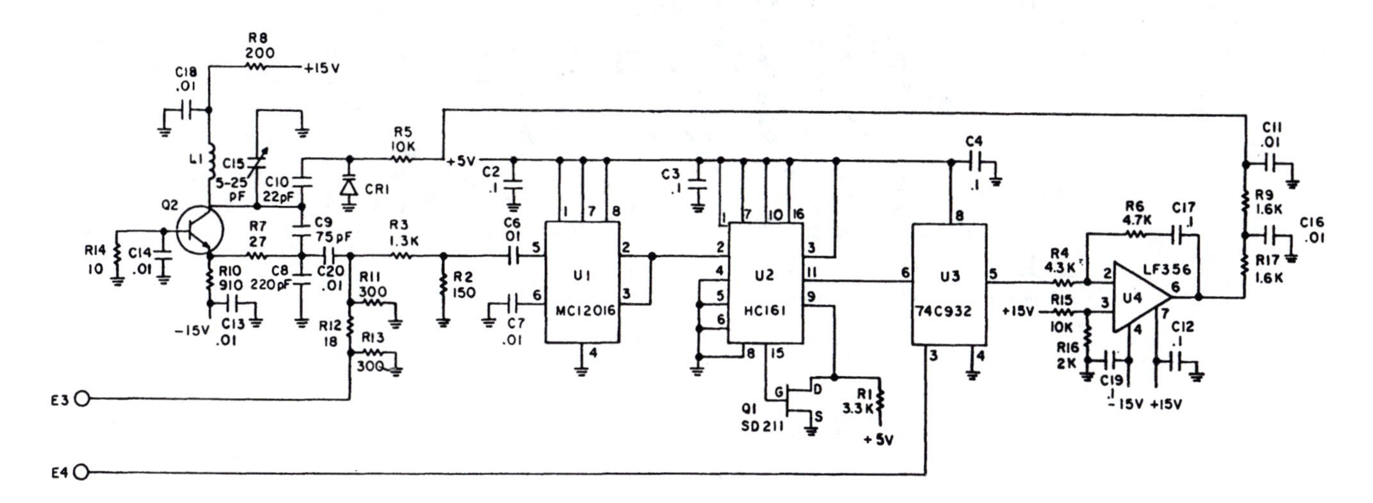

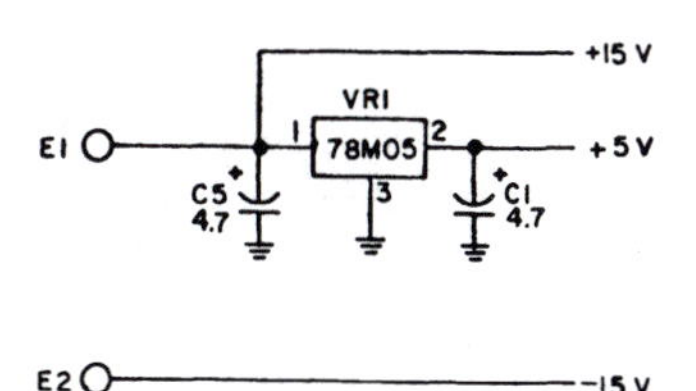

1. RESISTOR VALUES ARE IN OHMS, CAPACITOR VALUES IN uF.

NOTES: UNLESS OTHERWISE SPECIFIED

SCHEMATIC FOR EXERCISE 10-4

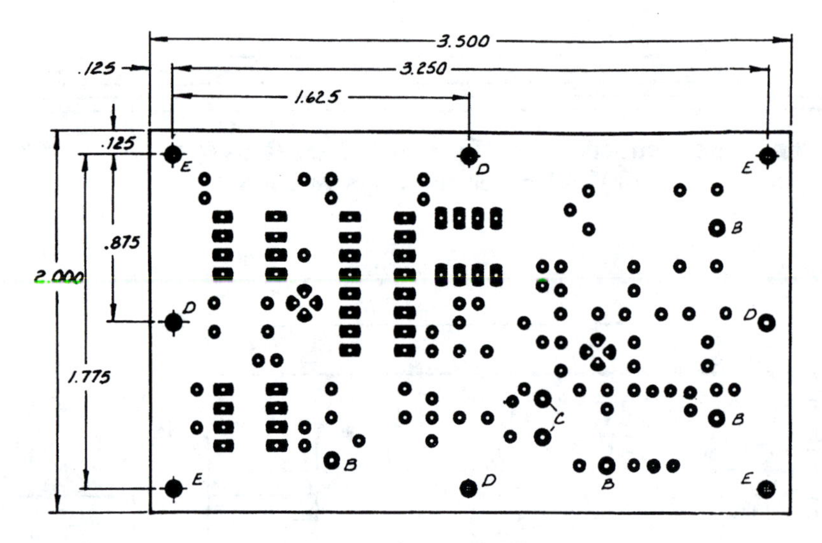

HOLE SCHEDULE			
LTR	SIZE	QTY	REMARKS
—	.032	—	ALL UNMARKED HOLES
A	.036	0	
B	.042	4	
C	.076	2	
D	.111	4	
E	.140	4	

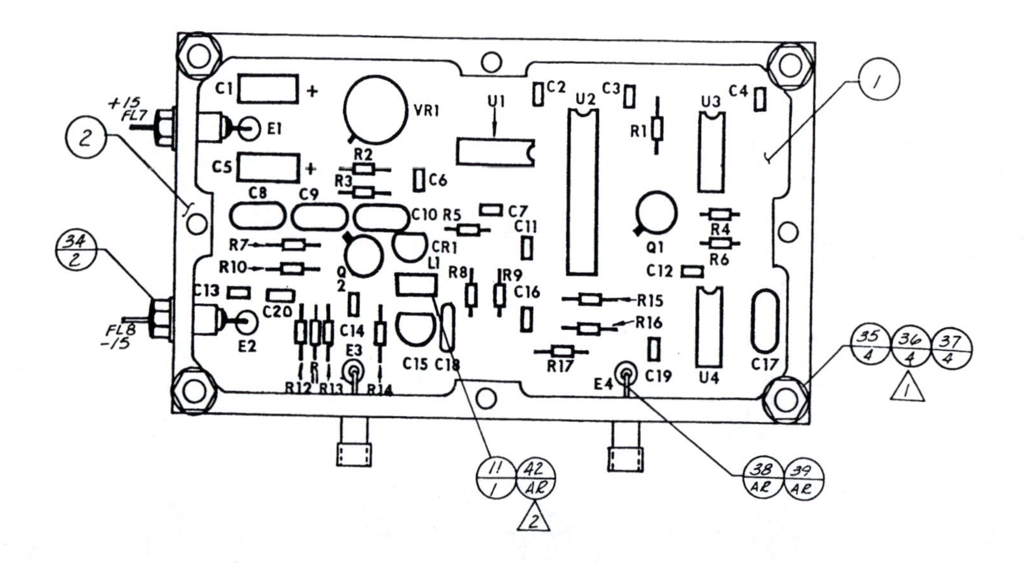

SPECIFICATIONS FOR EXERCISE 10-4

Chapter 11

Standard Racks, Panels, and Cases for Electronics

OBJECTIVES

After completing this chapter, you will be able to

- ✓ prepare a design layout for an electronics rack, given a set of RS–310 standard 19″ panels and cases.
- ✓ prepare detail drawings for the rack, using the design layout.
- ✓ design and detail a standard panel and cable support bracket from given engineering notes.

INTRODUCTION

The arrangement of electronic components into units which are easily accessible is a major part of mechanical packaging. Because this has been a requirement for many types of equipment for several years, the components of these packages have become standardized to a great extent. This chapter describes the major structural parts which have been standardized and the design considerations for arranging these parts.

The structure of these electronic housings consists of three basic elements: the rack, the panel, and the chassis or case.

RACK

A rack is a rectangular, skeleton structure composed of channel members at the top, bottom, and sides and secured to the bottom mounting plate by support angles. This is the basic unit onto which other units are mounted. Rack details, as standardized by Electronic Industries Association (EIA) in its standard RS–310, are shown in Figures 11–1 through 11–4.

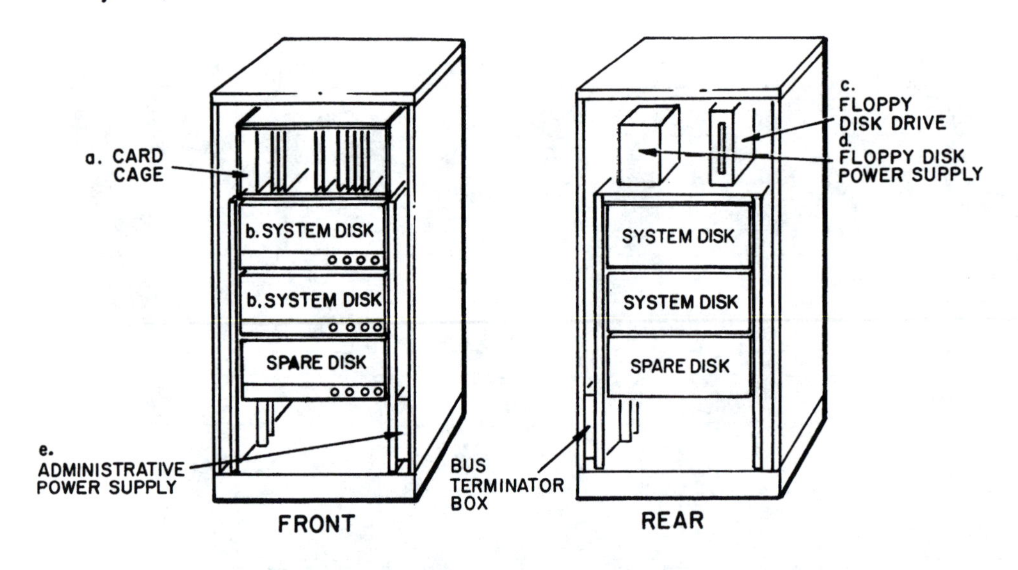

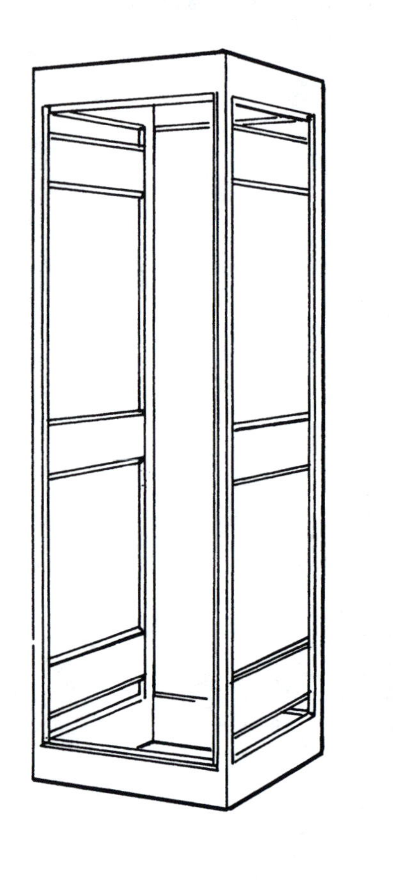

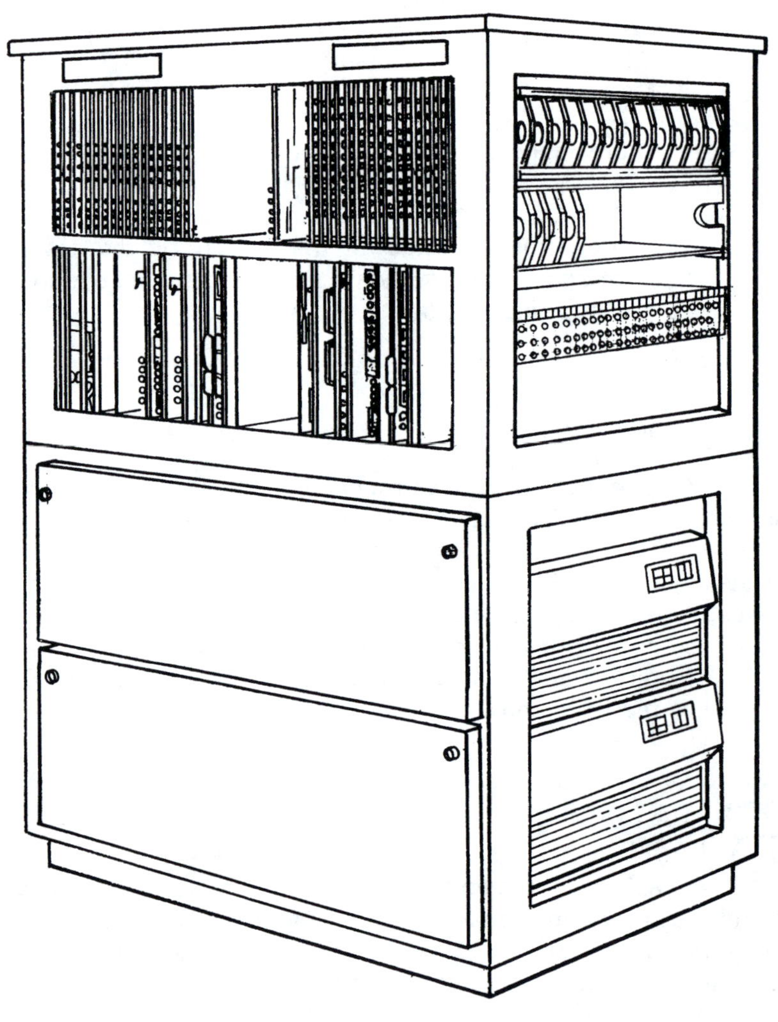

Courtesy of Electronic Industries Association

Figure 11–1 Equipment racks

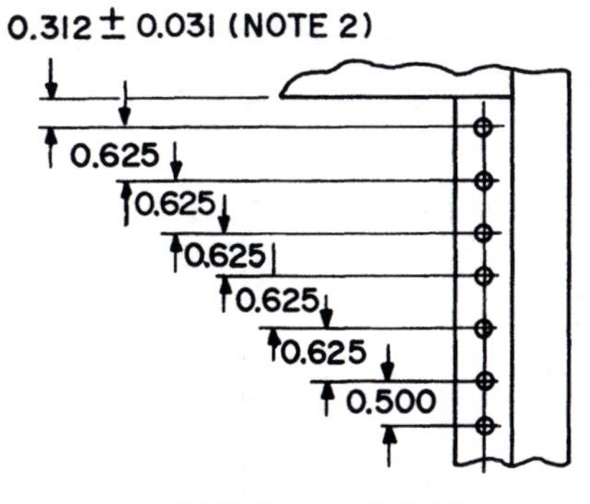

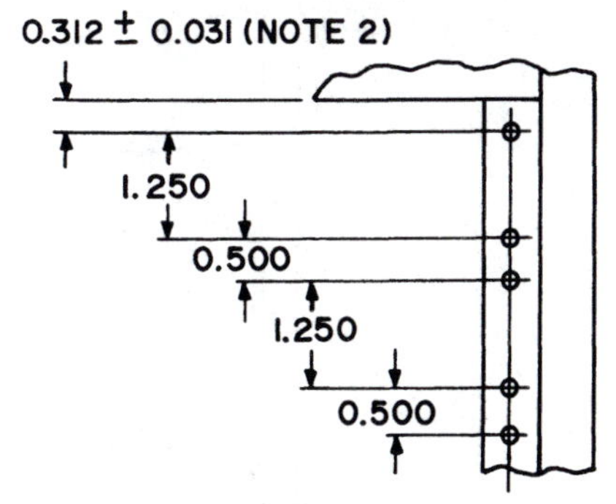

NOTES:
1. TOLERANCES TO BE NON-CUMULATIVE
2. TOLERANCE BETWEEN ANY TWO HOLES IS ±0.015
3. TYPICAL TOP AND BOTTOM

Courtesy of Electronic Industries Association

Figure 11–2 Rack mounting holes

WIDTH		
PANEL	B	C
19.000[1]	17.750 MIN	18.312
24.000	22.750 MIN	23.312
30.000	28.750 MIN	29.312

VERTICAL PANEL SPACE	
n[2]	A
12	21.125
16	28.125
18	31.625
21	36.875
38	66.625
41	71.875
44	77.125

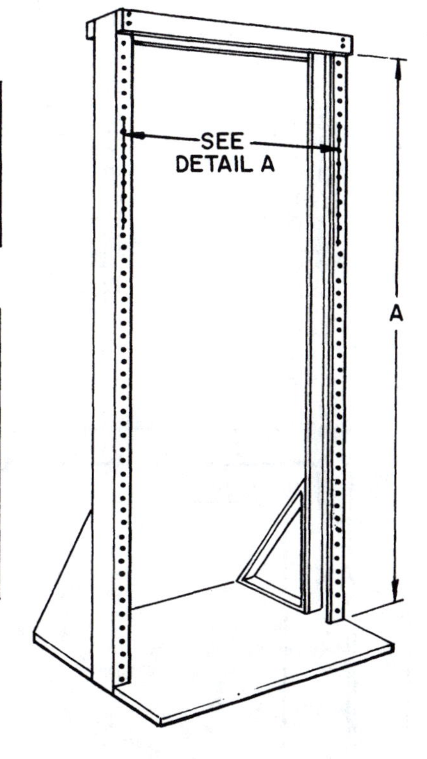

NOTES:
1. PREFERRED WIDTH
2. NUMBER OF 1.750 MODULES
3. TOLERANCE TO BE ±.062
UNLESS OTHERWISE SPECIFIED
TOLERANCES TO BE NON-CUMULATIVE

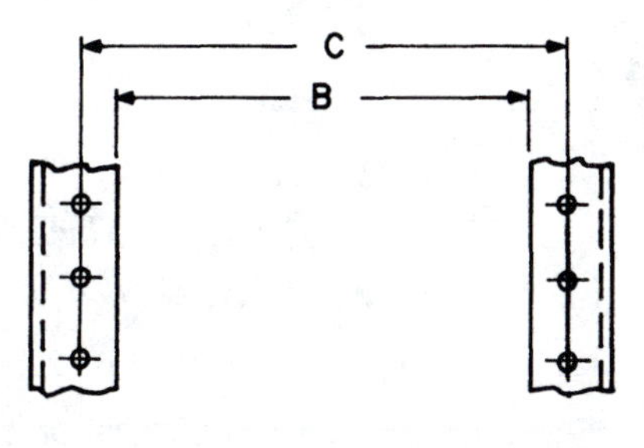

Courtesy of Electronic Industries Association

Figure 11–3 Open rack dimensions

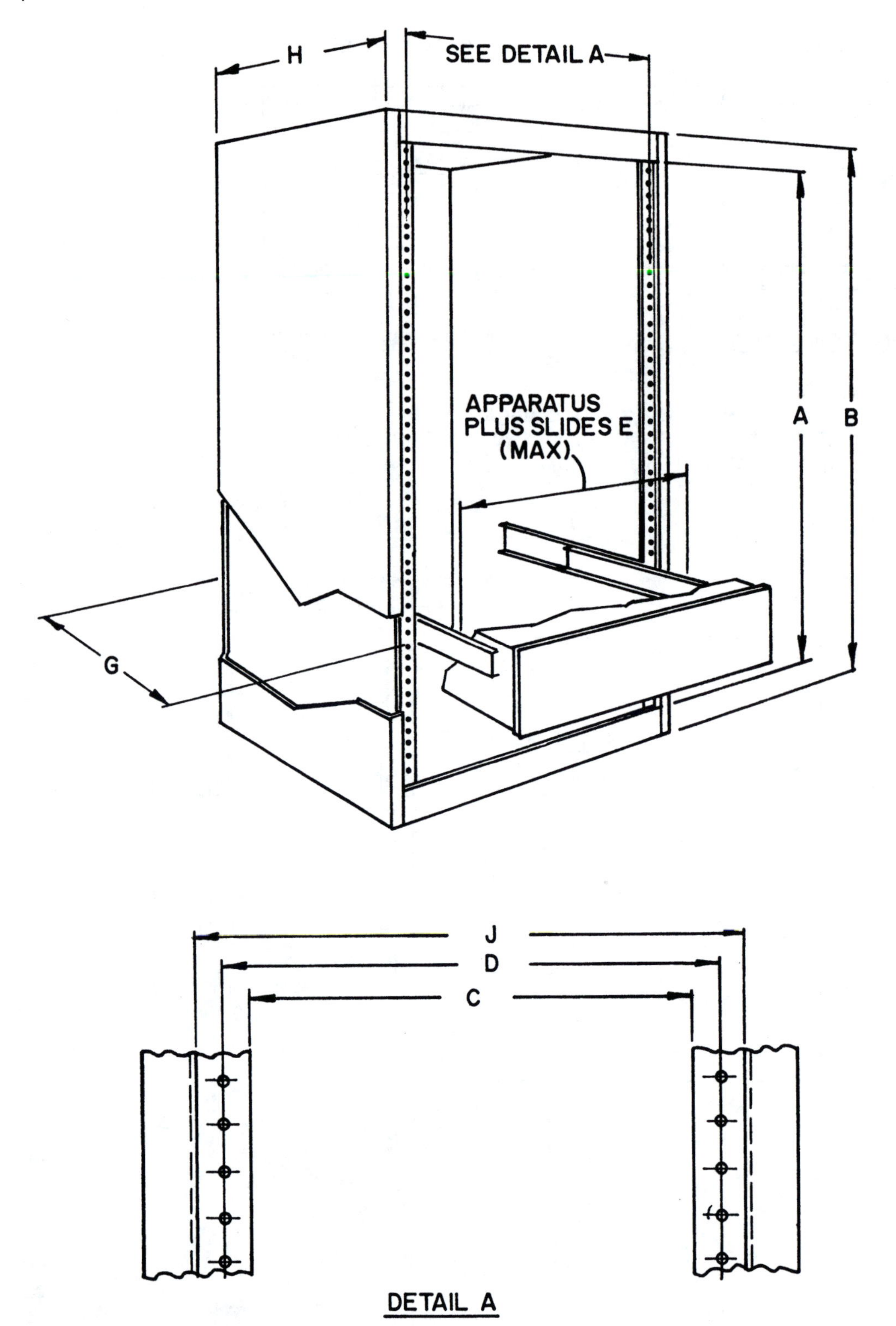

NOTES:

1. UNLESS OTHERWISE SPECIFIED TOLERANCE TO BE ±.062
2. TOLERANCES TO BE NON-CUMULATIVE

Courtesy of Electronic Industries Association

Figure 11–4 Closed rack dimensions

DEPTH	
H^5 ± .500	G^4
15.250	13.750
18.00	16.50
24.00	22.50
30.00	28.50

WIDTH				
PANEL	C	D	E	J ± .032
19.000[1]	17.750 Min.	18.312	17.625	19.062
24.000	22.750 Min.	23.312	22.625	24.062
30.000	28.750 Min.	29.312	28.625	30.062

		VERTICAL PANEL SPACE	RESULTING OVERALL HEIGHT
N^2	NAME	A	B^G ± .500
12	DESK	21.125	30.000
16	BENCH	28.125	36.000
20	COUNTER	35.125	42.000
30	LOOKOVER	52.625	60.00
N^2	NAME	A	B^6
35		61.375	See Note 3
40	BUILDING DOOR	70.125	80" Max.
44		77.125	See Note 3
45		78.875	See Note 3

NOTES:
1. PREFERRED PANEL WIDTH.
2. MINIMUM NUMBER OF 1.750 MODULES FOR VERTICAL PANEL SPACE.
3. IN ACCORDANCE WITH STRUCTURAL NEEDS.
4. MINIMUM CLEARANCE DEPTH BEHIND PANEL MOUNTING FLANGES APPLIES ONLY WHEN PANELS ARE MOUNTED FLUSH WITH THE FRONT OF THE CABINET. (APPLICABLE OVER FULL PANEL WIDTH.)
5. OVERALL DEPTH.
6. HEIGHT WITHOUT CASTERS.

Figure 11–4 (continued)

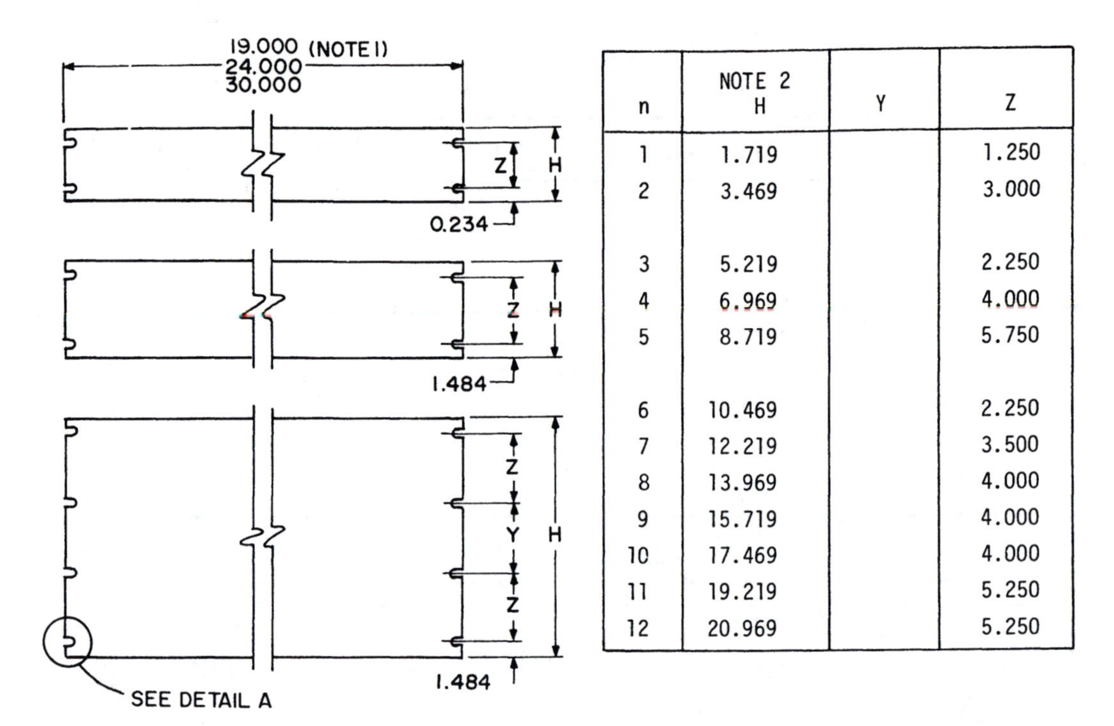

n	NOTE 2 H	Y	Z
1	1.719		1.250
2	3.469		3.000
3	5.219		2.250
4	6.969		4.000
5	8.719		5.750
6	10.469		2.250
7	12.219		3.500
8	13.969		4.000
9	15.719		4.000
10	17.469		4.000
11	19.219		5.250
12	20.969		5.250

NOTES:
1. PREFERRED WIDTH
2. DIMENSION H = $1.750n \; {}^{+0}_{-0.031}$
3. TOLERANCE TO BE ±0.015 UNLESS OTHERWISE SPECIFIED. TOLERANCES TO BE NON-CUMULATIVE. TOLERANCE BETWEEN ANY TWO SLOTS ±0.015
4. PANEL THICKNESS AND NO. OF SLOTS MAY VARY WITH LOAD REQUIREMENTS.

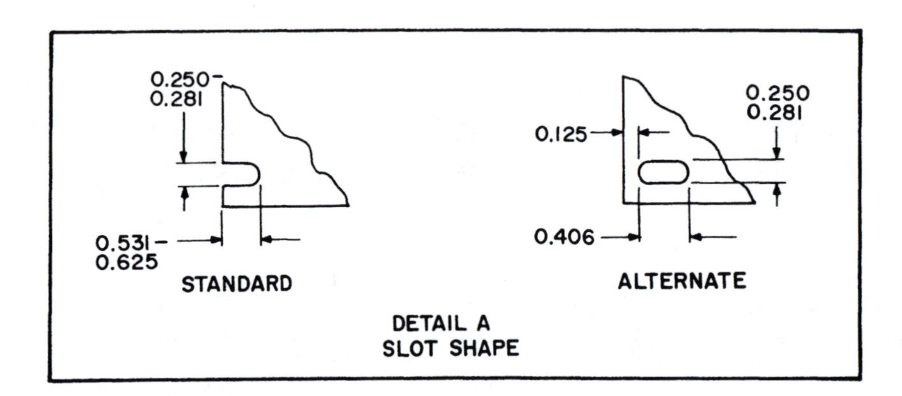

Figure 11–5 Rack panel details

Panel

The panel component is attached to the front of individual instruments, chassis, and cases to allow them to be mounted onto a rack. The 19″ wide rack panel is standard. All of the pertinent details of blank rack panels are shown in Figure 11–5. Holes are drilled in individual rack panels to allow switches, lights, meters, and knobs, and so forth, to be mounted on the face of the panel. Sometimes blank panels are built into the final rack assembly to cover up an unused space or to provide for the addition of future equipment.

Chassis or Case

Some differences exist between chassis manufactured for commercial use and those manufactured for use by the Army, Navy, and Air Force. Therefore, the two are described separately.

Commercial Chassis and Cases • Chassis of aluminum or steel are available commercially in standard sizes. These units may be attached to a standard rack panel, then mounted in a rack. They also may be used in a variety of other ways, such as for individual mounting and specialized rack mounting.

Usually these units are made of $\frac{1}{16}''$ material ($\frac{3}{32}''$ steel is used for larger sizes) and have connecting plugs or receptacles mounted on the rear of the case. They may be fastened vertically or horizontally on the rack and more than one chassis may be attached to a single rack panel. The two standard case size tables for commercially available equipment are shown in Tables 11–1 and 11–2.

TABLE 11–1 Sizes of commercially available chassis, in inches

Width	Depth	Height
7	5	2
$9\frac{1}{2}$	5	$1\frac{1}{2}, 2\frac{1}{2}$
10	5	3
14	6	3
7	7	2
9	7	2
11	7	2
12	7	3
13	7	2
15	7	3
17	7	3
12	8	2,3
17	8	2,3
12	10	3
14	10	3
17	10	2,3,4,5
23	10	3,4
17	11	2,3
17	12	2,3
17	13	2,3,4,5

TABLE 11–2 Electronic equipment case sizes, in inches (Aeronautical Radio, Inc.)

ATR case size	Width	Depth	Height
Short $\frac{1}{4}$	$2\frac{1}{4}$	$12\frac{5}{8}$	$7\frac{5}{8}$
Long $\frac{1}{4}$	$2\frac{1}{4}$	$19\frac{5}{8}$	$7\frac{5}{8}$
Short $\frac{3}{8}$	$3\frac{5}{8}$	$12\frac{5}{8}$	$7\frac{5}{8}$
Long $\frac{3}{8}$	$3\frac{5}{8}$	$19\frac{5}{8}$	$7\frac{5}{8}$
Short $\frac{1}{2}$	$4\frac{7}{8}$	$12\frac{5}{8}$	$7\frac{5}{8}$
Long $\frac{1}{2}$	$4\frac{7}{8}$	$19\frac{5}{8}$	$7\frac{5}{8}$
Short $\frac{3}{4}$	$7\frac{1}{2}$	$12\frac{5}{8}$	$7\frac{5}{8}$
Long $\frac{3}{4}$	$7\frac{1}{2}$	$19\frac{5}{8}$	$7\frac{5}{8}$
Long 1	$10\frac{1}{8}$	$19\frac{5}{8}$	$7\frac{5}{8}$
Long $1\frac{1}{2}$	$15\frac{3}{8}$	$19\frac{5}{8}$	$7\frac{5}{8}$

Military Chassis and Cases • There are standard case sizes and standard mounting bases for the cases shown in Figures 11–6 through 11–9. In addition to standard sizes, the specifications of military standard MIL-C-172 includes the following requirements.

1. All units must have handles to allow easier handling. When units are mounted on the front panel, the handles protect controls (switches, meters, and so forth) when the complete assembly is placed face downward for servicing.
2. Thumb or wing nuts used for fastening the equipment to the mounting base must be safety wired (wired so they will not vibrate and turn accidentally).
3. Four flat strips of beryllium copper or phosphor bronze must be used for grounding the base. Notice that Figures 11–8 and 11–9 indicate the allowable space beneath the mounting bases which may be used for a shock or vibration mount, if necessary.

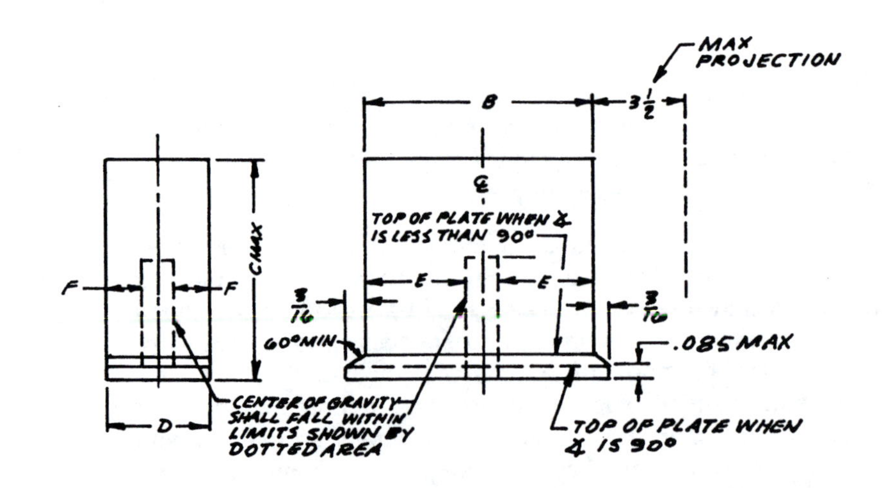

PART NUMBER	FIIW	DIMENSIONS (INCHES)					WEIGHT RANGE, POUNDS
		B	C	D	E	F	
M391402-51 M591402-52		9-7/8 13-1/8	7-5/8 7-5/8	5-1/8 4-7/8	4 5-1/2	2 2	6 to 12 10 to 22

Figure 11–6 Standard small-size cases

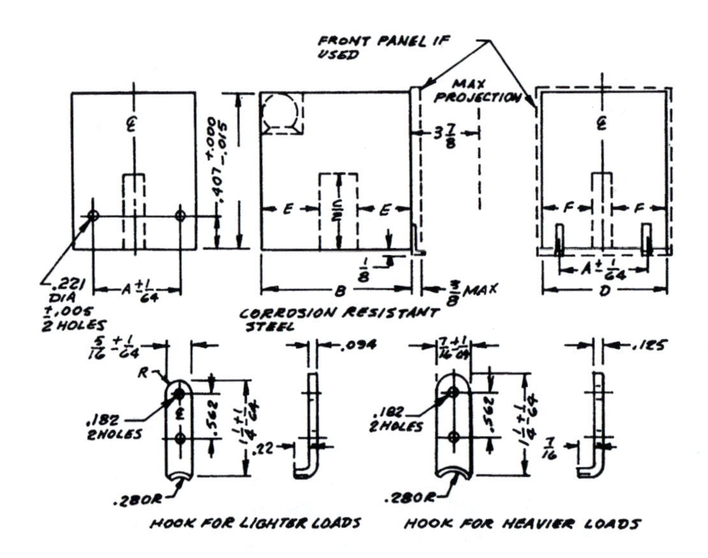

PART NUMBER	DIMENSIONS (INCHES)						WEIGHT RANGE, POUNDS
	A	B	C	D	E	F	
MS91403-A1B	4-1/8	12-9/16	7-5/8	4-7/8	5-1/4	2	8 to 18
MS91403-A1C	4-1/8	15-9/16	7-5/8	4-7/8	6-1/2	2	12 to 20
MS91403-A1D	4-1/8	19-9/16	7-5/8	4-7/8	8-3/16	2	18 to 40
MS91403-B1B	9-3/8	12-9/16	7-5/8	10-1/8	5-1/4	4-1/4	18 to 40
MS91403-B1C	9-3/8	15-9/16	7-5/8	10-1/8	6-1/2	4-1/4	18 to 40
MS91403-B1D1	9-3/8	19-9/16	7-5/8	10-1/8	8-3/16	4-1/4	25 to 50
MS91403-B1D2	9-3/8	19-9/16	7-5/8	10-1/8	8-3/16	4-1/4	40 to 80
MS91403-C1D	14-5/8	19-9/16	7-5/8	15-3/8	8-3/16	6-7/16	40 to 80
MS91403-C2D	14-5/8	19-9/16	10-5/8	15-3/8	8-3/16	6-7/16	40 to 80

Figure 11–7 Standard large-size cases

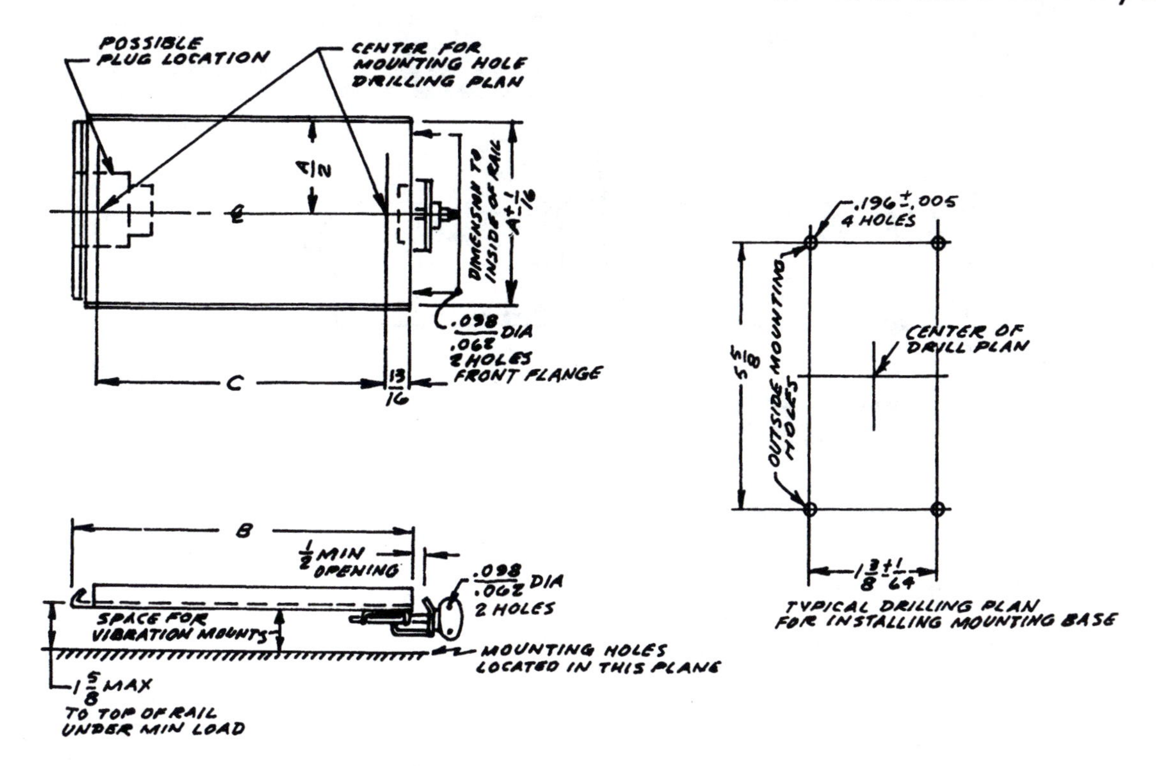

PART NUMBER	FIIW	DIMENSIONS (INCHES)			LOAD RANGE POUNDS
		A	B	C	
MS91404-51 MS91404-52		5-3/16 4-15/16	10-1/8 13-5/16	8-1/4 11-5/8	6 to 12 10 to 22

Figure 11–8 Mounting bases for small-size cases

DESIGN CONSIDERATIONS AND METHODS FOR ARRANGING CASES

The major elements which must be considered in designing a rack with several chassis are space, accessibility for servicing, equipment cooling, location of controls and indicators, and shielding.

Space

Often the overall dimensions of the equipment are established before the cases are designed. In this situation, several attempts may be needed to arrange the components into the final space.

One method often used is to draw a scaled outline of the rack and then arranging scaled cardboard cutouts of cases and rack panels to determine the best layout. Whenever possible, it is best to locate the heavier chassis near the bottom of the rack, both for ease and safety in handling and to reduce stress on the main structure. Space allowance is also required for shock mounts used on delicate components.

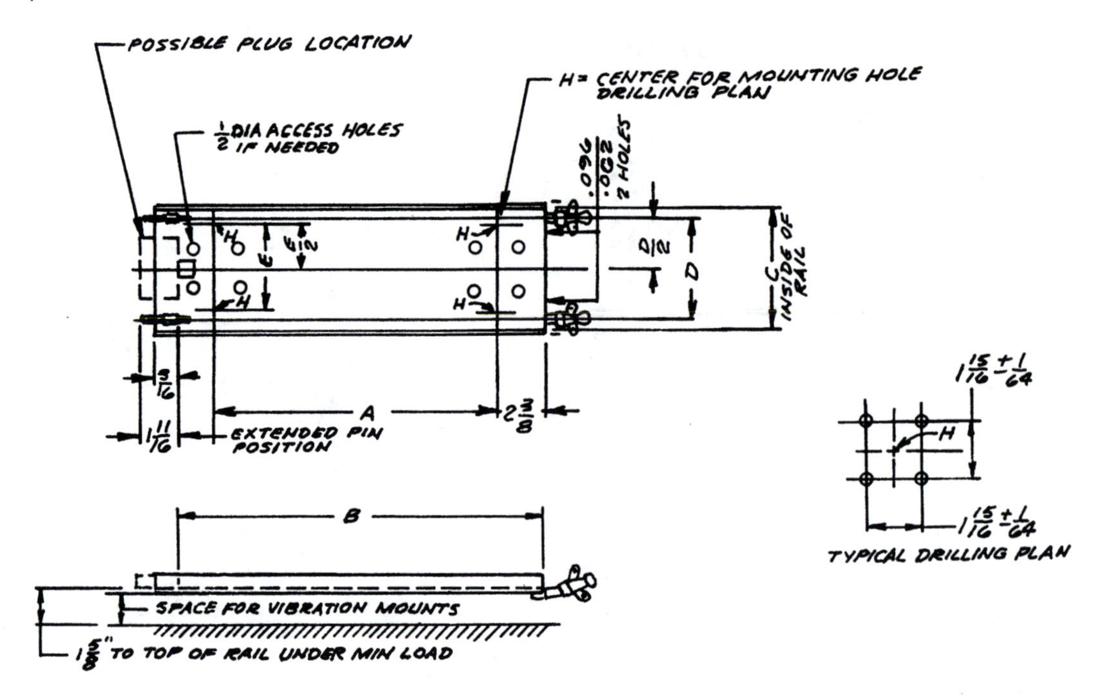

PART NUMBER	DIMENSIONS (INCHES)					LOAD RANGE, POUNDS
	A	B	C	D	E	
MS91405-A1B	9-1/16	12-5/8	5	4-1/8	3-1/2	8 to 18
MS91405-A1C	12-1/16	15-5/8	5	4-1/8	3-1/2	12 to 20
MS91405-A1D	16-1/16	19-5/8	5	4-1/8	3-1/2	18 to 40
MS91405-B1B	9-1/16	12-5/8	10-1/4	9-3/8	8-3/4	18 to 40
MS91405-B1C	12-1/16	15-5/8	10-1/4	9-3/8	8-3/4	18 to 40
MS91405-B1D1	16-1/16	19-5/8	10-1/4	9-3/8	8-3/4	25 to 50
MS91405-B1D2	16-1/16	19-5/8	10-1/4	9-3/8	8-3/4	40 to 80
MS91405-C1D	16-1/16	19-5/8	15-1/2	14-5/8	14	40 to 80
MS91405-C2D	16-1/16	19-5/8	15-1/2	14-5/8	14	40 to 80

Figure 11–9 Mounting bases for large-size cases

Accessibility for Servicing

Drawer slides, Figure 11–10, are required for cases that must be serviced frequently. These slides provide easy access, and sometimes have a tilting feature which allows the case to be held in a convenient position for servicing. Other means used are hinged chassis with removable doors. For individually housed chassis, easy removal from the case is provided by quick release fasteners which are attached to the front panel.

Equipment Cooling

A means must be provided to allow heat to be dissipated or removed from electronic components which generate heat in their operation. General specifications for electronic equipment manufactured for the military provide for such cooling, but give preference to nonpowered means, such as vents or louvers, before allowing blowers to be employed. Openings made to provide cooling are often screened to prevent the entrance of insects and some components must be enclosed to prevent dust from entering them.

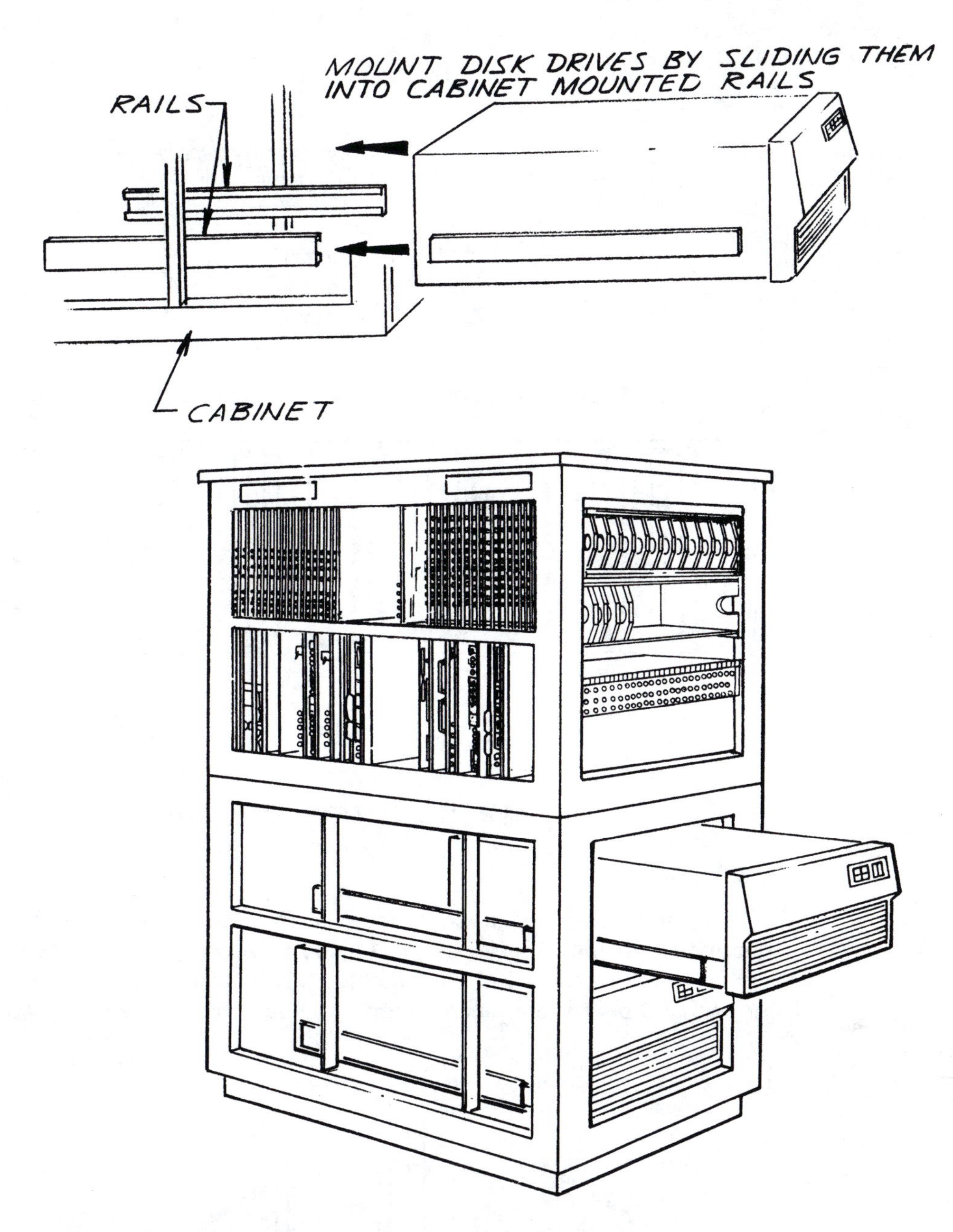

Figure 11–10 Drawer slides

Location of Controls and Indicators

Whenever possible, controls should be placed for the operator's convenience. The major controls should be placed near the center of the equipment and indicators should be located at eye level. A control and its indicator should be placed as close to each other as possible, and arranged so that the operation of the control does not obstruct a view of the indicator.

Shielding

Whenever the possibility of electrostatic or electromagnetic radiation exists, parts that can be affected by this radiation must be shielded. Metal is used as a shield in most cases.

In a chassis, metal partitions are often used to isolate sections within the chassis. Metallic runners of bronze or copper are used between removable shield parts to ensure adequate contact.

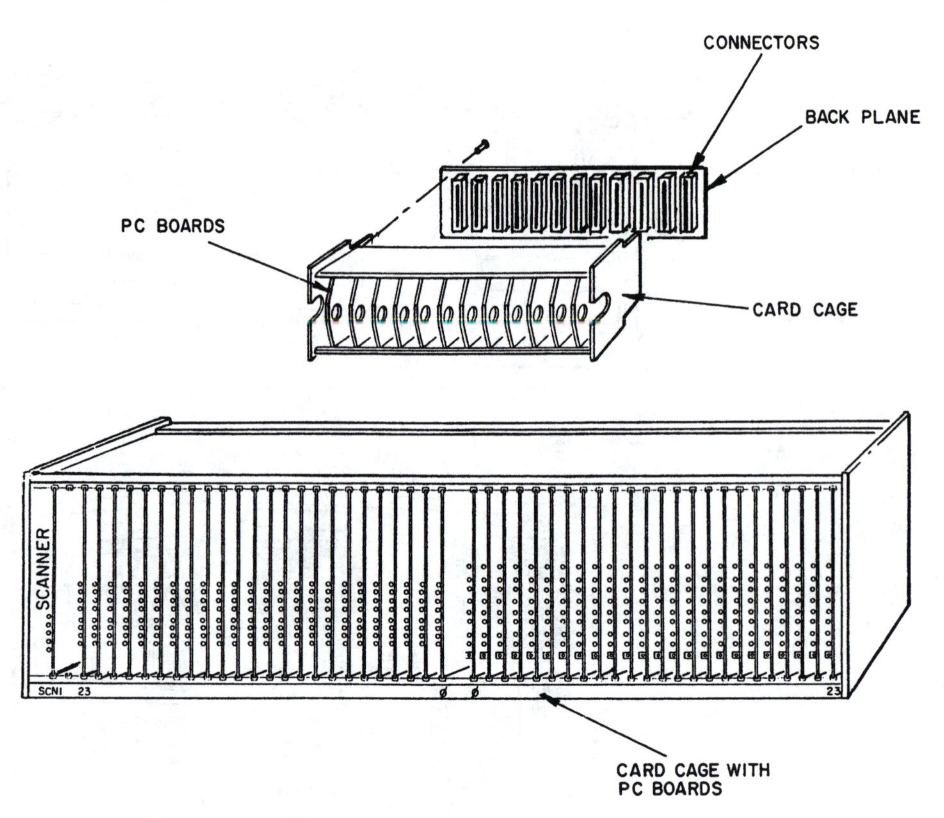

Courtesy of VMX, Inc.

Figure 11–11 Card cages and back planes

Card Cage

The purpose of a card cage is to hold printed circuit boards in electronics equipment. Figure 11–11 shows an example of a card cage and how it is mounted in the equipment. Individual plastic slides are provided for each PC board, and each PC board usually has a latch on it to keep the

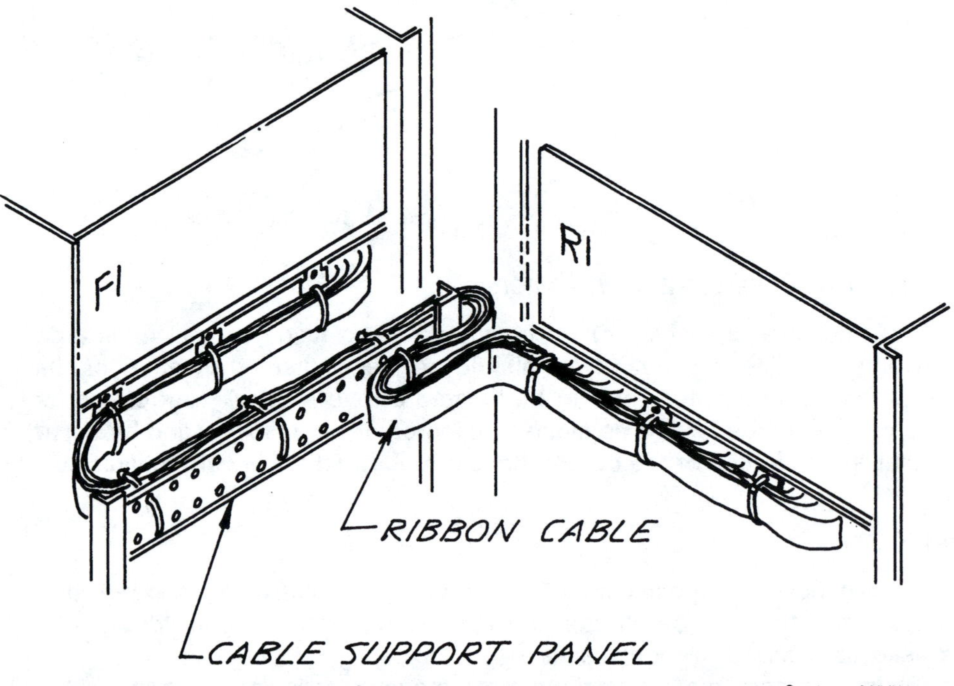

Courtesy of VMX, Inc.

Figure 11–12 Cable routing

board in place. Notice the back plane pointed out in this figure. The back plane is used to connect printed circuit boards with other printed circuit boards or other components. The back plane often has female connectors into which the PC board fingers are inserted to make connection between the PC board and the back plane.

Cabling

Cables for connecting electronic units in a rack often become a major problem. Brackets to support cables and to allow for efficient assembly are often needed. These brackets are designed to accommodate flat or round cables. Figure 11–12 shows examples of cable routing and support in a typical electronics rack.

EXERCISE 11-1

Design a rack using the standards described in this chapter, and the required panels shown here. Prepare a design layout showing the rack and panels to verify the design, and then complete a detailed shop drawing of the rack.

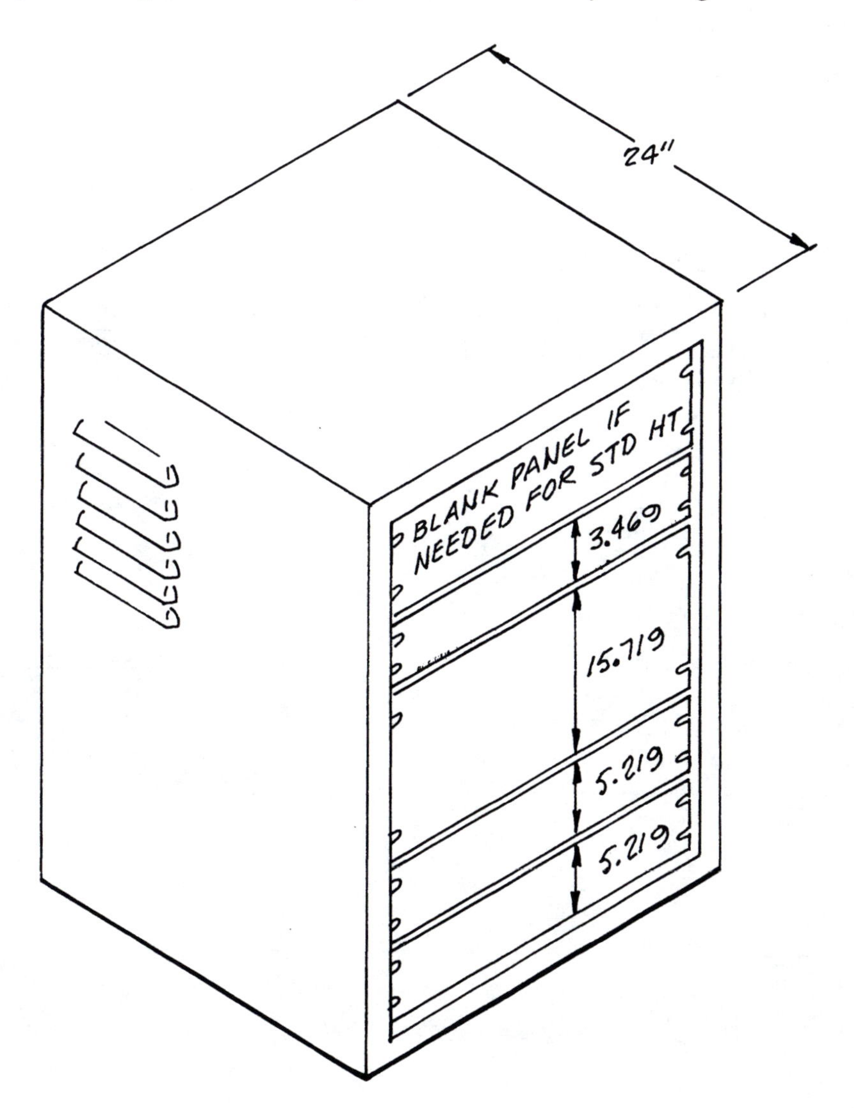

EXERCISE 11-2

Design a standard size panel using specifications in this chapter. Show holes in the panel and panel markings using rub-on letters or freehand lettering as your instructor indicates.

The following is an example drawing showing how your drawing should look. Draw the panel full scale on a C-size sheet.

Provide holes in the panel for mounting the components shown on the facing page. The height of the panel is your decision, but it should not take up more room than is necessary and should not be so small as to crowd components.

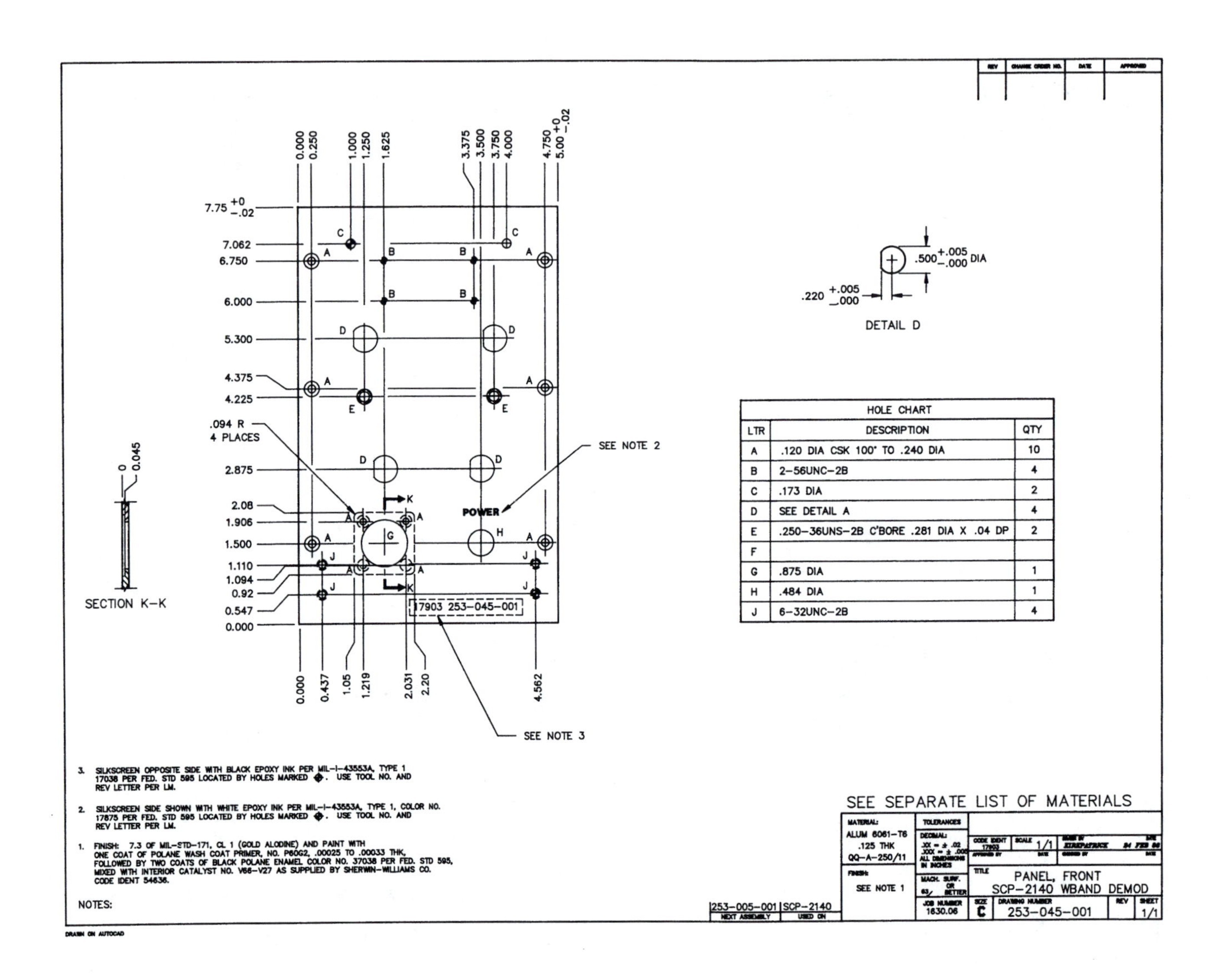

	HOLE CHART	
LTR	DESCRIPTION	QTY
A	.120 DIA CSK 100° TO .240 DIA	10
B	2–56UNC–2B	4
C	.173 DIA	2
D	SEE DETAIL A	4
E	.250–36UNS–2B C'BORE .281 DIA X .04 DP	2
F		
G	.875 DIA	1
H	.484 DIA	1
J	6–32UNC–2B	4

NOTES:

1. FINISH: 7.3 OF MIL–STD–171, CL 1 (GOLD ALODINE) AND PAINT WITH ONE COAT OF POLANE WASH COAT PRIMER, NO. P60G2, .00025 TO .00033 THK, FOLLOWED BY TWO COATS OF BLACK POLANE ENAMEL COLOR NO. 37038 PER FED. STD 595, MIXED WITH INTERIOR CATALYST NO. V66–V27 AS SUPPLIED BY SHERWIN–WILLIAMS CO. CODE IDENT 54636.
2. SILKSCREEN SIDE SHOWN WITH WHITE EPOXY INK PER MIL–I–43553A, TYPE 1, COLOR NO. 17875 PER FED. STD 595 LOCATED BY HOLES MARKED ◈. USE TOOL NO. AND REV LETTER PER LM.
3. SILKSCREEN OPPOSITE SIDE WITH BLACK EPOXY INK PER MIL–I–43553A, TYPE 1 17038 PER FED. STD 595 LOCATED BY HOLES MARKED ◈. USE TOOL NO. AND REV LETTER PER LM.

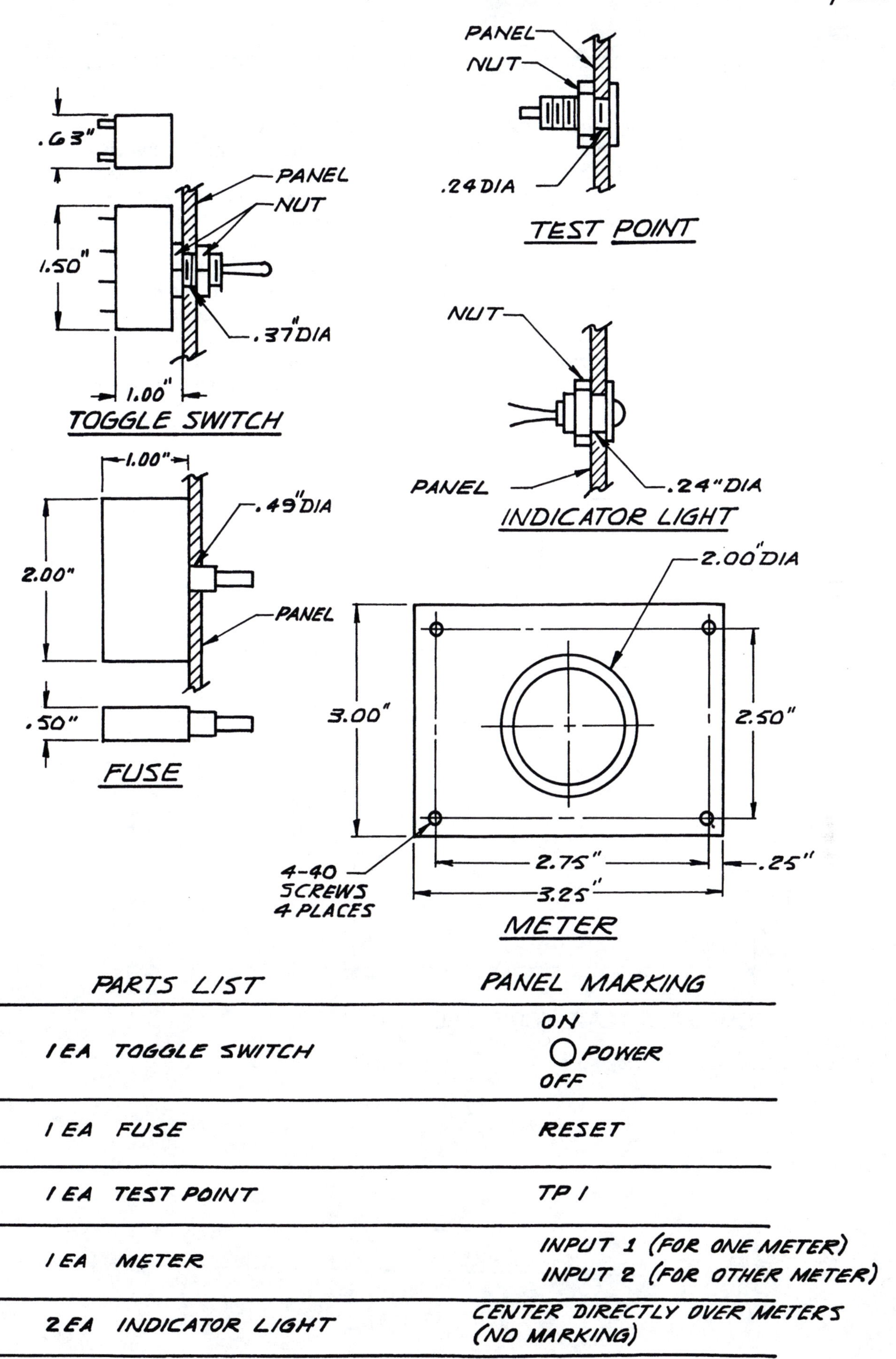

PARTS LIST	PANEL MARKING
1 EA TOGGLE SWITCH	ON ○ POWER OFF
1 EA FUSE	RESET
1 EA TEST POINT	TP 1
1 EA METER	INPUT 1 (FOR ONE METER) INPUT 2 (FOR OTHER METER)
2 EA INDICATOR LIGHT	CENTER DIRECTLY OVER METERS (NO MARKING)

EXERCISE 11-3

Arrange the components A2, AR1, PD1, and FL1 on the 6″ × 4″ plate. Arrange them so that all connectors (J1, J2, J3, and the small rectangles sticking out of PD1 and FL1) are clear. These connections will have coaxial cables attached to them. The plate is .125 thick aluminum. Approximate any dimensions not given. The components appear in your book at $\frac{1}{2}$ scale. After you have made an assembly layout full size similar to the top view of the drawing in Exercise 11–4, make a detail drawing of the plate. Locate all holes with respect to the upper left corner (similar to the figure in Exercise 11–2). Assume that all holes will be the same size as the mounting holes on the components (A2, AR1, PD1, and FL1). Show only the holes and the plate itself. Screws will be inserted through the plate into the components to mount them to the plate. Turn in both the design layout and the detail drawing of the plate.

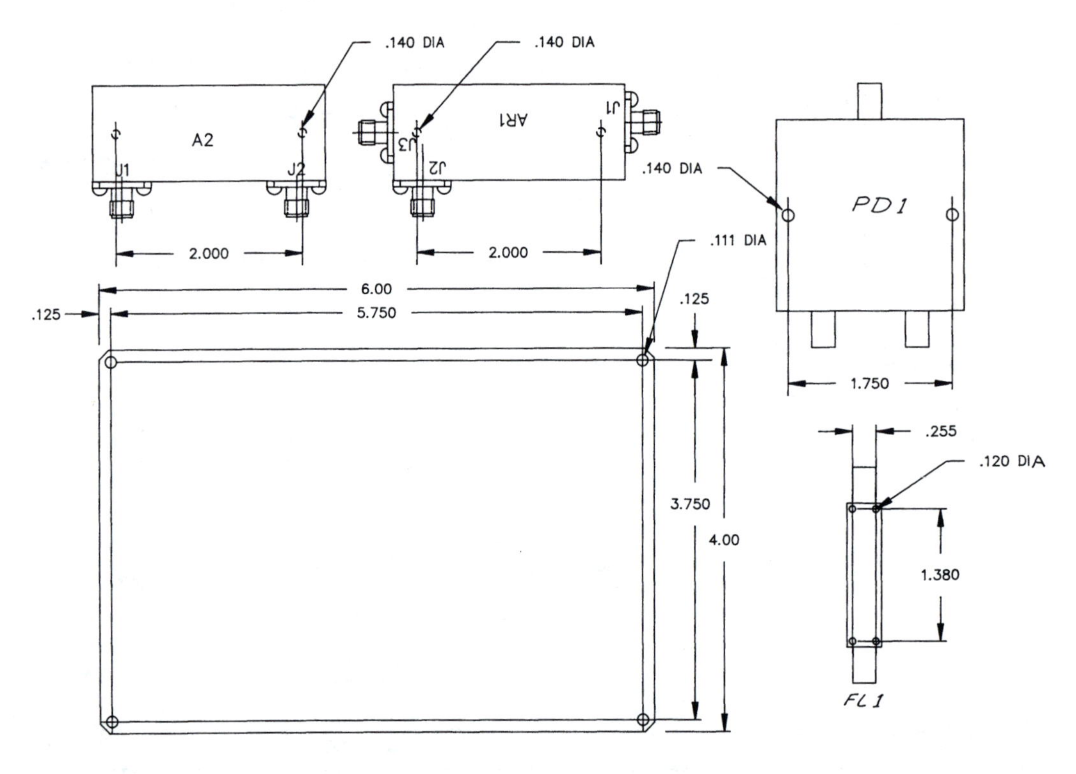

COMPONENT PLATE FOR EXERCISE 11–3

EXERCISE 11-4

Design a baseplate for the electronics unit shown on the next page. The drawing appears in your book half size. Notice that the component plate from Exercise 11–3 is mounted on spacers keeping it off of the baseplate. The holes for the spacers are .120 diameter. All other holes are as identified on the drawing. Countersink all holes on the bottom of the baseplate so that a flat-head screw may be used to mount all items from the bottom. Dimension all holes from the upper left corner. This will be a detail drawing of the baseplate similar to the drawing shown in Exercise 11–2. Show the plate and the holes only. The plate will be .25 thick aluminum.

ELECTRONICS UNIT FOR EXERCISE 11–4

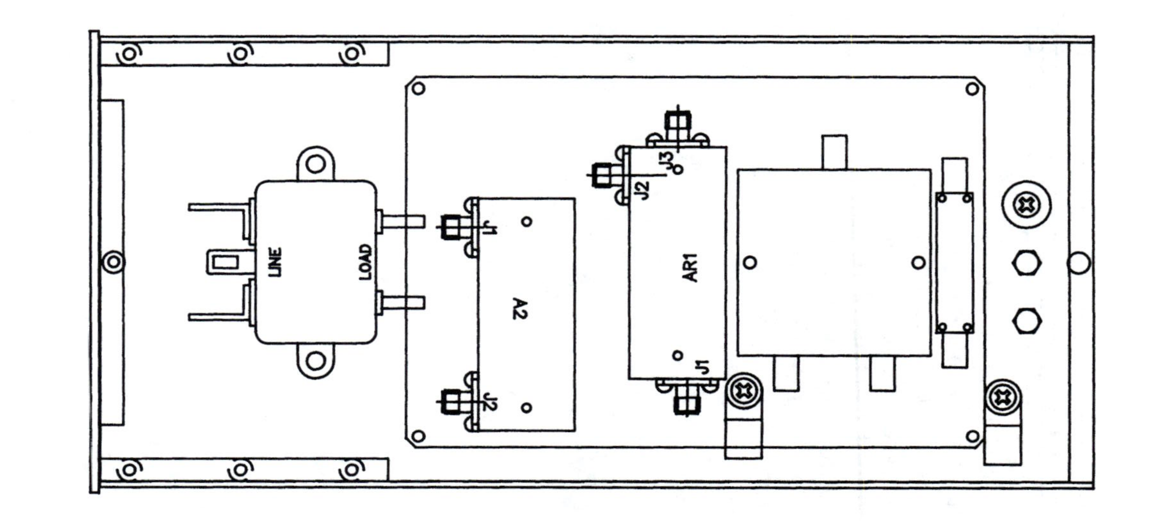

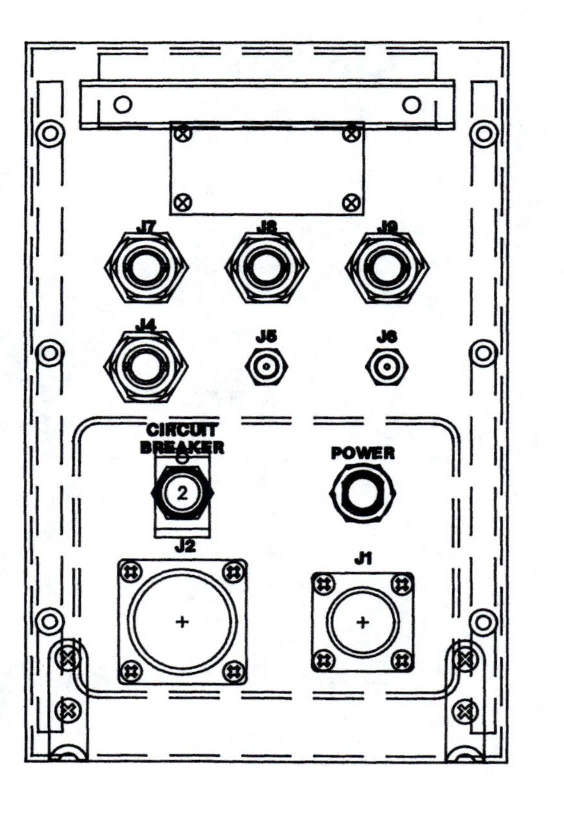

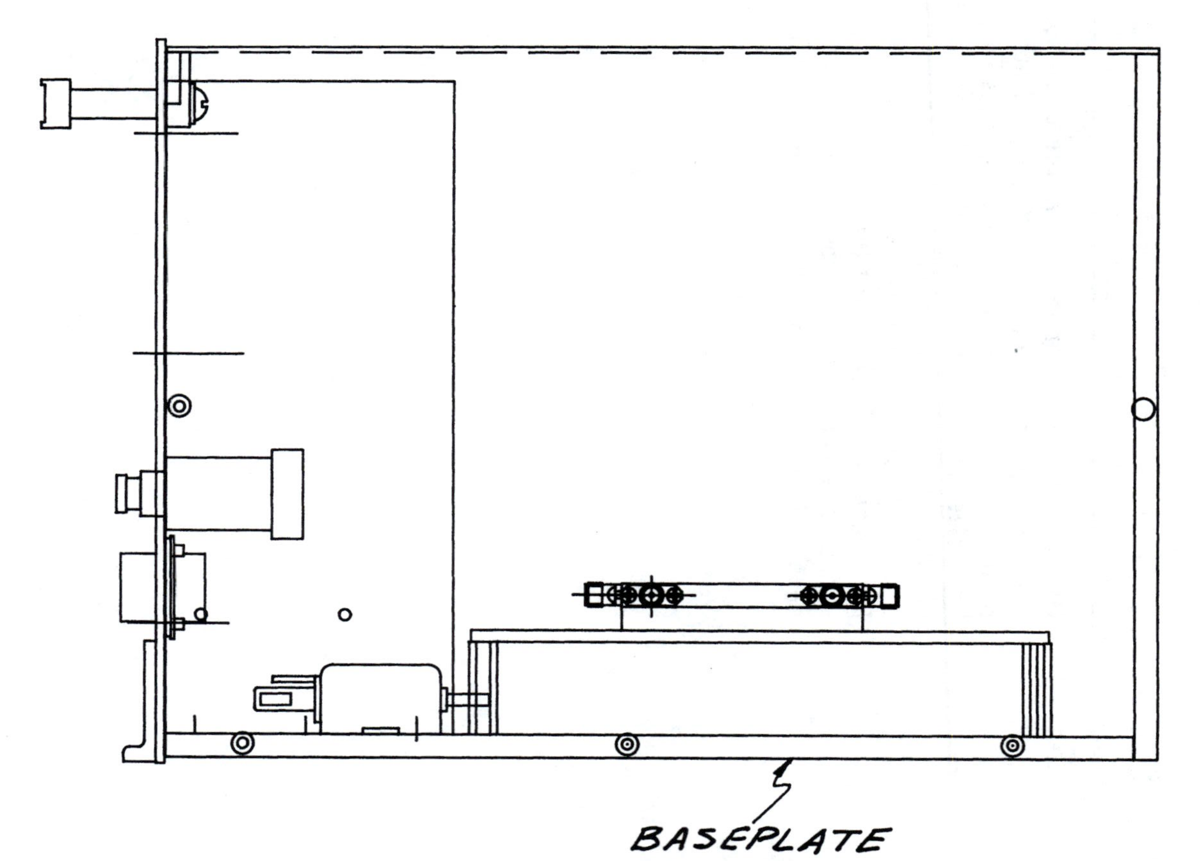

EXERCISE 11-5

Design a cable support panel to fit on standard mounting holes in your rack (the bracket should be approximately 19″ wide). This panel will be used to tie down 4″ ribbon cable with standard nylon cable ties as shown here. Make the height of the panel as short as possible, while still having room to accommodate cables comfortably. Fold over top and bottom edges to guard against cable damage from sharp edges. Draw the panel half scale using two views (front and right side on a B-size sheet). The choice of mounting screws and hole sizes is up to your own careful judgment.

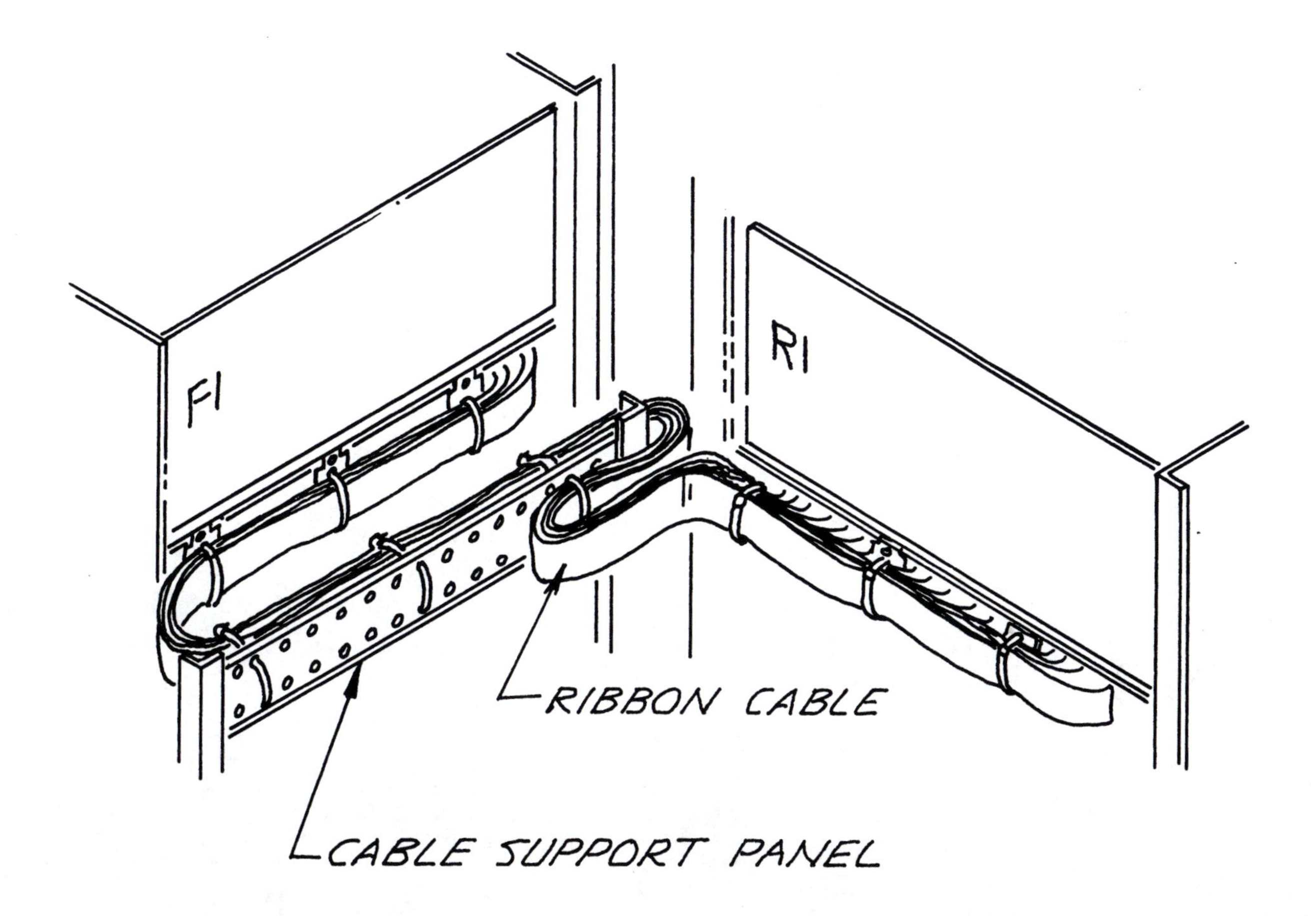

Chapter 12

Computer-Aided Design of Printed Circuits

OBJECTIVES

After completing this chapter, you will be able to

- ✓ list the major parts of a computer-aided design (CAD) system and describe the function of each part.
- ✓ list 10 of the best known software packages for printed circuit design.
- ✓ define the terms:
  - Schematic capture
  - Automatic component placement
  - Net list
  - Rat's nest
  - Automatic routing
- ✓ describe how a circuit may be designed using a CAD system from schematic to final artwork

INTRODUCTION

There are several different manufacturers who make computer-aided design (CAD) systems hardware and software. All of these systems have many similarities. The term *hardware* is used to describe the equipment. *Software* describes the programs for both microcomputers *(personal computers)* and minicomputers (larger systems which are dedicated to one software program).

PARTS OF THE SYSTEM

The parts of a typical personal computer system, Figure 12–1, are:

- The computer
- A video monitor
- A floppy disk drive
- A hard disk drive
- A keyboard
- A mouse, a tablet and stylus or puck, or a light pen
- A plotter

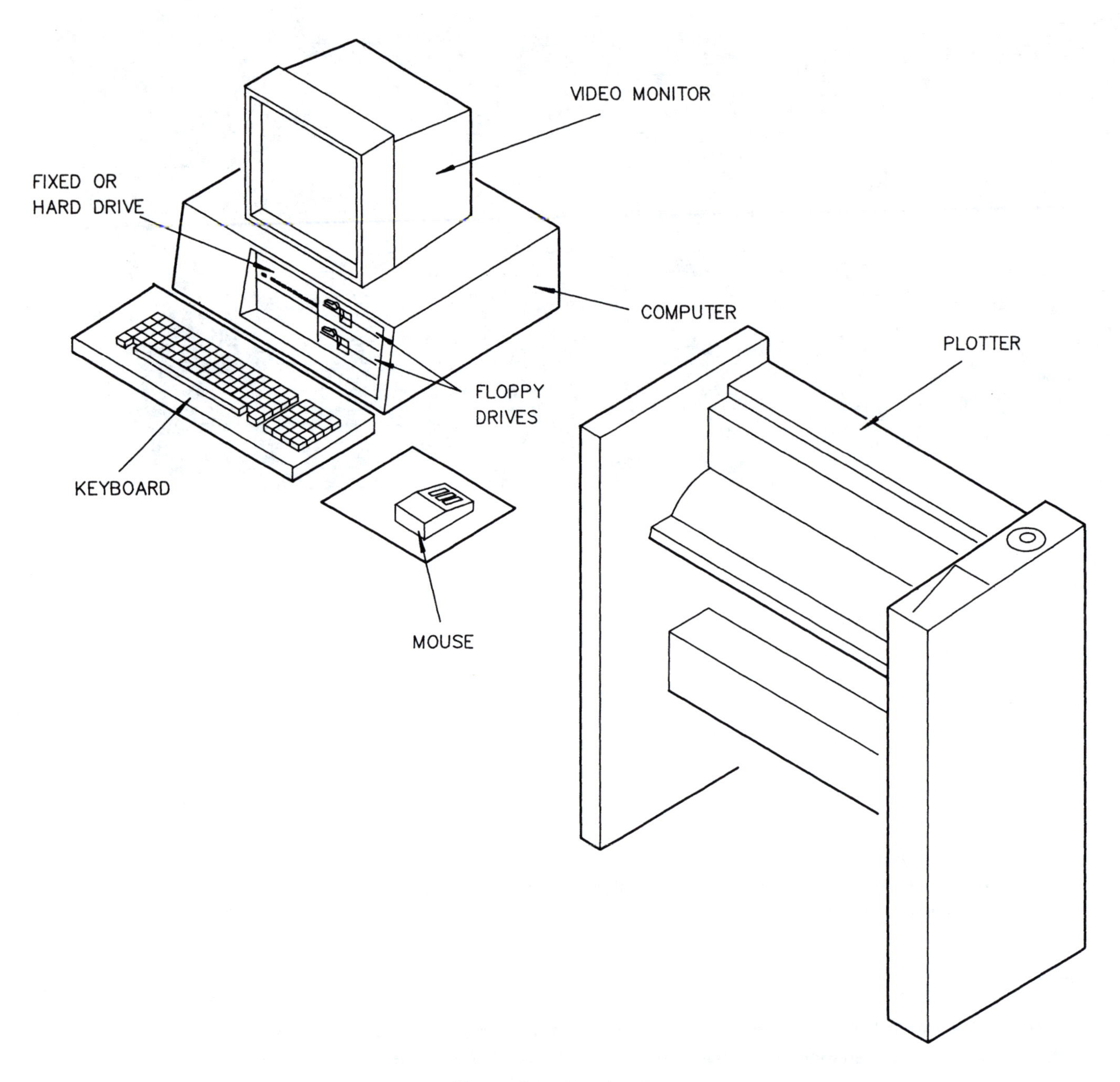

Figure 12–1 A workstation

The computer processes information and responds to commands from the keyboard or a menu which is activated by a mouse or a tablet stylus. The video monitor displays the drawing and any screen menu which is active. A floppy disk drive is used to move large amounts of information into and out of the computer. The hard drive is used to store software programs and drawings. The keyboard, tablet, and mouse are input devices used to activate commands telling the computer what to do. The plotter is sometimes used to produce the finished artwork and is often used to produce schematic, assembly, drill plans, and any detail drawings which do not require extreme accuracy. A photographic plotter is more often used to produce the final artwork.

STEPS TO PRODUCE A PC BOARD

In addition to the physical parts of the system there is stored information called software. The software packages used to create all of the documents necessary to produce a printed circuit board contain the following steps:

Step 1. Identify existing symbols or create new symbols called library figures or blocks.

To obtain such a symbol, the operator activates a command to insert a particular symbol. The CRT screen then displays a question asking where the symbol should be placed. The operator then moves the mouse or stylus on the graphics area of the tablet. This movement is shown on the screen, but the symbol is not located until the operator touches (digitizes) a location on the screen. The operator may locate as many of the same symbol as needed by digitizing these locations on the screen. After the operator touches the return key on the keyboard, the symbols appear on the screen.

If a symbol is needed that is not in the software package it may be drawn and stored for future use. These types of symbols are called *library figures* or blocks. These figures must be constructed so that they have the same features as the symbols that are already in the system. These features, shown in Figure 12–2, are described next.

Features of Symbols

Figure 12–2 shows a NAND gate symbol similar to those already studied. However, this symbol shows three diamonds, one on the end of each of the input and output lines. These diamonds are what the system recognizes as *connect nodes.* They are needed to connect one symbol with another. Near each of the diamonds is a triangle called a *text node.* The text node is used to place pin numbers on gate inputs and outputs. The figure shows two other text nodes: one inside the symbol and one outside. The one inside is the text node used to place the reference designator (such as UI) inside of the symbol. The one outside is used to place the device type (the part number of the IC, such as 7400) outside of the symbol. Many of these features are automatically produced by some software packages.

After all symbols have been identified or created, the schematic can be drawn.

Step 2. Create the schematic using standard symbols and connect them together. All pin numbers, reference designators, values, and part numbers must appear on the schematic. These are commonly inserted into the drawing when the symbol is placed into the drawing.

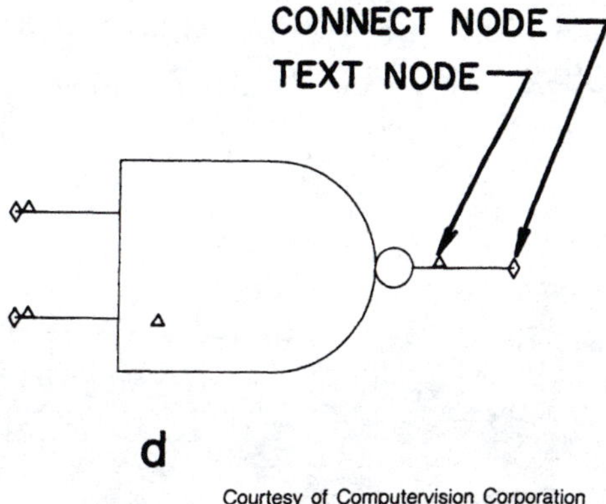

Courtesy of Computervision Corporation

Figure 12–2 Symbol features

Step 3. Create the printed circuit board outline and place components on the board. Some software programs will automatically place the components on the board. This process is appropriately called *automatic component placement.*

Step 4. Merge (combine) the information from the schematic and the printed circuit board so that all connecting points are correctly identified.

Step 5. Make the required connections between components to complete the design layout.

These steps will now be described in detail.

Step 1. Identify Symbols

All PC board software packages contain such symbols as resistors, capacitors, diodes, and integrated circuits that may be activated by typing a command or using a menu. Many of these symbol libraries are very large and contain most if not all of the parts needed for most printed circuit boards.

Step 2. Create the Schematic

A rough drawing of the schematic must be drawn by the engineer who designs the circuit. Using this rough drawing, the CAD operator then places symbols on the screen. After all symbols are placed on the screen

```
001   XSIG001   U3–1  P1–7  U2–1  U3–5
002   XSIG002   U2–9  U3–9  P1–5
003   XSIG003   U1–1  P1–1
004   XSIG004   U3–13  U2–2  U3–10  P1–9
005   XSIG005   U2–3  U3–12  P1–8
006   XSIG006   U2–13  U4–13  P1–10
007   XSIG007   U1–4  U1–11  P1–4
008   XSIG008   U1–12  U2–12
009   XSIG009   U4–1  U3–8
010   XSIG010   U1–5  U4–11
011   XSIG011   U4–12  U4–3
012   XSIG012   U4–2  U3–3
013   XSIG013   U3–2  U2–8
014   +5V       C1–1  C2–1  C3–1  U1–14  U2–14  U3–14  U4–14  P1–3
015   GND       C1–2  C2–2  C3–2  U1–7  U2–7  U3–7  U4–7  P1–6
016   XSIG014   P1–2  U1–3
017   XSIG015   U1–2  U1–6
018   XSIG016   U4–9  U3–6
019   XSIG017   U3–4  U2–4
020   XSIG018   U1–13  U4–8
021   XSIG019   U4–10  U3–11
```

Courtesy of Computervision Corporation

Figure 12-3 Creating a *Net List*

they are connected, as shown on the rough drawing. After all connections have been made, pin numbers, reference designators, values, and part numbers are added using text locations on each symbol. Much of this information is added automatically with the use of some software packages. When all of the correct information has been added to the connected symbols the schematic is complete.

A document called a *net list,* Figure 12–3, is then created using a single command. The software program identifies each connection on each symbol so that a list of connections similar to a point-to-point wire list can be generated. This list of connections is called a *net list.* The process of assembling all of this information is called *schematic capture.* Many software programs do this automatically.

Step 3. Create the Printed Circuit Board Outline and Place Components On the Board

Creating the printed circuit board begins by very accurately drawing the outline of the board, Figure 12–4. Next, a *keep-in area* is constructed. This produces a line that describes the area within which the leads between connecting points may be routed, Figure 12–5. Then, the connector or connectors are inserted onto the board outline.

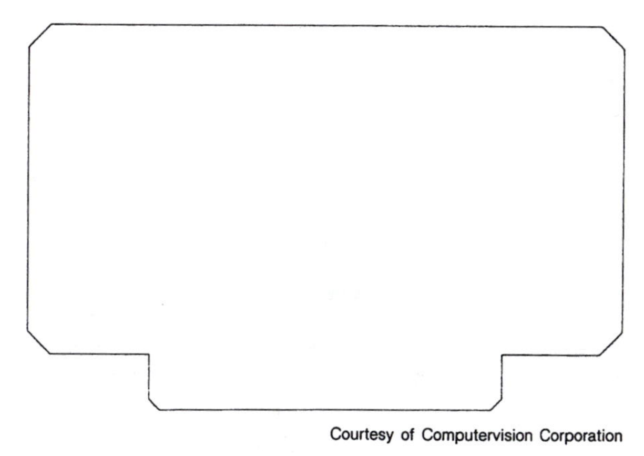

Courtesy of Computervision Corporation

Figure 12–4 Inserting the outline of the board

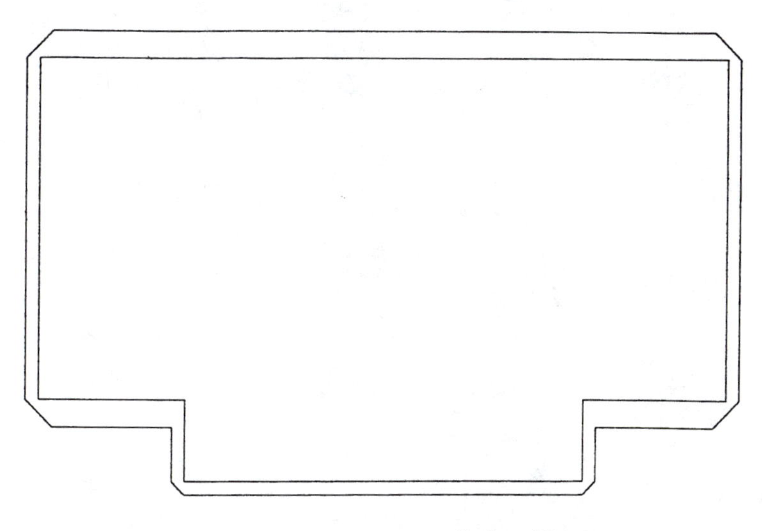

Courtesy of Computervision Corporation

Figure 12–5 Inserting a *Keep-In* boundary

Finally, all of the components of the circuit are inserted onto the board, using as the source a parts list which agrees with the schematic diagram (automatic placement is available for some board types). Notice that all of the four integrated circuits in Figure 12–6 have connect nodes or points for each pin, and text locating points for all pins, reference designators, and device types.

Figure 12–7 shows the board with all of its components in place and all pins, reference designators, and part numbers identified. At this stage the board is ready to be merged (combined with the connecting information from the schematic diagram: the net list).

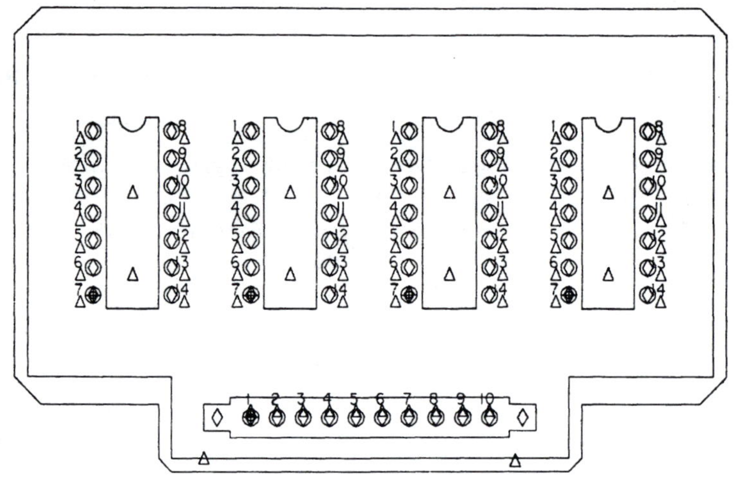

Courtesy of Computervision Corporation

Figure 12–6 Inserting components

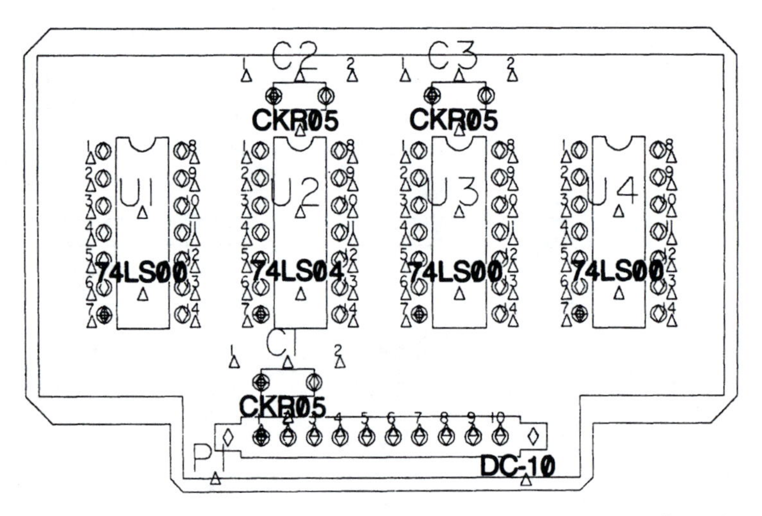

Courtesy of Computervision Corporation

Figure 12–7 Inserting *Text*

Step 4. Merge the Net List

A command called *Merge Net List* is used to supply the connecting information to the computer so that it may connect the correct components, Figure 12–8. If the components and all of their connect nodes have been correctly identified on both the schematic and PCB, the net list shows that all connecting points have successfully *merged.* If there is a mistake, the *Merge Net List* command will show where the mistake has been made. Any mistakes must be corrected at this time before the final step, "Automatic Routing" can be done.

Some software programs create what is called a *rat's nest.* This shows all of the connections without regard to layer or lead spacing or routing, Figure 12–8.

Step 5. Automatically Make All Connections (Automatic Routing)

Once the net list has been successfully combined with the PCB, a *Route* command is used to make all of the required connections on the board, Figure 12–9. This is the most rewarding of all commands if all previous steps have been done correctly. This routing is done very quickly and can be accomplished with a minimum of feed-thrus (VIAs) by repeating the *Route* command and limiting the number of feed-thrus.

XSIG001 MERGED
XSIG002 MERGED
XSIG003 MERGED
XSIG004 MERGED
XSIG005 MERGED
XSIG006 MERGED
XSIG007 MERGED
XSIG008 MERGED
XSIG009 MERGED
XSIG010 MERGED
XSIG011 MERGED
XSIG012 MERGED
XSIG013 MERGED
+5V MERGED
GND MERGED
XSIG014 MERGED
XSIG015 MERGED
XSIG016 MERGED
XSIG017 MERGED
XSIG018 MERGED
XSIG019 MERGED

A
NET LIST MERGED

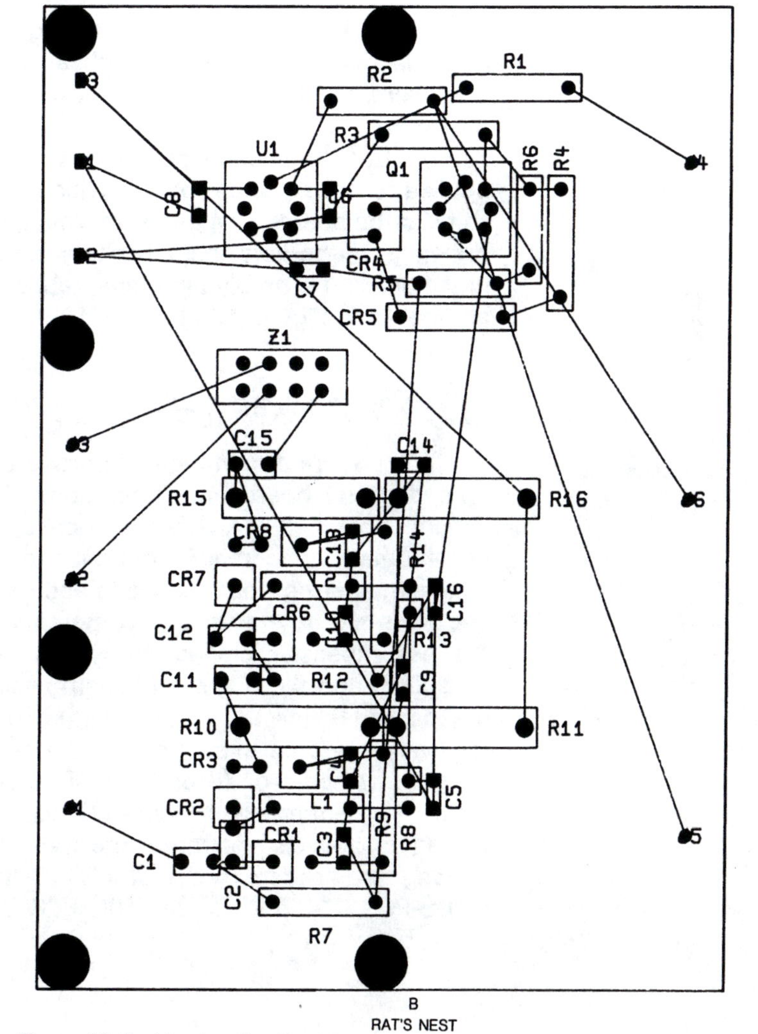

Courtesy of Computervision Corporation

B
RAT'S NEST

Figure 12–8 Merging the *Net List*

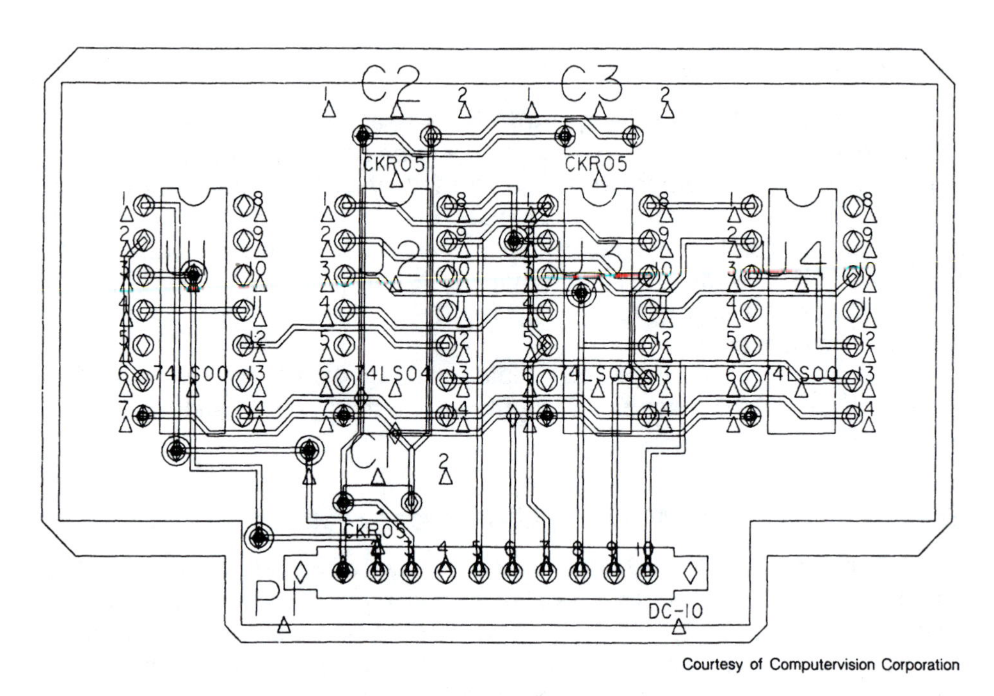

Courtesy of Computervision Corporation

Figure 12–9 Routing the board (connecting pads)

Not all connections can be made automatically with all software programs, because of some limiting factors which may have been programmed into the system, such as *do not cross wires* or *do not make diagonal lines*. Unconnected lines are then made manually with the mouse or stylus.

After the routing of the board is complete, the information may be applied to drive a photo plotter which is used to make the artwork necessary to build the board. This same information may be used to produce schematic drawings, parts lists, numerically controlled drill tapes for drilling the holes in the board, assembly drawings, fabrication drawings, and other documents as shown in Figure 12–10.

A SURVEY OF THE FIELD

There are a multitude of software packages that are used to produce printed circuit board drawings. Some are very limited in their capabilities. Others contain all of the features described in the preceeding example such as automatic placement, routing and checking of design rules (such as minimum spacing between leads and pads). Many of the more sophisticated packages are now available for personal computers such as the IBM AT or the less expensive clones. Many of these packages presently cost as little as $3000 and the number of them available is multiplying rapidly. Some of the packages for schematic capture only, cost as little as $400. These relatively low cost systems have made it possible for many PCB designers to work at home or to open small (often lucrative) offices.

A recent article in *Printed Circuit Design,* a magazine for printed circuit designers, reveals that there are more than 55 suppliers of PCB software. Among the leaders are COMPUTERVISION, REDAC, SCHEMA, P-CAPS, PADS-PCB, CALAY, CADNETIX, PCB DESIGNER, and McCAD PCB.

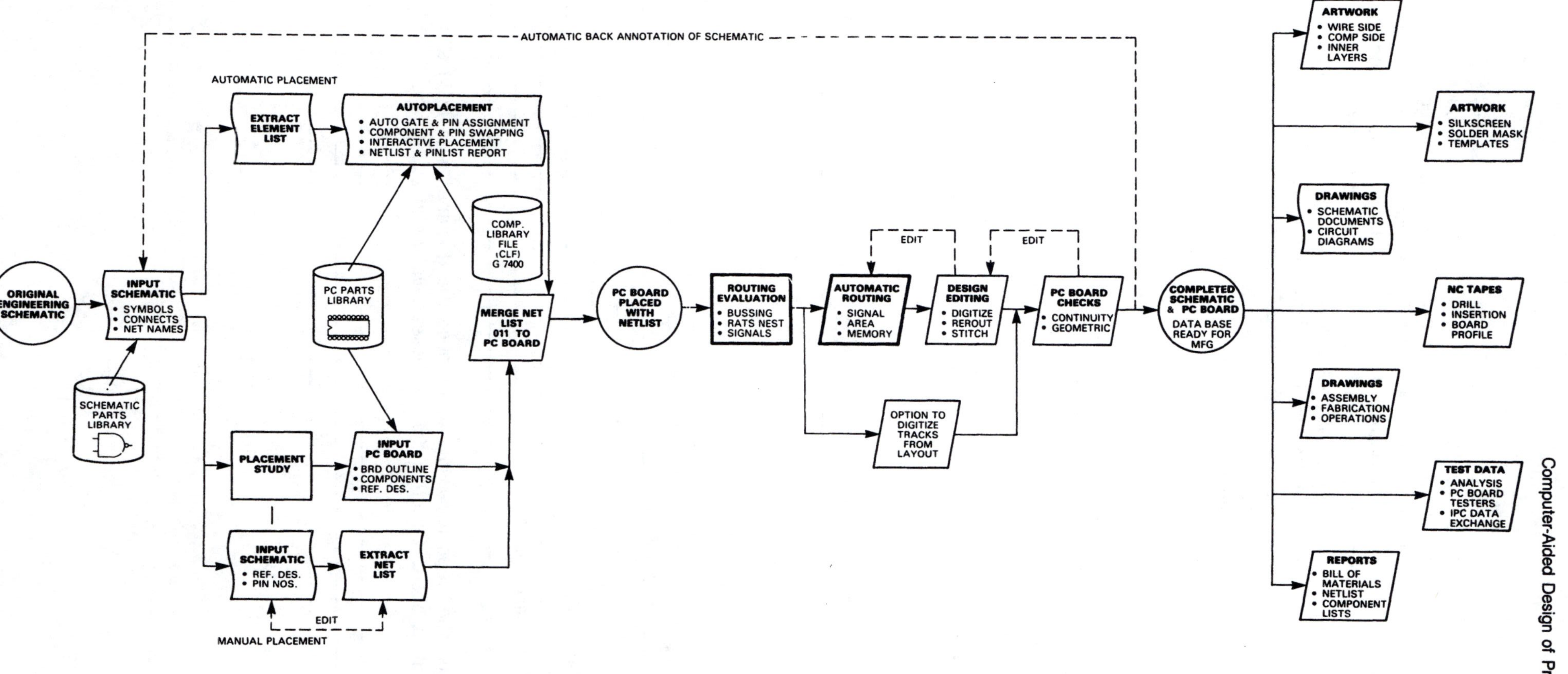

Courtesy of Computervision Corporation

Figure 12–10 Computervision CAD/CAM PC board design process

Some PCB designers use the less specialized CAD packages such as AutoCAD and customize their computer tablets or menus to allow this system to be used for PCB design. Automatic functions are not available for the more generalized CAD packages.

The job of the printed circuit designer is one that has many varied opportunities. Small companies and large use the services of the accomplished professional designer and many of the larger companies provide extensive training to the beginning designer. Job Shops (personnel agencies that provide temporary technical employees) often have openings for PCB designers that last from a few weeks to years, at salaries that are higher than many permanent jobs. This is an interesting job in an exciting period in the design field.

EXERCISE 12–1

On a separate sheet of paper, write your answers to the following test on the Objectives.

1. List the major parts of a CAD system, and describe the function of each.
2. List 10 of the best known software packages that are used to produce printed circuit boards.

______	______
______	______
______	______
______	______
______	______

3. Define the following terms.
   Schematic capture ______
   ______
   Automatic component placement ______
   ______
   Net list ______
   ______
   Rat's nest ______
   ______
   Automatic routing ______
   ______

4. List the steps necessary to produce the documents to build a printed circuit board. Describe the processes involved in each step.
   ______
   ______
   ______
   ______

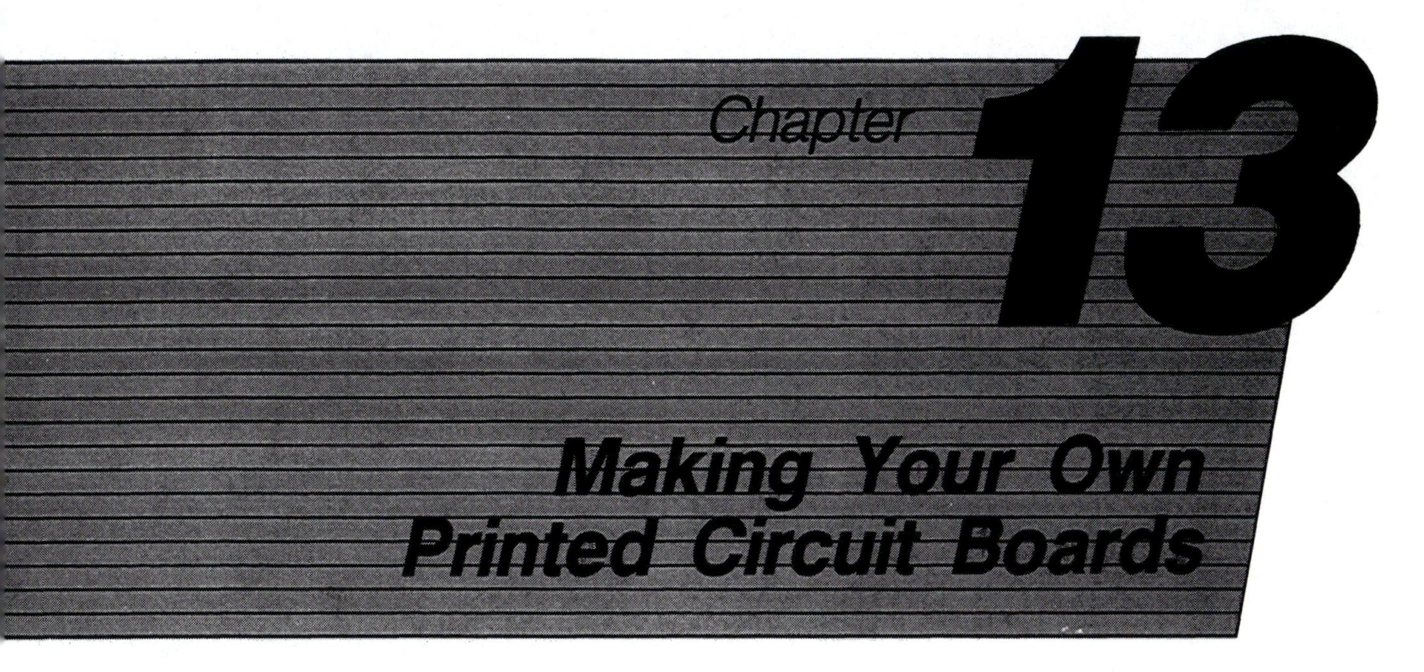

OBJECTIVES

After completing this chapter, you will be able to

- ✓ create physical specifications for a printed circuit board.
- ✓ produce an artwork which you can use to make a printed circuit board.
- ✓ expose and etch a printed circuit board from the artwork you have created.
- ✓ drill holes in the PC board.
- ✓ assemble components on the PC board and solder them in place.

INTRODUCTION

Making your own printed circuit boards can be an interesting and valuable project. Many electronic magazines and other project oriented publications have schematics, circuit artworks, and parts lists which can be used to manufacture electronic projects. Examples of these projects—such as testers, clocks, scramblers, amplifiers, flashing Christmas tree lights, radios, etc.,—can be found in every edition of these magazines.

This chapter will provide you with the information you need to take a construction project from the schematic diagram to the finished project, Figure 13–1.

The required steps are:

1. Find a schematic and a parts list and buy the parts.
2. Determine the size of the finished printed circuit board and how it is to be mounted.
3. Create or obtain an artwork for the circuit.
4. Expose the board.
5. Etch the board.
6. Drill holes in the board.
7. Assemble the components on the board.
8. Solder the components in place and trim the leads.
9. Clean and inspect the board.
10. Test the board.

SCHEMATIC

PARTS LIST

STEP 1
FIND SCHEMATIC
AND PARTS LIST

STEP 2
DETERMINE PACKAGE
AND BOARD SIZE

ARTWORK FROM
MAGAZINE

TAPED OR
PLOTTED ARTWORK

PAINT THE RESIST
DIRECTLY ON THE
BOARD

TAPE DIRECTLY
ON THE BOARD

STEP 3
CREATE ARTWORK

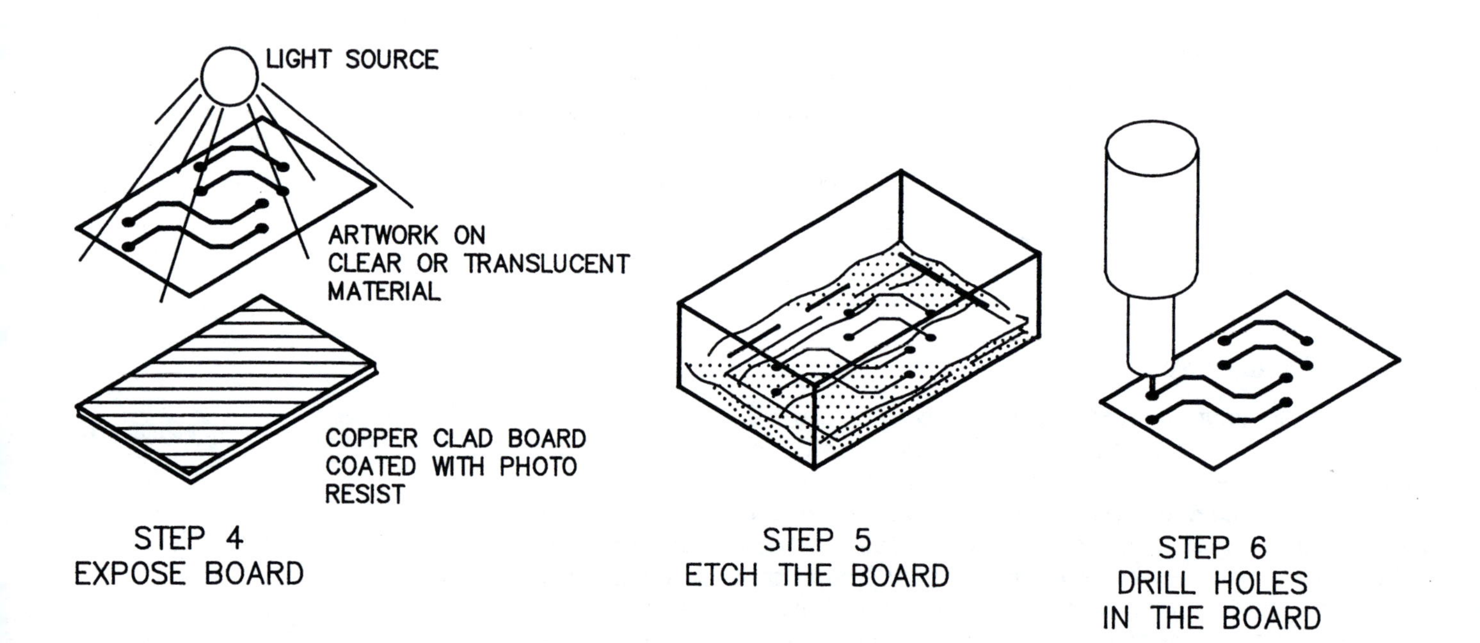

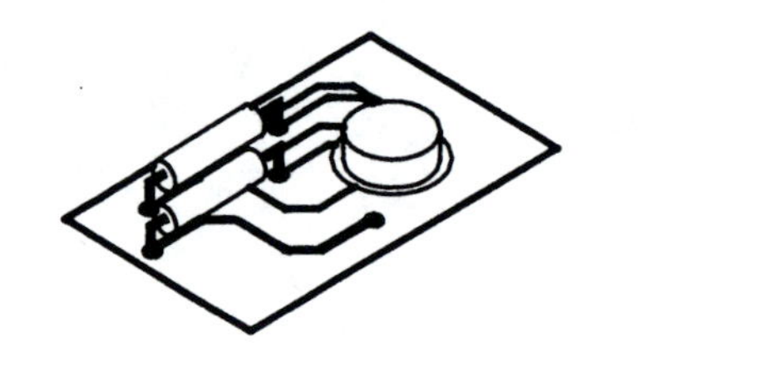

STEP 7
MOUNT COMPONENTS
ON THE BOARD

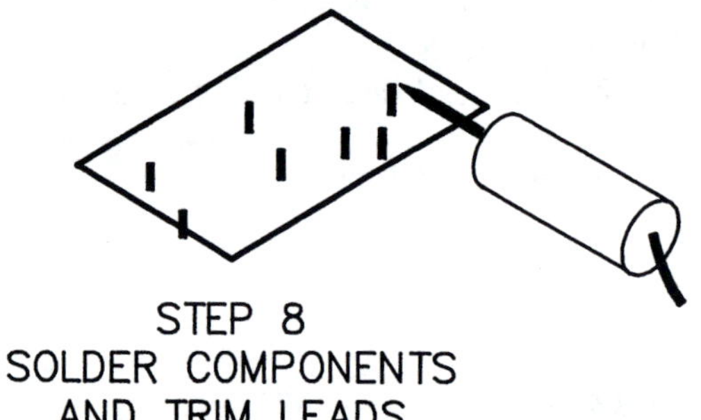

STEP 8
SOLDER COMPONENTS
AND TRIM LEADS

STEP 9
CLEAN AND
INSPECT THE
BOARD

STEP 10
TEST THE BOARD

Figure 13–1 Making a Printed Circuit Board

Step 1. Find a Schematic and a Parts List and Buy the Parts

Any technical library and many bookstores or magazine racks have detailed articles on construction projects. Those containing a parts list and a schematic diagram provide the basic information you will need. Articles that also contain an artwork with nothing printed on the back of the page are the most helpful. If the artwork is provided, it must be carefully inspected for weak lines or areas which are not dense black. All weak lines must be touched up with dense black ink such as India ink or its equivalent. If no artwork is provided, one must be created from the schematic diagram.

After the project has been identified, the parts to be mounted on the PC board must be purchased. Some parts substitutions may be necessary, or if parts are unavailable, a different project that provides the same end result may need to be selected. Be sure you have the parts in your possession before you proceed to the next step.

Step 2. Determine the Size of the Finished Printed Circuit Board and How it is to be Mounted, Figure 13–1

It is often easier and less expensive to buy a standard size box with standard covers than to form a box from sheet metal or plastic. These are available at retail electronic stores as well as surplus and wholesale outlets. Spacers are often needed to allow the PC board to be mounted a distance from the metal box or cover. All mounting features, including board size, should be determined before any new artwork is created. After the board size has been determined, the copper-clad, plastic fiber board may be purchased. These boards are found in many retail electronic or hobby shops. (Copper-clad means that one or both sides of the plastic is covered with a thin sheet of copper.)

Step 3. Create or Obtain an Artwork for the Circuit

An artwork for etching the circuit on the PC board may be obtained in at least four ways:

- Cut an existing artwork out of a magazine, inspect it carefully, and touch up any weak lines.
- Create an artwork on translucent paper using the principles described in the previous chapters of this book.
- Use a resist fluid (such as that found in the kits in retail electronic stores) to paint the circuit directly on the copper-coated plastic PC board. A resist fluid does not allow etching to take place in the areas where it resides; if this method is used, skip Step 4.
- Stick black PC tape and donuts directly on the copper-coated PC board to form the circuit; if this method is used, skip Step 4.

Step 4. Expose the Board

After the artwork has been created, the copper on the board must be coated with a substance which resists the etching solution in the areas where the circuit is to be formed. If one of the last two methods described in Step 3 is used, the board does not need to be exposed. If one of the first two methods in Step 3 is used, the following is necessary:

- Buy a copper-clad PC board material which is coated with a photo-resist or spray the photo-resist onto the copper surface (a photo-resist is an etch resisting material that is sensitive to light). Photo-resist is available in stores specializing in PCB or photographic supplies.

- Expose the board to ultraviolet light with the artwork attached firmly to the photo-resist copper-coated surface. Sunlight or commonly available exposure lamps can be used for the light source. The amount of exposure time should be controlled according to the photo-resist manufacturer's specifications.

Step 5. Etch the Board

After the resistive material is in place and the board has been exposed or otherwise prepared, immerse the board in an etching solution for the amount of time suggested by the manufacturer of the etching solution. *Observe any warnings on the bottle.* After the etching solution has done its work the copper will remain in only the areas coated with the resistive materials. Remove any resistive material and clean the board in accordance with the manufacturer's instructions.

Step 6. Drill Holes in the Board

Use the chart in Figure 9–24 to determine the correct size for all holes to be drilled in the centers of all pads. Carefully drill holes and remove any copper burrs or other foreign material.

Step 7. Assemble Components on the Board

Assemble components one at a time or tape them in place on the component side of the board before turning the board over for soldering.

Step 8. Solder Components in Place and Trim the Leads

Use good soldering techniques to solder components in place. A good solder joint has a minimum amount of solder to hold the lead firmly to the pad. Avoid excessive heat which can pull the circuit off the board. Carefully trim the excess length of each lead.

Step 9. Clean and Inspect the Board

Carefully clean the board and inspect it for good solder joints and possible shorts.

Step 10. Test the Board

Test the board, if possible, without placing it into its final assembly. Correct any malfunctions.

EXERCISE 13–1

List the ten steps in manufacturing a printed circuit board and describe how you plan to perform each step for your project.

Appendix

Specifications, Standards, and Manuals

STANDARDIZATION

Most companies make strong attempts to standardize their drawings. The reasoning behind the move toward standardization is that standardized drawings are often more economical to produce and use, present a better appearance to customers, are stored and reproduced more easily, and often provide more complete information to the manufacturer of the product.

IMPORTANCE OF STANDARDS

Governmental agencies and some companies have very strict and detailed standards. Other companies have more relaxed standards. One of the most important details that a drafter *must* attend to is to determine which standards apply to the work at hand, and then adhere to those standards thoroughly. Many drawings are redone every day because someone did not read a specification carefully or ignored a standard.

Types

There are standards for the drawings themselves, what they are drawn on, what they are drawn with, the lettering height and style, the parts represented by the drawings, how the drawings are reproduced and stored, and many other items that affect the drafter's job. The government even has a standard that standardizes which standards are to be used (DOD-STD-100C).

Classifications

Standard DOD-STD-100C classifies standard documents into three groups: specifications, standards, and manuals. Although these three groups seem to overlap, and the classifications have lost their distinction through many changes over a long period of time, it is probably important to use the correct group listing when ordering copies of these documents. A partial listing of the ones that apply to electronic drawings is as follows. Notice that specifications are divided into federal and military, standards into military and industry, and the manuals have no division.

Specifications

Federal

L-F-340	Film, Diazotype, Sensitized, Moist and Dry Process, Roll and Sheet
L-P-519	Plastic Sheet, Tracing, Glazed Matte Finish
UU-P-221	Paper, Direct-positive Sensitized, (Diazotype—Moist and Dry Process)
UU-P-561	Paper, Tracing
CCC-C-531	Cloth, Tracing

Military

DOD-D-1000	Drawings, Engineering and Associated Lists
MIL-D-5480	Data, Engineering and Technical Reproduction Requirements for
MIL-D-8510	Drawing, Undimensioned, Reproducibles, Photographic and Contact Preparation of (ASG)
MIL-M-38761	Microfilming and Photographing of Engineering/Technical Data and Related Documents: PCAM Card Preparation, Engineering Data Micro-Reproduction System, General Requirements for, Preparation of
DOD-STD-100	Types of Drawings
MIL-STD-454	Design Requirements

Standards

Military

MIL-STD-12	Abbreviations for Use on Drawings, Specifications, Standards and in Technical Documents
MIL-STD-17-1	Mechanical Symbols
MIL-STD-17-2	Mechanical Symbols for Aeronautical, Aerospacecraft, and Spacecraft Use
MIL-STD-34	Preparation of Drawings for Optical Elements and Optical Systems, General Requirements
MIL-STD-275	Printed Wiring for Electronic Equipment
DOD-STD-1476	Metric System, Application in New Design

Industry

ANSI B46.1–1978	Surface Texture (Surface Roughness, Waviness, and Lay)
ANSI X3.5–1970	Flowchart Symbols and Their Usage in Information Processing. Same as FIPS-24
ANSI Y14.1–1980	Drawing Sheet Size and Format
ANSI Y14.2M–1979	Line Conventions and Lettering
ANSI Y14.3–1975	Multi and Sectional View Drawings
ANSI Y14.5M–1982	Dimensioning and Tolerancing
ANSI Y14.6–1978	Screw Thread Representation and ANSI Y14.6aM (Metric Supplement)

ANSI Y14.7.1–1971	Gear Drawing Standards, Part 1 for Spur, Helical, Double Helical, and Rack
ANSI Y14.7.2–1978	Gear and Spline Drawing Standards—Part 2, Bevel and Hypoid Gears
ANSI Y14.13M–1981	Mechanical Spring Representation
ANSI Y14.15–1966	Electrical and Electronic Diagrams
ANSI Y14.15a–1971	Interconnection Diagrams
ANSI Y14.15b–1973	Supplement to ANSI Y14.15–1966 and ANSI Y14.15a–1970
ANSI Y14.26.3–1975	Computer-aided Preparation of Product Definition Data (including Engineering Drawings), Terms and Definitions
ANSI/IPC-D-350B–1978	Printed Board Description in Digital Form
ANSI/IPC-T-50B–1980	Terms and Definitions for Interconnecting and Packaging Electronic Circuits
AWS A3.0–1976	Welding Terms and Definitions
IEEE STD 91–1982	Reference Designations for Electric and Electronic Parts and Equipment. (Same as ANSI Y32.16–1975)
IEEE STD 315–1975	Graphic Symbols for Electrical and Electronic Diagrams (including Reference Designation Class Designation Letters). (Same as ANSI Y32.2–1975)

Manuals

Military

H4	Federal Supply Code for Manufacturers
H6	Federal Item Name Directory for Supply Cataloging
DOD 5220.22-M	Industrial Security Manual for Safeguarding Classified Information

Ordering

All professional drafters have access to the standards that affect their work. Often these standards may be obtained at a very low cost from a government or industrial agency. A new drafter is encouraged to order current copies of the required standards, including all of the latest changes (called addenda). To obtain copies of military specifications, call or write to the Government Printing Office in any large city. Copies of Industry Association Standards may be obtained from:

(ANSI) American National Standards Institute
1430 Broadway
New York, NY 10018

(ASTM) American Society for Testing and Materials
1916 Race Street
Philadelphia, PA 19103

The following is a typical letter for ordering copies of standards.

American National Standards Institute
1430 Broadway
New York, NY 10018

Dear Sir or Madam:

Please send copies of the following standards with all addenda to the address below:
ANSI B46.1–1978 Surface Texture
ANSI Y14.1–1980 Drawing Sheet Size and Format

Please send any required invoice to the same address.

Sincerely,

Name
Street address
City, State, Zip Code

Appendix

Printed Circuit Design Criteria

DIMENSION AND TOLERANCE CONSIDERATIONS

The information presented in the following chart is for reference only. It is a compilation of various Military Standards (MIL-STD) and Institute of Printed Circuits (IPC) specifications. For more detailed information, consult the referenced specifications.

The classes of Materials and Processes on the right of the chart indicate progressive degrees of sophistication. The use of one class for a specific characteristic does not mean that the class must be used for all other characteristics and tolerances. Whenever possible, dimensions and tolerances should be selected from the preferred data since this will typically result in the most producible part. Design requirements should always be discussed with individual circuit board manufacturers as process capabilities may vary considerably.

In some instances where a conflict between MIL-STD and IPC data for a specific feature or tolerance exists, both sets of data are listed. Note that the MIL-STD specifications generally represent a printed circuit board *user's* viewpoint while the IPC specifications typically emphasize the printed circuit board *manufacturer's* viewpoint.

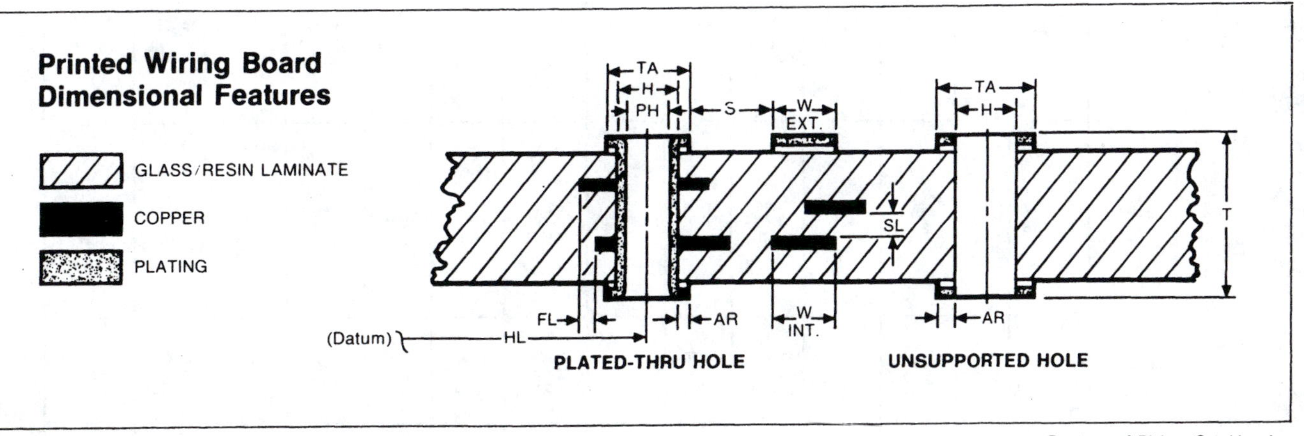

Courtesy of Bishop Graphics, Inc.

DESIGN CRITERIA & TOLERANCES		DATA FOR CLASSES OF MATERIALS & PROCESSES					
CODE LETTER	DESCRIPTION QUALIFIERS SUB-QUALIFIERS	SOURCE DOCUMENT	PREFERRED		STANDARD	REDUCED PRODUCIBILITY	
			CLASS 1	CLASS 2	CLASS 3	CLASS 4	CLASS 5
T	**BOARD THICKNESS (NOMINAL)**						
	SINGLE OR DOUBLE SIDED		.062" (1,57 mm)			.031" *(0,79 mm)*	
	MULTILAYER (MAX)	MIL-STD-1495	.100" *(2,54 mm)*		.150" *(3,81 mm)*	.200" *(5,08 mm)*	
T Tol.	**BOARD THICKNESS TOLERANCE (1)**						
	SINGLE OR DOUBLE SIDED						
	.031" *(0,79 mm)* THK	MIL-P-13949D	±.0065" *(0,17 mm)*		±.004" *(0,10 mm)*	±.003" *(0,08 mm)*	
	.062" *(1,57 mm)* THK		±.0075" *(0,19 mm)*		±.005" *(0,13 mm)*	±.003" *(0,08 mm)*	
	.093" *(2,36 mm)* THK		±.0090" *(0,23 mm)*		±.007" *(0,18 mm)*	±.004" *(0,10 mm)*	
	.125" *(3,18 mm)* THK		±.0120" *(0,30 mm)*		±.009" *(0,23 mm)*	±.005" *(0,13 mm)*	
	.250" *(6,35 mm)* THK		±.0220" *(0,56 mm)*		±.012" *(0,30 mm)*	±.006" *(0,15 mm)*	
	MULTILAYER	MIL-STD-1495, IPC-ML-910	±10% OF "T" NOMINAL OR ±.007" *(0,18 mm)* MINIMUM				
L	**MAXIMUM NUMBER OF MULTILAYER CONDUCTOR LAYERS (2)**	MIL-STD-1495	6		12	20	
W	**CONDUCTOR WIDTH (MINIMUM) (3)**						
	SINGLE OR DOUBLE SIDED	MIL-STD-275C	.010" *(0,25 mm)*				
	MULTILAYER						
	MIL-STD INTERNAL	MIL-STD-1495	.015" *(0,38 mm)*		.010" *(0,25 mm)*	.010" *(0,25 mm)*	
	MIL-STD EXTERNAL	MIL-STD-1495	.020" *(0,51 mm)*		.015" *(0,38 mm)*	.010" *(0,25 mm)*	
	IPC	IPC-ML-910	.020" *(0,51 mm)*		.010" *(0,25 mm)*	.005" *(0,13 mm)*	
W Tol.	**CONDUCTOR WIDTH TOLERANCE (4)**						
	IPC WITHOUT PLATING	IPC-D-300F	+.007" *(0,18 mm)* −.011" *(0,28 mm)*	+.005" *(0,10 mm)* −.006" *(0,15 mm)*	+.003" *(0,08 mm)* −.005" *(0,10 mm)*	+.002" *(0,05 mm)* −.003" *(0,08 mm)*	+.002" *(0,05 mm)* −.002" *(0,05 mm)*
	IPC WITH PLATING		+.016" *(0,41 mm)* −.018" *(0,28 mm)*	+.009" *(0,23 mm)* −.006" *(0,15 mm)*	+.005" *(0,10 mm)* −.005" *(0,10 mm)*	+.004" *(0,10 mm)* −.004" *(0,10 mm)*	+.003" *(0,08 mm)* −.003" *(0,08 mm)*
	MIL-STD WITHOUT PLATING	MIL-STD-1495	+.004" *(0,10 mm)* −.005" *(0,13 mm)*		+.002" *(0,05 mm)* −.004" *(0,10 mm)*	+.001" *(0,03 mm)* −.002" *(0,05 mm)*	
	MIL-STD WITH PLATING		±.0075" *(0,19 mm)*		±.005" *(0,13 mm)*	±.003" *(0,08 mm)*	
S	**CONDUCTOR SPACING (MINIMUM) (5)**						
	MIL-STD	MIL-STD-1495	.020" *(0,51 mm)*		.010" *(0,25 mm)*	.005" *(0,13 mm)* (6)	
	IPC	IPC-ML-910	.015" *(0,38 mm)*		.010" *(0,25 mm)*	.005" *(0,13 mm)*	

RECOMMENDED TERMINAL AREAS FOR UNSUPPORTED HOLES

All dimensions below are shown as standard inch with the metric equivalent in italic. Example: .0063" and *0,160*.

Component Lead Size		Recommended Finished Hole			Recommended Terminal Area: Preferred				Standard				Minimum			
AWG	Dia. in. *mm*	Dia. in. *mm*	Max. Allow Tol.	Drill	Term Dia.	Bishop Donut Pad/Dia. 4X	2X	1X	Term Dia.	Bishop Donut Pad/Dia. 4X	2X	1X	Term Dia.	Bishop Donut Pad/Dia. 4X	2X	1X
34	.0063" *0,160*	**.0145"** ***0,368***	±.003" *0,076*	#79	**.085"** ***2,16***	**D205** .343" *8,71*	**D307** .170" *4,32*	**D137** .093" *2,36*	**.075"** ***1,91***	**D111** .300" *7,62*	**D203** .150" *3,81*	**D311** .075" *1,91*	**.0625"** ***1,59***	**D109** .250" *6,35*	**D102** .125" *3,18*	**D136** .062" *1,57*
33	.0071" *0,180*															
32	.0080" *0,203*															
31	.0089" *0,226*															
30	.0100" *0,254*															
29	.0113" *0,287*															
28	.0126" *0,320*	**.026"** ***0,66***	±.003" *0,076*	#71	**.100"** *2,54*	**D115** .400" *10,16*	**D204** .200" *5,08*	**D101** .100" *2,54*	**.085"** *2,16*	**D205** .343" *8,71*	**D307** .170" *4,32*	**D137** .093" *2,36*	**.075"** *1,91*	**D111** .300" *7,62*	**D203** .150" *3,81*	**D311** .075" *1,91*
27	.0142" *0,361*															
26	.0159" *0,404*															
25	.0179" *0,455*															
24	.0201" *0,511*															
23	.0226" *0,574*															
22	.0253" *0,634*	**.036"** ***0,92***	±.004" *0,102*	#64	**.109"** ***2,77***	**D117** .437" *11,10*	**D143** .218" *5,54*	**D279** .111" *2,82*	**.093"** ***2,36***	**D114** .375" *9,53*	**D104** .187" *4,75*	**D137** .093" *2,36*	**.085"** ***2,16***	**D205** .343" *8,71*	**D307** .170" *4,32*	**D137** .093" *2,36*
21	.0258" *0,724*															
20	.0320" *0,813*															
19	.0359" *0,912*	**.052"** ***1,32***	±.004" *0,102*	#55	**.125"** ***3,18***	**D119** .500" *12,70*	**D109** .250" *6,35*	**D102** .125" *3,18*	**.109"** ***2,77***	**D117** .437" *11,10*	**D143** .218" *5,54*	**D279** .111" *2,82*	**.100"** ***2,54***	**D115** .400" *10,16*	**D204** .200" *5,08*	**D101** .100" *2,54*
18	.0403" *1,024*															
17	.0453" *1,151*															
16	.0508" *1,290*	**.0625"** ***1,59***	±.004" *0,102*	1/16	**.140"** ***3,56***	**D121** .562" *14,27*	**D110** .281" *7,14*	**D280** .140" *3,56*	**.125"** ***3,18***	**D119** .500" *12,70*	**D109** .250" *6,35*	**D102** .125" *3,18*	**.109"** ***2,77***	**D117** .437" *11,10*	**D143** .218" *5,54*	**D279** .111" *2,82*
15	.0571" *1,450*															

NOTE: This table was prepared on the basis of 2 oz./ft^2 copper with tin lead plate.
A .015" *(0,38 mm)* min. annular ring is assumed.

RECOMMENDED TERMINAL AREAS FOR PLATED-THRU HOLES

All dimensions below are shown as standard inch with the metric equivalent in italic. Example: .0063" and *0,160*

Recommended Terminal Area — Preferred

Component Lead Size AWG	Component Lead Size Dia in. *mm*	Recommended Finished Hole Dia in *mm*	Max. Allow. Tol	Preferred Term Dia	Preferred Bishop Donut Pad/Dia 4X	Preferred Bishop Donut Pad/Dia 2X	Preferred Bishop Donut Pad/Dia 1X
34 33 32 31 30 29	.0063" *0,160* .0071" *0,180* .0080" *0,203* .0089" *0,226* .0100" *0,254* .0113" *0,287*	**.020"** ***0,51***	± .005" *0,127*	**.085"** ***2,16***	**D205** .343" *8,71*	**D307** .170" *4,32*	**D137** .093" *2,36*
28 27 26 25 24 23	.0126" *0,320* .0142" *0,361* .0159" *0,404* .0179 *0,455* .0201" *0,511* .0226" *0,574*	**.030"** ***0,76***	± .005" *0,127*	**.093"** ***2,36***	**D114** .375" *9,53*	**D104** .187" *4,75*	**D137** .093" *2,36*
22 21 20	.0253" *0,634* .0258 *0,724* .0320" *0,813*	**.040"** ***1,02***	± .006" *0,152*	**.109"** ***2,77***	**D117** .437" *11,10*	**D143** .218" *5,54*	**D279** .111" *2,82*
19 18 17	.0359 *0,912* .0403 *1,024* .0453 *1,151*	**.052"** ***1,32***	± .006" *0,152*	**.115"** ***2,92***	**D224** .468" *11,89*	**D249** .236" *6,00*	**D265** .115" *2,92*
16 15	.0508" *1,290* .0571" *1,450*	**.067"** ***1,70***	± .007" *0,178*	**.140"** ***3,56***	**D121** .562" *14,27*	**D110** .281" *7,14*	**D280** .140" *3,56*

Recommended Terminal Area — Standard

Component Lead Size AWG	Component Lead Size Dia in. *mm*	Standard Term Dia	Standard Bishop Donut Pad/Dia 4X	Standard Bishop Donut Pad/Dia 2X	Standard Bishop Donut Pad/Dia 1X
34 33 32 31 30 29	.0063" *0,160* .0071" *0,180* .0080" *0,203* .0089" *0,226* .0100" *0,254* .0113" *0,287*	**.070"** ***1,79***	**D110** .281" *7,14*	**D280** .140" *3,56*	**D311** .075" *1,91*
28 27 26 25 24 23	.0126" *0,320* .0142" *0,361* .0159" *0,404* .0179 *0,455* .0201" *0,511* .0226" *0,574*	**.080"** ***2,03***	**D159** .320" *8,13*	**D247** .160" *4,06*	**D216** .080" *2,03*
22 21 20	.0253" *0,634* .0258 *0,724* .0320" *0,813*	**.093"** ***2,36***	**D114** .375" *9,53*	**D104** .187" *4,75*	**D137** .093" *2,36*
19 18 17	.0359 *0,912* .0403 *1,024* .0453 *1,151*	**.100"** ***2,54***	**D115** .400" *10,16*	**D204** .200" *5,08*	**D101** .100" *2,54*
16 15	.0508" *1,290* .0571" *1,450*	**.115"** ***2,92***	**D224** .468" *11,89*	**D247** .236" *6,00*	**D265** .115" *2,92*

Recommended Terminal Area — Minimum

Component Lead Size AWG	Component Lead Size Dia in. *mm*	Minimum Term Dia	Minimum Bishop Donut Pad/Dia 4X	Minimum Bishop Donut Pad/Dia 2X	Minimum Bishop Donut Pad/Dia 1X
34 33 32 31 30 29	.0063" *0,160* .0071" *0,180* .0080" *0,203* .0089" *0,226* .0100" *0,254* .0113" *0,287*	**.059"** ***1,50***	**D249** .236" *6,00*	**D317** .118" *3,00*	**D316** .059" *1,50*
28 27 26 25 24 23	.0126" *0,320* .0142" *0,361* .0159" *0,404* .0179 *0,455* .0201" *0,511* .0226" *0,574*	**.070"** ***1,79***	**D110** .281" *7,14*	**D280** .140" *3,56*	**D311** .075" *1,91*
22 21 20	.0253" *0,634* .0258 *0,724* .0320" *0,813*	**.080"** ***2,03***	**D159** .320" *8,13*	**D247** .160" *4,06*	**D216** .080" *2,03*
19 18 17	.0359 *0,912* .0403 *1,024* .0453 *1,151*	**.093"** ***2,36***	**D114** .375" *9,53*	**D104** .187" *4,75*	**D137** .093" *2,36*
16 15	.0508" *1,290* .0571" *1,450*	**.109"** ***2,77***	**D117** .437" *11,10*	**D143** .218" *5,54*	**D279** .111" *2,82*

NOTE: This table was prepared on the basis of 2 oz./ft² copper with tin lead plate. The plated-thru holes have a min. of .001" *(0,025 mm)* copper plating & .001" *(0,025 mm)* tin lead plating. A .010" *(0,25 mm)* min. annular ring is assumed.

TABLE III

CONDUCTOR WIDTH PROCESSING TOLERANCES

Final Conductor Width Dimensions After Etching Depends on the Following:

1. Photographic opacity of tape
2. Tape width tolerance
3. Over or underdevelopment in photo reduction
4. Copper laminate thickness
5. Manufacturing process (photo direct or silk-screened, and intensity of etch)

The following tolerances can generally be maintained (dimensions based upon reduced 1:1 artwork):

PLATING	CONDUCTIVE MATERIAL	CONDUCTOR WIDTH in.	CONDUCTOR WIDTH mm
unplated	2 oz. copper	±.005	±0.127
	1 oz copper	±.003	±0.076
	½ oz. copper	±.001	±0.025
panel plated copper .001 in. (0.025 mm) minimum	2 oz copper	+.005 −.008	+0.127 −0.203
	1 oz. copper	+.003 −.005	+0.076 −0.127
	½ oz. copper	+.001 −.003	+0.025 −0.076
pattern plated copper .001 in. (0.025 mm) minimum	2 oz. copper	±.005	±0.127
	1 oz. copper	±.003	±0.076
	½ oz copper	±.0015	±0.038

FIGURE I **CONDUCTOR THICKNESS AND WIDTH**

(For use in determining current carrying capacity and sizes of etched copper conductors for various temperature rises above ambient)

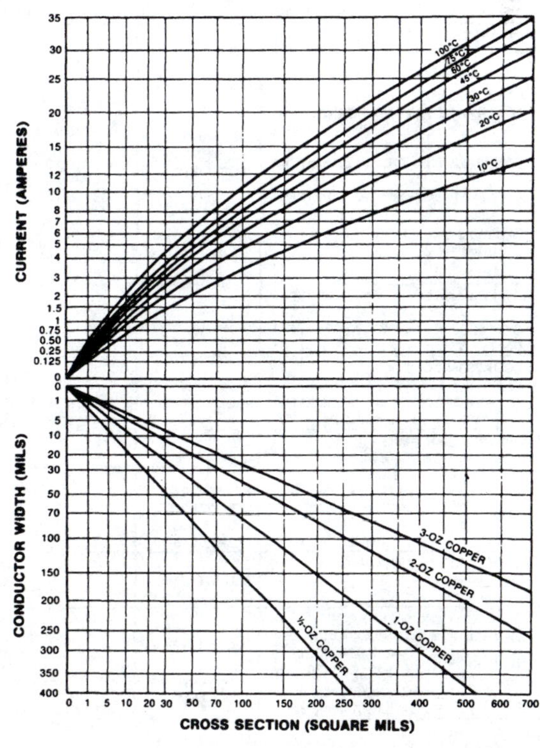

Courtesy of Bishop Graphics, Inc.

TABLE IV

CONDUCTOR SPACING MINIMUMS

<table>
<tr><th rowspan="3">Voltage Between Conductors DC or AC Peak (Volts)</th><th colspan="5">MIL-STD-275C & IPC-ML-910</th><th rowspan="3">MIL-STD-1495 Multilayer Printed Wiring</th><th rowspan="3">Voltage Between Conductors DC or AC Peak (Volts)</th></tr>
<tr><th colspan="2">Uncoated 0–10,000 Ft. Alt.</th><th>Uncoated Above 10,000 Ft. Alt.</th><th colspan="2">Coated & Internal Multilayer</th></tr>
<tr><th>MIL-STD-275C</th><th>IPC-ML-910</th><th>MIL-STD-275C & IPC-ML-910</th><th>MIL-STD-275</th><th>IPC-ML-910</th></tr>
<tr><td>0
9</td><td rowspan="6">.025" (0,64 mm)</td><td rowspan="4">.015" (0,38 mm)</td><td rowspan="4">.025" (0,64 mm)</td><td rowspan="3">.010" (0,25 mm)</td><td>.005" (0,13 mm)</td><td rowspan="2">.005" (1) (0,13 mm)</td><td>0
9</td></tr>
<tr><td>10
15</td><td rowspan="2">.010" (0,25 mm)</td><td>10
15</td></tr>
<tr><td>16
30</td><td>.010" (0,25 mm)</td><td>16
30</td></tr>
<tr><td>31
50</td><td colspan="2">.015" (0,38 mm)</td><td>.015" (0,38 mm)</td><td>31
50</td></tr>
<tr><td>51
100</td><td rowspan="2">.025" (0,64 mm)</td><td>.060" (1,52 mm)</td><td colspan="2">.020" (0,51 mm)</td><td>.020" (0,51 mm)</td><td>51
100</td></tr>
<tr><td>101
150</td><td rowspan="2">.125" (3,18 mm)</td><td colspan="2" rowspan="5">.030" (0,76 mm)</td><td rowspan="5">.030" (0,76 mm)</td><td>101
150</td></tr>
<tr><td>151
170</td><td colspan="2" rowspan="3">.050" (1,27 mm)</td><td>151
170</td></tr>
<tr><td>171
250</td><td>.250" (6,35 mm)</td><td>171
250</td></tr>
<tr><td>251
300</td><td rowspan="2">.500" (12,70 mm)</td><td>251
300</td></tr>
<tr><td>301
500</td><td colspan="2">.100" (2,54 mm)</td><td colspan="2">.060" (1,52 mm)</td><td>.060" (1,52 mm)</td><td>301
500</td></tr>
<tr><td>500+</td><td colspan="2">.0002 in/volt (0,0051 mm/volt)</td><td>.0010 in/volt (0,025 mm/volt)</td><td colspan="2">.00012 in/volt (0,0030 mm/volt)</td><td>.00012 in/volt (0,0030 mm/volt)</td><td>500+</td></tr>
</table>

(1) For Internal Layers Only.

Courtesy of Bishop Graphics, Inc.

DESIGN CRITERIA & TOLERANCES		DATA FOR CLASSES OF MATERIALS & PROCESSES					
CODE LETTER	DESCRIPTION QUALIFIERS SUB-QUALIFIERS	SOURCE DOCUMENT	PREFERRED CLASS 1	PREFERRED CLASS 2	STANDARD CLASS 3	REDUCED PRODUCIBILITY CLASS 4	REDUCED PRODUCIBILITY CLASS 5
S Tol.	CONDUCTOR SPACING TOLERANCE (4)						
	IPC WITHOUT PLATING	IPC-D-300F	+.012" (0,30 mm) −.008" (0,20 mm)	+.007" (0,18 mm) −.006" (0,15 mm)	+.006" (0,15 mm) −.004" (0,10 mm)	+.004" (0,10 mm) −.003" (0,08 mm)	+.003" (0,08 mm) −.003" (0,08 mm)
	IPC WITH PLATING		+.012" (0,30 mm) −.017" (0,43 mm)	+.007" (0,18 mm) −.010" (0,25 mm)	+.006" (0,15 mm) −.006" (0,15 mm)	+.005" (0,13 mm) −.005" (0,13 mm)	+.004" (0,10 mm) −.004" (0,10 mm)
	MIL-STD WITHOUT PLATING	MIL-STD-1495	+.005" (0,13 mm) −.004" (0,10 mm)		+.004" (0,10 mm) −.002" (0,05 mm)	+.002" (0,05 mm) −.001" (0,03 mm)	
	MIL-STD WITH PLATING		±.0075" (0,19 mm)		±.005" (0,13 mm)	±.003" (0,08 mm)	
SL	MINIMUM INTRALAYER CONDUCTOR SPACING (MULTILAYER)	IPC-ML-910	.004" (0,10 mm)		.003" (0,08 mm)	.002" (0,05 mm)	
H	HOLE DIA. MAXIMUM (UNSUPPORTED TERMINAL)	MIL-STD-275C	MINIMUM LEAD DIA. + .020" (0,51 mm) OR MINIMUM EYELET BARREL DIA. + .010" (0,25 mm)				
H Tol.	HOLE TOLERANCE (AS MACHINED) (7)	IPC-D-300F, MIL-STD-1495, IPC-ML-910					
	0"–.032" (0,81 mm) DIA.		±.003" (0,08 mm)	±.002" (0,05 mm)	±.001" (0,03 mm) −.002" (0,05 mm)	±.001" (0,03 mm)	±.001" (0,03 mm)
	.033" (0,82 mm)–.063" (1,60 mm) DIA.		±.004" (0,10 mm)	±.003" (0,08 mm)	±.002" (0,05 mm)	+.001" (0,03 mm) −.002" (0,05 mm)	±.001" (0,03 mm)
	.064" (1,61 mm)–.188" (4,78 mm) DIA.		±.005" (0,13 mm)	±.004" (0,10 mm)	±.003" (0,08 mm)	±.002" (0,05 mm)	+.001" (0,03 mm) −.002" (0,05 mm)
PH	PLATED-THRU HOLE DIA. (8) MINIMUM	MIL-STD-1495, IPC-ML-910	1/3 T MAX		1/4 T MAX	1/5 T MAX	
	MAXIMUM	MIL-STD-275C	MINIMUM LEAD DIA. + .028" (0,71 mm)				
PH Tol.	PLATED-THRU HOLE DIA. TOLERANCE (8) (9)	IPC-D-300F					
	0"–.032" (0,81 mm) DIA.		±.005" (0,13 mm)	±.003" (0,08 mm)	±.002" (0,05 mm)	±.001" (0,03 mm)	±.001" (0,03 mm)
	.033" (0,82 mm)–.063" (1,60 mm) DIA.		±.006" (0,15 mm)	±.004" (0,10 mm)	±.003" (0,08 mm)	±.002" (0,05 mm)	±.001" (0,03 mm)
	.064" (1,61 mm)–.188" (4,78 mm) DIA.		±.007" (0,18 mm)	±.005" (0,13 mm)	±.004" (0,10 mm)	±.004" (0,10 mm)	±.002" (0,05 mm)
HL Tol.	HOLE LOCATION TOLERANCE (RADIUS TRUE POSITION VARIANCE)	IPC-D-300F, MIL-STD-1495					
	<6.00" (15,2 cm) FROM DATUM		.010" (0,25 mm)	.007" (0,18 mm)	.005" (0,13 mm)	.003" (0,08 mm)	.002" (0,05 mm)
	>6.00" (15,2 cm) FROM DATUM		.014" (0,36 mm)	.010" (0,25 mm)	.007" (0,18 mm)	.005" (0,13 mm)	.003" (0,08 mm)
AR	MINIMUM ANNULAR RING						
	UNSUPPORTED HOLE	MIL-P-55110B	.015" (0,38 mm)				
	PLATED-THRU HOLE						
	EXTERNAL (MIL-STD)	MIL-STD-1495	.005" (0,13 mm)				
	INTERNAL (MIL-STD)	MIL-STD-1495	.002" (0,05 mm)				
	EXTERNAL (IPC)	IPC-ML-910	.002" (0,05 mm)				
	INTERNAL (IPC)	IPC-ML-910	.001" (0,03 mm)				
TA	MINIMUM TERMINAL AREA (10)	MIL-STD-275C	MAX. EYELET OR TERMINAL FLANGE DIA. + .020" (0,51 mm), OR MAX. DIA. OF UNSUPPORTED HOLE + .040" (1,02 mm), OR MAX. DIA. OF PLATED-THRU HOLE + .020" (0,51 mm)				
FL Tol.	FEATURE LOCATION TOLERANCE (RADIUS TRUE POSITION VARIANCE)						
	MIL-STD <12.000" (30,5 cm)	MIL-STD-1495	.008" (0,20 mm)		.007" (0,18 mm)	.006" (0,15 mm)	
	MIL-STD >12.000" (30,5 cm)		.010" (0,25 mm)		.009" (0,23 mm)	.008" (0,20 mm)	
	IPC < 6.000" (15,2 cm) (11)	IPC-D-300F	.021" (0,53 mm)	.014" (0,36 mm)	.010" (0,25 mm)	.007" (0,18 mm)	.0045" (0,11 mm)
	IPC > 6.000" (15.2 cm) (11)		.025" (0,64 mm)	.018" (0,46 mm)	.014" (0,36 mm)	.010" (0,25 mm)	.007" (0,18 mm)

Courtesy of Bishop Graphics, Inc.

NOTES:

(1) Board thickness tolerances apply only to copper-clad laminated glass/resin based materials.

(2) The number of conductor layers should be the optimum for the required board function and good producibility.

(3) The thickness and width of the conductors should be determined on the basis of the current-carrying capacity required. The allowable temperature rise should be determined in accordance with Figure I. Maximum size, consistent with minimum spacing requirements, should be maintained for ease of manufacture and durability in usage.

(4) The conductor width and spacing tolerances listed represent process tolerances that can be expected with normal processing. (Specific process tolerances should be discussed with the supplier.) These tolerances are based on 2 oz/ft^2 copper. For 1 oz/ft^2 copper, a .001″ (0,025 mm) reduction in variation can be expected per conductor edge. Final product drawings and specifications should specify only minimums for conductor width and spacing.

(5) The minimum spacing dimensions listed are limited by voltage rating, altitude and coatings. See Table IV for more detailed information.

(6) For internal layers only.

(7) The tolerances listed are for nominal base material thicknesses up to and including 0.0625″ (1,59 mm). For nominal base material thickness over 0.0625″ (1,59 mm), add ±0.001″ (0,025 mm) to each tolerance. Tolerances indicate total spread and may be varied from the nominal to satisfy design requirements.

(8) All references to plated-thru-hole diameters indicate finished hole diameters (after plating).

(9) The tolerances listed are for holes no less than one-third of the nominal overall board thickness. Tolerances indicate total spread, and may be varied from the nominal to satisfy design requirements. For hole diameter to nominal board thickness ratios greater than ⅓ add ±0.001″ (0,025 mm) to these tolerances. For ratios greater than ¼ add ±0.002″ (0,05 mm).

(10) MIL-STD minimum terminal areas are listed for reference only. For a discussion and examples on how to determine terminal area, see section on Design "Terminal Areas (Pads)."

(11) Tolerances apply when a registration datum is used.

Courtesy of Bishop Graphics, Inc.

Appendix III

Resistor and Capacitor Standard Color Code

The standard color code provides the necessary information required to properly identify color coded resistors and capacitors. Refer to the color code for numerical values and the number of zeros (or multiplier) assigned to the colors used. A fourth color band on resistors determines the tolerance rating. Absence of the fourth band indicates a 20 percent tolerance rating (REF MIL-STD-1285).

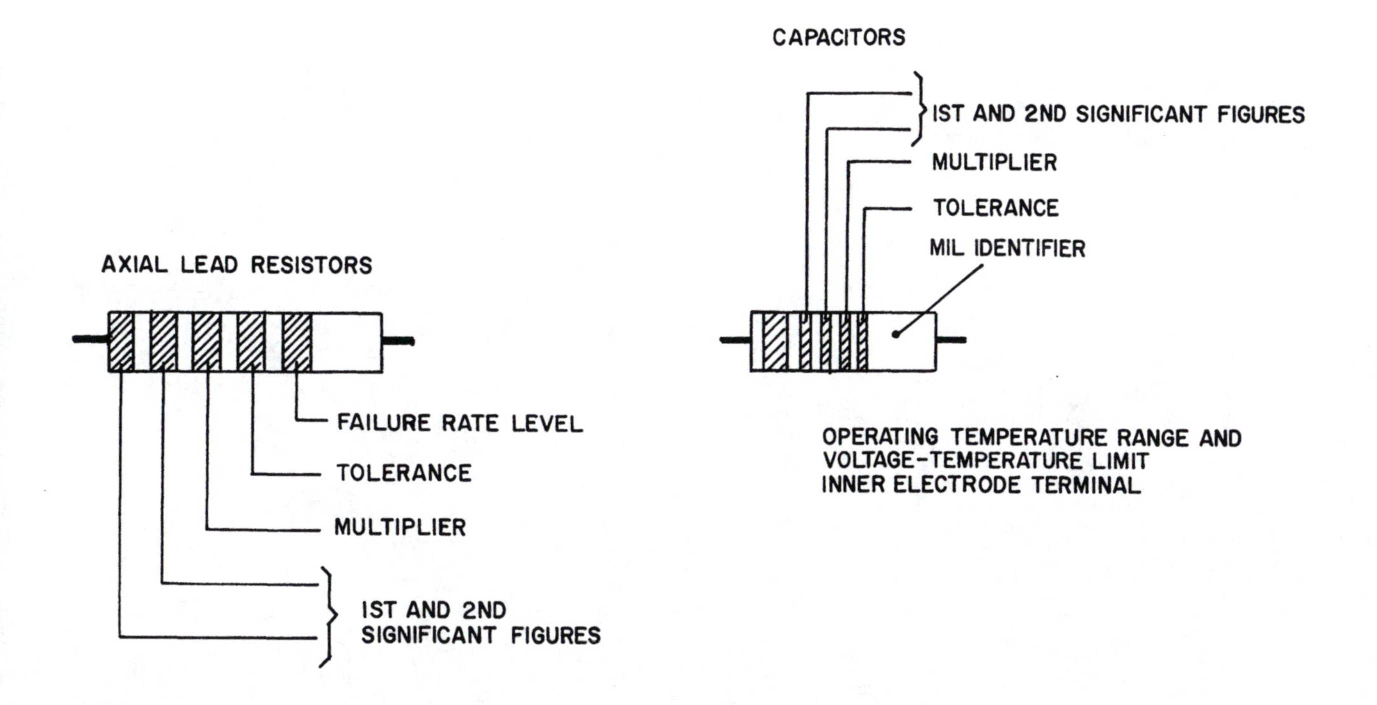

Courtesy of Texas Instruments, Inc.

Color Code

Color	ABV	First Figure	Second Figure	Multiplier	Tolerance		Failure Rate Level
					Percent	LTR	
Black	BL	0	0	1	±20	M	L
Brown	BR	1	1	10	±1	F	M
Red	R	2	2	100	± 2	G	P
Orange	O	3	3	1,000			R
Yellow	Y	4	4	10,000			S
Green	G	5	5	100,000			T
Blue	BL	6	6	1,000,000			
Purple (Violet)	V	7	7				
Gray	GY	8	8				
White	W	9	9				
Silver	SIL			0.01	±10	K	
Gold	GLD			0.1	± 5	J	

Courtesy of Texas Instruments, Inc.

Appendix IV Quick Reference to Electronic Symbols

1. Qualifying Symbols

1.1 Adjustability Variability

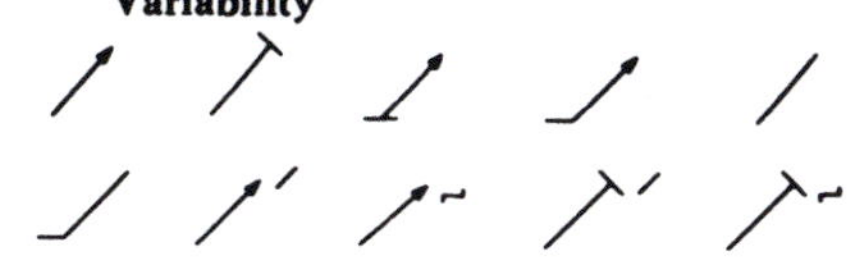

1.2 Special-Property Indicators

1.3 Radiation Indicators

1.4 Physical State Recognition Symbols

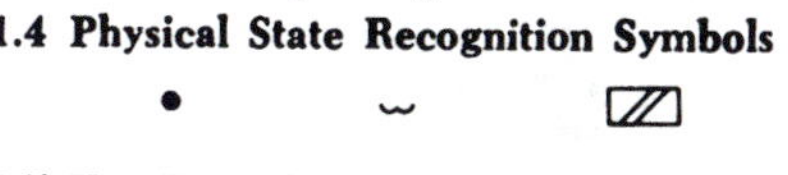

1.5 Test-Point Recognition Symbol

1.6 Polarity Markings

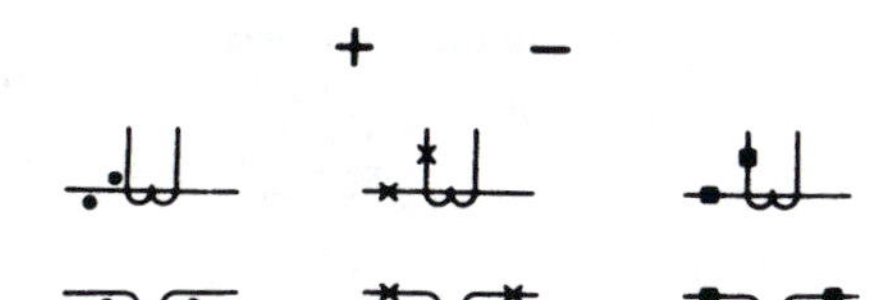

1.7 Direction of Flow of Power, Signal, or Information

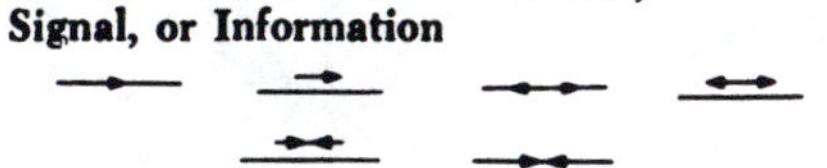

1.8 Kind of Current

1.9 Connection Symbols

1.10 Envelope Enclosure

1.11 Shield Shielding

1.12 Special Connector or Cable Indicator

1.13 Electret

2. Fundamental Items

2.1 Resistor

2.2 Capacitor

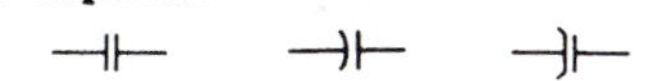

2.3 Antenna

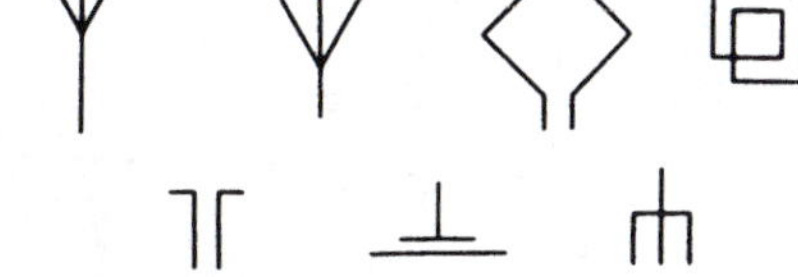

2.4 Attenuator

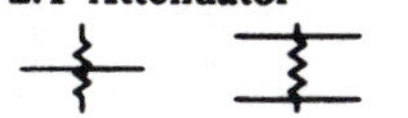

2.5 Battery

2.6 Delay Function Delay Line Slow-Wave Structure

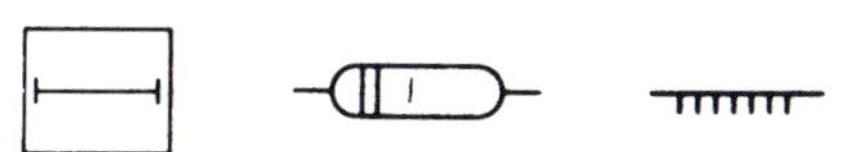

2.7 Oscillator Generalized Alternating-Current Source

2.8 Permanent Magnet

2.9 Pickup Head

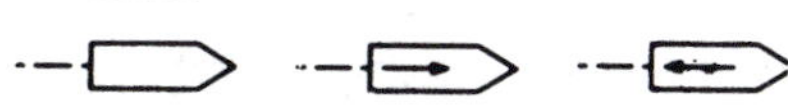

2.10 Piezoelectric Crystal Unit

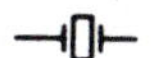

2.11 Primary Detector Measuring Transducer

2.12 Squib, Electrical

2.13 Thermocouple

Courtesy of The Institute of Electrical and Electronics Engineers, Inc.

2.14 Thermal Element
Thermomechanical
Transducer

2.15 Spark gap
Igniter gap

2.16 Continuous Loop Fire Detector (temperature sensor)

2.17 Ignitor Plug

3. Transmission Path

3.1 Transmission Path
Conductor
Cable
Wiring

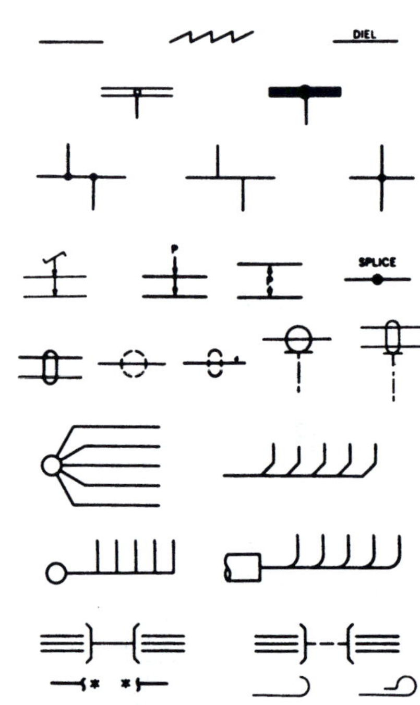

3.2 Distribution lines
Transmission lines

F S T V

3.3 Alternative or Conditioned Wiring

3.4 Associated or Future

3.5 Intentional Isolation of Direct-Current Path in Coaxial or Waveguide Applications

3.6 Waveguide

3.7 Strip-Type Transmission Line

3.8 Termination

3.9 Circuit Return

3.10 Pressure-Tight Bulkhead Cable
Gland
Cable Sealing End

4. Contacts, Switches, Contactors, and Relays

4.1 Switching Function

4.2 Electrical Contact

4.3 Basic Contact Assemblies

4.4 Magnetic Blowout Coil

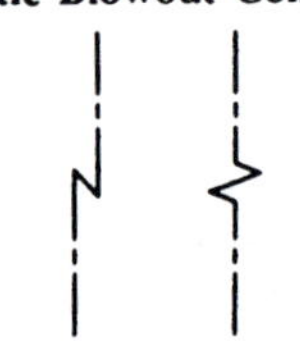

4.5 Operating Coil
Relay Coil

4.6 Switch

4.7 Pushbutton, Momentary or Spring-Return

4.8 Two-Circuit, Maintained or Not Spring-Return

4.9 Nonlocking Switch, Momentary or Spring-Return

4.10 Locking Switch

4.11 Combination Locking and Nonlocking Switch

4.12 Key-Type Switch
Lever Switch

4.13 Selector or Multiposition Switch

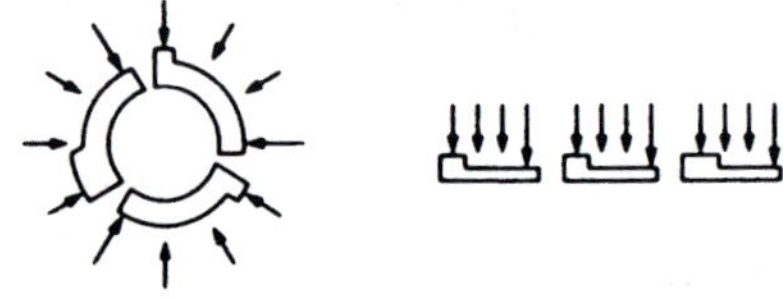

Courtesy of The Institute of Electrical and Electronics Engineers, Inc.

4.14 Limit Switch
Sensitive Switch

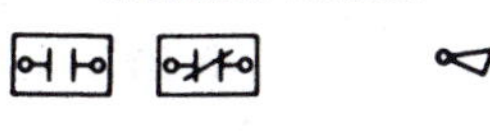
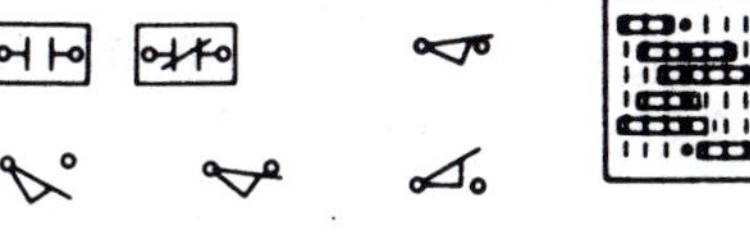

4.15 Safety Interlock

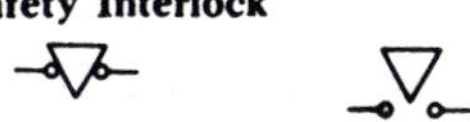

4.16 Switches with Time-Delay Feature

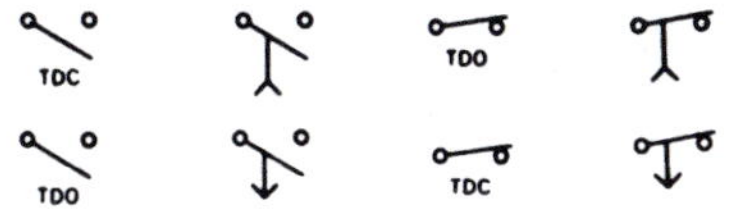

4.17 Flow-Actuated Switch

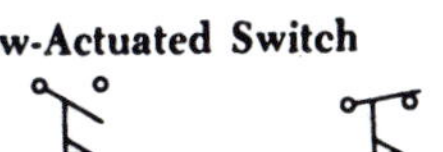

4.18 Liquid-Level-Actuated Switch

4.19 Pressure- or Vacuum-Actuated Switch

4.20 Temperature-Actuated Switch

4.21 Thermostat

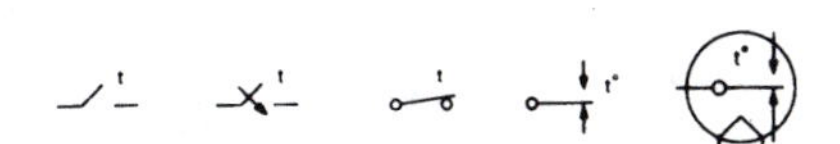

4.22 Flasher
Self-interrupting switch

4.23 Foot-Operated Switch
Foot Switch

4.24 Switch Operated by Shaft Rotation and Responsive to Speed or Direction

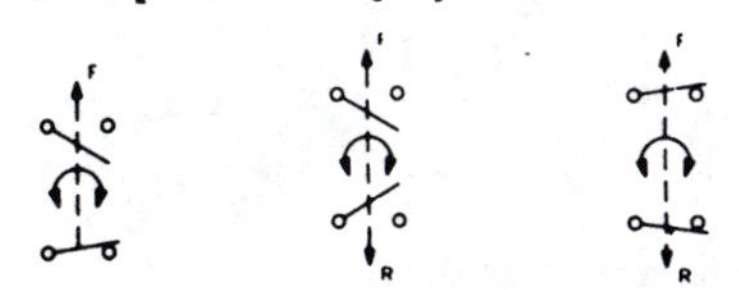

4.25 Switches with Specific Features

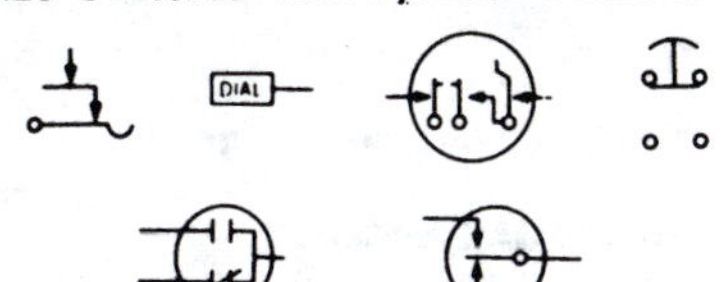

4.26 Telegraph Key

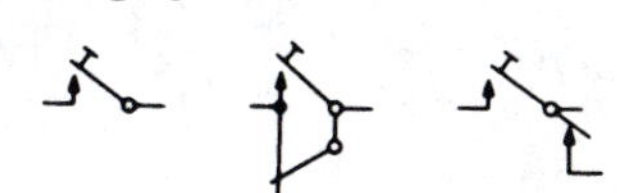

4.27 Governor
Speed Regulator

4.28 Vibrator
Interrupter

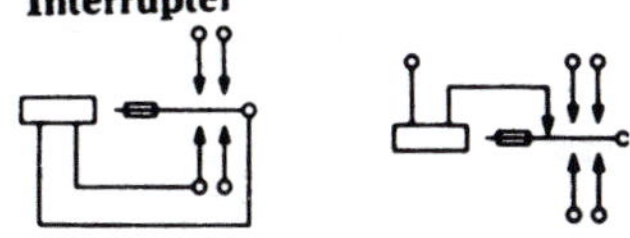

4.29 Contactor

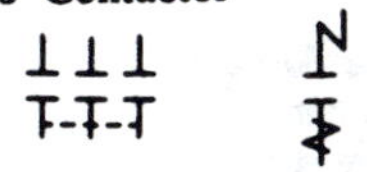

4.30 Relay

AC
P
SO
SR

D	DP	MG
DB	EP	NB
SA	SW	NR
L	ML	FO
		FR

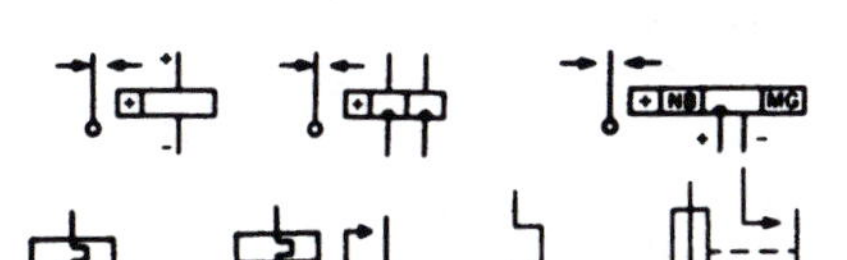

4.31 Inertia Switch

4.32 Mercury Switch

4.33 Aneroid Capsule

5. Terminals and Connectors

5.1 Terminals

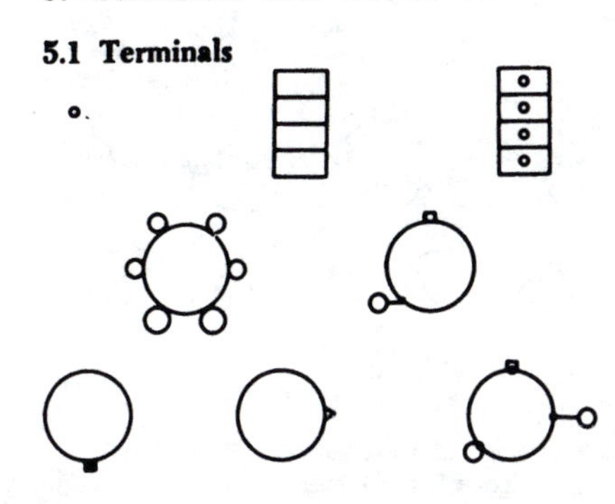

5.2 Cable Termination

5.3 Connector
Disconnecting Device

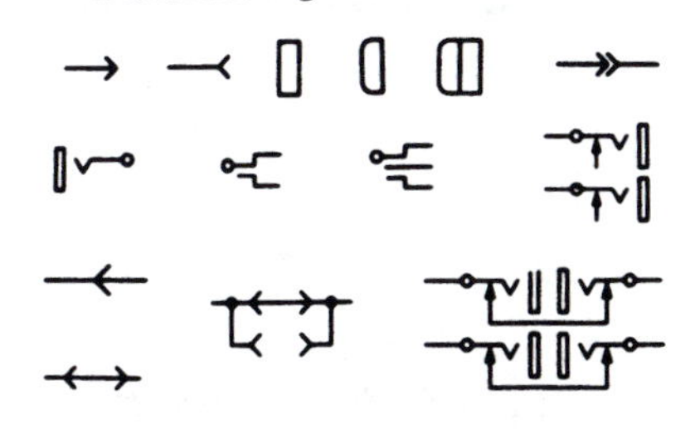

5.4 Connectors of the Type Commonly Used for Power-Supply Purposes

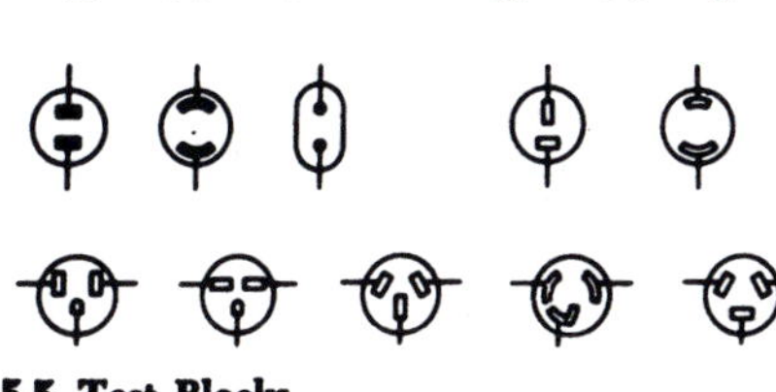

5.5 Test Blocks

5.6 Coaxial Connector

5.7 Waveguide Flanges
Waveguide junction

6. Transformers, Inductors, and Windings

6.1 Core

6.2 Inductor
Winding
Reactor
Radio frequency coil
Telephone retardation coil

6.3 Transductor

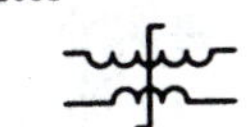

6.4 Transformer
Telephone induction coil
Telephone repeating coil

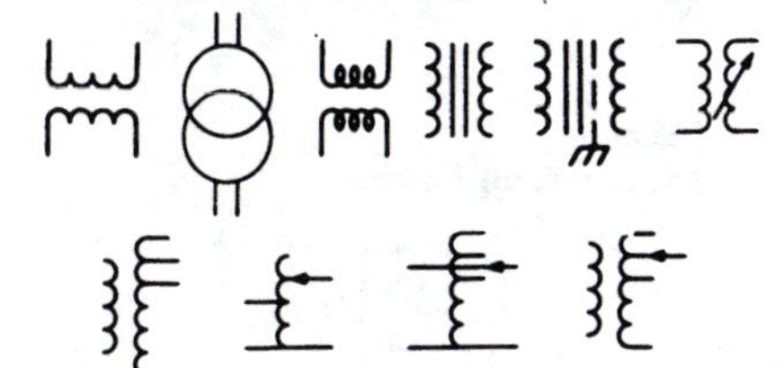

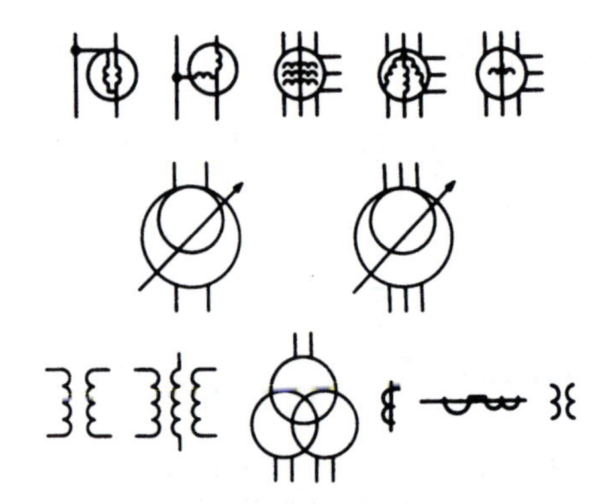

6.5 Linear Coupler

7. Electron Tubes and Related Devices

7.1 Electron Tube

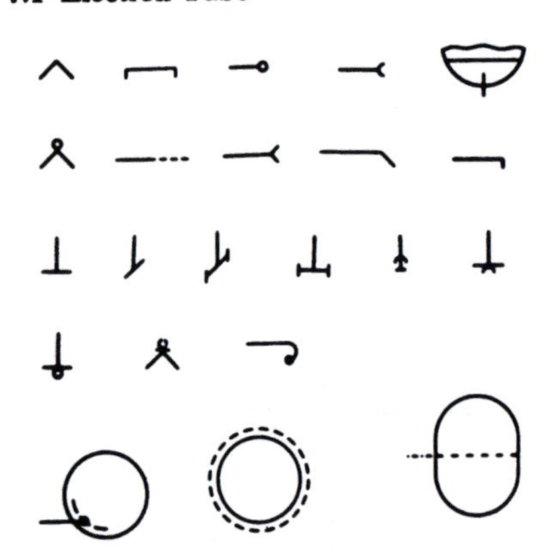

7.2 General Notes

7.3 Typical Applications

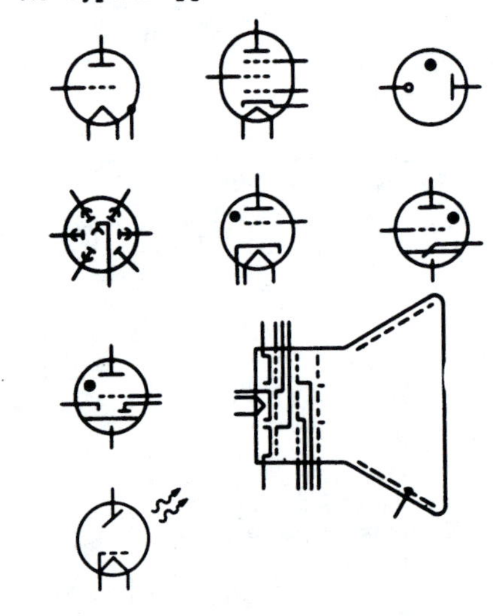

7.4 Solion
Ion-Diffusion Device

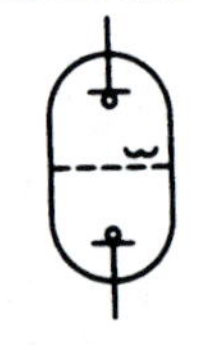

7.5 Coulomb Accumulator
Electrochemical Step-Function Device

7.6 Conductivity cell

7.7 Nuclear-Radiation Detector
Ionization Chamber
Proportional Counter Tube
Geiger-Müller Counter Tube

8. Semiconductor Devices

8.1 Semiconductor Device
Transistor
Diode

8.2 Element Symbols

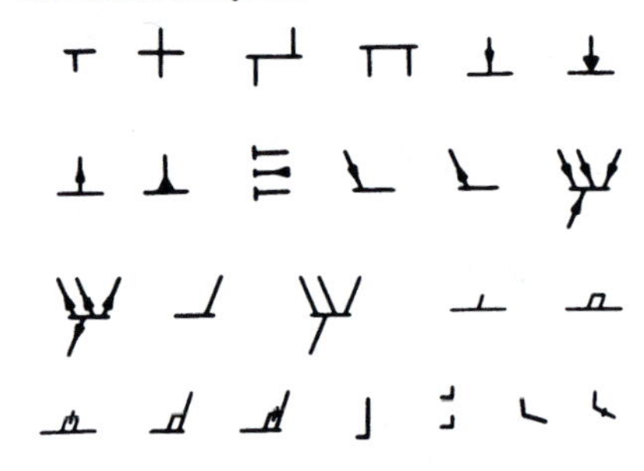

8.3 Special Property Indicators

8.4 Rules for Drawing Style 1 Symbols

8.5 Typical Applications: Two-Terminal Devices

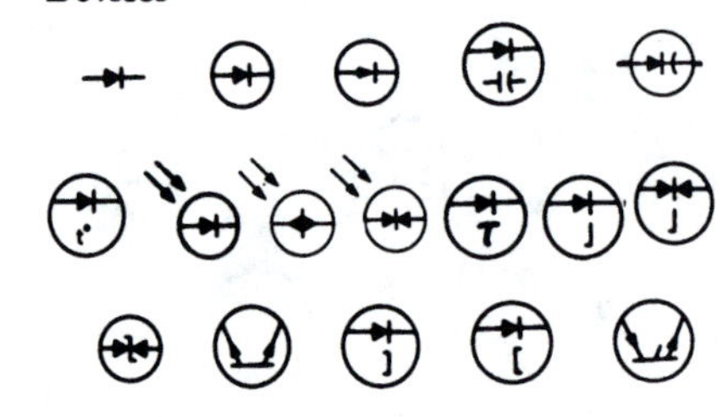

8.6 Typical Applications: Three- (or More) Terminal Devices

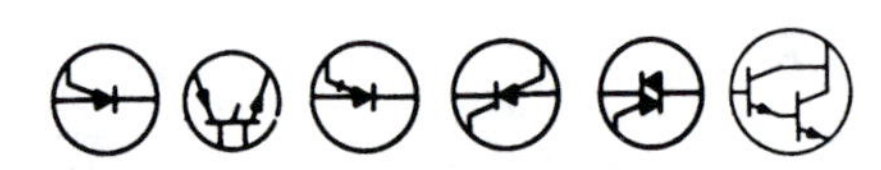

8.7 Photosensitive Cell

8.8 Semiconductor Thermocouple

8.9 Hall Element
Hall Generator

8.10 Photon-coupled isolator

8.11 Solid-state-thyratron

9. Circuit Protectors

9.1 Fuse

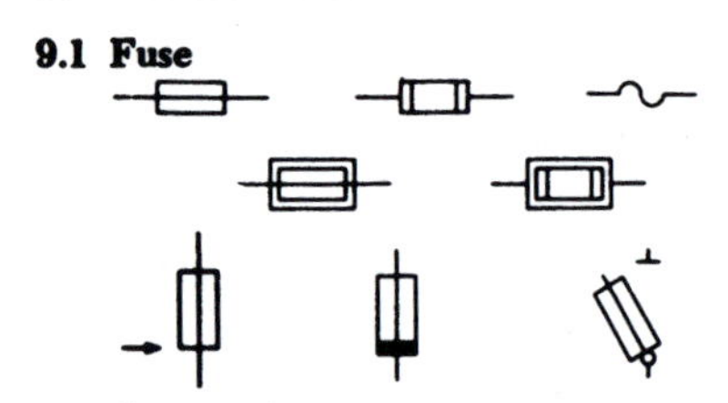

9.2 Current Arrester

9.3 Lightning Arrester
Arrester
Gap

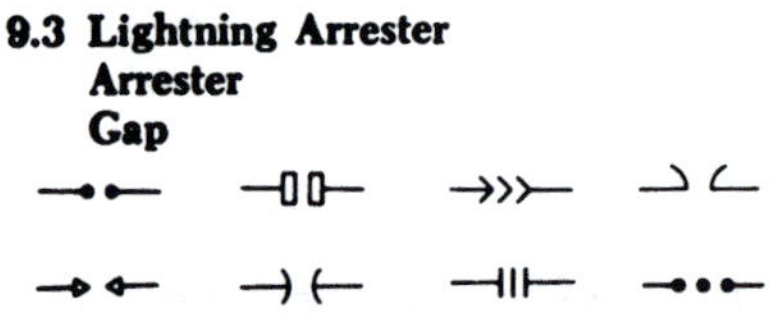

9.4 Circuit Breaker

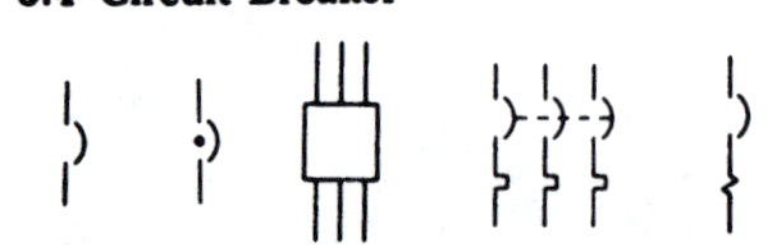

9.5 Protective Relay

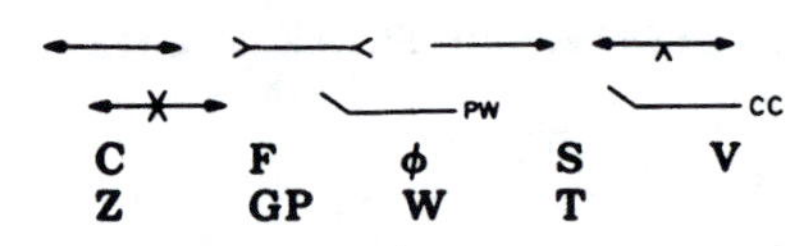

10. Acoustic Devices

10.1 Audible-Signaling Device

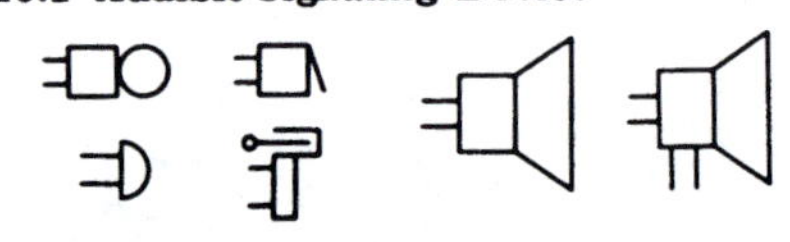

10.2 Microphone

Courtesy of The Institute of Electrical and Electronics Engineers, Inc.

10.3 Handset
Operator's Set

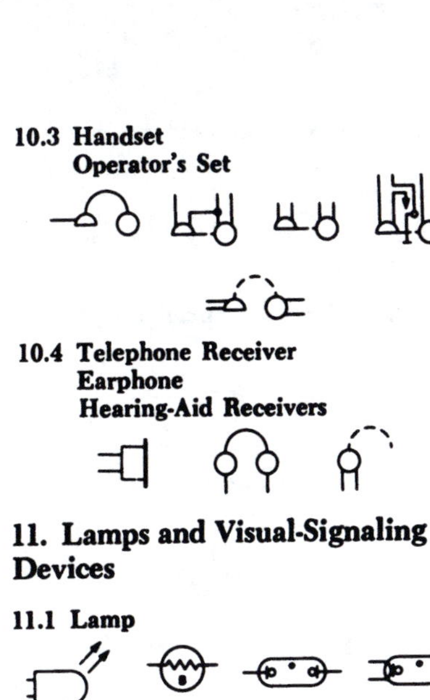

10.4 Telephone Receiver
Earphone
Hearing-Aid Receivers

11. Lamps and Visual-Signaling Devices

11.1 Lamp

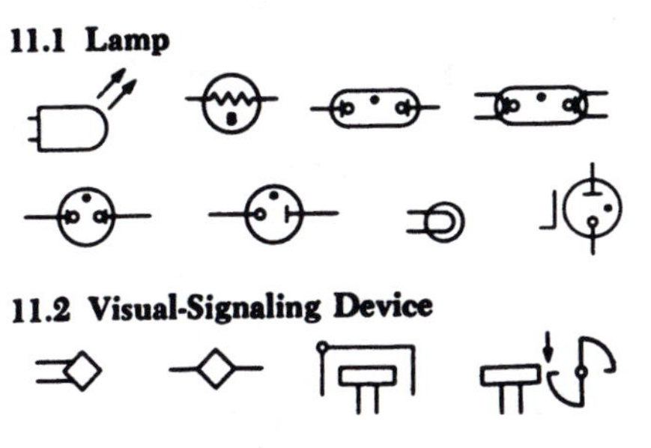

11.2 Visual-Signaling Device

12. Readout Devices

12.1 Meter
Instrument

A	DB	I	OP	RF	VA
AH	DBM	INT	OSCG	SY	VAR
C	DM	μA	PH	TLM	VARH
CMA	DTR	UA	PI	t°	VI
CMC	F	MA	PF	THC	VU
CMV	G	NM	RD	TT	W
CRO	GD	OHM	REC	V	WH

12.2 Electromagnetically Operated
Counter
Message Register

13. Rotating Machinery

13.1 Rotating Machine

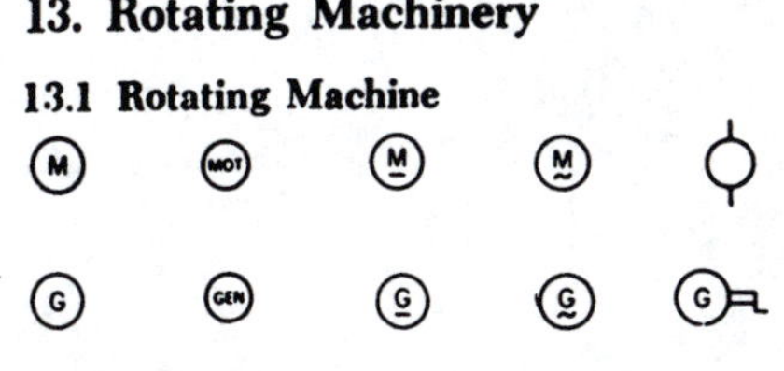

13.2 Field, Generator or Motor

13.3 Winding Connection Symbols

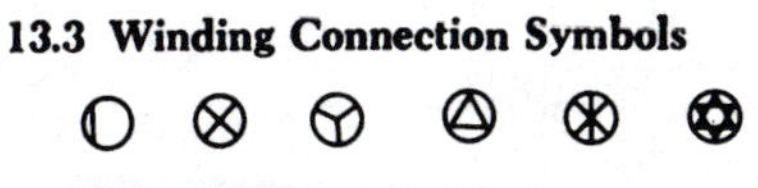

13.4 Applications: Direct-Current Machines

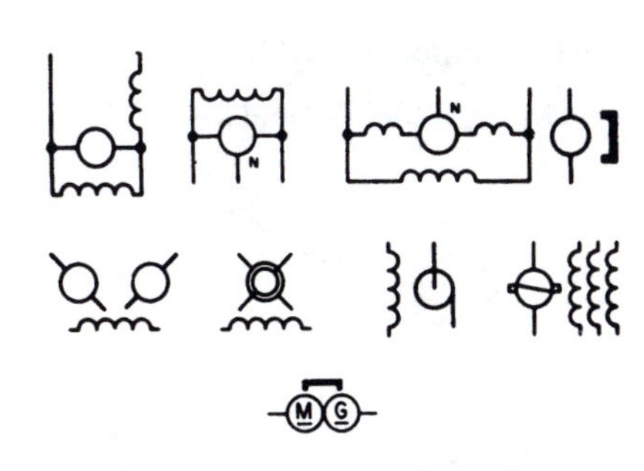

13.5 Applications: Alternating-Current Machines

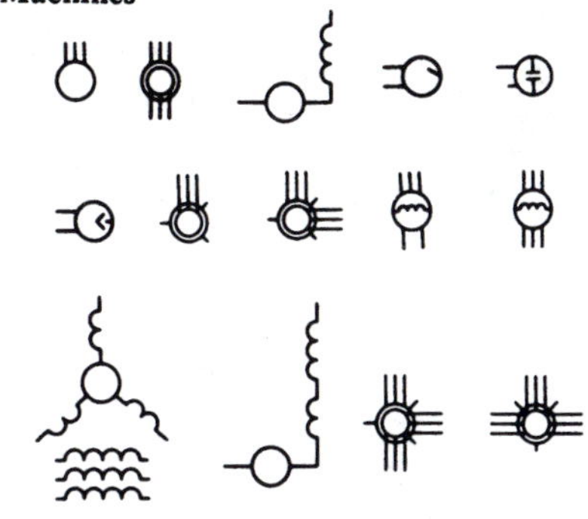

13.6 Applications: Alternating-Current Machines with Direct-Current Field Excitation

13.7 Applications: Alternating- and Direct-Current Composite

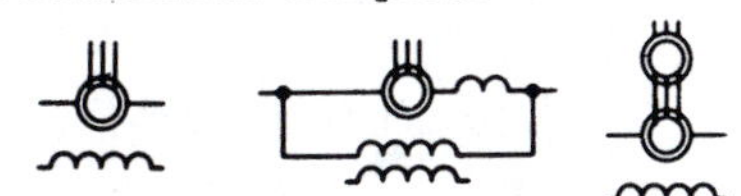

13.8 Synchro

CDX	TDX
CT	TR
CX	TX
TDR	RS

14. Mechanical Functions

14.1 Mechanical Connection
Mechanical Interlock

14.2 Mechanical Motion

14.3 Clutch
Brake

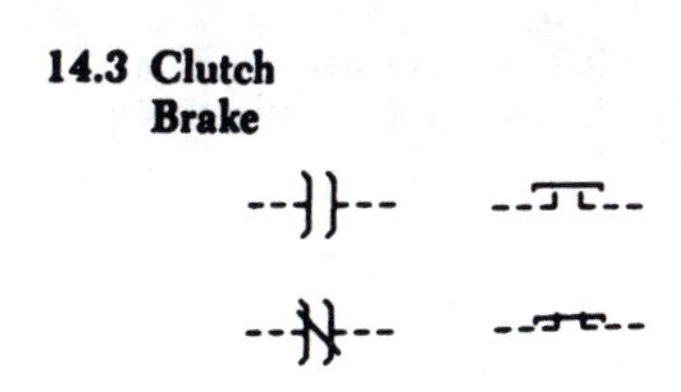

14.4 Manual Control

16. Composite Assemblies

16.1 Circuit assembly
Circuit subassembly
Circuit element

EQ	FL-BP	RG	TPR
FAX	FL-HP	RU	TTY
FL	FL-LP	DIAL	CLK
FL-BE	PS	TEL	IND
ST-INV			

16.2 Amplifier

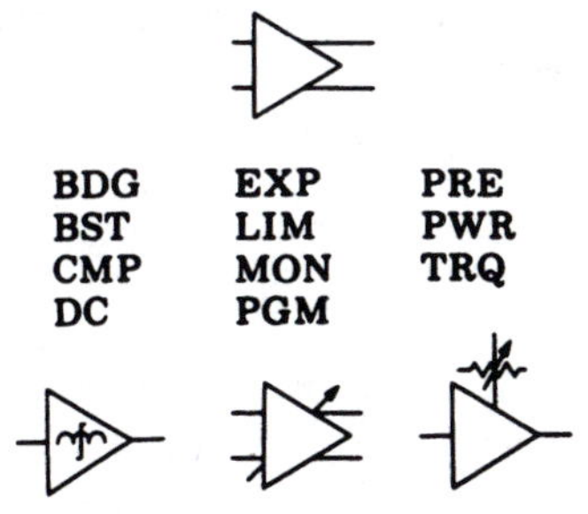

BDG	EXP	PRE
BST	LIM	PWR
CMP	MON	TRQ
DC	PGM	

16.3 Rectifier

16.4 Repeater

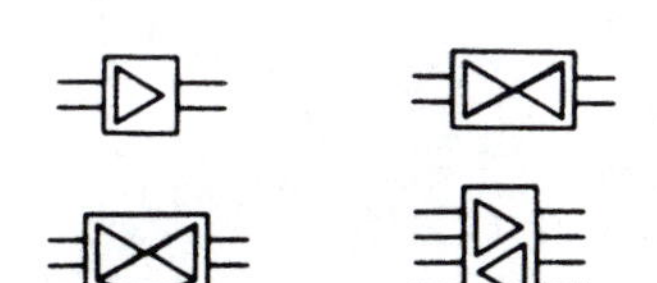

16.5 Network

16.6 Phase Shifter
Phase-Changing Network

16.7 Chopper

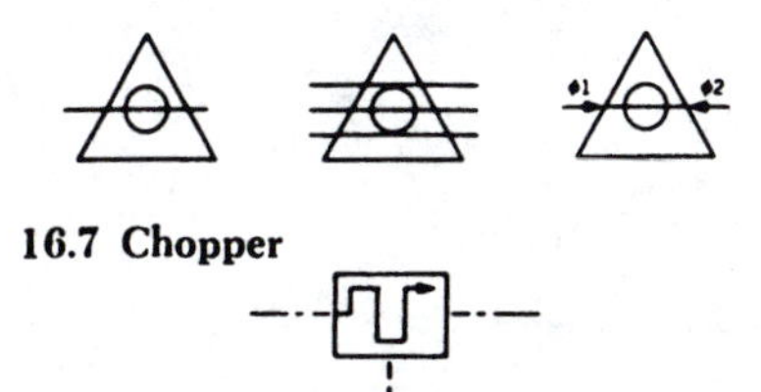

16.8 Diode-type ring demodulator
Diode-type ring modulator

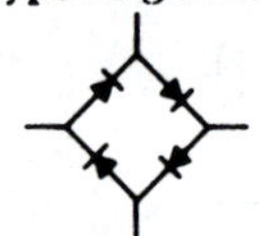

16.9 Gyro
Gyroscope
Gyrocompass

16.10 Position Indicator

16.11 Position Transmitter

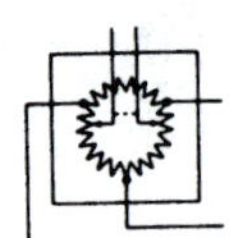 

16.12 Fire Extinguisher Actuator Head

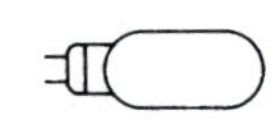 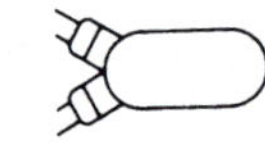

17. Analog Functions

17.1 Operational Amplifier

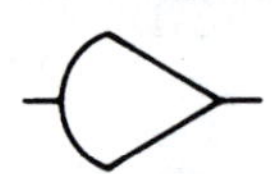

17.2 Summing Amplifier

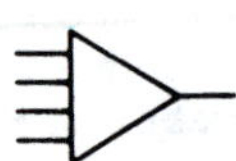 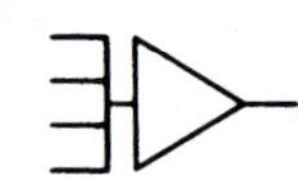

17.3 Integrator

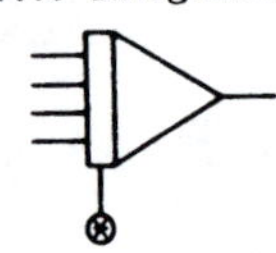 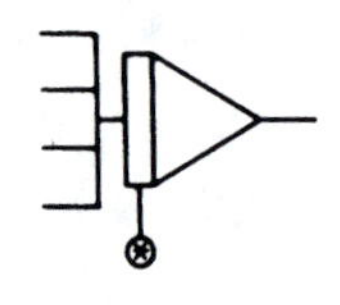

17.4 Electronic Multiplier

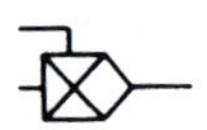

17.5 Electronic Divider

17.6 Electronic Function Generator

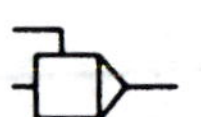

17.7 Generalized Integrator

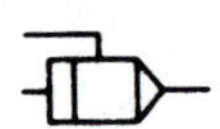

17.8 Positional Servo-mechanism

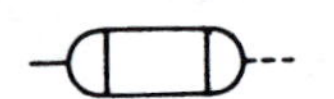

17.9 Function Potentiometer

Courtesy of The Institute of Electrical and Electronics Engineers, Inc.

Appendix V

Quick Reference to Logic Symbols

1. General Element Symbols

1.1, 1.2 Basic Symbols

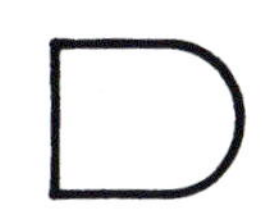
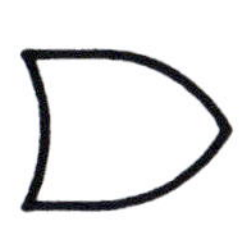

1.3 Contiguous Block

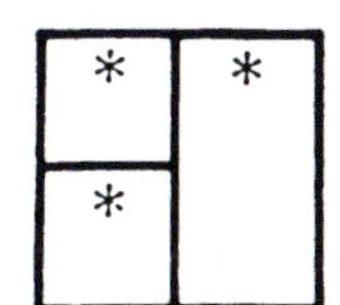
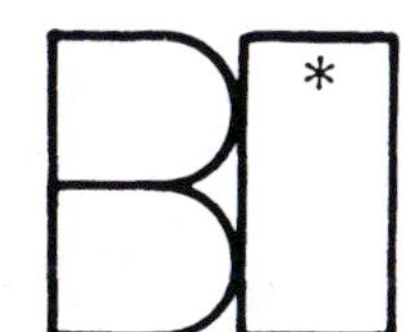

1.4 Common Control Block

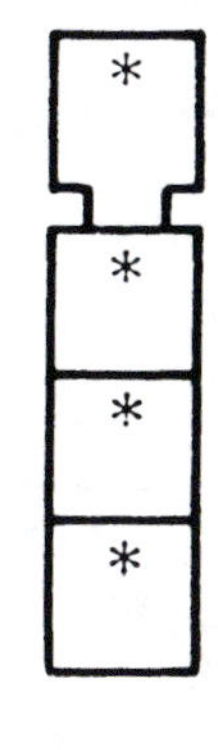
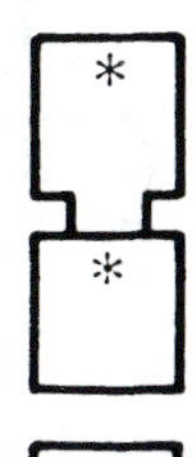

1.5 Bundling, Grouping

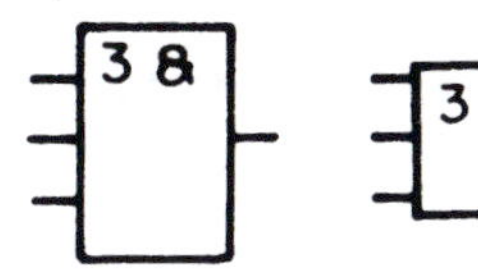

2. Indicator Symbols

2.1 Negation Indicator

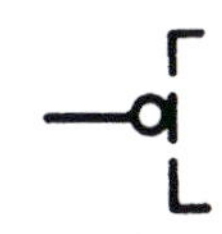

2.2 Polarity Indicator

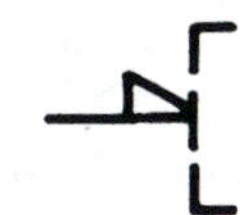

2.3 Dynamic Indicator

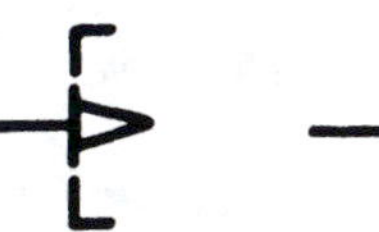

Courtesy of The Institute of Electrical and Electronics Engineers, Inc.

2.4 Nonlogic Indicator

2.5 Extender-Connection Indicator

2.6 Inhibiting-Input Indicator

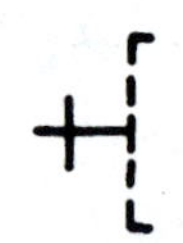

2.7 Output-Delay Indicator

3. Combinational Logic Symbols

3.1 AND Function

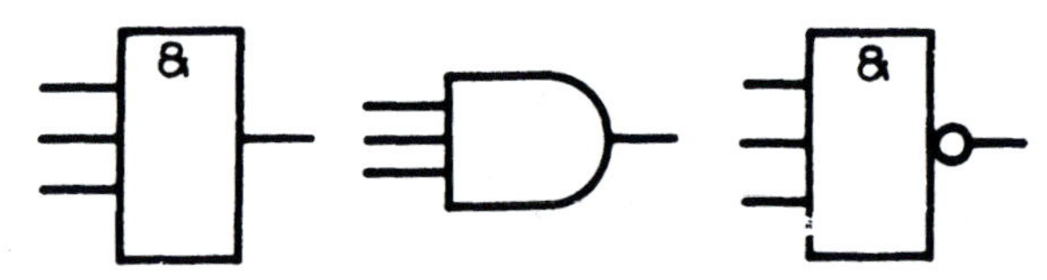

3.2 OR Function

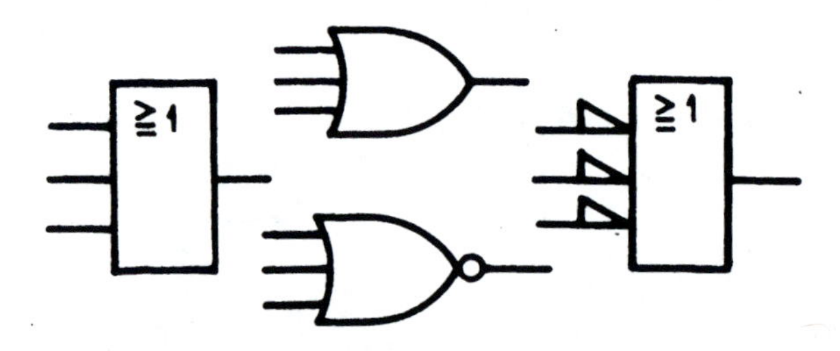

3.3 Exclusive OR Function

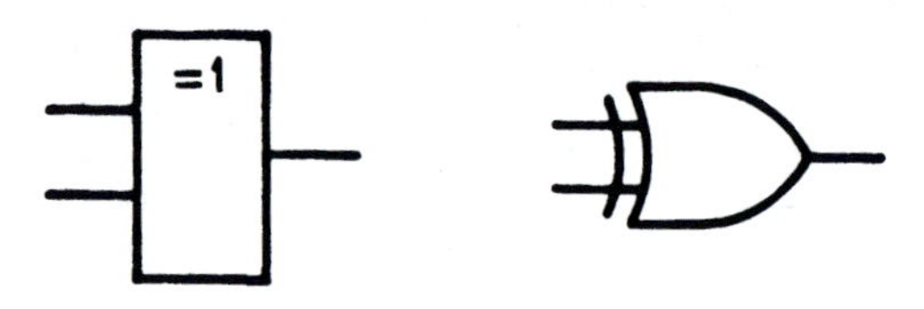

3.4 Logic Threshold Function

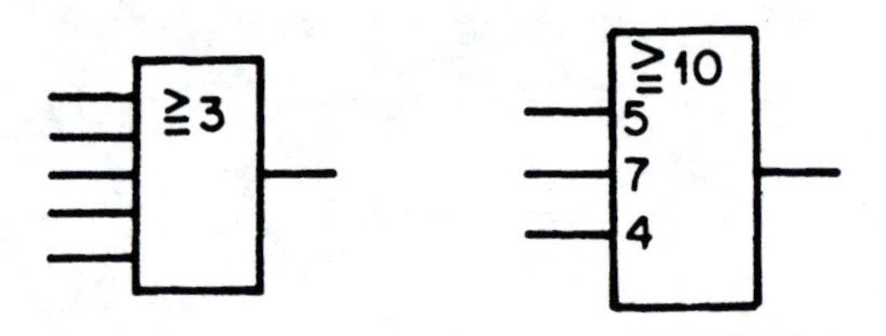

3.5 Majority Function

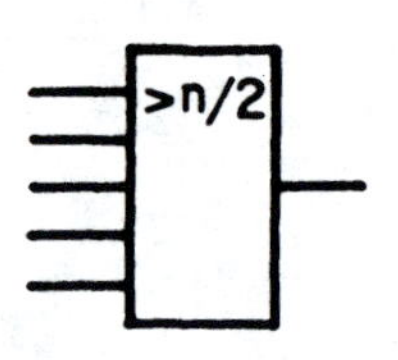

3.6 m and only m Function

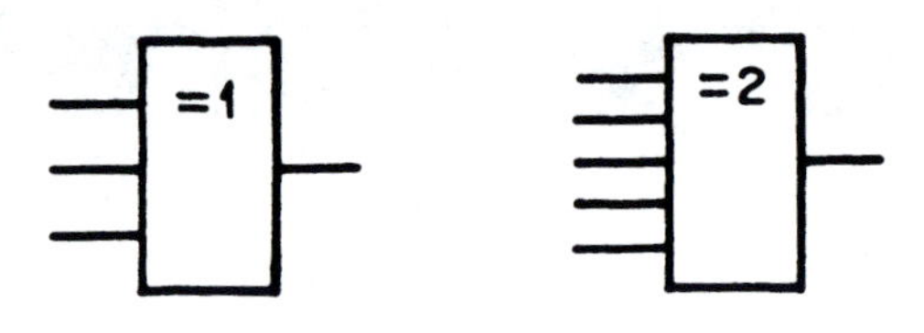

3.7 Even Function

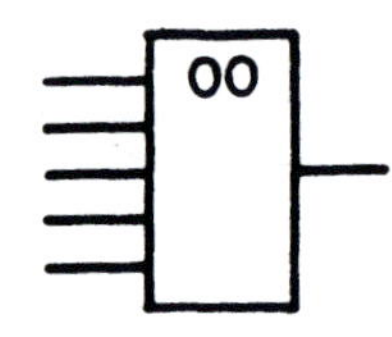

3.8 Addition Modulo 2 (ODD) Function

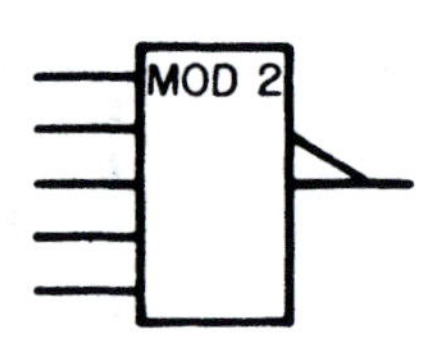

3.9 Logic Identity Function

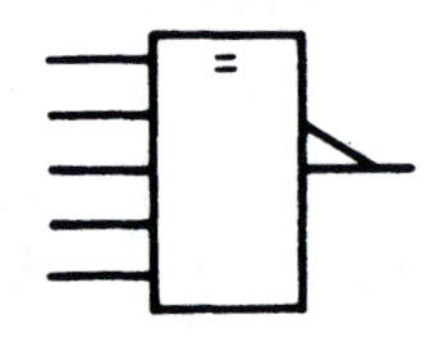

3.10 Coder

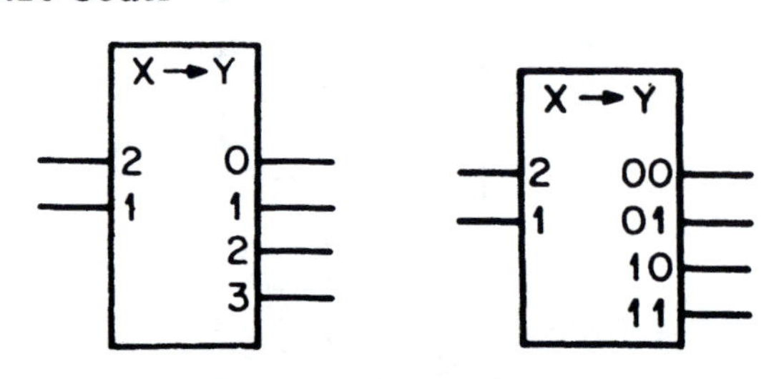

Courtesy of The Institute of Electrical and Electronics Engineers, Inc.

3.11 Distributed Connection (DOT-AND, DOT-OR)

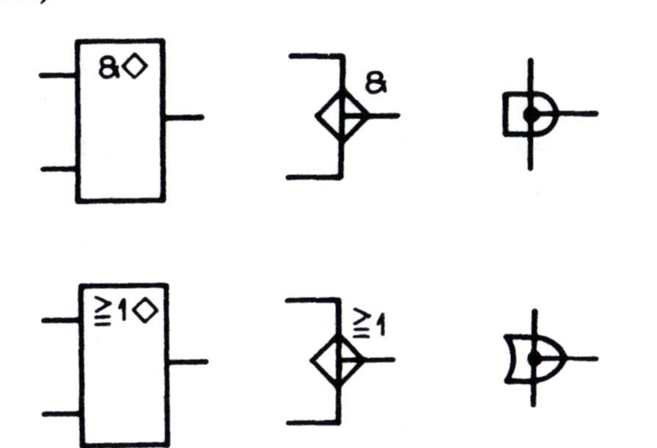

3.12 Amplifier Function

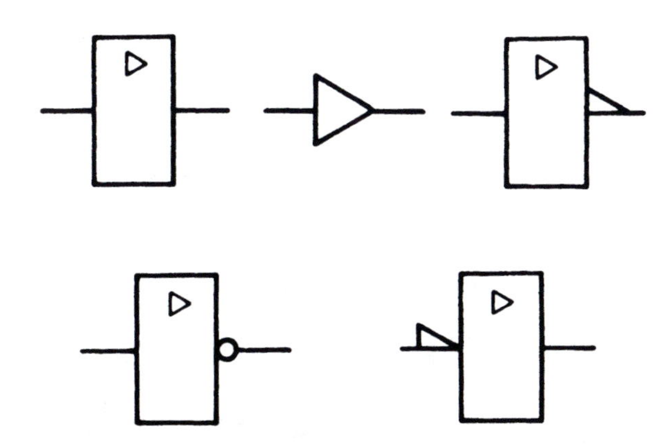

3.13 Inverter Function (Logic Negation Indicator)

3.14 Inverter Function (Polarity Indicator)

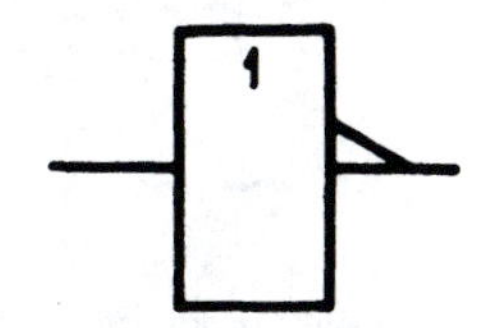

3.15 Signal Level Converter

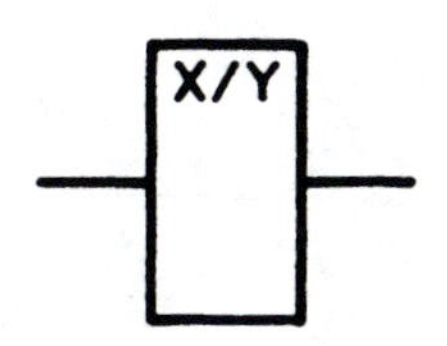

3.16 Oscillator

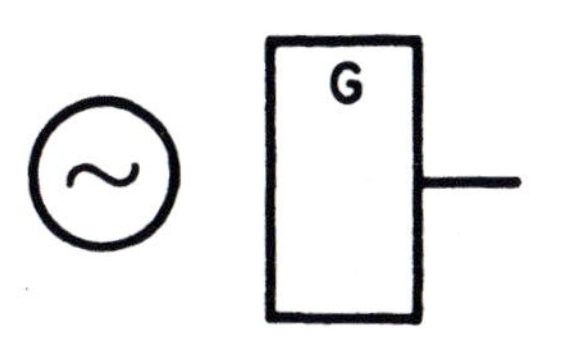

4. Sequential Logic Functions

4.1 Monostable Element (Single Shot)

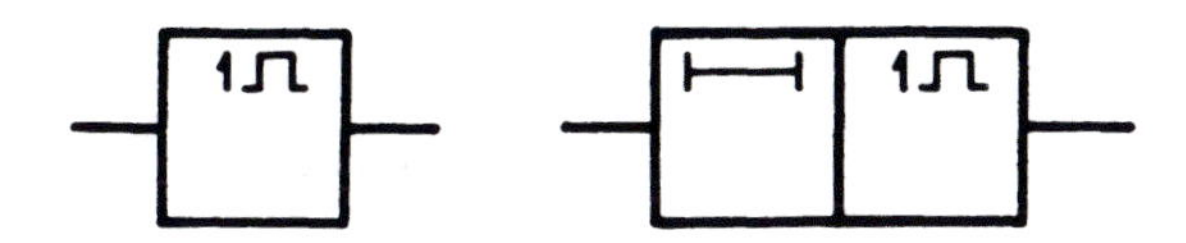

4.2 Delay Element

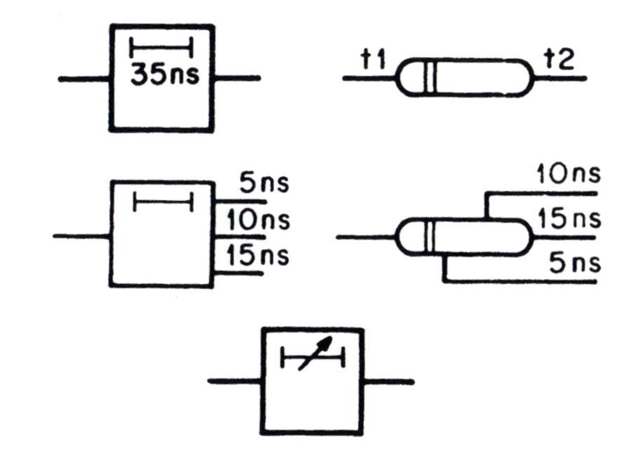

4.3 Flip-Flop

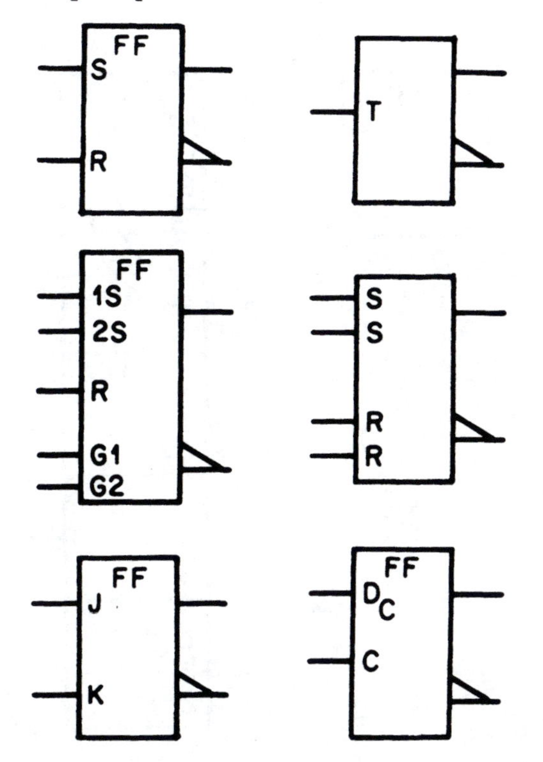

5. Control Block Functions

5.1 Register Control Block

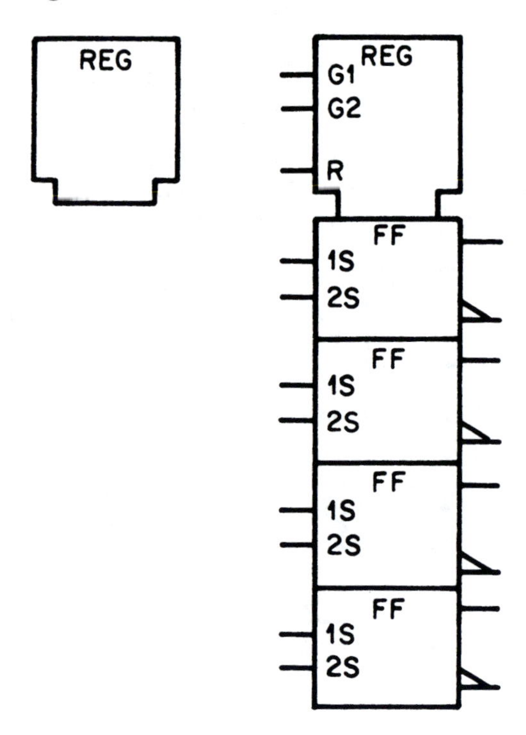

5.2 Selector Control Block

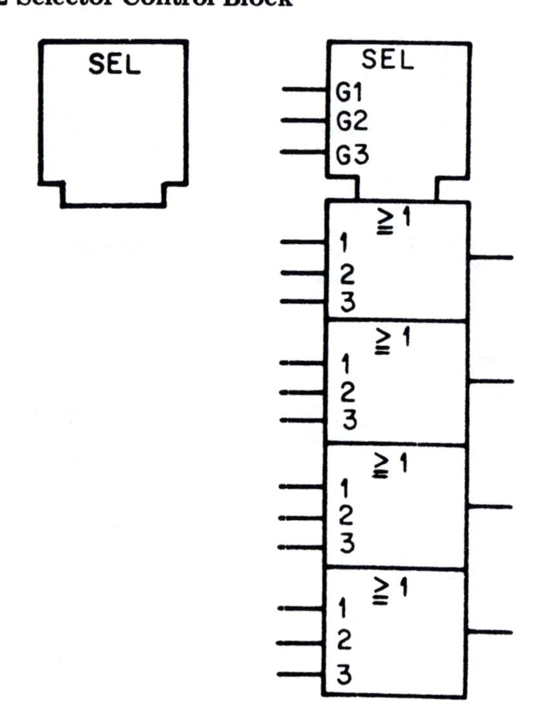

5.3 Counter Control Block

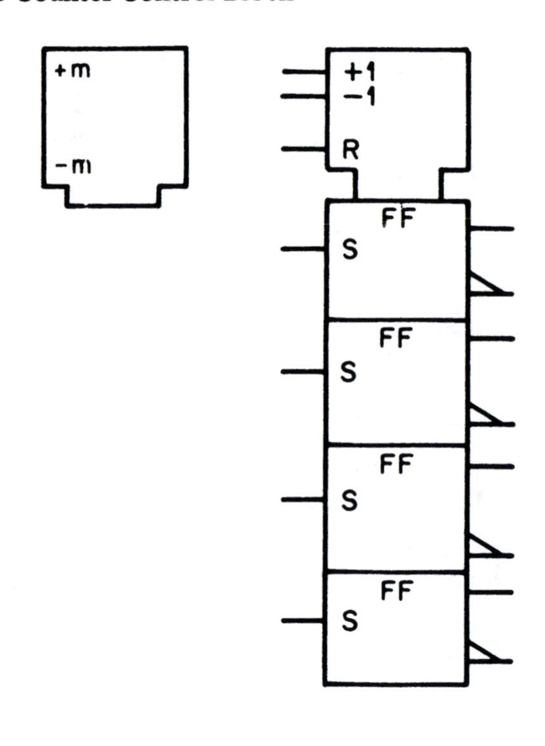

5.4 Shift Register Control Block

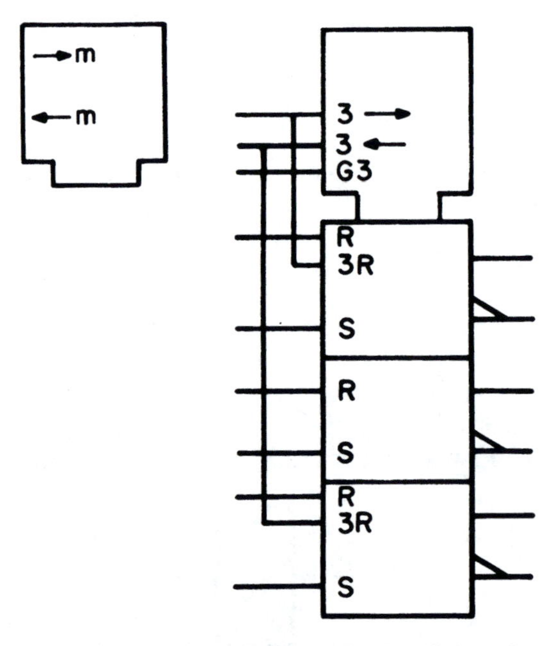

Courtesy of The Institute of Electrical and Electronics Engineers, Inc.

6. Miscellaneous Functions

6.1 Schmitt Trigger

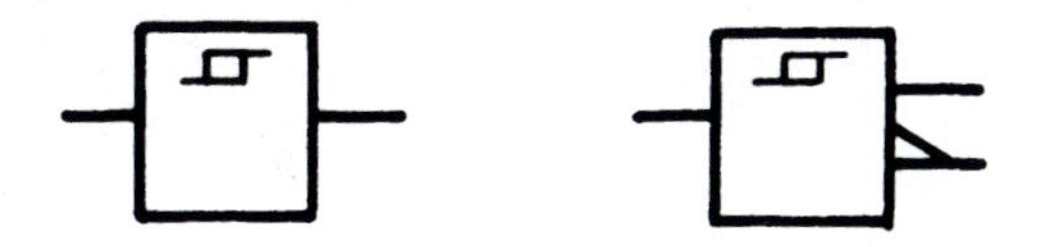

6.2 Nonreciprocal Directional Coupler

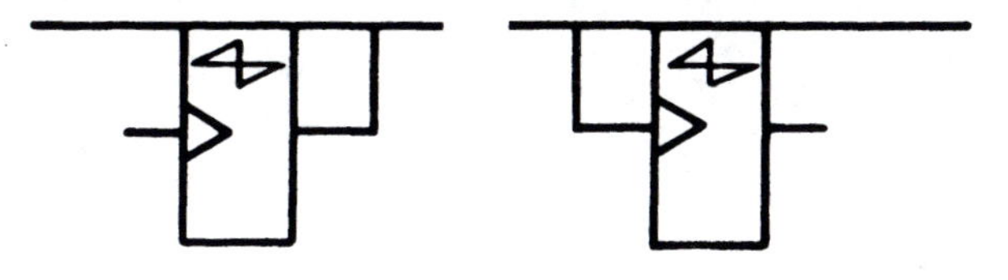

Courtesy of The Institute of Electrical and Electronics Engineers, Inc.

Appendix VI

Mechanical Design Data

INCH/METRIC – EQUIVALENTS					
Fraction	Decimal Equivalent Customary (in.)	Decimal Equivalent Metric (mm)	Fraction	Decimal Equivalent Customary (in.)	Decimal Equivalent Metric (mm)
1/64	.015625	0.3969	33/64	.515625	13.0969
1/32	.03125	0.7938	17/32	.53125	13.4938
3/64	.046875	1.1906	35/64	.546875	13.8906
1/16	.0625	1.5875	9/16	.5625	14.2875
5/64	.078125	1.9844	37/64	.578125	14.6844
3/32	.09375	2.3813	19/32	.59375	15.0813
7/64	.109375	2.7781	39/64	.609375	15.4781
1/8	.1250	3.1750	5/8	.6250	15.8750
9/64	.140625	3.5719	41/64	.640625	16.2719
5/32	.15625	3.9688	21/32	.65625	16.6688
11/64	.171875	4.3656	43/64	.671875	17.0656
3/16	.1875	4.7625	11/16	.6875	17.4625
13/64	.203125	5.1594	45/64	.703125	17.8594
7/32	.21875	5.5563	23/32	.71875	18.2563
15/64	.234375	5.9531	47/64	.734375	18.6531
1/4	.250	6.3500	3/4	.750	19.0500
17/64	.265625	6.7469	49/64	.765625	19.4469
9/32	.28125	7.1438	25/32	.78125	19.8438
19/64	.296875	7.5406	51/64	.796875	20.2406
5/16	.3125	7.9375	13/16	.8125	20.6375
21/64	.328125	8.3384	53/64	.828125	21.0344
11/32	.34375	8.7313	27/32	.84375	21.4313
23/64	.359375	9.1281	55/64	.859375	21.8281
3/8	.3750	9.5250	7/8	.8750	22.2250
25/64	.390625	9.9219	57/64	.890625	22.6219
13/32	.40625	10.3188	29/32	.90625	23.0188
27/64	.421875	10.7156	59/64	.921875	23.4156
7/16	.4375	11.1125	15/16	.9375	23.8125
29/64	.453125	11.5094	61/64	.953125	24.2094
15/32	.46875	11.9063	31/32	.96875	24.6063
31/64	.484375	12.3031	63/64	.984375	25.0031
1/2	.500	12.7000	1	1.000	25.4000

Courtesy *Drafting for Trades & Industry*, Nelson. Delmar Publishers Inc.

U.S. STANDARD GAUGES OF SHEET METAL

GAUGE	THICKNESS		WT. PER SQ. FT.		GAUGE
10	.1406″	3.571 MM	5.625 LBS	2.551 Kg.	10
11	.1250″	3.175 MM	5.000 LBS	2.267 Kg.	11
12	.1094″	2.778 MM	4.375 LBS	1.984 Kg.	12
13	.0938″	2.383 MM	3.750 LBS	1.700 Kg.	13
14	.0781″	1.983 MM	3.125 LBS	1.417 Kg.	14
15	.0703″	1.786 MM	2.813 LBS	1.276 Kg.	15
16	.0625″	1.588 MM	2.510 LBS	1.134 Kg.	16
17	.0563″	1.430 MM	2.250 LBS	1.021 Kg.	17
18	.0500″	1.270 MM	2.000 LBS	0.907 Kg.	18
19	.0438″	1.111 MM	1.750 LBS	0.794 Kg.	19
20	.0375″	0.953 MM	1.500 LBS	0.680 Kg.	20
21	.0344″	0.877 MM	1.375 LBS	0.624 Kg.	21
22	.0313″	0.795 MM	1.250 LBS	0.567 Kg.	22
23	.0280″	0.714 MM	1.125 LBS	0.510 Kg.	23
24	.0250″	0.635 MM	1.000 LBS	0.454 Kg.	24
25	.0219″	0.556 MM	0.875 LBS	0.397 Kg.	25
26	.0188″	0.478 MM	0.750 LBS	0.340 Kg.	26
27	.0172″	0.437 MM	0.687 LBS	0.312 Kg.	27
28	.0156″	0.396 MM	0.625 LBS	0.283 Kg.	28
29	.0141″	0.358 MM	0.563 LBS	0.255 Kg.	29
30	.0120″	0.318 MM	0.500 LBS	0.227 Kg.	30

Courtesy *Drafting for Trades & Industry,* Nelson. Delmar Publishers Inc.

AMERICAN WIRE GAUGE

B & S Gauge Number	Diameter in Mils	Area in Circular Mils	B & S Gauge Number	Diameter in Mils	Area in Circular Mils
0000	460	211 600	21	28.5	810
000	410	167 800	22	25.4	642
00	365	133 100	23	22.6	510
0	325	105 500	24	20.1	404
1	289	83 690	25	17.9	320
2	258	66 370	26	15.9	254
3	229	52 640	27	14.2	202
4	204	41 740	28	12.6	160
5	182	33 100	29	11.3	126.7
6	162	26 250	30	10.0	100.5
7	144	20 820	31	8.93	79.7
8	128	16 510	32	7.95	63.2
9	114	13 090	33	7.08	50.1
10	102	10 380	34	6.31	39.8
11	91	8 234	35	5.62	31.5
12	81	6 530	36	5.00	25.0
13	72	5 178	37	4.45	19.8
14	64	4 107	38	3.96	15.7
15	57	3 257	39	3.53	12.5
16	51	2 583	40	3.15	9.9
17	45.3	2 048	41		
18	40.3	1 624	42	2.50	6.3
19	35.9	1 288	43		
20	32.0	1 022	44	1.97	3.9

Courtesy *Introduction to Electricity and Electronics,* Loper, Ahr, & Clendenning. Delmar Publishers Inc.

BEND RADII FOR STRAIGHT BENDS IN SHEET METALS, MINIMUM

ALUMINUM							
MATERIAL THICKNESS	ALLOY AND TEMPER						
	2024 T3 or T4	5052 0	5052 H32	5052 H34	6061 0	6061 T4	6061 T6
.025	.06	.03	.03	.03	.03	.03	.03
.032	.09	.03	.03	.03	.03	.03	.06
.040	.12	.03	.03	.03	.03	.04	.06
.050	.12	.03	.03	.06	.03	.06	.12
.063	.19	.03	.06	.06	.03	.06	.12
.080	.25	.06	.09	.09	.06	.09	.19
.090	.31	.06	.09	.12	.06	.09	.25
.100	.38	.09	.12	.12	.09	.12	.25
.125	.50	.12	.12	.12	.12	.16	.38
.160	.75	.16	.16	.19	.16	.19	.62
.190	1.00	.19	.19	.25	.29	.25	.87

STEEL			
Material Thickness	Corrosion Resistant		Plain Carbon
	Types 301 302 304 (Annealed)	Types 301 302 304 (1/4H)	
.020	.03	.03	.03
.025	.03	.03	.03
.032	.03	.03	.03
.040	.03	.03	.03
.050	.06	.06	.06
.063	.06	.06	.06
.080	.09	.09	.09
.090	.09	.09	.09
.100	.09	.09	.12
.125	.12	.25	.12
.160	.12	.25	.19
.190	.12	.25	.19

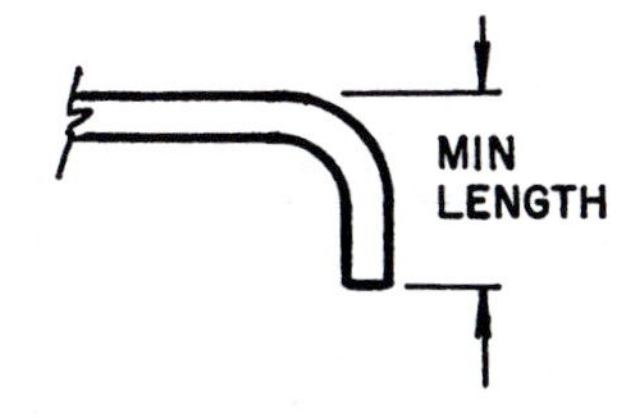

Minimum Length = 2T + 2R
T = Material Thickness
R = Bend Radius

Preferred Options would be to:

1. Reduce Material Thickness
2. Reduce Bend Radius
3. Both of the above
4. Bend Flange Long and Machine to Length

NOTE: The recommended radii shown on this chart are considered minimum based on the economics of using standard tooling and forming methods. Design restrictions that require smaller radii than shown on chart may be accomplished by various methods of annealing, heat treating, tooling and forming equipment. However, due to the increased cost of these processes, each deviation should be given careful consideration.

Courtesy of Texas Instruments, Inc.

DIMENSIONS OF CAP SCREW HEADS

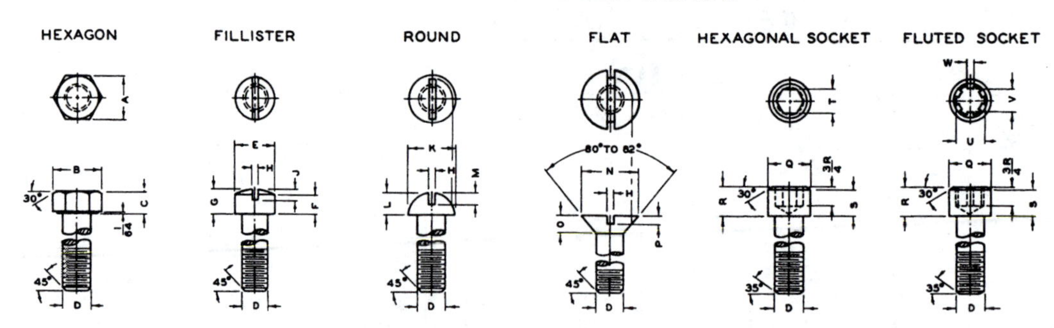

Diam.	A	B	C	E	F	G	H	J	K	L	M	N	O	P	Q	R	S	T	U	V	W	
2-.086															.140 .136	.086 .083	.079	1/16	.074 .073	.064 .063	.016 .015	Max. Min.
3-.099															.161 .157	.099 .096	.091	5/64	.098 .097	.082 .080	.022 .021	Max. Min.
4-.112															.183 .178	.112 .109	.103	5/64	.098 .097	.082 .080	.022 .021	Max. Min.
5-.125															.205 .200	.125 .122	.115	3/32	.115 .113	.098 .096	.025 .023	Max. Min.
6-.138															.226 .221	.138 .134	.127	3/32	.115 .113	.098 .096	.025 .023	Max. Min.
8-.164															.270 .265	.164 .160	.150	1/8	.149 .147	.128 .126	.032 .030	Max. Min.
10-.190															5/16	.190 .185	.174	5/32	.188 .186	.163 .161	.039 .037	Max. Min.
12-.216															11/32	.216 .211	.198	5/32	.188 .186	.163 .161	.039 .037	Max. Min.
1/4	7/16	.505	3/16	3/8	.172 .157	.216 .195	.075 .064	.097 .077	7/16	.191 .175	.117 .097	1/2	.140	.069 .046	3/8	1/4	.229	3/16	.221 .219	.190 .188	.050 .048	Max. Min.
5/16	1/2	.577	15/64	7/16	.203 .186	.253 .230	.084 .072	.115 .090	9/16	.246 .226	.151 .126	5/8	.176	.086 .057	7/16	5/16	.286	7/32	.256 .254	.221 .219	.060 .058	Max. Min.
3/8	9/16	.649	9/32	9/16	.250 .229	.314 .285	.094 .081	.143 .113	5/8	.273 .252	.168 .138	3/4	.210	.103 .069	9/16	3/8	.344	5/16	.380 .377	.319 .316	.092 .089	Max. Min.
7/16	5/8	.722	21/64	5/8	.297 .274	.368 .337	.094 .081	.168 .133	3/4	.328 .302	.202 .167	13/16	.210	.103 .069	5/8	7/16	.401	5/16	.380 .377	.319 .316	.092 .089	Max. Min.
1/2	3/4	.866	3/8	3/4	.328 .301	.412 .376	.106 .091	.188 .148	13/16	.355 .328	.219 .179	7/8	.210	.103 .069	3/4	1/2	.458	3/8	.463 .460	.386 .383	.112 .109	Max. Min.
9/16	13/16	.938	27/64	13/16	.375 .347	.466 .428	.118 .102	.214 .169	15/16	.410 .379	.253 .208	1	.245	.120 .080	13/16	9/16	.516	3/8	.463 .460	.386 .383	.112 .109	Max. Min.
5/8	7/8	1.010	15/32	7/8	.422 .392	.521 .480	.133 .116	.240 .190	1	.438 .405	.270 .220	1 1/8	.281	.137 .092	7/8	5/8	.573	1/2	.604 .601	.509 .506	.138 .134	Max. Min.
3/4	1	1.155	9/16	1	.500 .466	.612 .566	.149 .131	.283 .233	1 1/4	.547 .506	.337 .277	1 3/8	.352	.171 .115	1	3/4	.688	9/16	.631 .627	.535 .531	.149 .145	Max. Min.
7/8	1 1/8	1.299	21/32	1 1/8	.594 .556	.720 .669	.167 .147	.334 .264				1 5/8	.423	.206 .139	1 1/8	7/8	.802	9/16	.709 .705	.604 .600	.168 .164	Max. Min.
1	1 5/16	1.516	3/4	1 5/16	.656 .613	.802 .744	.188 .166	.372 .292				1 7/8	.494	.240 .162	1 5/16	1	.917	5/8	.801 .797	.685 .681	.189 .185	Max. Min.
1 1/8	1 1/2	1.732	27/32												1 1/2	1 1/8	1.031	3/4	.970 .966	.828 .824	.231 .227	Max. Min.
1 1/4	1 11/16	1.949	15/16												1 3/4	1 1/4	1.146	3/4	.970 .966	.828 .824	.231 .227	Max. Min.
1 3/8															1 7/8	1 3/8	1.260	3/4	.970 .966	.828 .824	.231 .227	Max. Min.
1 1/2															2	1 1/2	1.375	1	1.275 1.271	1.007 1.003	.298 .294	Max. Min.

Length of Thread on Hexagon, Fillister, Round and Flat = 2 D + 1/4"

Length of Thread on Hex. and Fluted Socket, For NC Thread = 2D + 1/2", or 1/2 Bolt Length } Whichever is Greater
For NF Thread = 1 1/2D + 1/2, or 3/8 Bolt Length

Courtesy *Basic Drafting Technology,* Rotmans, Horton, Good. Delmar Publishers Inc.

DIMENSIONS OF MACHINE SCREW HEADS

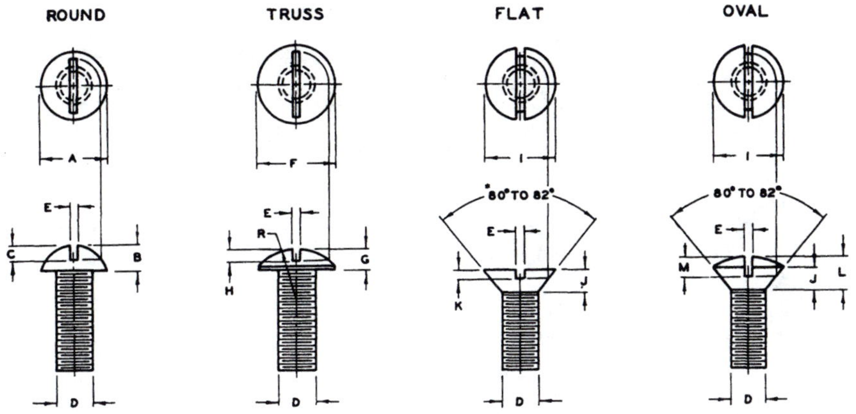

ROUND TRUSS FLAT OVAL

Diam.	A	B	C	E	F	G	H	H[1]	I	J	K	L	M	N	
0-.060	.113	.053	.039	.023					.119	.035	.015	.056	.030	.096	Max.
	.099	.043	.029	.016					.105	.026	.010	.041	.025	.083	Min.
1-.073	.138	.061	.044	.026					.146	.043	.019	.068	.038	.118	Max.
	.122	.051	.033	.019					.130	.033	.012	.052	.031	.104	Min.
2-.086	.162	.069	.048	.031	.194	.053	.031	.050	.172	.051	.023	.080	.045	.140	Max.
	.146	.059	.037	.023	.180	.044	.022	.040	.156	.040	.015	.063	.037	.124	Min.
3-.099	.187	.078	.053	.035	.226	.061	.036	.055	.199	.059	.027	.092	.052	.161	Max.
	.169	.067	.040	.027	.211	.051	.026	.044	.181	.048	.017	.073	.043	.145	Min.
4-.112	.211	.086	.058	.039	.257	.069	.040	.060	.225	.067	.030	.104	.059	.183	Max.
	.193	.075	.044	.031	.241	.059	.030	.049	.207	.055	.020	.084	.049	.166	Min.
5-.125	.236	.095	.063	.043	.289	.078	.045	.070	.252	.075	.034	.116	.067	.205	Max.
	.217	.083	.047	.035	.272	.066	.034	.058	.232	.062	.022	.095	.055	.187	Min.
6-.138	.260	.103	.068	.048	.321	.086	.050	.080	.279	.083	.038	.128	.074	.226	Max.
	.240	.091	.051	.039	.303	.074	.037	.067	.257	.069	.024	.105	.060	.208	Min.
8-.164	.309	.120	.077	.054	.384	.102	.058	.110	.332	.100	.045	.152	.088	.270	Max.
	.287	.107	.058	.045	.364	.088	.045	.096	.308	.084	.029	.126	.072	.250	Min.
10-.190	.359	.137	.087	.060	.448	.118	.068	.120	.385	.116	.053	.176	.103	.313	Max.
	.334	.123	.065	.050	.425	.103	.053	.105	.359	.098	.034	.148	.084	.292	Min.
12-.216	.408	.153	.096	.067	.511	.134	.077	.155	.438	.132	.060	.200	.117	.357	Max.
	.382	.139	.072	.056	.487	.118	.061	.139	.410	.112	.039	.169	.096	.334	Min.
1/4	.472	.175	.109	.075	.573	.150	.087	.190	.507	.153	.070	.232	.136	.414	Max.
	.443	.160	.082	.064	.546	.133	.070	.172	.477	.131	.046	.197	.112	.389	Min.
5/16	.590	.216	.132	.084	.698	.183	.106	.230	.635	.191	.088	.290	.171	.518	Max.
	.557	.198	.099	.072	.666	.162	.085	.208	.600	.165	.058	.249	.141	.490	Min.
3/8	.708	.256	.155	.094	.823	.215	.124	.295	.762	.230	.106	.347	.206	.622	Max.
	.670	.237	.117	.081	.787	.191	.100	.270	.722	.200	.070	.300	.170	.590	Min.
7/16	.750	.328	.196	.094	.948	.248	.142		.812	.223	.103	.345	.210	.625	Max.
	.707	.307	.148	.081	.907	.221	.116		.771	.190	.066	.295	.174	.589	Min.
1/2	.813	.355	.211	.106	1.073	.280	.161		.875	.223	.103	.354	.216	.750	Max.
	.766	.332	.159	.091	1.028	.250	.131		.831	.186	.065	.299	.176	.710	Min.
9/16	.938	.410	.242	.118	1.198	.312	.179		1.000	.260	.120	.410	.250	.812	Max.
	.887	.385	.183	.102	1.149	.279	.146		.950	.220	.077	.350	.207	.768	Min.
5/8	1.000	.438	.258	.133	1.323	.345	.196		1.125	.298	.137	.467	.285	.875	Max.
	.944	.411	.195	.116	1.269	.309	.162		1.069	.253	.088	.399	.235	.827	Min.
3/4	1.250	.547	.320	.149	1.573	.410	.234		1.375	.372	.171	.578	.353	1.000	Max.
	1.185	.516	.242	.131	1.511	.368	.182		1.306	.319	.111	.497	.293	.945	Min.

* Flat Head Machine Screws Also Are Made With 100° Angle

Courtesy *Basic Drafting Technology*, Rotmans, Horton, Good. Delmar Publishers Inc.

DIMENSIONS OF MACHINE SCREW HEADS – Continued

FILLISTER BINDING PAN HEXAGON

Diam.	O	P	Q	R	R[1]	S	T	U	V	W	X	Y	Z	
0-.060	.045 .037	.059 .043	.025 .015											Max. Min.
1-.073	.053 .045	.071 .055	.031 .020											Max. Min.
2-.086	.062 .053	.083 .066	.037 .025	.129	.035	.181 .171	.046 .041	.018 .013	.030 .024	.167 .155	.053 .045	.033 .023	.125 .120	Max. Min.
3-.099	.070 .061	.095 .077	.043 .030	.151	.037	.208 .197	.054 .048	.022 .016	.036 .029	.193 .180	.060 .051	.037 .027	.187 .181	Max. Min.
4-.112	.079 .069	.107 .088	.048 .035	.169	.042	.235 .223	.063 .056	.025 .018	.042 .034	.219 .205	.068 .058	.041 .030	.187 .181	Max. Min.
5-.125	.088 .078	.120 .100	.054 .040	.191	.044	.263 .249	.071 .064	.029 .021	.048 .039	.245 .231	.075 .065	.045 .032	.187 .181	Max. Min.
6-.138	.096 .086	.132 .111	.060 .045	.211	.046	.290 .275	.080 .071	.032 .024	.053 .044	.270 .256	.082 .072	.050 .038	.250 .244	Max. Min.
8-.164	.113 .102	.156 .133	.071 .054	.254	.052	.344 .326	.097 .087	.039 .029	.065 .054	.322 .306	.096 .085	.058 .043	.250 .244	Max. Min.
10-.190	.130 .118	.180 .156	.083 .064	.283	.061	.399 .378	.114 .102	.045 .034	.077 .064	.373 .357	.110 .099	.067 .050	.312 .305	Max. Min.
12-.216	.148 .134	.205 .178	.094 .074	.336	.078	.454 .430	.130 .117	.052 .039	.089 .074	.425 .407	.125 .112	.077 .060	.312 .305	Max. Min.
1/4	.170 .155	.237 .207	.109 .087	.375	.087	.513 .488	.153 .138	.061 .046	.105 .088	.492 .473	.144 .130	.087 .070	.375 .367	Max. Min.
5/16	.211 .194	.295 .262	.137 .110	.457	.099	.641 .609	.193 .174	.077 .059	.134 .112	.615 .594	.178 .162	.109 .092	.500 .491	Max. Min.
3/8	.253 .233	.355 .315	.164 .133	.538	.143	.769 .731	.234 .211	.094 .071	.163 .136	.740 .716	.212 .195	.130 .113	.562 .552	Max. Min.
7/16	.265 .242	.368 .321	.170 .135	.619										Max. Min.
1/2	.297 .273	.412 .362	.190 .151	.701										Max. Min.
9/16	.336 .308	.466 .410	.214 .172	.783										Max. Min.
5/8	.375 .345	.521 .461	.240 .193	.863										Max. Min.
3/4	.441 .406	.612 .542	.281 .226	1.024										Max. Min.

Length of Thread: Screws 2" or Shorter are Threaded to Head. Screws over 2" are Threaded 1 3/4" Minimum. Screw Heads Also are Made With Cross Recess Slots.

Courtesy *Basic Drafting Technology,* Rotmans, Horton, Good. Delmar Publishers Inc.

DIMENSIONS OF MACHINE SCREW AND STOVE BOLT NUTS

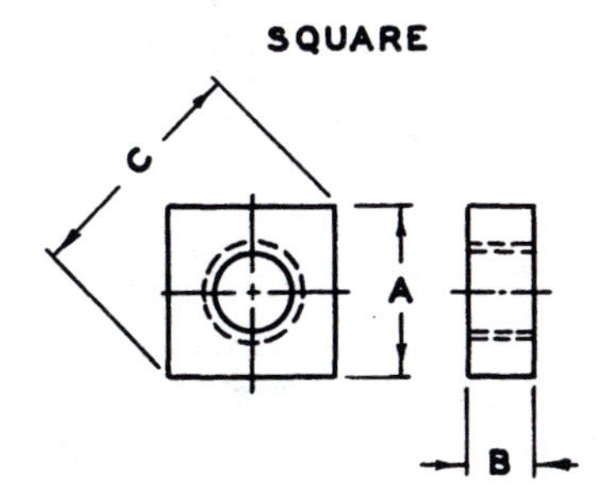

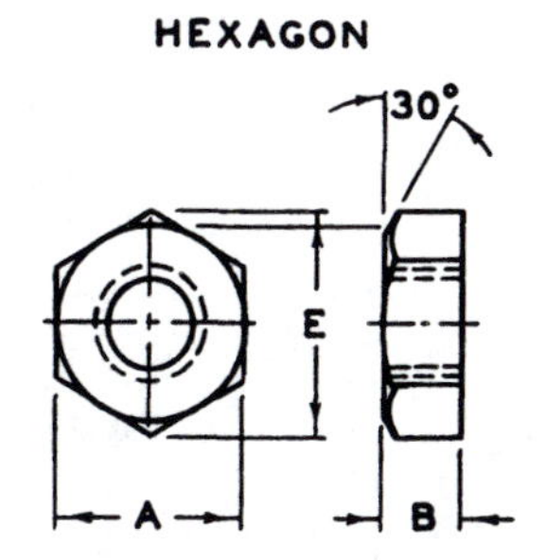

Diameter of Thread	A			B			C	E
		Max.	Min.		Max.	Min.		
0-.060	5/32	.1562	.150	3/64	.050	.043	.2209	.1804
1-.073	5/32	.1562	.150	3/64	.050	.043	.2209	.1804
2-.086	3/16	.1875	.180	1/16	.066	.057	.2651	.2165
3-.099	3/16	.1875	.180	1/16	.066	.057	.2651	.2165
4-.112	1/4	.250	.241	3/32	.098	.087	.3535	.2886
5-.125	5/16	.3125	.302	7/64	.114	.102	.4419	.3608
6-.138	5/16	.3125	.302	7/64	.114	.102	.4419	.3608
8-.164	11/32	.3438	.332	1/8	.130	.117	.4861	.3968
10-.190	3/8	.375	.362	1/8	.130	.117	.5303	.4329
12-.216	7/16	.4375	.423	5/32	.161	.148	.6187	.5051
1/4	7/16	.4375	.423	3/16	.193	.178	.6187	.5051
5/16	9/16	.5625	.545	7/32	.225	.208	.7955	.6494
3/8	5/8	.625	.607	1/4	.257	.239	.8839	.7216

Courtesy *Basic Drafting Technology,* Rotmans, Horton, Good. Delmar Publishers Inc.

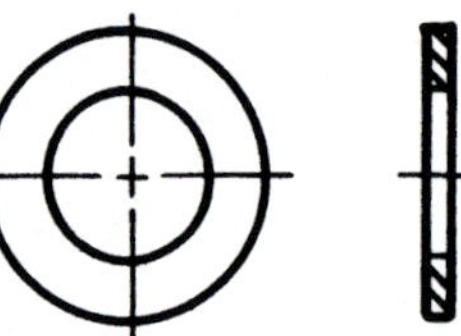

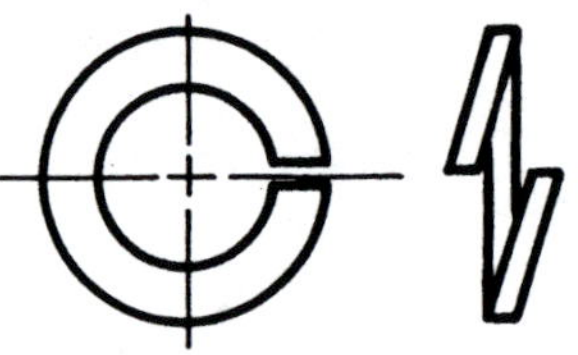

Nominal Screw Size		Flat Washer			Lockwasher		
Number or Fraction	Decimal	Inside Dia A	Outside Dia B	Thickness C	Inside Dia A	Outside Dia B	Thickness C
6	.138	.16	.38	.05	.14	.25	.03
8	.164	.19	.44	.05	.17	.29	.04
10	.190	.22	.50	.05	.19	.33	.05
12	.216	.25	.56	.07	.22	.38	.06
1/4	.250 N	.28	.63	.07	.26	.49	.06
1/4	.250 W	.31	.73	.07			
5/16	.312 N	.34	.69	.07	.32	.59	.08
5/16	.312 W	.38	.88	.08			
3/8	.375 N	.41	.81	.07	.38	.68	.09
3/8	.375 W	.44	1.00	.08			
7/16	.438 N	.47	.92	.07	.45	.78	.11
7/16	.438 W	.50	1.25	.08			
1/2	.500 N	.53	1.06	.10	.51	.87	.12
1/2	.500 W	.56	1.38	.11			
5/8	.625 N	.66	1.31	.10	.64	1.08	.16
5/8	.625 W	.69	1.75	.13			
3/4	.750 N	.81	1.47	.13	.76	1.27	.19
3/4	.750 W	.81	2.00	.15			
7/8	.875 N	.94	1.75	.13	.89	1.46	.22
7/8	.875 W	.94	2.25	.17			
1	1.000 N	1.06	2.00	.13	1.02	1.66	.25
1	1.000 W	1.06	2.50	.17			
1 1/8	1.125 N	1.25	2.25	.13	1.14	1.85	.28
1 1/8	1.125 W	1.25	2.75	.17			
1 1/4	1.250 N	1.38	2.50	.17	1.27	2.05	.31
1 1/4	1.250 W	1.38	3.00	.17			
1 3/8	1.375 N	1.50	2.75	.17	1.40	2.24	.34
1 3/8	1.375 W	1.50	3.25	.18			
1 1/2	1.500 N	1.62	3.00	.17	1.53	2.43	.38
1 1/2	1.500 W	1.62	3.50	.18			

N–SAE Sizes (Narrow)
W–Standard Plate (Wide)

Common Washer Sizes

Courtesy *Interpreting Engineering Drawings,* Third Edition, Jensen & Hines. Delmar Publishers Inc.

DIMENSIONS OF SQUARE HEAD AND HEADLESS SET SCREWS

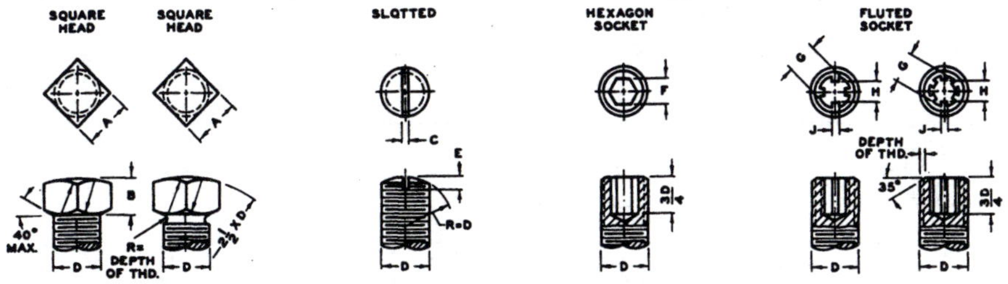

Diam.	A	B	C	E	F	G	H	J	K	L	M	N	
5-.125			.023	.031	1/16	.071 .070	.053 .052	.022 .021	1/16	.083 .078	.06	.03	Max. Min.
6-.138			.025	.035	1/16	..079 .078	.056 .055	.022 .021	.069	.092 .087	.07	.03	Max. Min.
8-.164			.029	.041	5/64	.098 .097	.082 .080	.022 .021	5/64	.109 .103	.08	.04	Max. Min.
10-.190			.032	.048	3/32	.115 .113	.098 .096	.025 .023	3/32	.127 .120	.09	.04	Max. Min.
12-.216			.036	.054	3/32	.115 .113	.098 .096	.025 .023	7/64	.144 .137	.11	.06	Max. Min.
1/4	1/4	3/16	.045	.063	1/8	.149 .147	.128 .126	.032 .030	1/8	5/32	1/8	1/16	Max. Min.
5/16	5/16	15/64	.051	.078	5/32	.188 .186	.163 .161	.039 .037	11/64	13/64	5/32	5/64	Max. Min.
3/8	3/8	9/32	.064	.094	3/16	.221 .219	.190 .188	.050 .048	13/64	1/4	3/16	3/32	Max. Min.
7/16	7/16	21/64	.072	.109	7/32	.256 .254	.221 .219	.060 .058	15/64	19/64	7/32	7/64	Max. Min.
1/2	1/2	3/8	.081	.125	1/4	.298 .296	.254 .252	.068 .066	9/32	11/32	1/4	1/8	Max. Min.
9/16	9/16	27/64	.091	.141	1/4	.298 .296	.254 .252	.068 .066	5/16	25/64	9/32	9/64	Max. Min.
5/8	5/8	15/32	.102	.156	5/16	.380 .377	.319 .316	.092 .089	23/64	15/32	5/16	5/32	Max. Min.
3/4	3/4	9/16	.129	.188	3/8	.463 .460	.386 .383	.112 .109	7/16	9/16	3/8	3/16	Max. Min.
7/8	7/8	21/32			1/2	.604 .601	.509 .506	.138 .134	33/64	21/32	7/16	7/32	Max. Min.
1	1	3/4			9/16	.631 .627	.535 .531	.149 .145	19/32	3/4	1/2	1/4	Max. Min.
1 1/8	1 1/8	27/32			9/16	.709 .705	.604 .600	.168 .164	43/64	27/32	9/16	9/32	Max. Min.
1 1/4	1 1/4	15/16			5/8	.801 .797	.685 .681	.189 .185	3/4	15/16	5/8	5/16	Max. Min.
1 3/8	1 3/8	1 1/32			5/8	.869 .865	.744 .740	.207 .203	53/64	1 1/32	11/16	11/32	Max. Min.
1 1/2	1 1/2	1 1/8			3/4	.970 .966	.828 .824	.231 .227	29/32	1 1/8	3/4	3/8	Max. Min.
1 3/4					1	1.275 1.271	1.007 1.003	.298 .294	1 1/16	1 5/16	7/8	7/16	Max. Min.
2					1	1.275 1.271	1.007 1.003	.298 .294	1 7/32	1 1/2	1	1/2	Max. Min.

SET SCREW POINTS

*Angle Y = 118° When Length Equals Diam. or Less, Y = 90° When Length Exceeds Diam.

Courtesy *Basic Drafting Technology*, Rotmans, Horton, Good. Delmar Publishers Inc.

TAP DRILL SIZES FOR AMERICAN STANDARD THREADS

Diam. of Thread	Threads per Inch	Drill*	Decimal Equiv.
No. 0–.060	80 NF	3/64	.0469
1–.073	64 NC	1.5 MM	.0591
	72 NF	53	.0595
2–.086	56 NC	50	.0700
	64 NF	50	.0700
3–.099	48 NC	5/64	.0781
	56 NF	45	.0820
4–.112	40 NC	43	.0890
	48 NF	42	.0935
5–.125	40 NC	38	.1015
	44 NF	37	.1040
6–.138	32 NC	36	.1065
	40 NF	33	.1130
8–.164	32 NC	29	.1360
	36 NF	29	.1360
10–.190	24 NC	25	.1495
	32 NF	21	.1590
12–.216	24 NC	16	.1770
	28 NF	14	.1820
1/4	20 NC	7	.2010
	28 NF	3	.2130
	32 NEF	7/32	.2188
5/16	18 NC	F	.2570
	24 NF	I	.2720
	32 NEF	9/32	.2812
3/8	16 NC	5/16	.3125
	24 NF	Q	.3320
	32 NEF	11/32	.3438
7/16	14 NC	U	.3680
	20 NF	25/64	.3906
	28 NEF	Y	.4040
1/2	12 N	27/64	.4219
	13 NC	27/64	.4219
	20 NF	29/64	.4531
	28 NEF	15/32	.4687
9/16	12 NC	31/64	.4844
	18 NF	33/64	.5156
	24 NEF	33/64	.5156
5/8	11 NC	17/32	.5312
	12 N	35/64	.5469
	18 NF	14.5 MM	.5709
	24 NEF	37/64	.5781
11/16	12 N	39/64	.6094
	24 NEF	16.5 MM	.6496
3/4	10 NC	16.5 MM	.6496
	12 N	17 MM	.6693
	16 NF	17.5 MM	.6890
	20 NEF	45/64	.7031
13/16	12 N	18.5 MM	.7283
	16 N	3/4	.7500
	20 NEF	49/64	.7656
7/8	9 NC	49/64	.7656
	12 N	20 MM	.7874
	14 NF	20.5 MM	.8071
	16 N	13/16	.8125
	20 NEF	21 MM	.8268
15/16	12 N	55/64	.8594
	16 N	7/8	.8750
	20 NEF	22.5 MM	.8858
1	8 NC	7/8	.8750
	12 N	59/64	.9219
	14 NF	23.5 MM	.9252
	16 N	15/16	.9375
	20 NEF	61/64	.9531
1 1/16	12 N	25 MM	.9843
	16 N	1	1.0000
	18 NEF	25.5 MM	1.0040
1 1/8	7 NC	63/64	.9844
	8 N	25.5 MM	1.0039
	12 NF	26.5 MM	1.0433
	16 N	1 1/16	1.0625
	18 NEF	1 5/64	1.0781
1 3/16	12 N	28 MM	1.1024
	16 N	1 1/8	1.1250
	18 NEF	1 9/64	1.1406
1 1/4	7 NC	1 7/64	1.1094
	8 N	1 1/8	1.1250
	12 NF	29.5 MM	1.1614
	16 N	1 3/16	1.1875
	18 NEF	30.5 MM	1.2008
1 5/16	12 N	1 15/64	1.2344
	16 N	1 1/4	1.2500
	18 NEF	32 MM	1.2598

*To produce approximately 75% full thread

Courtesy *Basic Drafting Technology,* Rotmans, Horton, Good. Delmar Publishers Inc.

TAP DRILL SIZES FOR AMERICAN STANDARD THREADS – Continued

Diam. of Thread	Threads per Inch	Drill*	Decimal Equiv.
1 3/8	6 NC	1 13/64	1.2031
	8 N	1 1/4	1.2500
	12 NF	1 19/64	1.2969
	16 N	1 5/16	1.3125
	18 NEF	33.5 MM	1.3189
1 7/16	12 N	34.5 MM	1.3583
	16 N	1 3/8	1.3750
	18 NEF	35 MM	1.3780
1 1/2	6 NC	1 21/64	1.3281
	8 N	1 3/8	1.3750
	12 NF	36 MM	1.4173
	16 N	1 7/16	1.4375
	18 NEF	1 29/64	1.4531
1 9/16	16 N	1 1/2	1.5000
	18 NEF	1 33/64	1.5156
1 5/8	8 N	1 1/2	1.5000
	12 N	39 MM	1.5354
	16 N	1 9/16	1.5625
	18 NEF	40 MM	1.5748
1 11/16	16 N	1 5/8	1.6250
	18 NEF	41.5 MM	1.6339
1 3/4	5 NC	1 35/64	1.5469
	8 N	1 5/8	1.6250
	12 N	1 43/64	1.6719
	16 NEF	1 11/16	1.6875
1 13/16	16 N	1 3/4	1.7500
1 7/8	8 N	1 3/4	1.7500
	12 N	45.5 MM	1.7913
	16 N	1 13/16	1.8125
1 15/16	16 N	1 7/8	1.8750
2	4 1/2 NC	1 25/32	1.7812
	8 N	1 7/8	1.8750
	12 N	1 59/64	1.9219
	16 NEF	1 15/16	1.9375
2 1/16	16 N	2	2.0000
2 1/8	8 N	2	2.0000
	12 N	2 3/64	2.0469
	16 N	2 1/16	2.0625
2 3/16	16 N	2 1/8	2.1250
2 1/4	4 1/2 NC	2 1/32	2.0312
	8 N	2 1/8	2.1250
	12 N	55 MM	2.1654
	16 N	2 3/16	2.1875
2 5/16	16 N	2 1/4	2.2500
2 3/8	12 N	2 19/64	2.2969
	16 N	2 5/16	2.3125
2 7/16	16 N	2 3/8	2.3750
2 1/2	4 NC	2 1/4	2.2500
	8 N	2 3/8	2.3750
	12 N	61.5 MM	2.4213
	16 N	2 7/16	2.4375
2 5/8	12 N	64.5 MM	2.5394
	16 N	2 9/16	2.5625
2 3/4	4 NC	2 1/2	2.5000
	8 N	2 5/8	2.6250
	12 N	2 43/64	2.6719
	16 N	2 11/16	2.6875
2 7/8	12 N	71 MM	2.7953
	16 N	2 13/16	2.8125
3	4 NC	2 3/4	2.7500
	8 N	2 7/8	2.8750
	12 N	74 MM	2.9134
	16 N	2 15/16	2.9375
3 1/8	12 N	3 1/16	3.0625
	16 N	3 1/16	3.0625
3 1/4	4 NC	3	3.0000
	8 N	3 1/8	3.1250
	12 N	3 3/16	3.1875
	16 N	3 3/16	3.1875
3 3/8	12 N	3 5/16	3.3125
	16 N	3 5/16	3.3125
3 1/2	4 NC	3 1/4	3.2500
	8 N	3 3/8	3.3750
	12 N	3 7/16	3.4375
	16 N	3 7/16	3.4375
3 3/4	4 NC	3 1/2	3.5000

* To produce approximately 75% full thread

Courtesy *Basic Drafting Technology*, Rotmans, Horton, Good. Delmar Publishers Inc.

DRILL SIZES—DECIMAL AND METRIC EQUIVALENTS

Size	Equivalent		
	Decimal	3 Pl	MM
80	.0135	.014	.3429
79	.0145	.014	.3683
1/64	.015625	.016	.3969
78	.0160		.4064
77	.0180		.4572
76	.0200		.5080
75	.0210		.5334
74	.0225	.022	.5715
73	.0240		.6096
72	.0250		.6350
71	.0260		.6604
70	.0280		.7112
69	.0292	.029	.7417
68	.0310		.7874
1/32	.03125	.031	.7938
67	.0320		.8128
66	.0330		.8382
65	.0350		.8890
64	.0360		.9144
63	.0370		.9398
62	.0380		.9652
61	.0390		.9906
60	.0400		1.0160
59	.0410		1.0414
58	.0420		1.0668
57	.0430		1.0922
56	.0465	.046	1.1811
3/64	.046875	.047	1.1906
55	.0520		1.3208
54	.0550		1.3970
53	.0595	.060	1.5113
1/16	.0625	.062	1.5875
52	.0635	.064	1.6129
51	.0670		1.7018
50	.0700		1.7780
49	.0730		1.8542
48	.0760		1.9304
5/64	.078125	.078	1.9844
47	.0785	.078	1.9939
46	.0810		2.0574
45	.0820		2.0828
44	.0860		2.1844
43	.0890		2.2606
42	.0935	.094	2.3749
3/32	.09375	.094	2.3813
41	.0960		2.4384
40	.0980		2.4892

Size	Equivalent		
	Decimal	3 Pl	MM
39	.0995	.100	2.5273
38	.1015	.102	2.5781
37	.1040		2.6416
36	.1065	.106	2.7051
7/64	.109375	.109	2.7781
35	.1100		2.7940
34	.1110		2.8194
33	.1130		2.8702
32	.1160		2.9464
31	.1200		3.0480
1/8	.125		3.1750
30	.1285	.128	3.2639
29	.1360		3.4544
28	.1405	.140	3.5687
9/64	.140625	.141	3.5719
27	.1440		3.6576
26	.1470		3.7338
25	.1495	.150	3.7973
24	.1520		3.8608
23	.1540		3.9116
5/32	.15625	.156	3.9688
22	.1570		3.9878
21	.1590		4.0386
20	.1610		4.0894
19	.1660		4.2164
18	.1695	.170	4.3053
11/64	.171875	.172	4.3656
17	.1730		4.3942
16	.1770		4.4958
15	.1800		4.5720
14	.1820		4.6228
13	.1850		4.6990
3/16	.1875	.188	4.7625
12	.1890		4.8006
11	.1910		4.8514
10	.1935	.194	4.9149
9	.1960		4.9784
8	.1990		5.0546
7	.2010		5.1054
13/64	.203125	.203	5.1594
6	.2040		5.1816
5	.2055	.206	5.2197
4	.2090		5.3086
3	.2130		5.4102
7/32	.21875	.219	5.5563
2	.2210		5.6134
1	.2280		5.7912

Courtesy of Texas Instruments, Inc.

DRILL SIZES—DECIMAL AND METRIC EQUIVALENTS (Continued)

	Equivalent		
Size	Decimal	3 Pl	MM
A	.2340		5.9436
15/64	.234375	.234	5.9531
B	.2380		6.0452
C	.2420		6.1468
D	.2460		6.2484
1/4	.250		6.3500
E	.2500		6.3500
F	.2570		6.5278
G	.2610		6.6294
17/64	.265625	.266	6.7469
H	.2660		6.7564
I	.2720		6.9088
J	.2770		7.0358
K	.2810		7.1374
9/32	.28125	.281	7.1438
L	.2900		7.3660
M	.2950		7.4930
19/64	.296875	.297	7.5406
N	.3020		7.6708
5/16	.3125	.312	7.9375
O	.3160		8.0264
P	.3230		8.2042
21/64	.328125	.328	8.3344
Q	.3320		8.4328
R	.3390		8.6106
11/32	.34375	.344	8.7313
S	.3480		8.3892
T	.3580		9.0932
23/64	.359375	.359	9.1281
U	.3680		9.3472
3/8	.375		9.5250
V	.3770		9.5758
W	.3860		9.8044
25/64	.390625	.391	9.9219
X	.3970		10.0838
Y	.4040		10.2616
13/32	.40625	.406	10.3188
Z	.4130		10.4902

	Equivalent		
Size	Decimal	3 Pl	MM
27/64	.421875	.422	10.7156
7/16	.4375	.438	11.1125
29/64	.453125	.453	11.5094
15/32	.46875	.469	11.9063
31/64	.484375	.484	12.3031
1/2	.5000		12.7000
33/64	.515625	.516	13.0969
17/32	.53125	.531	13.4938
35/64	.546875	.547	13.8906
9/16	.5625	.562	14.2875
37/64	.578125	.578	14.6844
19/32	.59375	.594	15.0813
39/64	.609375	.609	15.4781
5/8	.625		15.8750
41/64	.640625	.641	16.2719
21/32	.65625	.656	16.6688
43/64	.671875	.672	17.0656
11/16	.6875	.688	17.4625
45/64	.703125	.703	17.8594
23/32	.71875	.719	18.2563
47/64	.734375	.734	18.6531
3/4	.750		19.0500
49/64	.765625	.766	19.4469
25/32	.78125	.781	19.8438
51/64	.796875	.797	20.2406
13/16	.8125	.812	20.6375
53/64	.828125	.828	21.0344
27/32	.84375	.844	21.4313
55/64	.859375	.859	21.8281
7/8	.875		22.2250
57/64	.890625	.891	22.6219
29/32	.90625	.906	23.0188
59/64	.921875	.922	23.4156
15/16	.9375	.938	23.8125
61/64	.953125	.953	24.2094
31/32	.96875	.969	24.6063
63/64	.984375	.984	25.0031
1 IN.	1.000		25.4000

NOTES:

1. Example of hole callout on drawings:
   .040 DIA (#60)
   - (#60) — Drill size if applicable
   - .040 — Round off to three places per ANSI Z25.1-1940.
2. When standard hole tolerance (in drawing title block) does not apply, omit drill size and indicate required tolerance.
3. Metric equivalents are given for general information only and are based upon 1 inch = 25.4 mm.

Courtesy of Texas Instruments, Inc.

ISO METRIC STANDARD SIZES

Nominal Size Diam. (mm) Column[a]			Pitches (mm)														Nom-inal Size Diam. (mm)
			Series With Graded Pitches		Series With Constant Pitches												
1	2	3	Coarse	Fine	6	4	3	2	1.5	1.25	1	0.75	0.5	0.35	0.25	0.2	
0.25			0.075	–	–	–	–	–	–	–	–	–	–	–	–	–	0.25
0.3			0.08	–	–	–	–	–	–	–	–	–	–	–	–	–	0.3
	0.35		0.09	–	–	–	–	–	–	–	–	–	–	–	–	–	0.35
0.4			0.1	–	–	–	–	–	–	–	–	–	–	–	–	–	0.4
	0.45		0.1	–	–	–	–	–	–	–	–	–	–	–	–	–	0.45
0.5			0.125	–	–	–	–	–	–	–	–	–	–	–	–	–	0.5
	0.55		0.125	–	–	–	–	–	–	–	–	–	–	–	–	–	0.55
0.6			0.15	–	–	–	–	–	–	–	–	–	–	–	–	–	0.6
	0.7		0.175	–	–	–	–	–	–	–	–	–	–	–	–	–	0.7
0.8			0.2	–	–	–	–	–	–	–	–	–	–	–	–	–	0.8
	0.9		0.225	–	–	–	–	–	–	–	–	–	–	–	–	–	0.9
1			0.25	–	–	–	–	–	–	–	–	–	–	–	–	0.2	1
	1.1		0.25	–	–	–	–	–	–	–	–	–	–	–	–	0.2	1.1
1.2			0.25	–	–	–	–	–	–	–	–	–	–	–	–	0.2	1.2
	1.4		0.3	–	–	–	–	–	–	–	–	–	–	–	–	0.2	1.4
1.6			0.35	–	–	–	–	–	–	–	–	–	–	–	–	0.2	1.6
	1.8		0.35	–	–	–	–	–	–	–	–	–	–	–	–	0.2	1.8
2			0.4	–	–	–	–	–	–	–	–	–	–	–	0.25	–	2
	2.2		0.45	–	–	–	–	–	–	–	–	–	–	–	0.25	–	2.2
2.5			0.45	–	–	–	–	–	–	–	–	–	–	0.35	–	–	2.5
3			0.5	–	–	–	–	–	–	–	–	–	–	0.35	–	–	3
	3.5		0.6	–	–	–	–	–	–	–	–	–	–	0.35	–	–	3.5
4			0.7	–	–	–	–	–	–	–	–	–	0.5	–	–	–	4
	4.5		0.75	–	–	–	–	–	–	–	–	–	0.5	–	–	–	4.5
5			0.8	–	–	–	–	–	–	–	–	–	0.5	–	–	–	5
		5.5	–	–	–	–	–	–	–	–	–	–	0.5	–	–	–	5.5
6			1	–	–	–	–	–	–	–	–	0.75	–	–	–	–	6
		7	1	–	–	–	–	–	–	–	–	0.75	–	–	–	–	7
8			1.25	1	–	–	–	–	–	–	1	0.75	–	–	–	–	8
		9	1.25	–	–	–	–	–	–	–	1	0.75	–	–	–	–	9
10			1.5	1.25	–	–	–	–	–	1.25	1	0.75	–	–	–	–	10
		11	1.5	–	–	–	–	–	–	–	1	0.75	–	–	–	–	11
12			1.75	1.25	–	–	–	–	1.5	1.25	1	–	–	–	–	–	12
	14		2	1.5	–	–	–	–	1.5	1.25[b]	1	–	–	–	–	–	14
		15	–	–	–	–	–	–	1.5	–	1	–	–	–	–	–	15
16			2	1.5	–	–	–	–	1.5	–	1	–	–	–	–	–	16
		17	–	–	–	–	–	–	1.5	–	1	–	–	–	–	–	17
	18		2.5	1.5	–	–	–	2	1.5	–	1	–	–	–	–	–	18
20			2.5	1.5	–	–	–	2	1.5	–	1	–	–	–	–	–	20
	22		2.5	1.5	–	–	–	2	1.5	–	1	–	–	–	–	–	22
24			3	2	–	–	–	2	1.5	–	1	–	–	–	–	–	24
		25	–	–	–	–	–	2	1.5	–	1	–	–	–	–	–	25
		26	–	–	–	–	–	–	1.5	–	1	–	–	–	–	–	26
	27		3	2	–	–	–	2	1.5	–	1	–	–	–	–	–	27
		28	–	–	–	–	–	2	1.5	–	1	–	–	–	–	–	28
30			3.5	2	–	–	(3)	2	1.5	–	1	–	–	–	–	–	30
		32	–	–	–	–	–	2	1.5	–	–	–	–	–	–	–	32
	33		3.5	2	–	–	(3)	2	1.5	–	–	–	–	–	–	–	33
		35[c]	–	–	–	–	–	–	1.5	–	–	–	–	–	–	–	35[c]
36			4	3	–	–	–	2	1.5	–	–	–	–	–	–	–	36
		38	–	–	–	–	–	–	1.5	–	–	–	–	–	–	–	38
	39		4	3	–	–	–	2	1.5	–	–	–	–	–	–	–	39
		40	–	–	–	–	3	2	1.5	–	–	–	–	–	–	–	40
42			4.5	3	–	4	3	2	1.5	–	–	–	–	–	–	–	42
	45		4.5	3	–	4	3	2	1.5	–	–	–	–	–	–	–	45

a Thread diameter should be selected from columns 1, 2 or 3; with preference being given in that order.
b Pitch 1.25 mm in combination with diameter 14 mm has been included for spark plug applications.
c Diameter 35 mm has been included for bearing locknut applications.
The use of pitches shown in parentheses should be avoided wherever possible.
The pitches enclosed in the bold frame, together with the corresponding nominal diameters in Columns 1 and 2, are those combinations which have been established by ISO Recommendations as a selected "coarse" and "fine" series for commercial fasteners. Sizes 0.25 mm through 1.4 mm are covered in ISO Recommendation R 68 and, except for the 0.25 mm size, in AN Standard ANSI B1.10.

CLEARANCE HOLES FOR THREADED FASTENERS

Thread Size	Tolerance Between Hole Centers		
	±.005	±.010	±.02
.060 [0]	.076 Dia (#48)	.089 Dia (#43)	
.086 [2]	.102 Dia (#38)	.116 Dia (#32)	.144 Dia (#27)*
.112 [4]	.125 Dia (#1/8)	.140 Dia (#28)	.166 Dia (#19)*
.138 [6]	.150 Dia (#25)	.166 Dia (#19)	.191 Dia (#11)*
.164 [8]	.177 Dia (#16)	.189 Dia (#12)	.219 Dia (#7/32)
.190 [10]	.199 Dia (#8)	.213 Dia (#3)	.242 Dia (C)
.250 [1/4]	.257 Dia (F)	.272 Dia (I)	.302 Dia (N)

* Exclude Fillister and socket head cap screws.
FLAT WASHERS WILL BE REQUIRED WITH ± .02 TOLERANCE.

1. This chart applies to the following:
   a. Where a row of holes is on the same centerline.
   b. Where the holes are on more than one centerline.
   c. Where clearance holes are in both parts.
   d. Where clearance holes are in one part and tapped holes or studs are in the other part.
2. This chart does not apply to the following:
   a. Clearance holes for rivets.
   b. One fastener used independent of any other.
   c. Clearance holes for flat head machine screws.
3. The hole sizes listed have been formulated for the most severe conditions for screw fasteners and represent a noninterference probability of approximately 98%.

All dimensioning shall be from an established base hole or fixed point.

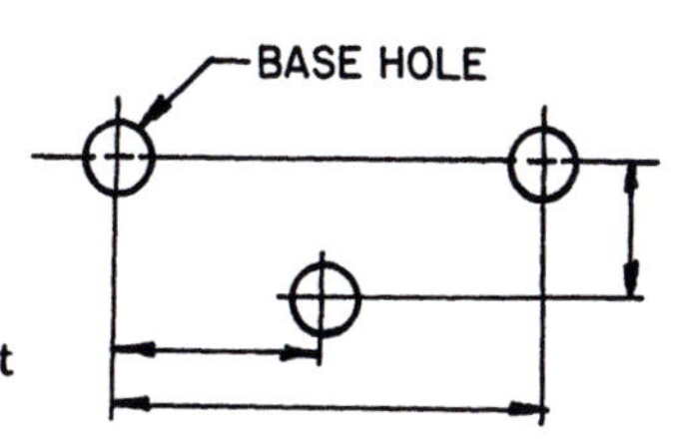

GRIP LENGTH CHART

Screw Length	SCREW SIZE						
	.060-80	.086-56	.112-40	.138-32	.164-32	.190-32	.250-28
.250	.098	.075					
.312	.160	.137	.074				
.375	.223	.200	.137	.086			
.438		.263	.200	.149	.119	.112	
.500		.365	.262	.211	.181	.174	
.625		.490	.387	.336	.306	.299	.162
.750		.615	.512	.461	.431	.424	.287
.875		.740	.637	.586	.556	.549	.412
1.00			.762	.711	.681	.674	.537
1.25					.837	.830	.693
1.50						.986	.849

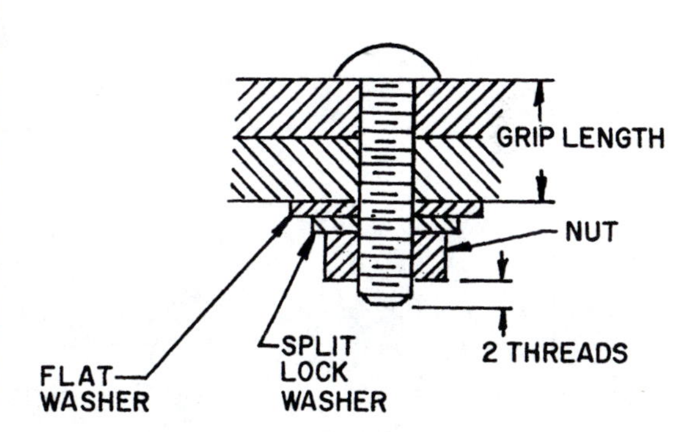

With grip length known (under screw size), read required screw length in left hand vertical column. Grip lengths are maximum.

Courtesy of Texas Instruments, Inc.

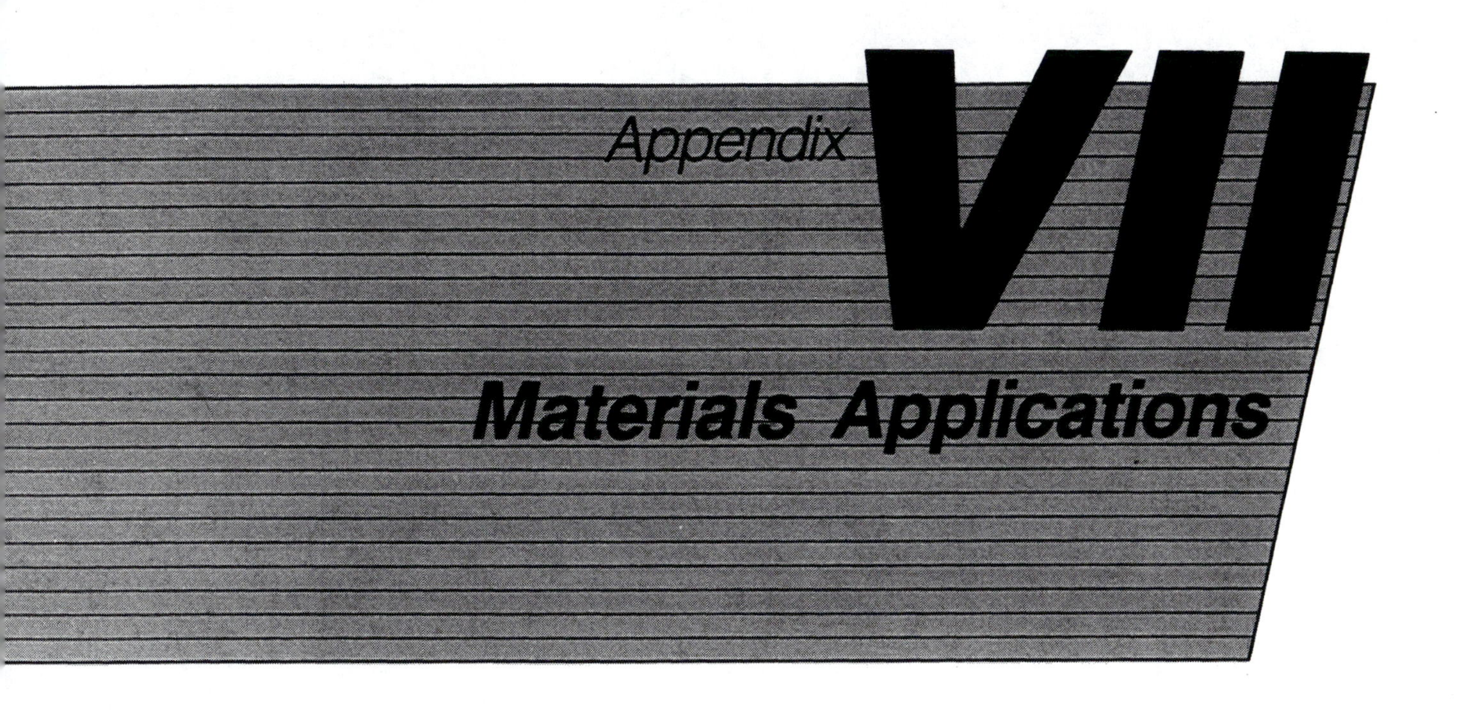

DISSIMILAR METALS

Without proper protective coating, do not use a metal from one group in contact with a metal listed in another group:

1. Tin, cadmium, and zinc may be used with all metals in Groups II and III.
2. Stainless steel may be used with all metals in Groups II, III, and IV.

Use metals within each group together.
MIL-STD-454

Group I	Group II	Group III	Group IV
Magnesium Alloys	Aluminum	Zinc	Copper and Its Alloys
Tin	Aluminum Alloys	Cadmium	Nickel and Its Alloys
Al Alloy (5052)	Zinc	Steel	Chromium
Al Alloy (5056)	Cadmium	Lead	Stainless steel
Al Alloy (5356)	Tin	Tin Tin Lead (solder)	Gold
Al Alloy (6061)	Stainless Steel	Stainless Steel	Silver
Al Alloy (6063)	Tin Lead (solder)	Nickel and Its Alloys	

Courtesy of Texas Instruments, Inc.

Index